AF718973

WITHDRAWN

Engineering Plastics

Engineering Plastics

A Handbook of Polyarylethers

Robert J. Cotter
Durham Technology Consultants, Inc.
Durham, New Hampshire

GORDON AND BREACH PUBLISHERS
Australia • Austria • Belgium • France • Germany • India • Japan • Malaysia • Netherlands • Russia • Singapore • Switzerland • Thailand • United Kingdom • United States

Copyright © 1995 by OPA (Overseas Publishers Association) Amsterdam B.V. Published under licence by Gordon and Breach Science Publishers S.A.

All rights reserved.

No part of this book may be reproduced or utilized in any form or by any means, electronic or mechanical, including photocopying and recording, or by any information storage or retrieval system, without permission in writing from the publisher.
Printed in the United States of America.

Gordon and Breach Science Publishers S.A.
Postfach
4004 Basel
Switzerland

BRITISH LIBRARY CATALOGUING IN PUBLICATION DATA

Cotter, Robert J.
Engineering Plastics: Handbook of Polyarylethers
I. Title
668.423

ISBN 2-88449-112-0

TP
1180
.P553
C65
1995

To my wife, Barbara Rigg Cotter, for her supportive and loving help over the decades.

Contents

Tables

Preface

The primary purpose of this work is to provide plastics industry professionals with a comprehensive review and compilation of those polymers that contain the aryl ether group in their structure. The development of processes for the synthesis of high molecular weight polyarylethers has led to the commercial introduction of many members of this polymer family. In fact, it can be said that it provided the basis for the emergence of the class of plastics known as high-performance engineering thermoplastics. Few, if any, other polymer families have spawned as many discrete molecular entities for use in such a wide variety of demanding applications.

Industry personnel ranging from the synthetic polymer chemist to the process engineer to marketing and sales personnel will find the abundance of data in this book invaluable. It is organized into two sections. Section one (chapters 1–3, tables 1–17) presents the synthetic methods that have been used for polyarylether synthesis, along with the history of their development. The methods are assigned designations according to their intrinsic mechanism, i.e., nucleophilic (N) or electrophilic (E). An additional category (A) is employed for those cases that do not fall neatly into the nucleophilic versus electrophilic classification. These method letters are used in the tables to designate the route by which a specific polymer was synthesized in the cited table reference. An experimental procedure is presented for each of the methods described in the first three chapters.

This work is intended to be comprehensive and complete for those homopolymers that contain the aryl ether group. The alert reader will no doubt find exceptions. A few polymers that do not contain this group are covered because they are close relatives to some of the included polyarylethers. Their inclusion will facilitate structure-property correlations. Although cross-checking tests do indicate a high degree of success in achieving comprehensive coverage, it is hoped that the omissions that undoubtedly have occurred are not major ones. As noted above, the table entries are for those polyarylether homopolymers that have been synthesized by the designated methods. No attempt was made to include the various types of copolymeric polyarylethers because of the ease of reliably predicting copolymer properties from homopolymer data. The hierarchy that is used throughout the tables and the book generally is in the order of polyarylethersulfones, polyaryletherketones, polyaryletherimides and polyarylethers. All polymers with a backbone sulfone group are included in the first group of tables (1–5); all polymers with a backbone ketone group are included in the second group (6–10); all polymers with a backbone imide group are included in the third group (11–13); and all other polyarylethers are included in the last group (14–17). Each of these groups has at least one member that is currently of commercial significance. The organization of the tables is described in more detail in the Guide to the Tables (p. 99).

For each polymer entry in the tables, the glass and melting transition temperatures are recorded, where available. The synthetic method that was used and a summary of the conditions are also included. These data can serve also as a ready key to the literature for the monomers that are useful for polyarylether synthesis. Application of a little retro-synthetic analysis to the repeat unit and the synthetic method will indicate which monomers were employed. Thus, a reference where a high molecular weight polymer was made from a given monomer will most likely describe the monomer purification procedures. Quickly locating a reference where the polymerization of a particular monomer has been studied can be a time saver for those faced with choosing one synthetic path from among many to that monomer. Knowing that other workers have found that a particular synthetic route yields "polymerization grade" monomers should prove extremely valuable. The organization at which the research was conducted is also listed. This is always of particular interest to the industrial community. The authors of the cited publications are listed in the References for Tables (p. 337).

Section two of the handbook is comprised of the Guide to the Tables, chapters 4–7, tables 18–22, and chapter 8. Chapters 4–7 cover polyarylethers containing Bisphenol A moieties, diphenyl sulfone moieties, crystalline polyarylethers and heterocyclic ring-containing polyarylethers. The latter two topics represent the more recent research thrusts in the field. The data in tables 18–22 were sorted and taken from the basic tables 1–17. Chapter 8 presents the basic property data for all of the polyarylethers that have ever been commercialized. The tables in this chapter will allow one to easily survey the property ranges that are available from these commercial products. Selection of candidate products for an application can then be pursued with the manufacturers. Chapter 8 does not attempt to present the wealth of all types of data that resin manufacturers can supply about their products. Working closely with a resin manufacturer to develop an application is recommended.

As is the case with most endeavors of this type, many people have helped to make it possible. The author wishes to acknowledge and extend his sincere appreciation to them. Many of the programs carried out by my colleagues at Union Carbide's Bound Brook (New Jersey) laboratories were seminal in opening up the field of polyarylethers. Most notable was the discovery of Polysulfone by A.G. Farnham and R.N. Johnson. The manuscript of the book was reviewed in detail by L.E. Brydia, D.R. Kelsey, and R.P. Kurkjy.

Many helpful suggestions were received for which the author is thankful. Suggestions and/or help with the documentation were received from R.A. Clendinning, P.T. McGrail, J.E. McGrath, D.G. Parker, and S. Savariar. The resources and staff of the Chemistry Department's library at the University of New Hampshire provided numerous invaluable assists to the author.

Introduction

Polyarylethers, or aromatic polyethers, comprise one of the major polymer families that gave rise to the high-performance levels that are achievable from plastics today. They are known for their superior performance characteristics in an unsurpassed number of application areas. The terms "high-performance plastics" and "engineering plastics" are used to convey their attributes in a qualitative way. But chemists have long recognized that the fundamental basis for their performance resides in the aromatic ether linkages that are present in their molecular structure.

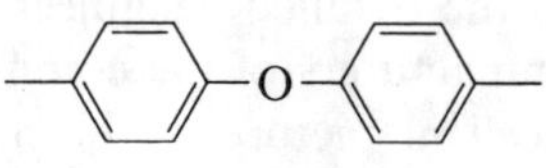

Aryl ether group

The desirability of incorporating this group into polymer backbones was recognized long before it was actually achieved. Many earlier studies on the thermal and chemical properties of low molecular weight aromatic polyethers highlighted their high thermal stability and hydrolytic resistance. Thus, incorporation of this functionality into polymers transferred these desirable attributes to a large number of commercially valuable, thermoplastic polyarylethers.

Many commercially important sub-categories exist under the polyarylether heading. Starting with polyarylethersulfones, the list includes polyaryletherketones, polyetherimides and polyphenylene ethers. This handbook is designed to provide a comprehensive database for these polymer families. Additionally, it provides plastics industry professionals, ranging from chemist to engineer to marketing personnel, with a handy reference work for the structures and properties that have been achieved from this polymer class. It also provides a facile entry to the synthetic routes that have been used for their preparation. This information will be especially valuable for new participants in the field of aromatic polyethers.

One key feature of this book is the tabular databases for the polyarylethers that have been synthesized. Only homopolymers are listed in these tables because of the complexity of the added task of cataloguing the various types of copolymers. The ease of predicting copolymer properties from homopolymer data also argued against their inclusion. The detailed, published synthetic procedures that have been used for polyarylether synthesis are a second valuable component of the book. They were selected by the author to illustrate the different methods and because they were reported to give high molecular weight polymers. They are presented in this book exactly as reported in the original publications. Since the synthetic procedures were not experimentally re-checked for inclusion herein, no guarantees are offered or implied by their republication. To the synthetic polymer chemist, they do represent an excellent starting point for the synthesis of a sample of a particular polymer. The chemical engineer who is concerned with process scale-up should also find these experimental procedures valuable. They will allow for early visualization of what an ultimate process will entail, while providing useful guidance in a laboratory process development program. The earlier the considerations that must be met for the commercial scale preparation of a polymer are defined and communicated, the higher the chances for eventual success.

This book also includes the detailed "engineering properties" that have been published by the manufacturers of the commercialized polyarylethers. These data provide an excellent picture of the properties that have made this polymer family so valuable. The author has not attempted to include the mechanical property data that are published sporadically on research samples of the aromatic polyethers. In the writer's experience, all aromatic polyethers can be assumed to be "tough" unless proven otherwise. That is to say that if one can prepare a sample of a given polymer at molecular weights above the threshold level (or entanglement molecular weight) for mechanical properties to appear, one can assume that it will be tough. Exceptions to this statement do and will exist in the future. But inspection of the repeat unit structure by the trained eye of the polymer scientist will forewarn when to expect brittle polymers. Excessive amounts of rigid structures in the polymer backbone or pendant to it will induce brittleness in polyethers, as it will in any polymer class. Data for products that are no longer commercially available have also been included for their value in structure/property correlation studies.

The author attempted to include at least one reference for each polymer structure shown in the tables. An attempt was made to use the original reference or at least one from the group that made the seminal contribution. Simply knowing that a specific structure has been synthesized can give greater impetus and confidence to a subsequent, exhaustive literature search, if such is needed.

It is hoped that the wealth of data presented in this volume will facilitate the ongoing worldwide research, development, and commercialization efforts on aromatic polyethers.

CHAPTER 1

Nucleophilic Synthetic Methods

Introduction

The advance of polyarylethers onto the commercial stage was precipitated by the development of nucleophilic routes for their synthesis. The pioneering researches of Farnham and Johnson at the Bound Brook Laboratories of the Union Carbide Corporation led to the development of the polyarylether from Bisphenol A and 4,4′-dichlorodiphenyl sulfone (*1*, *2*, *3*, *4*, *5*). This polymer, now referred to simply as Polysulfone was introduced commercially in 1965. The decision to commercialize it was based on the polymer's high glass transition (190°C) and its excellent impact strength. It also possessed excellent stability for long periods in air at elevated temperatures and outstanding hydrolytic stability. The basic patent (*2*) for Polysulfone was originally applied for in 1963, but did not issue until 1978 because of continuations-in-part and interferences. Since it will not expire until 1995, it has provided a legal monopoly for Polysulfone to the inventors (Union Carbide/Amoco) for thirty-two years!

The process chemistry that was originally developed for the commercial production of Polysulfone involved the formation of the bisphenate salt from sodium hydroxide and bisphenol A; dehydration of the reaction mixture; and polymerization in a solvent mixture containing dimethyl sulfoxide. The arrival of dimethyl sulfoxide on the commercial scene at about that time played a key role in the success of this chemistry. It was one of the first of several dipolar, aprotic solvents to be exploited in a variety of monomer and polymer chemistries. It is believed that this technology is still in

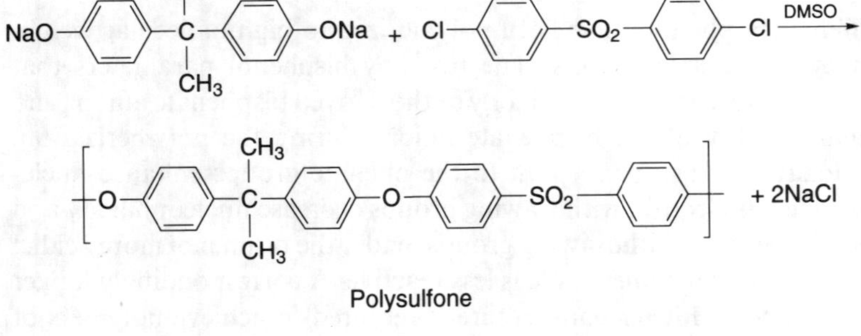

(I-1)

use today at the Polysulfone commercial plant now owned and operated by Amoco Performance Products in Marietta, Ohio. Amoco sells Polysulfone under its UDEL™ trademark. No other producer of polyarylethers is known to use this "caustic process" on a commercial scale.

A. POLY(ETHER SULFONES)

1. Polyetherification

N-1. Diphenols and Activated Aromatic Dihalides: Caustic Process

The original process that was developed for Polysulfone is now commonly referred to as the "caustic process". This is a "textbook" example of a condensation or step-growth polymerization.

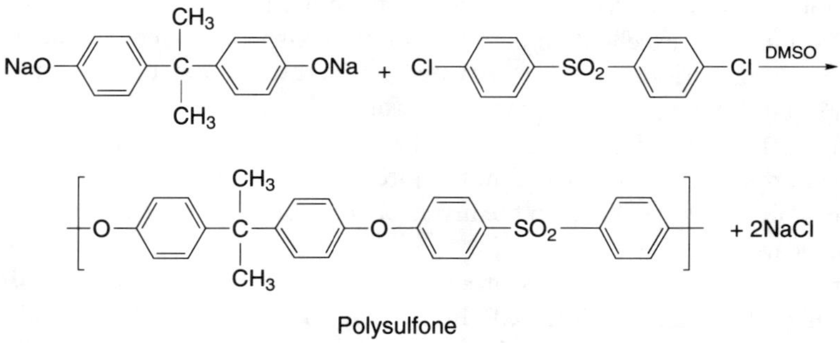

(I-2)

The scope of this chemistry and the variables that need to be controlled were discussed recently by the original developers (*3*).

Reaction Scope and Variables

Bisphenol Reactivity

Many different bisphenols have been polymerized to high molecular weight polymers by the caustic process. The two key bisphenol parameters that determine success are the nucleophilicity of the derived bisphenate anions and the thermal stability of the bisphenate anions during the polymerization. Electron donating substituents *para* to the phenate groups enhance nucleophilicity, whereas electron withdrawing groups decrease nucleophilicity and reactivity. The electron withdrawing groups render the bisphenol more acidic; hence the corresponding phenoxide is less reactive. A correspondingly longer reaction time and/or higher temperature is required to achieve polymers of

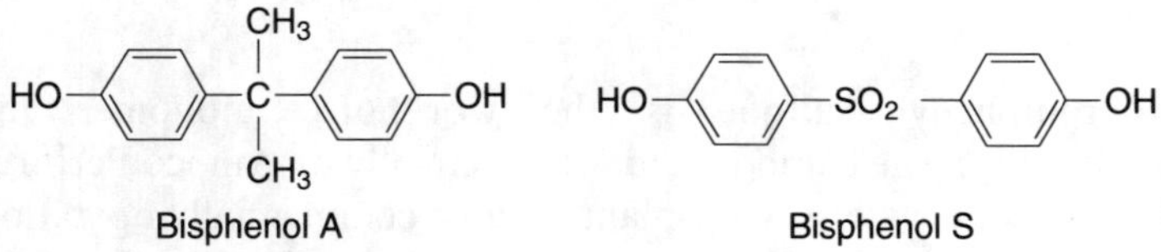

(I-3)

useful molecular weight from the more acidic bisphenols. Bisphenol A which has an alkylidene connecting group yields very high molecular weight polymer with 4,4′-dichlorodiphenyl sulfone in an hour at 160°C. In contrast, Bisphenol S, which has a sulfonyl group connecting the phenol rings, gave only low molecular weight polymer after ten hours (*3*).

The thermal stability of the phenoxide anions is another consideration for a successful polymerization. For example, the mono anion of Bisphenol A undergoes a cleavage reaction as shown in the accompanying equation. This tendency imposes an upper temperature limit of about 165°C for polymerization of Bisphenol A by any process involving its anions. Side reactions such as

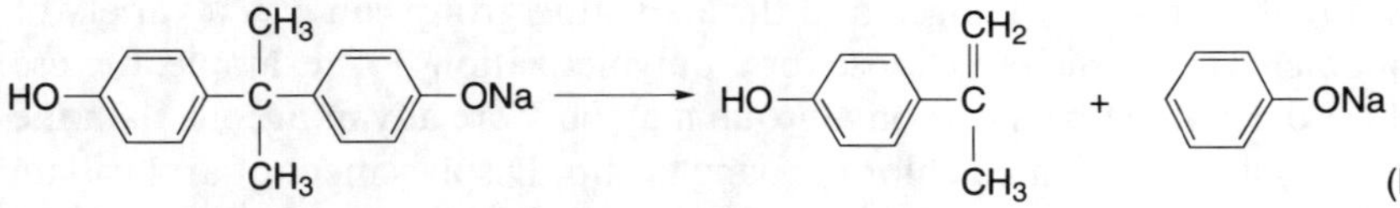

(I-4)

this one not only disrupt the polymerization stoichiometry, but also adversely affect the final product quality. Anions of bisphenols with electron withdrawing groups are more thermally stable and can be polymerized at higher temperatures.

In addition to maintaining an anhydrous system for these polymerizations, it is also necessary that they be air-free. Reactions of phenoxides with oxygen are well known and very rapid. Oxygen leads to severely discolored polymers and to difficulty in achieving high molecular weight. This is especially true for polymerization of hydroquinone by the caustic process. Badly discolored, low molecular weight products are obtained if adequate precautions are not taken.

The purity of the bisphenol is a very important factor that affects the polymerization stoichiometry and the quality of the polymer. Two quality parameters that are of prime importance for Polysulfone products are the polymer color and its melt stability. Very high purity Bisphenol A is required in order to obtain a high quality Polysulfone. Fortunately, it is readily available commercially. The grades of Bisphenol A that were developed for Polycarbonate manufacture are used commercially for Polysulfone manufacture.

Dihalide Reactivity

4,4′-Dichlorodiphenyl sulfone is the workhorse monomer in the polyarylethersulfone field and is made commercially by Amoco Performance Products, Inc. A new world-scale plant became commercially operational in 1991 in Augusta, Georgia. This new process for preparing diaryl sulfones (*6*) has the environmental benefit of low emissions, is energy efficient, and has replaced an older thionyl chloride-based process. The purity of this monomer is also very critical and must be controlled and managed in order to produce a high quality polymer.

Activation of the aromatic halo groups by electron withdrawing groups is essential for a facile polymerization. As expected from earlier model studies, halogen reactivity decreases in the order F >> Cl > Br, I when the activating group is a sulfone. Additionally, the relative effectiveness of the group activating the halogen has been found to be in the order

$$-NO_2 \sim -SO_2- > -\overset{\overset{\displaystyle O}{\|}}{C}- > -N{=}N- \qquad \text{(I-5)}$$

Varying the choice of halogen and the activating group can give a variety of monomers from which to choose for a polymerization. For example, the use of fluoro activated by a carbonyl group may be more advantageous than the use of a sulfone-activated chloro leaving group. Displacement of aryl triflate and brosylate groups by bis-phenoxides can also yield high molecular weight polymer in some cases (*7*).

A major side reaction involving the dihalide that must be controlled is hydrolysis of halo groups in incompletely dehydrated polymerization reactions. Water in the system can lead to hydrolysis of the bisphenol salt which produces sodium hydroxide. Caustic can then react with polymer end groups or with dichlorodiphenyl sulfone as shown in the accompanying

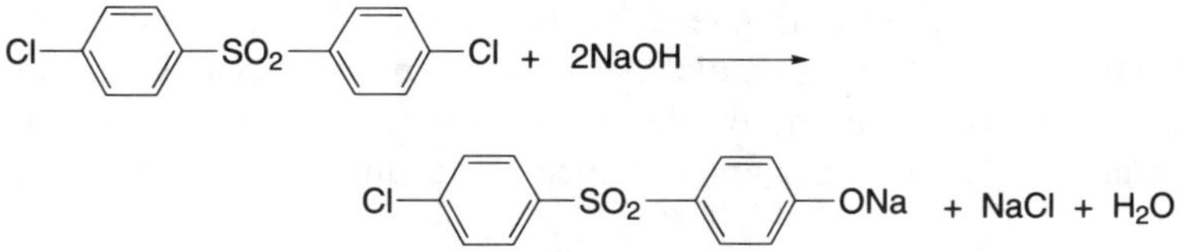

(I-6)

equation. The sodium salt produced above is not nucleophilic enough to react further in DMSO at 160°C, thereby acting as a terminator and upsetting the stoichiometry of the polymerization.

The reactivity of 4,4′-dichlorodiphenyl sulfone and related model compounds has been studied extensively by researchers at ICI (*8*, *9*, *10*, *11*). Their interest in commercializing Polyethersulfone (PES) by a nucleophilic route prompted these studies. A key practical result was the finding that

4-chloro-4′-hydroxydiphenylsulfone was obtainable in 95% yield by partial hydrolysis of 4,4′-dichlorodiphenyl sulfone. This work showed that with diphenyl sulfone derivatives the electronic effects of the sulfone group on the reactivity of a functional group, e.g. halogen or phenoxide, attached to one phenylene ring is modified by substituents in the other. Thus, the rate constants for reaction of the first and second chlorine atoms of 4,4′-dichlorodiphenyl sulfone with KOH are 179 and 3.2 ($\times 10^{-5}$ l/mole/sec) respectively (*11*). The first chloro group is removed about 1.5 times faster than is the single chlorine from 4-chlorophenyl phenyl sulfone ($k = 112 \times 10^{-5}$ l/mol/sec) (*10*). However, when one chlorine has been displaced by hydroxide ion, the

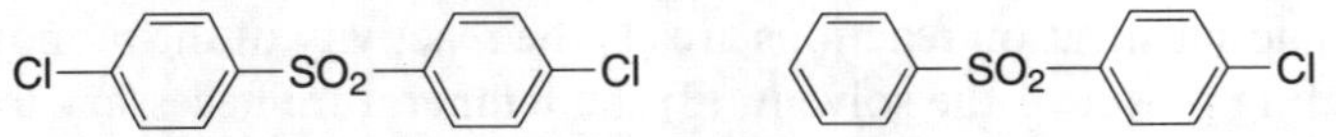

(I-7)

second chlorine reacts at only one-thirtieth the rate of the first chlorine. The electronic effects of a substituent in one ring is transmitted to the other via the sulfone group. Halogen reactivity in one ring is enhanced by the presence of a *para*-chloro substituent in the other, while the strongly electron donating *para*-phenoxide substituent reduces halogen reactivity in the adjacent ring. These effects are the basis for obtaining the high yields of 4-chloro-4′-hydroxydiphenyl sulfone from hydrolysis of 4,4′-dichlorodiphenyl sulfone rather than the expected statistical mixture. The ready availability of this monomer served as the basis for the initial commercialization of PES by a nucleophilic route by ICI (*12*). The potassium halophenoxides polymerize rapidly on melting at ca. 300°C. Solution polymerization in sulfolane was significantly slower. This is an example of the "caustic process" in that a preformed, alkali

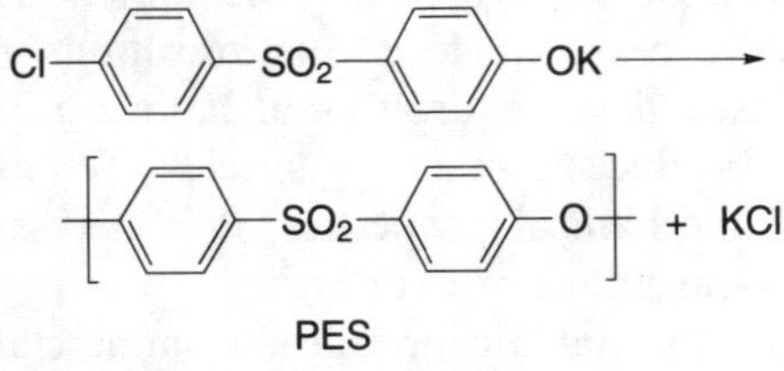

(I-8)

metal, anhydrous salt of the monomer is formed and then polymerized to high molecular weight polymer. It is also an example of a self-polymerizable monomer wherein exact 1:1 stoichimetry is built into the monomer.

Choice of Base

Economic considerations dictate that the base of choice in the "caustic process" is caustic or sodium hydroxide. Potassium hydroxide functions as well and may offer better reactivity and/or solubility of the phenate salts in some cases. The use of inorganic carbonates to prepare polysulfones is described below under the "carbonate process".

Solvent Selection

Dimethyl sulfoxide is favored and is believed to be used commercially for Polysulfone preparation by the caustic process. Its purity must be controlled and care must be taken to avoid conditions during the polymerization that could lead to deleterious side reactions involving the solvent. Some of the factors that dictate the choice of a dipolar aprotic solvent for aromatic nucleophilic substitution reactions are (1) the reactivity of the monomers, (2) the solvating power of the solvent, (3) the temperature necessary to achieve high molecular weight in a reasonable time, (4) the solubility of the polymer under the reaction conditions, and (5) the choice of base. The carbonate process, which is discussed in a later section, can utilize a wider variety of solvents than the caustic process.

Stoichiometry and Molecular Weight Control

The polymerization of a bisphenol with an activated aromatic dihalide is an excellent example of a step-growth or condensation polymerization (*13*). The purity of the reactants, reagents, and solvents are crucial to obtaining high molecular weight polymer. Highest molecular weights have been shown to be achieved at 1:1 stoichiometry of the bisphenate and dihalide reactants in accord with theory. The sodium hydroxide stoichiometry is also very crucial in these reactions. Small excesses can lead to side reactions such as ether interchange between chains and the hydrolysis of halo groups that can upset stoichiometry and/or produce terminating species. To obtain Polysulfone polymers with useful mechanical properties, it is necessary to achieve a reduced viscosity of 0.4 (in chloroform) or higher.

Thermoplastic polymers that are produced commercially must be processable from the melt in a variety of fabrication systems. Hence, they must be produced so as to possess a range of melt viscosities that allow fabrication in these systems. And, moreover, they must be stable against degradation or crosslinking in the melt so that the initial melt viscosities are preserved during fabrication. Because melt viscosity is most directly related to polymer molecular weight, a means of stopping the polymerization at the desired point is essential. For Polysulfone prepared by the caustic process, this can be accomplished by the addition of a monofunctional reagent such as methyl chloride at the appropriate point. Methyl chloride reacts with the growing phenoxide end groups to form methyl ether end groups.

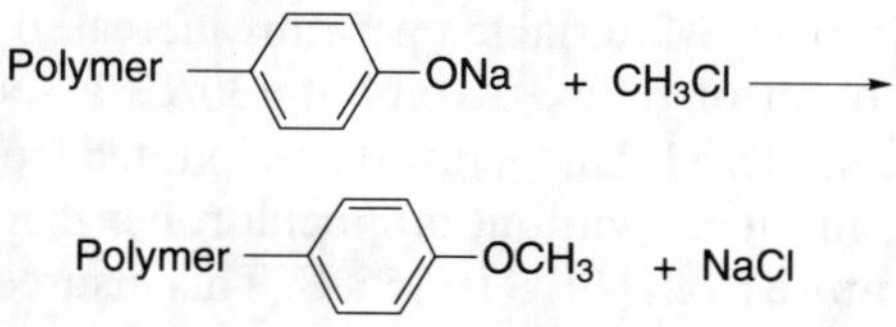

(I-9)

This not only stops the polymerization but also provides a more color and thermally stable polymer. Other monofunctional halo compounds can also be used for termination but they offer no practical advantages on a commercial scale. To produce the most thermally stable polymers (with respect to color and molecular weight), the elimination of metal phenate end groups is essential. Keep in mind that polyarylethers are fabricated from melt temperatures in the range of 350–425°C. These extreme conditions mandate that precise control be maintained over the species that are present in the polymer in order to insure a quality end product. The literature on the degradation and stabilization of polysulfones has been reviewed (*14*).

EXAMPLE I: PREPARATION OF THE POLYSULFONE FROM BISPHENOL A AND 4,4′-DICHLORODIPHENYL SULFONE BY THE “CAUSTIC PROCESS” (*15*)

This example is taken from a patent (*15*) belonging to the commercial producer of Udel™ Polysulfone. This procedure incorporates techniques for producing polymer possessing high melt stability and low color.

A four-neck 500 ml round bottom distilling flask was equipped with a mechanical stirrer, two three-way parallel side-arm adapters, a thermometer, two dropping funnels, a dry nitrogen stainless steel inlet tube, a Liebig condenser, and a distillation receiver. Into the flask was charged 205.44 g (1.825 moles) of monochlorobenzene, 102.72 g (1.315 moles) of dimethylsulfoxide and 51.36 g (0.225 moles) of Bisphenol A; 120 g of monochlorobenzene was charged to the distillation receiver. The mixture was purged with dry nitrogen for 30 minutes at about 25°C and then heated to 40°C. A continuous nitrogen sweep was maintained on the flask during the heating period and throughout the remainder of the reaction. From an addition funnel, 17.92 g (0.45 moles) of sodium hydroxide was charged to the flask as a 50% solution in distilled water. Bisphenol A and sodium hydroxide react to form the disodium salt of bisphenol A and two moles of water.

The water of reaction plus that entering with the sodium hydroxide, was removed from the reaction mixture by co-distillation with monochlorobenzene. Reflux began at 115°C. Water, monochlorobenzene, and dimethyl sulfoxide were boiled up together, condensed, and separated in the distillation receiver. Monochlorobenzene was recycled from the bottom of the receiver to the vapor space in the flask. Material temperature increased to 140°C in about one hour after the onset of reflux. At about 130°C, a sodium bisphenate-dimethyl sulfoxide complex (1/2 mole ratio) precipitated from solution.

Distillation was continued, without monochlorobenzene recycle, to bring the material temperature from 140°C to 155°C. This insured solubilization of the bisphenate-dimethyl sulfoxide complex and complete water removal from the reaction system. The total distillate collected was approximately 272 ml of monochlorobenzene plus 30 ml of water and dimethyl sulfoxide. From a dropping funnel, 92% of a 50% solution of 4,4′-dichlorodiphenyl sulfone (64.62 g) in monochlorobenzene (64.62 g) was charged to the flask at a temperature of about 120°C over a 10 minute period. 4,4′-Dichlorodiphenyl sulfone reacts exothermically with the sodium bisphenate to form polysulfone oligomers and sodium chloride.

Monochlorobenzene distillation (46 ml) was then continued to raise the material temperature to about 160°C. Prepolymerization was complete after 30 minutes at about 160°C, at which time the remaining 8% of the 4,4′-dichlorodiphenyl sulfone/monochlorobenzene solution was charged to the flask. Polymerization was then conducted at about 160°C, until the desired molecular weight was achieved (15 to 30 minutes). The reaction was then quenched to about 135°C with 166 g of anhydrous monochlorobenzene, terminated by sparging methyl chloride through the viscous polymer solution for about 30 minutes, and then reterminating with 0.17 g of 50% sodium hydroxide to cause conversion of bisphenol A hydroxyl end-groups to phenates. Methyl chloride addition was then continued for about 30 minutes, the batch cooled to about 90°C, acidified with 0.50 g of oxalic acid to a pH of 4, and further diluted with monochlorobenzene to a volume of about 500 ml. At about 25°C, the polymer solution was filtered through a medium porosity sintered-glass funnel to remove sodium chloride.

Approximately 50 ml of the clear filtrate was removed. Percent polymer was determined from a 1.0000 g sample heated for 10 minutes on an Ohaus Moisture Determination Scale. The 50 ml sample was then adjusted to 10% solids by addition of monochlorobenzene and pressure filtered through a fine porosity sintered glass funnel. The batch color factor was determined by comparing the sample's % transmittance (94.6) to monochlorobenzene in a Fisher Electrophotometer II using a blue (425 nm) filter. Percent transmittance was converted to absorbance (0.0238) using the Beer-Lambert equation, i.e.,

$$\text{Absorbance} = \log \frac{100}{\%\ \text{Transmittance}}$$

and the color factor defined by the following equation;

$$\text{Color factor} = 1000(\text{Absorbance})$$

Polymer recovery from the salt-free reaction mixture was obtained by methanol coagulation (5/1 methanol/polymer solution ratio) in a Waring blender. The coagulant was then washed in the blender with additional methanol (350 ml) to ensure removal of dimethyl sulfoxide and monochlorobenzene, filtered, and then dried at about 120°C and 30 mm Hg vacuum. A polymer was obtained having a reduced viscosity of 0.50 as measured in chloroform at 25°C.

EXAMPLE II: PREPARATION OF POLYSULFONE FROM BISPHENOL A AND 4,4′-DICHLORODIPHENYL SULFONE BY THE "CAUSTIC PROCESS" (*1*)

This example is taken from the original publication (*1*) of the inventors of Polysulfone. Additional details and refinements have appeared in a later publication (*16*).

Into a 1-liter stainless steel resin kettle, fitted with an inert gas sparge tube, thermocouple, mechanical stirrer, dropping funnel, and take-off to a six-plate (glass helix) fractionating column connected to a moisture trap and condenser, is placed 51.36 g (0.225 mole) of ultra high purity Bisphenol A [2,2-bis(4-hydroxyphenyl)propane], 115 g of dimethyl sulfoxide (DMSO), and 330 g of chlorobenzene. The mixture is heated to 60–80°C, whereupon a clear solution is obtained. Air is displaced from the system by flushing with nitrogen or argon and exactly 0.450 mole of 50.00% aqueous sodium hydroxide (35.86 g of a 50.20% solution) is added with good stirring over about 10 min. In so doing, two liquid phases appear: one predominantly chlorobenzene, the other disodium salt dissolved in aqueous DMSO. The system is brought to reflux with inert gas sparging through the reaction mixture. Water is removed from the system, the chlorobenzene that codistils being returned continuously. In so doing the temperature of the contents rises from about 120°C initially to 140°C at the conclusion of this step. When this point is reached, most of the water originally present has distilled, and the disodium salt of Bisphenol A appears as a precipitate.

Excess azeotrope solvent is distilled from the system until the temperature of the contents reaches 155–160°C. At this point the precipitate will redissolve with the formation of a very viscous solution. It is believed that at this point only traces of water remain.

A 50% solution of 64.61 g (0.225 mole) of 4,4′-dichlorodiphenyl sulfone (DCDPS) in dry chlorobenzene maintained at 110°C is added over a period of about 10 min., the excess solvent being allowed to distil at a rate sufficient to hold the material temperature at about 160°C. When all the sulfone has been added, polymerization is continued until the desired degree of polymerization is reached. A reduced viscosity as measured in chloroform solution (0.2 g/100 ml) at 25°C. of about 1.0 or higher is reached in about 1 hr. at 160°C. after addition of all the sulfone.

The temperature must not drop below about 150°C. until the polymerization is well along, as sodium-terminated low polymer may precipitate on the walls of the vessel. This material, which is polymeric, is very difficult to redissolve upon heating owing to the familiar problem of solvent migration. Too high a temperature during addition of the sulfone monomer and subsequent polymerization is to be avoided, as the reaction is mildly exothermic, extremely rapid above 160°C, and excessive solvent decomposition and/or discoloration or even gelation of the reaction mass may occur. To prevent the danger of excessive overheating of the reaction mass, use of oil baths for heating is recommended. Upon addition of the sulfone, the reaction mixture becomes vividly colored, generally orange to yellow, but often deep greens are observed. The color is believed due to the formation of the Meisenheimer type sigma-complex or possibly a charge-transfer complex composed of the reactants and the DMSO solvent.

The polymerization may be terminated in a variety of ways; one of which is to pass methyl chloride into the polymerizing mixture when the desired degree of polymerization is reached. The color of the reaction mixture fades in so doing to a light amber, signalling the completion of the termination. The methyl chloride reaction, which is very rapid and efficient, involves the conversion of aryl oxide ends to their respective methyl ethers, thus preventing further polymerization. The color change is usually complete within about 5 min. when gaseous methyl chloride is passed through the polymerizing mixture at about 120–160°C. That termination has occurred is indicated not only by arrest of the viscosity rise but also by the appearance of a peak in the NMR at 211.3 cps (relative to tetramethylsilane in chlorobenzene solution) not found in polymer terminated by treatment with acids. The dimethyl ether of Bisphenol A also gives a peak at 211.3 cps not found in the spectrum of bisphenol A. The viscous polymer solution is then cooled and diluted with about 700 g of chlorobenzene which provides a solution convenient for workup. The byproduct sodium chloride is removed by simple pressure or vacuum filtration, and the clear light amber solution is coagulated in three or four volumes of ethanol. The resulting fluff is dried in a vacuum oven at 135°C for several hours. The yield is about 90% of theory or higher depending on the severity of losses during the workup.

Films may be obtained by casting or by pressing at about 280°C; the polymer softens at about 200°C. Fibers may be obtained by pulling from the melt. The polymer is soluble in dipolar aprotic solvents, cyclic ethers such as

THF, polar aromatic solvents, most chlorinated hydrocarbons, and some carbonyls such as cyclohexanone. Hazy solutions may be noted upon dissolving in certain of these solvents. The haze which may not develop immediately is due to the precipitation of cyclic oligomers which are crystalline; the polymer is amorphous.

EXAMPLE III: PREPARATION OF POLYETHERSULFONE IN DIPHENYL SULFONE (*26*)

Hydrated dipotassium salt of 4,4′-dihydroxydiphenyl sulfone was finely ground to pass through a sieve having a mesh opening of 500 μm. A sample of the ground salt was estimated by titration against a standard solution of sulfuric acid. Accordingly a sample (37.03 g, corresponding to 0.100 mol) of the hydrated salt was weighed into a glass flask of capacity 250 ml fitted with a stirrer, nitrogen purge, and an air condenser. Diphenyl sulfone (50 g) was added and powder blended with the hydrated salt. The flask was flushed with nitrogen and heated on a solder bath at 230°C. As the diphenyl sulfone melted, the stirrer was started, nitrogen purge commenced and the pressure in the flask reduced to 50 torr. Water distilled from the mixture and a slurry of the dipotassium salt in diphenyl sulfone remained in the flask. Care was taken to ensure that the diphenyl sulfone did not boil and hence did not splash the dipotassium salt on to the upper walls of the flask. The pressure in the apparatus was increased to atmospheric by addition of nitrogen via the nitrogen inlet. 4,4′-Dichlorodiphenylsulfone (28.64 g; 0.0997 mol) and diphenyl sulfone (19.6 g) were added to the flask and the temperature increased with stirring to 285°C. After 16 hours polymerization at 285°C, the polymerization was terminated by bubbling methyl chloride gas through the reaction mixture via a syringe needle for 30 minutes while the mixture was maintained at 285°C. The resultant reaction mixture was cooled and milled with dry-ice to a powder of particle size less than 1 mm. The solvent (diphenyl sulfone) and reaction byproduct (potassium chloride) were extracted by means of a Soxhlet type extractor, first with a mixture of water (200 ml) and methanol (700 ml) for six hours; then with water (700 ml) for four hours; and finally with a mixture of acetone (300 ml) and methanol (600 ml) for two hours. The resultant polymer powder was then dried at 140°C in an air oven for 24 hours at 100 torr. The resultant off-white colored powder had a reduced viscosity of 0.48 (measured at 25°C on a solution in dimethyl formamide containing 1 g of polymer in 100 ml of solution). A sample (3.5 g) of the polymer was molded into film about 0.25 mm thick by pressing at 320°C for five minutes. The film was tough, transparent and pale amber in color.

N-2. Diphenols and Activated Aromatic Dihalides: Carbonate Process

The use of alkali metal carbonates as bases for the synthesis of aromatic polyethers was originally developed in the laboratories of the Union Carbide Corporation (*17*). All of the monomers, reagents and solvents are charged at the start of the reaction. This process does not require the separate formation and dehydration of the bisphenate salt. Formation of high molecular weight polymer without precipitation of salts is achievable even when using bisphenols whose dialkali metal salts are very insoluble.

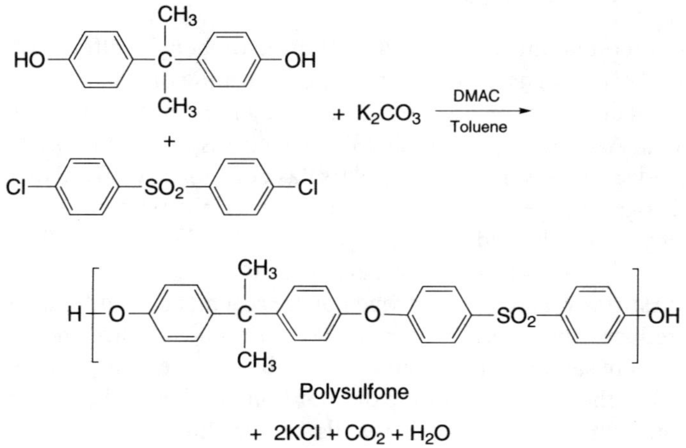

(I-10)

Because of this feature and others, it is now the preferred laboratory method for the synthesis of new and/or small samples of polyarylethersulfones (*18*). It is believed to be used commercially for the preparation of certain non Bisphenol A-containing polyarylethersulfones. Polyethersulfone (PES), which has been sold commercially as Victrex™ PES by ICI, is believed to be produced by the carbonate process. The following equations depict the chemistry. PES is an amorphous polymer which requires higher polymeriza-

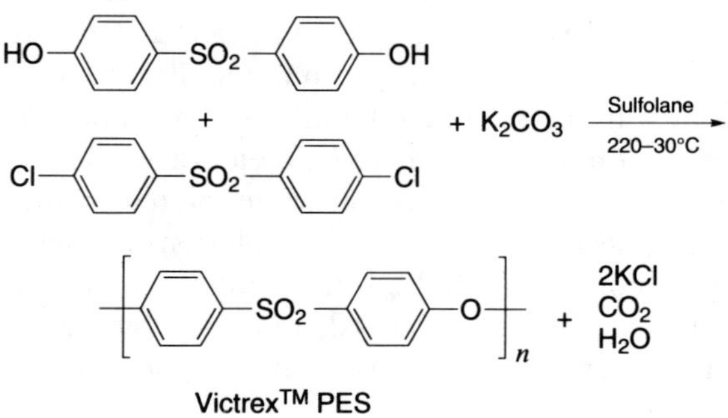

(I-11)

tion temperatures because of the less reactive Bisphenol S monomer. Diphenyl sulfone may also be used as the solvent for this polymerization.

Reaction Scope and Variables

Bisphenol Reactivity

As in the caustic process, the phenolic groups in this polymerization react via their phenate derivatives. Thus, they should be acidic enough to be converted to their phenates in an essentially anhydrous medium. A wide range of diphenols have been successfully polymerized by this technique. The polymerization can be conducted over a very wide temperature range depending on the choice of solvent. For amorphous polymers, a temperature of between 150° and 250°C is generally used. Crystalline polymers (especially polyketones) that need to be kept in solution during polymerization have been prepared at temperatures up to 335°C!

The major consideration regarding the bisphenol is its thermal stability at the polymerization temperature. The tendency of Bisphenol A and other alkylidene bisphenols to cleave at temperatures upwards of 160°C can be a limitation of this process. Thus, the thermal stability of a specific bisphenol and its phenate salts must be kept in mind when choosing a solvent (and polymerization temperature) for polymerization by the carbonate process. A variety of polyarylethersulfone compositions useful for molding into circuit boards have been prepared by this process (*19*).

Dihalide Reactivity

The same considerations about this parameter that were discussed earlier under the caustic process also apply to the carbonate process. Recent research has shown the amide groups in the *para* positions of aromatic fluorides can activate their displacement by phenoxides (*20*). Several polyaryletheramides have been prepared from *N*,*N'*-bis(4-fluorobenzoyl)-*m* (and *p*-)-phenylene diamine using the carbonate process. Polymerization of oxazole and thiazole monomers containing trifluoromethyl groups to activate adjacent fluorine atoms with diphenoxides has produced high molecular weight polymers (*21*). Triethyl amine was used as the base. Azine groups that are *para* to aromatic fluoro groups will activate them toward displacement polymerization in the carbonate process (*22*).

Choice of Base

Potassium carbonate is the base of choice because it is more soluble than sodium carbonate. Additionally, the resulting potassium phenoxides are also more reactive than the analogous sodium compounds. The mechanism that has been postulated for this process involves formation of the potassium monophenate and potassium bicarbonate as the first step (*3*). The monosalt then reacts with the dihalo compound to initiate chain growth. The potassium bicarbonate disproportionates to potassium carbonate producing water and carbon dioxide as shown in the following equations.

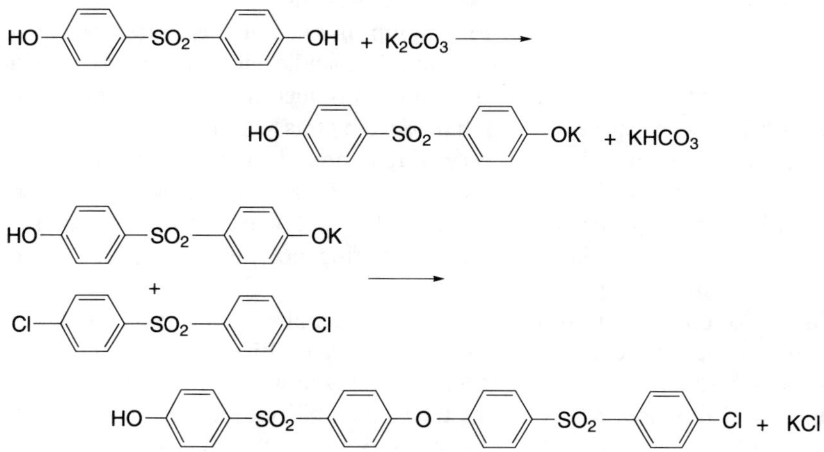

$$2KHCO_3 \longrightarrow K_2CO_3 + H_2O + CO_2 \qquad \text{(I-12)}$$

An azeotrope-forming solvent is included in the polymerization to remove the water as it is formed. When weaker bases such as potassium carbonate are used for the polymerization of diphenols there is a finite concentration of phenolic hydroxyl groups present during the polymerization. This is important because it enhances the solubility of the intermediate oligomeric salts which facilitates selection of the polymerization solvent. Both factors increase the probability for a successful polymerization. Additional studies on the mechanism of this polymerization have shown that there is a competition between polar and single electron transfer (SET) pathways under certain conditions (*23*).

Solvent Selection

The major factors that influence the choice of a solvent for nucleophilic substitution reactions were noted above under the caustic process. The dipolar solvents that are most frequently used in the carbonate process to prepare polyarylethersulfones are *N,N*-dimethylacetamide (*24*), *N*-methylpyrolidinone (NMP) and sulfolane. They provide progressively higher polymerization temperatures (boiling points), which enhance the bisphenol reactivity and the solubility of the polymers as they are being formed. Certain solvents, such as DMSO, can decompose in this process because of side reactions caused by the acidic phenolic groups that are present during the polymerization. This can lead to stoichiometry upsets and products of low molecular weight and quality. Sulfolane and NMP are preferred solvents for polymerization at higher temperatures, viz., 200–230°C. Polymerization of the more acidic bisphenol, 4,4′-dihydroxydiphenyl sulfone (Bisphenol S), with 4,4′-dichlorodiphenyl sulfone requires temperatures of 220–230°C in sulfolane to produce high molecular weight polymer.

Stoichiometry and Molecular Weight Control

The polymerization of a bisphenol with an activated aromatic dihalide by the carbonate process is a step-growth or condensation polymerization (*13*). The purity of the reactants, reagents, and solvents is critical to obtaining high molecular weight polymer. One to one stoichiometry of the diphenol and the dihalo compound will give the highest molecular weight. Maintaining the exact stoichiometric amount of the inorganic carbonate is less critical, in that small excesses (1–5 mole %) can be tolerated.

As in the case of polymers prepared by the caustic process, the polymer end groups play a determining role in the polymer melt stability. To generalize, aromatic halo groups are more melt stable than phenolic groups which are in turn more stable than phenate groups. Molecular weight control can be achieved in this polymerization by unbalancing the stoichiometry by a few percent. This is preferably done by using excess dihalo compound because it results in the incorporation of more aromatic halo groups as end groups. Monofunctional terminators can also be used to incorporate other specific end groups, but provide no real advantage in terms of thermal stability. Dendritic polyarylethers have been synthesized by this process in one step (*25*).

EXAMPLE I: PREPARATION OF POLYSULFONE FROM BISPHENOL A AND 4,4′-DICHLORODIPHENYL SULFONE BY THE CARBONATE PROCESS IN DMAC (*24*)

This example illustrates the use of the carbonate/DMAC process for the preparation of hydroxyl terminated oligomers of Polysulfone (*24*). The use of a 1:1 stoichiometry of the reactants would give high molecular weight polymer. This reference also contains extensive information about the purification of monomers and solvents for use in these polymerizations.

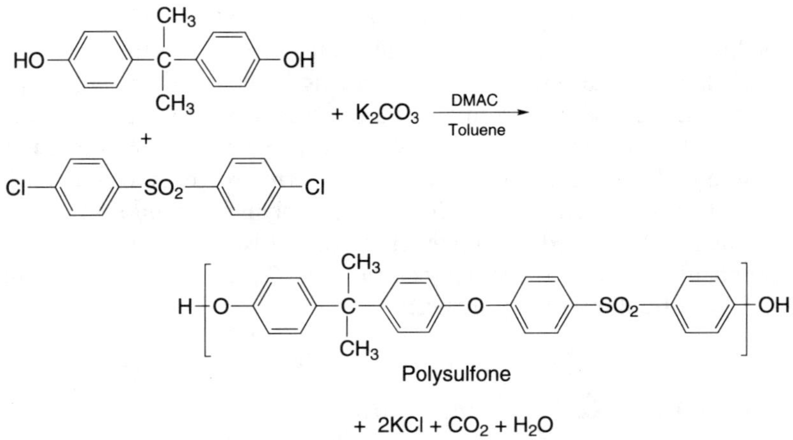

(I-13)

The 500 ml reaction assembly was purged with nitrogen, then 150 ml of *N,N*-dimethylacetamide (freshly purified) was added along with accurately weighed Bisphenol A (22.8 g, 0.1 mole) and 4,4′-dichlorodiphenyl sulfone (23.88 g, 0.083 mole). The aluminium weighing pans and powder funnel were rinsed with 75 ml toluene. The reaction mixture was stirred vigorously with a constant purge of nitrogen and heated to reflux. It is maintained at reflux until no more water droplets from the reaction are observed. Toluene is removed continuously from the trap until the temperature rises to about 155°C. The reaction mixture appears lightly colored and is maintained at this temperature for nearly 10–12 hr. At this time the reaction is assumed to be complete. The reaction mixture is cooled to about 100°C and about 75 ml of chlorobenzene is added to dilute the solution and precipitate the inorganic salts. The mixture was filtered through a medium pore size sintered glass funnel. The filtrate was neutralized with acetic acid. The clear neutralized solution (in 25 ml portions) is coagulated in a blender containing 250 ml (1:1) water and methanol. The precipitate was filtered and washed with methanolic water and finally with water. Next, it was boiled in distilled water for 1 hr. to remove any trapped salts, filtered and dried in a vacuum oven at 100°C; yield was >90%.

EXAMPLE II: SYNTHESIS OF A 50:50 "RANDOM" HYDROQUINONE: BIPHENOL POLYSULFONE (*27, 28*)

The reaction assembly is a 500 ml, 4-neck, round bottom flask equipped with a stirrer, thermometer, Dean Stark trap and condenser and it is purged with argon prior to the addition of the monomers or solvents. Next, 9.3105 gms or 0.05 moles of 4,4′-dihydroxybiphenyl was accurately weighed into Teflon coated aluminum pans. The monomer was delivered into the flask through a funnel and the delivery pan was washed several times with distilled *N*-methylpyrolidinone. In a similar fashion, 5.0055 gms or 0.05 moles of hydroquinone were accurately weighed out and placed in the flask. Then the 4,4′-dichlorodiphenyl sulfone was weighed out and added (28.716 gms or 0.1 moles). Each of the monomers was carefully washed down into the flask with distilled NMP. The final volume of NMP should be approximately 200 mls. To this solution, approximately 100 ml of toluene was added as an azeotroping solvent. Finally, approximately 18 gms or 0.135 moles of K_2CO_3 base was added. Note that K_2CO_3 was used in 30 to 40% excess. This reaction mixture was then heated until the toluene began to reflux. An optimum reflux temperature range appears to be about 140–150°C. The Dean Stark trap should be calibrated to hold approximately 40 mls of fluid. The initial distillate collected in the Dean Stark trap will be very cloudy. This is due to the water azeotroped from the system, as well as that released during phenoxide formation. The actual amount of water collected is related to the moles of phenol groups. After 1–2 hours the liquid in the Dean Stark trap will separate into two layers, the top layer is relatively pure toluene and the bottom layer is the azeotrope mixture. Refluxing must proceed until no more water is observed; which may take 4–6 hours. During this period the joints around the thermometer, gas inlet and mechanical stirrer must be kept fairly dry. This can be accomplished by utilizing a heat gun and heating the joints until the toluene and water particles coalesce and collect back into the solution; this, in fact, is a key step in the dehydration. During this stage of the reaction the solution may undergo several color changes. For example, during the initial formation of the phenolate, a yellow-brown color can be observed. As the refluxing proceeds, the color changes to a dark brown, and the solution may possibly have a dark green tint.

Once the water has been removed, the temperature is increased slowly. This may be accomplished by removing toluene through the Dean Stark trap. The temperature should increase to 180°C after removing toluene for about one hour. The Dean Stark trap must remain full and also a small amount of toluene should be left in the reaction vessel. This serves two purposes; (i) any water that is formed will be removed by the slowly refluxing toluene, (ii) the refluxing serves as an excellent method for temperature control. These reactions vary in time from 1.5 to 4 hours; completion or near completion may be estimated by the point the viscosity increases drastically; the polymer may even appear to come out of solution with very high molecular weight materials. At this point more solvent can be removed and the reaction temperature increased to 190–195°C. The reaction is allowed to continue for an additional one hour.

The high molecular weight product is then filtered hot through a 2 liter coarse filter under a water aspirator. The solution must be diluted with an additional 200 ml of NMP. This will aid in filtration and in the cleanliness of the final product. The filtered solution is cooled and acetic acid should be added to neutralize the phenate. Acetic acid is added, usually 10–20 mls, until the color changes from a dark green-brown to a transparent brownish color. The polymer solution is coagulated first by adding it to a one liter separatory funnel and then allowing the solution to drip slowly into a stirred blender containing about 10x volume of methanol and water in a 50/50 mixture. The coagulated polymer is filtered and boiled in water (pH approximately 5) for one hour to extract any trapped salts. Once again the polymer is filtered and dried under vacuum at 80°C to a constant weight. The final product should be a white, fibrous appearing polymer. This poly(arylene ether sulfone) copolymer possessed an intrinsic viscosity of 0.71–0.85 (in NMP at 25°C) and a glass transition temperature of 217°C.

The above procedure can also be used for the preparation of the homopolymers of hydroquinone or biphenol and 4,4′-dichlorodiphenyl sulfone using either NMP or sulfolane as the reaction solvent. The biphenol/4,4′-dichlorodiphenyl sulfone polymer is sold commercially under the Radel™ R tradename.

N-3. Self Polymerization of Halophenylsulfonyl Phenoxides

A wide variety of alkali metal halophenylsulfonyl phenoxides have been synthesized and their polymerization studied (*29, 30*). The ready availability of potassium 4-chlorophenylsulfonyl-4′-phenoxide by partial hydrolysis of 4,4′-dichlorodiphenyl sulfone was mentioned earlier. It polymerizes to usefully high molecular weight polymer at 305°C in thirty minutes as shown in the accompanying equation.

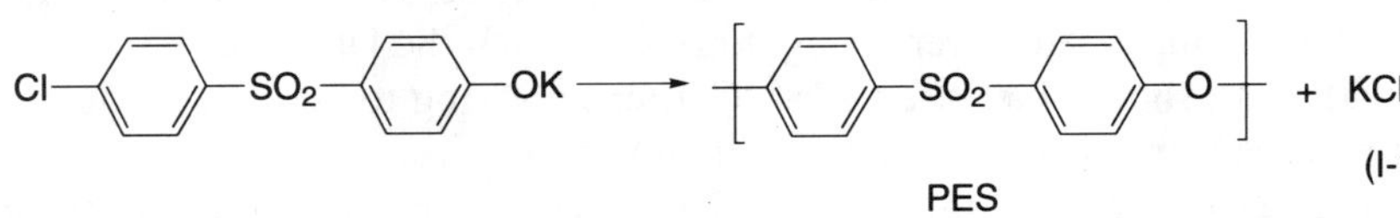

(I-14)

On the other hand, solution polymerization in sulfolane at 230°C required 24 hours to produce polymer with a reduced viscosity of only 0.45. The sodium or potassium salts of halophenylsulfonyl phenols give the best results when the dehydrated salts are melted in continuously evacuated tubes. The degree of polymerization is determined by the purity of the salt employed. Reaction temperatures below 300°C are preferred to minimize gel formation. High molecular weight polymers were obtained only from those halophenoxides where the halogen is *para* to sulfone, and the phenoxide group is in either the *meta* or the *para* position in the other ring. Fluorides gave better results than chlorides. Kinetic studies (*30*) have shown that the reactivities lie in the orders *p*-F > *p*-Cl > *o*-Cl >> *m*-Cl and *m*-OK > *p*-OK > *o*-OK in these systems.

The kinetics of the polycondensation of potassium 4-chlorophenylsulfonyl-4′-phenoxide has been studied and compared to the kinetics of the polycondensation of 4,4′-dichlorodiphenyl sulfone with the dipotassium salt of 4,4′-dihydroxydiphenyl sulfone. Both routes, of course, give Polyethersulfone (*31*). The initial rate constant for the self polymerization was 1500 times smaller than that for the two monomer case. This was due to the different types and reactivities of the functional groups that were present at the start of the polymerization. As the reaction progressed, both routes proceeded at the same rate in accord with theory. The analogous fluorine-containing monomers have also been studied in these two polymerizations with similar results (*11*).

EXAMPLE I: PREPARATION OF THE POTASSIUM SALT OF 4-FLUORO-4′-HYDROXYDIPHENYL SULFONE AND ITS MELT POLYMERIZATION (*30*)

4-Fluoro-4′-hydroxydiphenyl sulfone was dissolved in methanol and the calculated amount of 4N potassium hydroxide solution in aqueous methanol added from a burette. The resulting solution was evaporated to dryness using a rotary evaporator at 14 mm Hg pressure. The flask was heated by a water bath, the temperature being gradually increased to 100°C. The potassium salt was removed from the flask, ground up and further dried under high vacuum (100°–120°C, 0.001 mm Hg). The salts were generally prepared using a 0.25% excess of the halogenophenol; this stoichiometry was shown to give the highest RV melt polymer. The potassium salt was placed in a test tube which was evacuated by means of an oil pump to about 1 mm Hg. The tube was then immersed in a molten metal bath (usually for 45 min/280°C). The tube was kept evacuated by the oil pump during the polymerization. The tube was then cooled, the polymer removed, dissolved in a suitable solvent (e.g. *N,N*-dimethylformamide) and the solution precipitated into water or methanol in a laboratory blender. The fine polymer powder obtained was boiled with water to remove potassium halide and solvent, and then dried.

EXAMPLE II: SOLUTION COPOLYCONDENSATION OF POTASSIUM 4-(4-FLUOROPHENYLSULFONYL) PHENOXIDE WITH THE 4,2′-ISOMER (*30*)

The potassium salt of 4-fluoro-4′-hydroxydiphenyl sulfone (60 g, 0.2066 mol) and the potassium salt of 4-fluoro-2′-hydroxydiphenyl sulfone (60 g, 0.2066 mol) were stirred under dry nitrogen in sulfolane (120 g) at 200°C, (in an oil bath) for 44.5 hour. The solution was then cooled to 140°C, and methyl chloride passed through the stirred solution for 2 hours. The polymer solution was cooled to 20°C and diluted with *N,N*-dimethylformamide. This solution was poured into methanol in a laboratory blender. The polymer was filtered off and refluxed once with methanol and twice with water and dried (yield 92 g).

N-4. Diphenols and Activated Aromatic Dinitro Compounds

Activated aromatic nitro compounds can also be polymerized with bisphenates to yield polyarylethersulfones (*32*). The preparation of polysulfone by this route is shown in the accompanying equation.

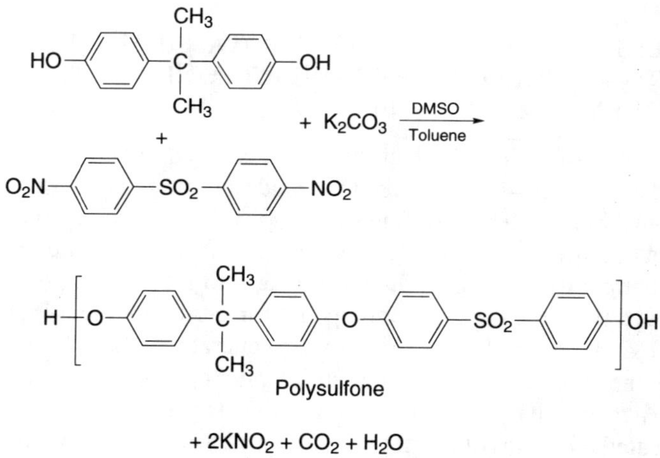

(I-15)

EXAMPLE I: POLYCONDENSATION OF BISPHENOL A WITH 4,4′-DINITRODIPHENYL SULFONE USING POTASSIUM CARBONATE (*32*)

4,4′-Dinitrodiphenyl sulfone, 30.83 g (0.10 mol), 2,2-*bis*(4-hydroxyphenyl)propane, 22.83 g (0.10 mol) and anhydrous potassium carbonate, 13.82 g (0.10 mol) were reacted in a mixed solvent of dimethyl sulfoxide, 200 ml, and toluene, 100 ml, at 130°C over 4 hours and further at 150°C over 2 hours. The by-product water was removed azeotropically. 42.41 gm of polymer was obtained by pouring the reaction mixture into excess methanol. Polyarylene ether with an intrinsic viscosity of 0.56 was obtained.

Aromatic nitro groups that are activated solely by ortho-trifluoromethyl groups can be polymerized by the carbonate process (*33*). The condensation of the bisphenate salt of Bisphenol S with 4,4′-dinitrobenzophenone has also been reported (*34*). The product, a poly(ether ketone sulfone), possessed a relative viscosity of 1.07 (NMP). A "caustic type" process

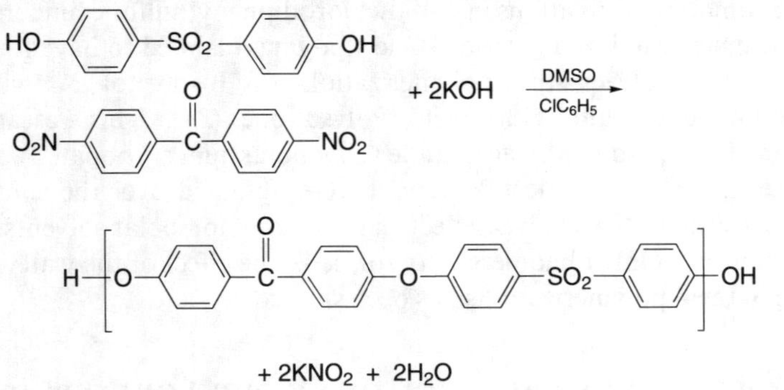

(I-16)

using potassium hydroxide gave the best results. Extensive model compound studies were also reported in this publication (*34*). The use of the nitro monomers offers no advantage over the halo monomers, especially when cost and availability factors are weighed. Additionally, the release of nitrite ion into the polymerization mixture may adversely affect quality factors, such as color.

EXAMPLE II: POLYCONDENSATION OF THE DIPOTASSIUM SALT OF 4,4′-DIHYDROXYDIPHENYL SULFONE WITH 4,4′-DINITROBENZOPHENONE (*34*)

A solution of 50.0 g of 4,4′-dihydroxydiphenylsulfone (0.2 mol) in a mixture of 450 ml dimethylsulfoxide and 200 ml chlorobenzene was mixed under nitrogen with a solution of 25.5 g of potassium hydroxide (87.9%)(0.4 mol) in 30 ml of water. After azeotropic removal of the water, the chlorobenzene was removed by distillation under vacuum and 54.4 g (0.2 mol.) of 4,4′-dinitrobenzophenone was added under nitrogen. The reaction mixture was then condensed at 140°C for 20 hours. The yellow colored polycondensate was precipitated in water, ground in a blender, filtered, washed with methanol and dried in vacuum at 50°C. Yield 57.4 g (67% of theory). Relative viscosity = 1.07 in NMP @ 25°C.

N-5. Phase Transfer Catalyzed Systems

The synthesis of polyarylethersulfones under solid-liquid phase transfer or interfacial conditions has the potential to offer economic advantages over the original solution process. These would lie primarily in the areas of polymer isolation and purification, and the solvent purification technologies that are now required to produce a commercial product. The relatively low reactivity of the aromatic chloro atoms in 4,4′-dichlorodiphenyl sulfone hinders a facile phase transfer-catalyzed reaction. One study has claimed to have found that 18-crown-6 catalyzes the polymerization of Bisphenol A with 4,4′-dichlorodiphenyl sulfone to yield Polysulfone (*35*). This research was performed in *N,N*-dimethylacetamide using potassium carbonate. Very small increases in polymer molecular weight were obtained over the control non catalyzed system. No catalytic effect was found in non polar solvents, such as chlorobenzene. Only oligomers were formed. The effect of the water concentration in these polymerizations has been studied.

EXAMPLE I: POLYMERIZATION OF 4,4′-DICHLORODIPHENYL SULFONE AND BISPHENOL-A CATALYZED BY 18-CROWN-6 (*35*)

To a solution of 5.00 mmol of 4,4′-dichlorodiphenyl sulfone and 5.00 mmol. of Bisphenol A in 20 ml of DMAC and 20 ml of toluene was added 20.0 mmol. of K_2CO_3. The reaction mixture was purged with nitrogen for 1 hr. and then refluxed for 3 hour to remove water as an azeotrope with the solvent through the use of a Dean Stark apparatus. After cooling to room temperature, the Dean Stark trap was removed and 1.25 mmol of 18-crown-6 catalyst and/or incremental aliquots of water were added. The reaction mixture was allowed to reflux for 17 hour, cooled to room temperature, and slowly added to 300 ml of methanol vigorously stirred in a blender. The filtered solids were then washed three times with 300 ml portions of distilled water, air dried, and placed in a vacuum oven for 24 hour. Polymer viscosity was determined in chloroform (30°C). At a mole fraction of water/catalyst of 0.2, the polymer possessed an inherent viscosity of 0.27 ($CHCl_3$ @ 30°C).

More recently, cryptands have been employed successfully as catalysts to prepare Polysulfone in non polar solvents (*36*). Very high molecular weight polymer has been obtained from Bisphenol A and 4,4′-dichlorodiphenyl sulfone in chlorinated benzenes and other inert solvents. The free bisphenols or their phenate salts can be used with the activated aromatic halo compounds.

EXAMPLE II: PREPARATION OF POLYSULFONE IN A (2.2.2)-CRYPTAND CATALYZED PROCESS (*36*)

A 250 ml, four-neck, round bottom flask was equipped with an overhead stainless steel stirrer through a stuffing box, a nitrogen inlet, a Claisen adapter and a stopper. The Claisen adapter was attached to a thermocouple thermometer and a two-way addition funnel and a condenser. The flask was charged with aqueous potassium hydroxide (12.25 g, 45.8%, 100 mmol), water (10 ml) and bisphenol A (11.4140 g, 50 mmol). Prepurified nitrogen was allowed to bubble into the reaction mixture and was maintained continuously throughout the reaction. After 0.5 hour at room temperature, while being stirred slowly, bisphenol A dissolved completely. Toluene (100 ml) was added. After another 0.5 hour at room temperature, the reaction flask was immersed in an oil bath. The temperature was set at 115°C and the batch was heated. Water that was collected at the two way addition funnel as an azeotrope of toluene was removed and measured. Toluene was returned to the reactor. During the dehydration, a pressure equalizing addition funnel was charged with dichlorodiphenyl sulfone (14.3590 g, 50 mmol), cryptand (0.94 g, 2.5 mmol) and dichlorobenzene (60 ml). Flexible electric heating tape was wrapped around the addition funnel. A nitrogen inlet was inserted from the top to the bottom of the addition funnel. The slurry was heated while being degassed to the point that the monomer dissolved and kept at that temperature. When the total amount of water removed was above 15.7 ml, the above hot solution of sulfone monomer was added. Immediately, a large amount of the disalt dissolved and the reaction became slightly yellowish. The temperature was now set to 150°C and toluene was removed from the system. After 15 minutes, the temperature was set to 180°C and all toluene was removed. The reaction was maintained at 180°C for 3 hours. The viscous polymer solution thus obtained was cooled to room temperature and diluted with chlorobenzene (200 ml) containing g-acetic acid (5 ml). The salts from the reaction were filtered through a sintered filter funnel (medium). The clear filtrate was coagulated into methanol (1.5 l) in a large high speed blender. The isolated polymer was dried at 110°C and at 50 mm vacuum. Reduced viscosity = 0.46.

N-6. Silylated Diphenols and Activated Aromatic Dihalides

The polymerization of silylated diphenols with activated aromatic dihalides was developed by Kricheldorf and coworkers (*37*, *38*, *39*). It is now referred to as the "silyl method"(*40*). First, a bisphenol is silylated using hexamethyldisilazane or chlorotrimethylsilane/triethylamine and purified by distillation. The former method is preferable because it reduces the risk of contaminating the product with chloride ion (*39*). The *bis*-(*O*-trimethylsilylated) bisphenol and 4,4′-difluorodiphenyl sulfone are polymerized in bulk using a catalytic amount of cesium fluoride. It is believed that the polymerization is initiated by fluoride ion attack on the *O*-trimethylsilyl group to generate a phenoxide anion. The volatile fluorotrimethylsilane is easily removed from the reaction mixture. The preparation of Polysulfone by this method is shown in the accompanying equations.

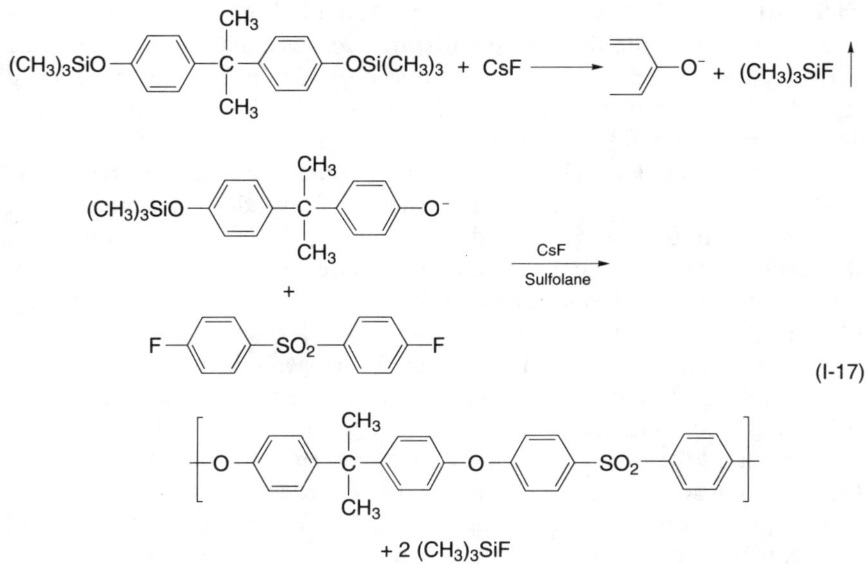

(I-17)

The temperatures required to effect this polymerization vary with the nucleophilicity of the phenate anion since it is basically a polyetherification method. For example, silylated Bisphenol A polymerizes at 220–230°C whereas silylated Bisphenol S required 300°C (*39*).

A general deficiency of the conventional metal bisphenate routes to polysulfones is that the resulting polymers need to be purified from the metal salt by-products and also from the expensive high-boiling point solvents. In contrast, the silylation method has the advantage in that the molten polymers do not need purification from large amounts of metal salts and solvents prior to processing, although a catalytic amount of fluoride ion exists in the molten polymers. On the other hand, the need to use the expensive aromatic difluorides makes this method unattractive for commercialization.

Recent work (*38*) has focussed on the use of 4,4′-dichlorodiphenyl sulfone (DCDPS) in this polymerization. In order to get polymerization of DCDPS with silylated Bisphenol A to occur in bulk, a molar excess of potassium fluoride or an equimolar amount of KF with a trace amount of CsF was necessary. This requirement immediately negates the advantage of this type of process. It was found that high molecular weight Polysulfone was obtainable from DCDPS and silylated Bisphenol A by conducting the polymerization in NMP or DMAC using an equimolar amount of potassium carbonate. The trimethylsilyl moieties are converted to hexamethyldisiloxane. Since DCDPS and free bisphenols give high polymer under these conditions, the use of silylated bisphenols under these conditions offers no advantage. The successful development of a catalytic process of this type that used 4,4′-dichlorodiphenyl sulfone would be of commercial interest. An excellent review on the use of silylated nucleophilic monomers for the synthesis of a wide variety of condensation polymers is available (*41*). Linear and star-branched poly(arylene ethers) have been synthesized from a variety of silylated intermediates (*42*).

EXAMPLE I: POLYMERIZATION OF SILYLATED BISPHENOL A WITH 4,4′-DIFLUORODIPHENYL SULFONE (*37*)

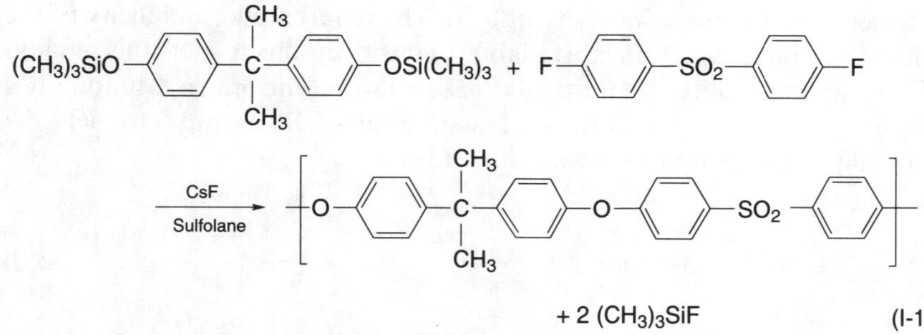

(I-18)

0.1 Mol of *bis*(*O*-trimethylsilyl) Bisphenol A and 0.1 mol 4,4′-difluorodiphenyl sulfone were heated in a glass flask under a slow stream of nitrogen to 180°C. Then 50 mg of dry, finely powdered cesium fluoride was added and heating at 180°C was continued for 0.5 hour, when the evolution of fluorotrimethylsilane nearly ceased. Then the reaction temperature was raised to 320°C for 2.5 hour, when a viscous melt was obtained. After cooling, the polymer was dissolved in 300–500 ml boiling methylene chloride, precipitated from 1–2 litre. cold methanol, and dried at 80°C/12 mbar. The inherent viscosity of the product was 1.4 (in chloroform @ 20°C).

Bis(carbamates) of bisphenols have been polymerized with sulfone-activated dihalides to yield polysulfones in a process that bears a formal similarity to the use of silylated bisphenols (*43*).

2. Disulfinates and Activated Aromatic Dihalides

N-7. Disulfinates and Activated Aromatic Dihalides

Disodium 4,4′-diphenyl ether disulfinate has been condensed with a doubly activated aromatic dihalide to produce an aromatic polysulfone (*44*). The polymer, which is shown in the accompanying equation, was obtained at a

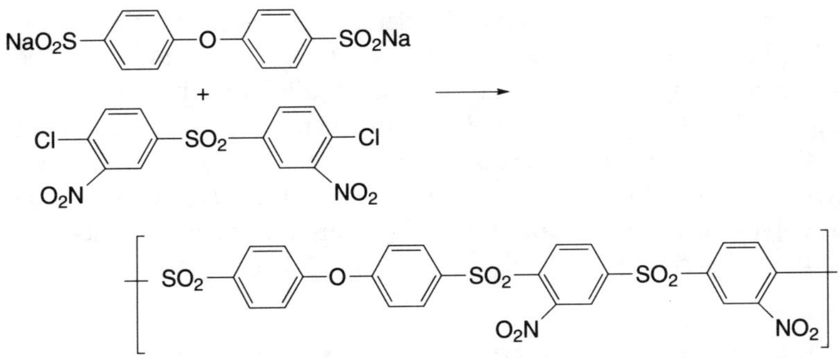

(I-19)

very low molecular weight. A phase transfer catalyst, tetrabutylammonium chloride, was used in a water-nitrobenzene system to give a 93% yield of polymer with a reduced viscosity of 0.16. The reduced nucleophilicity of the sulfinate anion and monomer availability considerations renders this method of very limited utility. Interestingly, *bis*(4-chloro-3-nitrophenyl)sulfone has also been polymerized with the dipotassium salt of Bisphenol S to yield the nitrated polyethersulfone shown below (*45*).

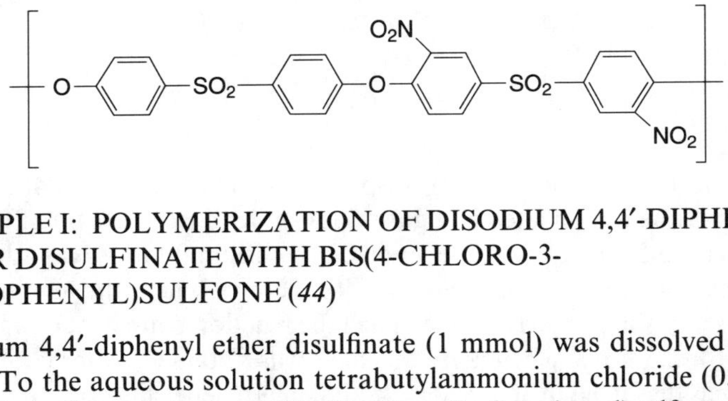

(I-20)

EXAMPLE I: POLYMERIZATION OF DISODIUM 4,4′-DIPHENYL ETHER DISULFINATE WITH BIS(4-CHLORO-3-NITROPHENYL)SULFONE (*44*)

Disodium 4,4′-diphenyl ether disulfinate (1 mmol) was dissolved in water (4 ml). To the aqueous solution tetrabutylammonium chloride (0.5 mmol) was added. Then a solution of *bis*(4-chloro-3-nitrophenyl)sulfone (1 mmol) in nitrobenzene (6 ml) was placed into the aqueous solution. The reaction mixture was stirred for 24 hours at 80°C. After the completion of the reaction, the upper aqueous layer was removed by decantation and the nitrobenzene solution of the polymer was poured into an excess of diethyl ether to give a precipitate. The polymer obtained was filtered off, washed with water and diethyl ether, and dried in vacuo. Yield: 93%. Reduced viscosity: 0.16 (DMAC).

B. POLY(ETHER KETONES)

Polyaryletherketones, or PAEKS, comprise a commercially valuable class of aromatic polyethers. The major intrinsic property difference between polyaryletherketones and polyarylethersulfones is that many of the former are crystalline polymers. The two major property advantages that crystallinity confers on polymers are higher use temperatures and improved solvent resistance. PAEKS are improved in both of these respects over the polysulfones, but these advantages come at a cost. PAEKS are somewhat more difficult to synthesize because they are less soluble in the usual reaction solvents due to their crystallinity.

The accompanying table illustrates some of the PAEKS that are or have been of commercial interest (*46*, *47*). The acronyms depicted are widely used in referring to PAEKS. For example, poly(aryl ether ether ketone) is referred to as PEEK. The E's and K's refer to the arrangement of ether and keto groups in the repeat units of the polymers. The aromatic groups in the polymers that have been commercialized are all 1,4-phenylene moieties.

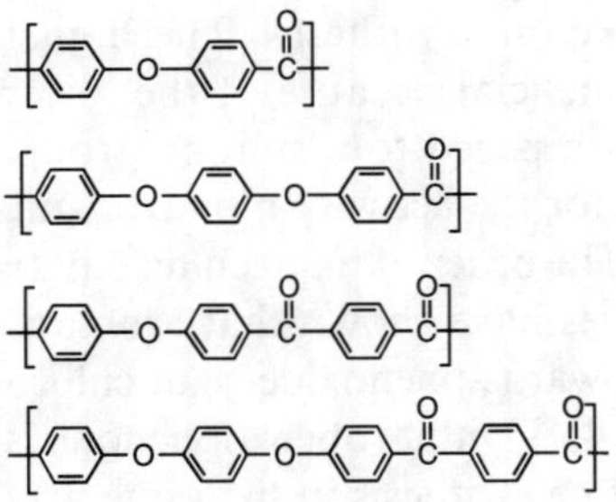

Name	Tg, °C	Tm, °C
PEK	155	365
PEEK	143	334
PEKK	165–175	385
PEEKK	155	365

(I-21)

1. Polyetherification

N-8. Diphenols and Activated Aromatic Dihalides: Caustic Process

Attempts were made to prepare polyaryletherketones by the "caustic process" that was discussed earlier in detail under Method *N*-1(1,2). High molecular weight polymer was not achieved because of premature crystallization and precipitation of the growing polymer chain. Polymerization of 4,4′-difluorobenzophenone with the potassium salts of hydroquinone or 4,4′-dihydroxybenzophenone in either dimethyl sulfoxide or sulfolane yielded polymer below the useful molecular weight range (RV=0.27). A key breakthrough was the discovery that diphenyl sulfone was a uniquely useful solvent for the synthesis of PAEKS. Polymerizations of potassium phenates with 4,4′-difluorobenzophenone at 300°C in diphenyl sulfone gave very high molecular weight polymer (*46*, *48*). An example is shown in the accompanying equation. In general, the potassium phenates are prepared and isolated separately as

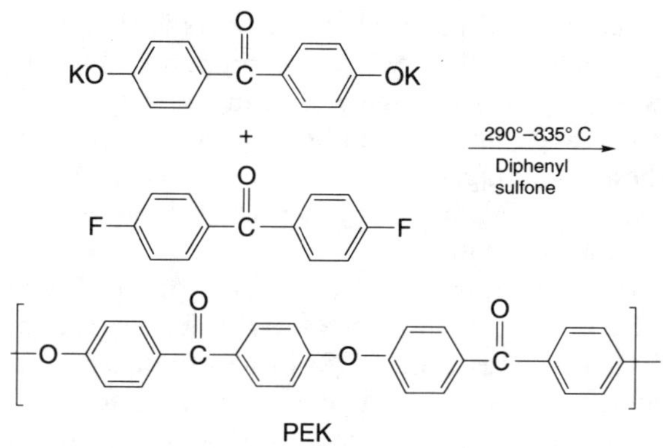

(I-22)

contrasted to the one-pot procedure that can be used for the synthesis of amorphous polysulfones.

The order of reactivity for phenates derived from alkali metal cations is Cs > K > Na > Li. On a reactivity/cost basis, potassium is preferred. The fluoroaromatics are generally used for polyketone syntheses. The higher reactivity of the aromatic fluoro group is beneficial because of the lower activating power of the carbonyl group as compared to a sulfone group. The power of the electron withdrawing group for the activation of aromatic halides is $NO_2 \sim SO_2$ > —CO— > —N=N—. The order of halogen reactivity is F>>> Cl > Br > I. Model compound studies have shown that aromatic fluorine is one hundred times more reactive towards phenoxide than chloro when both are activated by a keto group (*49*). 4,4′-Difluorobenzophenone is the workhorse monomer for polyaryletherketone synthesis in the same way that 4,4′-dichlorodiphenyl sulfone is for polyarylethersulfone synthesis.

The use of 4,4′-dichlorobenzophenone for poly(aryletherketone) synthesis has been a desirable research goal because of the economic advantages it would provide. It has been extensively investigated for this purpose but high quality polymers have not been obtained reproducibly. Polymers containing gel and/or branching with attendant poor melt processability are produced from the chloro monomers (*48*). Some research results have indicated that the nucleophilic aromatic substitution reaction on the chloro monomers may have a significant amount of radical character (*50, 51*). This factor could lead to side reactions that detract from the polymer's properties.

This discussion points up the fact that polyketones can differ significantly in properties when synthesized by slightly different routes (*48*). For example, partially crystalline films of PEK synthesized by the two monomer route shown above are tough at reduced viscosities of >= 1.03. The same polymer when prepared by the self polymerization of the potassium salt of 4-fluorophenyl-4-hydroxyphenyl ketone requires a threshold RV of about 1.9 for toughness.

The difference in the RV vs. toughness relationship found for the two polymers is due to some structural irregularity in the polymer from the fluorophenoxide route. This structural defect is likely to involve chain branching, a precursor to gel formation (*48*). Another route to branching in these polymers is via polyfunctional impurities in either of the monomers. Very small amounts of impurities can lead to deleterious effects in the melt viscosity properties of the polymers. Monomer purity must be very carefully monitored and controlled in order to produce a quality polymer.

The use of preformed, anhydrous phenates to produce polyketones has largely been supplanted by the "Carbonate" process which is discussed in the next section.

EXAMPLE I: POLYMERIZATION OF THE POTASSIUM SALT OF 4,4′-DIHYDROXYBENZOPHENONE WITH 4,4′-DIFLUOROBENZOPHENONE (*48*)

4,4′-Dihydroxybenzophenone, 2 equivalents of 4N KOH and just sufficient methanol to dissolve the whole were placed in a rotary evaporator flask and shaken until the pale yellow solution was clear. The methanol/water was then removed by rotary evaporation under reduced pressure for several hours. The pale yellow cake was then scraped from the flask and ground until it passed through a 500 μ sieve. Two samples of the salt were then titrated against 0.1 N sulfuric acid to determine the water content (ca. 10 wt. %) precisely. The partially dehydrated bis-potassium salt was mixed with half the required quantity of diphenyl sulfone by shaking in a bottle. This mixture was placed in a flask blanketed with nitrogen, heated to 220–240°C and stirred while the flask was gradually evacuated to 30 mm. giving a slurry of dehydrated bis-phenoxide. A slight excess (to control molecular weight) of 4,4′-difluoro-benzophenone was added with the rest of the diphenyl sulfone solvent and then the mixture was heated to 335°C for 2–3 hours. On cooling, the viscous melt set to a hard solid which was ground to a powder in a hammer mill and then extracted (3x each) with boiling methanol, boiling water, and methanol. The polymer was dried in a vacuum oven at 140°C overnight. Yields of polymer were usually >98% of theory. Reduced viscosities were run in 98% sulfuric acid at 25°C.

N-9. Diphenols and Activated Aromatic Dihalides: Carbonate Process

The carbonate process has been extremely useful for the preparation of aromatic polyethers (*18*). In contrast to the "caustic process", one does not have to prepare the anhydrous bisphenate, either *in situ* or in a separate step. This removes the necessity of carefully controlling the base stoichiometry during the phenate preparation and its subsequent reaction with the dihalo compound. Additionally, bisphenates are quite insoluble in organic media and easily oxidized by air to colored by-products that adversely affect product quality. In the carbonate process, diphenates are not formed. Rather, the monophenates react rapidly giving growing polyether chains that have enhanced solubility in the medium. This is particularly true of phenates derived from hydroquinone, a reactant used in the commercial synthesis of poly(ether ether ketone), or PEEK. However, application of the carbonate process for the preparation of PEEK is not straightforward, because of the insolubility of this crystalline polymer in conventional solvents.

The ICI group found that aryl sulfones, especially diphenyl sulfone, were uniquely useful for the preparation of aromatic polyethers by the carbonate process (*52*, *53*, *54*). (Needless to say, amorphous polyaryletherketones can be prepared in the more conventional solvents like DMAC, NMP, and sulfolane.) Their initial work was with poly (ether ketone sulfones) prepared from aromatic chloro monomers. When these products could not be scaled up successfully, their attention turned to polyaryletherketones, e.g. PEEK, that were derived from fluoro monomers. High molecular weight PEEK was prepared for the first time in 1977 by adding a slight excess of potassium carbonate to a mixture of hydroquinone and 4,4′-difluorobenzophenone in diphenyl sulfone. This is shown in the accompanying equation.

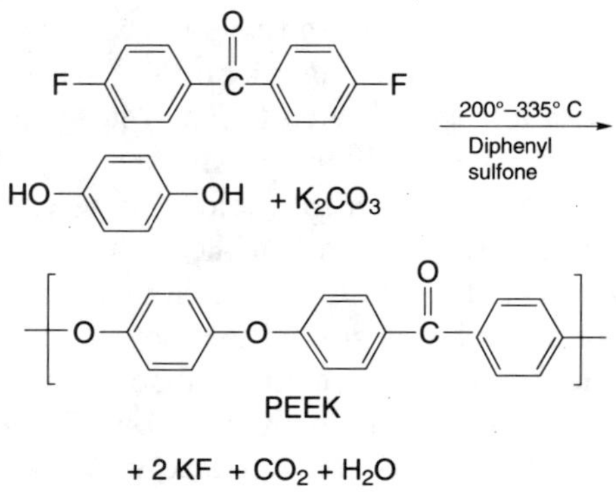

(I-23)

A programmed rise in reaction temperature is used for these polymerizations in order to keep the growing, crystallizable chains in solution (*53*). An excess of the fluoro monomer can be used to control molecular weight.

As the above equation shows, potassium fluoride, a strong base, is formed in these polymerizations. It can induce degradative ether interchange reactions at the high terminal polymerization temperatures. A way to prevent polymer degradation has been described in a recent patent (*55*). Addition of an alkali metal salt, such as lithium chloride, at the terminal polymerization conditions stops the polymerization and stabilizes the polymer against degradation. Alkaline earth salts such as magnesium chloride are also effective. This patent also teaches the use of the less expensive sodium carbonate as the base in conjunction with catalytic amounts of potassium fluoride in these polymerizations. Polyaryletherketone oligomers have been prepared and utilized to make products (Kadel™) that are very similar to PEEK (*56*).

EXAMPLE I: PREPARATION OF PEEK BY THE CARBONATE PROCESS (*54*)

4,4′-Difluorobenzophenone (21.82 g, 0.10 m); hydroquinone (11.01 g, 0.10 m); and diphenylsulphone (60 g) were charged to a 3-necked glass flask equipped with a stirrer, nitrogen inlet and air condenser. These materials were heated with stirring to 180°C to form a nearly colorless solution and, while maintaining a nitrogen blanket, anhydrous sodium carbonate (10.7 g, 0.10 m.) was added. The temperature was raised to 200°C and maintained there for 1 hour. The temperature was then raised to 250°C and maintained there for 15 minutes and then finally raised to 320°C and maintained there for 2.5 hours. The reaction mixture, which had become rather dark, was cooled and the resulting solid product milled so as to pass through a 500 μm sieve. Diphenylsulfone and inorganic salts were removed by washing successively with acetone (twice), water (thrice), and acetone/methanol (twice). The resulting solid polymer was dried at 140°C. under vacuum. The polymer had a reduced viscosity of 0.60 [RV measured at 25°C. on a solution of the polymer in concentrated sulfuric acid of density 1.84 g cm^{-3}, said solution containing 1 g of polymer per 100 cm^{3} of solution].

EXAMPLE II: PREPARATION OF AN AMORPHOUS POLYKETONE (*57*)

A three-neck flask equipped with a nitrogen inlet, thermometer, Dean-Stark trap and condenser was charged with 3 g (4.79 mmol.) of 1,2-bis(4-fluoro-

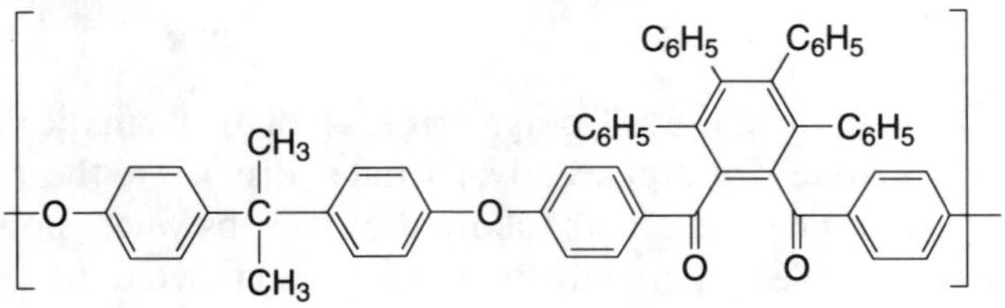

(I-24)

benzoyl)-3,4,5,6-tetraphenylbenzene, Bisphenol A (4.79 mmol, 1.09 g), and *N,N*-dimethylacetamide (16 ml). Toluene (8 ml) was added, followed by potassium carbonate (9.58 mmol, 1.32 g). The reaction mixture was then heated until the toluene began to reflux. An optimum reflux temperature range was achieved between 135° and 140°C. The reaction mixture was maintained at this range until the presence of water was no longer observed in the Dean-Stark trap, which took approximately 1 hour. Upon dehydration and slow removal of toluene from the reaction mixture, the temperature gradually increased to 165°C. The reaction was refluxed at this temperature for 1 hour. Completion or near completion of polymerization was qualitatively estimated by the point where the viscosity increased dramatically. The reaction mixture was cooled and diluted with DMAC (8 ml), and several drops of weak acid (e.g, acetic acid) were added to neutralize phenoxide end groups. The polymer solution was then coagulated in about 30 ml of methanol and filtered, redissolved in chloroform, and filtered hot through a thin layer of Celite. The chloroform solution was concentrated and then coagulated in methanol (30 ml). The white fibrous polymer was filtered and dried in a vacuum oven (120°C) for 48 hour to give the polyaryletherketone in almost quantitative yield with an inherent viscosity of 0.47 in chloroform at 25°C.

N-10. Self Polymerization of Activated Mono- and Dihalides

Polyaryletherketones can be synthesized by the self polymerization of monomers containing both alkali metal phenate and activated halo groups. The potassium salt of 4-hydroxy-4′-fluorobenzophenone polymerizes to high molecular weight PEK in diphenyl sulfone (*48*). This is shown in the accompanying equation. The PEK obtained from this route was found to be inferior in toughness at equivalent RV's to that obtained via the two monomer route discussed above. Branching in the polymer was found to be the cause of this property deficiency (*48*).

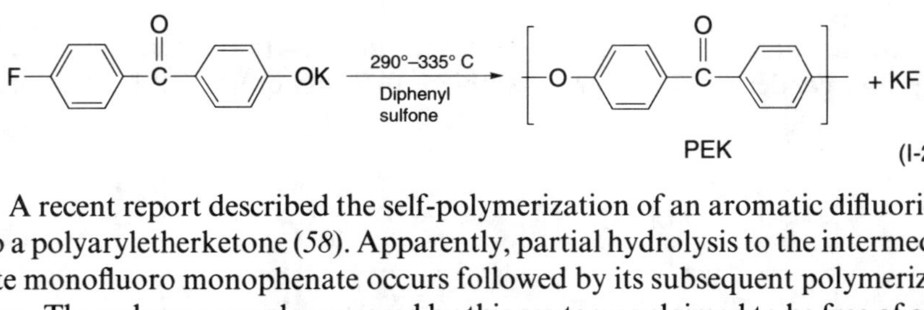

(I-25)

A recent report described the self-polymerization of an aromatic difluoride to a polyaryletherketone (*58*). Apparently, partial hydrolysis to the intermediate monofluoro monophenate occurs followed by its subsequent polymerization. The polymer sample prepared by this route was claimed to be free of gel.

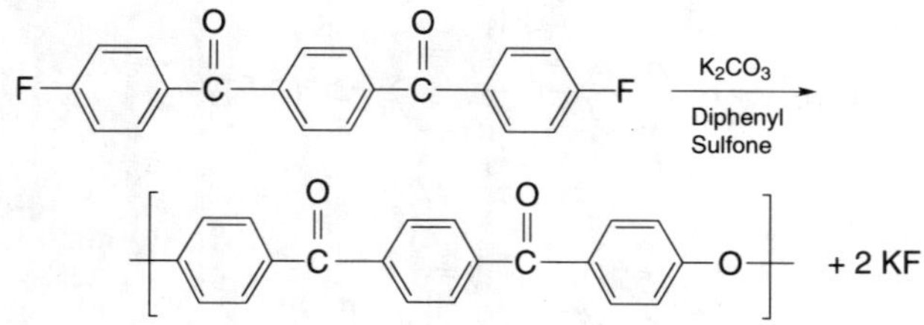

(I-26)

EXAMPLE I: POLYMERIZATION OF THE POTASSIUM SALT OF 4-HYDROXY-4-FLUOROBENZOPHENONE (*48*)

Precise neutralization of a methanolic solution of the fluorophenol with 4N KOH followed by rotary evaporation gave the bright yellow potassium salt which was ground to a powder and dried at 80–90°C at < 0.01 mm pressure. The fluorophenoxide plus a small quantity of bis-4-fluorophenyl ketone to control molecular weight, was placed in a flask with sufficient diphenyl sulfone to give the desired concentration of polymer (generally 30–50%) and heated with stirring under nitrogen at 335°C for 2–3 hour. On cooling the viscous melt set to a hard solid which was ground to a powder in a hammer mill and then extracted (3 × each) with boiling methanol, boiling water, and methanol. The polymer was dried in a vacuum oven at 140°C overnight. Yields of polymers were usually > 98% of theory. Reduced viscosities were measured in conc. H_2SO_4 @ 25°C.

EXAMPLE II: POLYMERIZATION OF 4,4′-DIFLUOROTEREPHTHALOPHENONE (*58*)

4,4′-Difluoroterephthalophenone (10.73 g), Na_2CO_3 (4.24 g), silica (2.0 g), and diphenylsulfone (40.0 g) are charged into a 100 ml flask, and nitrogen replaced the air, agitated and heated up to 280°C in 30 min. The mixture is kept for 1.5 hrs. at that temperature, heated to 310°C and reacted for 3 hrs., then heated up to 335°C within 10 minutes and reacted for 8.5 hrs. A polymer with an intrinsic viscosity of 1.15 (25°C) was obtained by removal of the salts and diphenyl sulfone by washing with water and acetone.

N-11. Silylated Diphenols and Activated Aromatic Dihalides and Diaryl Ethers

Polyaryletherketones can be prepared from silylated diphenols in a manner analogous to that described in Method N-6 for polyarylether sulfones (*39*, *59*). Fluoro monomers, specifically 4,4′-difluorobenzophenone (DFBP), must also be used in this case. Catalysis by fluoride ion (0.1 wt% CsF relative to the sum of both monomers) is also essential. The polyketone from Bisphenol A

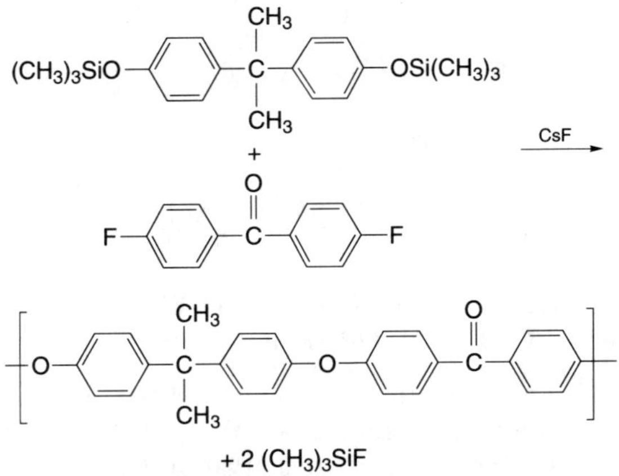

(I-27)

and DFBP is shown in the accompanying equation. The polymerization was conducted at various temperature schedules over the range of 220–320°C.

Most polyaryletherketones that are of commercial interest are crystalline and possess melting points greater than 300°C. The best known example is PEEK or poly(ether ether ketone). The utility of this method for the synthesis of this type of polymer is limited because crystallization during polymerization limits molecular weight growth. Temperatures greater than 320–330°C cannot be employed because of polymer crosslinking reactions. The absence of a solvent and the volatile nature of the by-product, fluorotrimethylsilane, are advantageous factors. Most of the polyaryletherketones that have been synthesized by this method have been amorphous (*39*).

EXAMPLE I: POLYMERIZATION OF BIS(O-TRIMETHYLSILYL) BISPHENOL A WITH 4,4′-DIFLUOROBENZOPHENONE (*39*)

The silylated Bisphenol A (0.1 mol), 4,4′-difluorobenzophenone (0.1 mol) and cesium fluoride (~100 mg) were heated in a three necked flask with stirring under a slow stream of nitrogen. The temperature at which a moderate and steady evolution of fluorotrimethylsilane occurred (220–230°C) was maintained for ~30 min. When the reaction temperature was raised to 320°C, a viscous foam was formed. After cooling, the polymer was dissolved in ~600 ml methylene chloride. The solution was filtered, and the filtrate precipitated into 4.0 l of ethanol. The polymer was dried at 80°C/12 mm. For elemental analyses small samples were dried at 185°C/12 mm. Polymer with relative viscosities of 1.1 (in conc. H_2SO_4 @ 25°C) was obtained. The Tg of the polymer was 152°C.

A variant of the "silyl" method that uses a diaryl ether instead of a dihalide has been reported (*60*). Heating a bis(silylated) diphenol with a bis(phenoxy) aromatic compound leads to exchange of the aryloxy moieties and formation of trimethylsilyloxybenzene and polymer. An example is shown in the accompanying equation

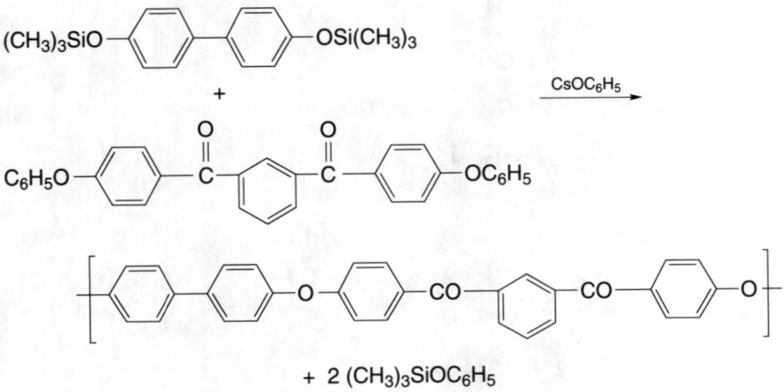

(I-28)

EXAMPLE II: POLYMERIZATION OF BIS(O-TRIMETHYLSILYLOXY)BIPHENYL WITH 1,3-(BIS(4-PHENOXYBENZOYL) BENZENE (*60*)

Heating 11.76 g of 1,3-bis[4-phenoxybenzoyl]benzene, 9.08 g of bis[4,4′-trimethylsilyloxy]biphenyl, and 90.8 mg Cs phenate, while removing trimethylsilyloxybenzene, under argon at 300°C for 1 hr., at 315°C for 150 min., and at 340°C for 50 min provided a polyether with inherent viscosity (0.50 g/dL in conc. H_2SO_4) of 1.05 dL/g.

Bis(carbamates) of bisphenols have been polymerized with keto-activated dihalides to yield polyketones in a process that bears a formal similarity to the use of silylated bisphenols (43).

N-12. Diphenols and Activated Aromatic Dinitro Compounds

Several diphenols have been condensed with 4,4′-dinitrobenzophenone to produce poly(arylether ketones)(*34*). The following equation illustrates the use of Bisphenol A in this synthesis. Polymers with relative viscosities of 1.07–1.38 were obtained from Bisphenol A, biphenol and 4,4’-dihydroxydiphenyl sulfone. The use of nitro monomers offers no advantage over the analogous halo monomers, especially when cost and availability factors are weighed. Additionally, the release of nitrite ion into the polymerization mixture may adversely affect quality factors, such as color.

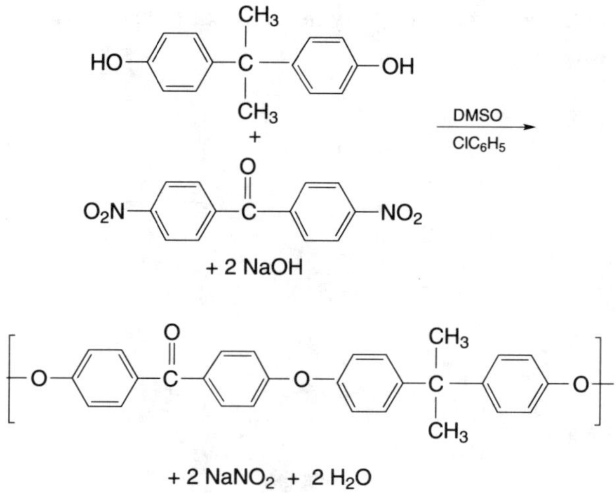

(I-29)

EXAMPLE 1: POLYCONDENSATION OF THE DISODIUM SALT OF 2,2-BIS(4-HYDROXYPHENYL) PROPANE WITH 4,4′-DINITROBENZOPHENONE (*34*)

A solution of 16.0 g (0.4 mol) of sodium hydroxide in 25 ml of water was added under nitrogen with stirring to a solution of 45.6 g (0.2 mol) of 2,2-bis-(4-hydroxyphenyl) propane in a mixture of 390 ml of dimethylsulfoxide and 210 ml of chlorobenzene. After azeotropic removal of water from the system under nitrogen, 54.4 g of 4,4′-dinitrobenzophenone (0.2 mol) was added. Reaction was conducted for 15 hours at 65°C. A part of the polymer precipitates towards the end of the reaction. The whole contents of the flask were then poured into excess methanol. The light yellow product was ground up in a blender with methanol, then thoroughly washed with water, and dried in vacuum at 50°C. Yield 78.2 g (94 % of theory). Relative viscosity of 1.38 (NMP @ 25°C).

N-13. Synthesis of Poly(ether ketones) via Amorphous Precursors

A novel strategy for preparing "defect-free" polyaryletherketones has been reported and patented (*61*). The synthesis involved the preparation of an amorphous precursor poly(ketal ketone) under mild reaction conditions by nucleophilic displacement polycondensation of an acetal monomer [2,2-bis(4-hydroxyphenyl)-1,3-dioxolane] with 4,4′-difluorobenzophenone in DMAC at 150°C or sulfolane at 220°C. The amorphous poly(ketal ketone) was then quantitatively converted to high molecular weight, crystalline, insoluble

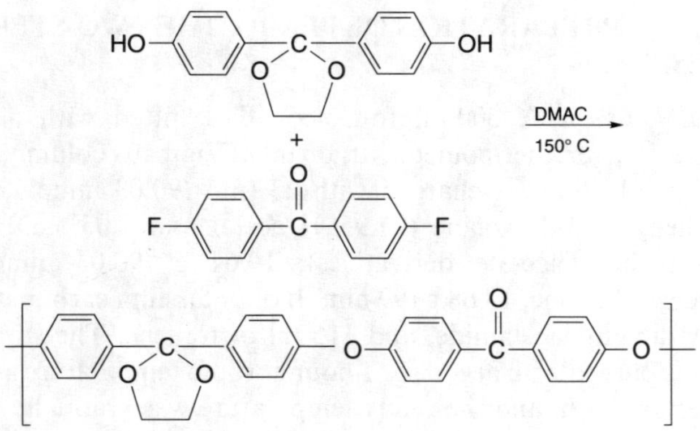

(I-30)

poly(ether ketone) [PEK] by acid-catalyzed hydrolysis under heterogeneous conditions. Homogeneous hydrolysis in concentrated sulfuric acid (25°C) was

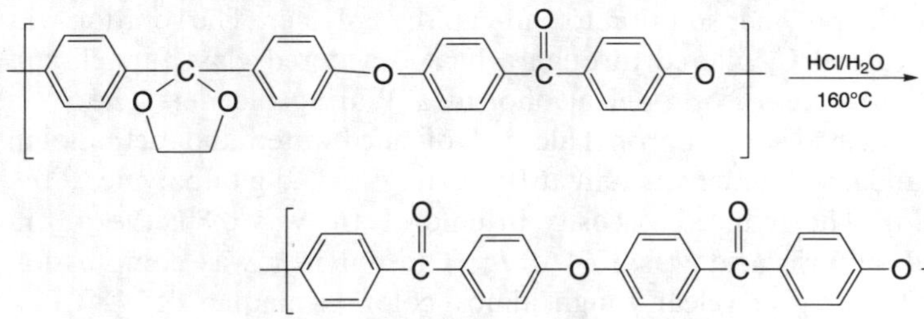

(I-31)

PEK

also effective. The PEK samples prepared by this route were more crystalline than those prepared by earlier electrophilic or nucleophilic routes. This effect was attributed to the absence of defects, such as position isomers or branching sites, that could have been introduced in the polymer structure by the conditions or reagents used in the other routes. As a result of the higher level of crystallinity, the polymer was improved in solvent resistance and had a higher modulus plateau between Tg and Tm. An improved synthesis of the novel cyclic acetals of 4,4′-dihydroxybenzophenone was developed which employs a combination of glycol, trialkylorthoformate, and a montmorillonite clay catalyst. This monomer was necessary because it was found that the acyclic dimethyl acetal of 4,4′-DHBP underwent an unusual degradation reaction under the basic polymerization conditions.

Imine groups have also been used to produce keto-protected polyaryletherketones (*62*). 4,4′-Difluoro(N-benzohydroxylidene aniline) was polymerized with hydroquinone by the carbonate process. The ketimine groups were then hydrolyzed to produce the polyaryletherketone. More recently, polyketones have been prepared via soluble poly(aminonitriles) (*63*).

EXAMPLE I: PREPARATION OF PEK BY THE TWO-STEP SYNTHESIS (*61*)

Poly(ketal ketone). A 500-ml four-neck flask fitted with a mechanical stainless steel stirrer, thermometer, argon inlet, Vigreaux column, Dean-Stark trap, and condenser was charged with 23.16 g (90.03 mmol) of 2,2-bis(4-hydroxyphenyl)-1,3-dioxolane (97.95 % acetal and 2.05% DHBP by GC analysis of the diacetate derivatives), 19.64 g (90.03 mmol) of 4,4′-difluorobenzophenone, 13.68 g (99 mmol) of potassium carbonate, 160 ml of dried *N,N*-dimethylacetamide, and 115 ml of toluene. The stirred reaction mixture was purged with argon for 1 hour at room temperature and heated to reflux in an oil bath, and the reflux temperature was gradually increased to 150°C by removing distillate from the trap. After 5.5 hour, 0.02 g of 4,4′-difluorobenzophenone (in 2 ml of DMAC) was added to the viscous polymer solution to assure stoichiometry. After 30 min, 135 ml of DMAC was added and the reaction temperature adjusted to 110°C. Methyl chloride was sparged for 1 hr. into the polymer solution to end-cap the polymer. The solution was then heated to 150°C, filtered through a heated sintered glass funnel, and coagulated into excess isopropyl alcohol in a Waring blender. The solid polymer was washed with isopropyl alcohol, distilled water, and methanol in the blender and dried under vacuum at 100°C to give 35.5 g of polymer (90% isolated yield). The reduced viscosity in chloroform was 0.80; the μ_{red} in concentrated sulfuric acid was 1.64 dL/g. The polymer was compression molded at 250°C to give a clear, tough, almost colorless plaque. Tg 145°C.

Heterogeneous Hydrolysis. The glass liner of a Parr rocker bomb was charged with poly(ketal ketone) fluff and dilute aqueous acid (10–100 ml/g of polymer), capped with the ground glass stopper (with pinhole to allow pressure equalization), and sealed in the bomb. The bomb was placed in the heater jacket and adjusted to rock gently in an almost vertical position. The temperature of the bomb (thermocouples) was initially adjusted to 150°C; temperatures between 160–200°C were then used. The polymer (PEK) was recovered by filtration and the polymer washed with water, stirred twice with boiling water, washed with methanol, and dried under vacuum at 100–120°C.

C. POLY(ETHER IMIDES)

Polyimides are one of the best known, large families of engineering polymers. Research in the mid-1950's at the DuPont Company led to the development and commercialization of the polyimide based on pyromellitic dianhydride and 4,4′-oxydianiline. This polymer was introduced in 1961 under several different trade names for different application areas. The major deficiency of this "super" performance polymer was that it was not thermoplastic and,

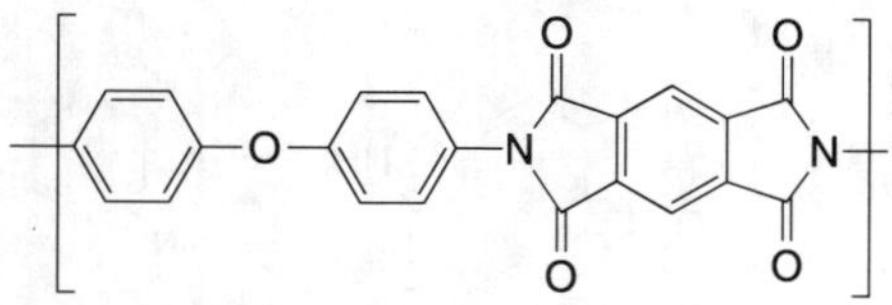

(I-32)

Kapton™, Pyre M L™, Vespel™, Pyralin™

hence, not injection moldable. Methods for the synthesis of polyimides and an extensive database of polyimides were included in a 1972 survey of ring-forming polymerizations (*64*). A large body of literature, which is beyond the scope of this book, exists on polyimides of this type (*65, 66, 67, 68*).

The focus of this section in this book is on thermoplastic, injection-moldable polyetherimides of the type represented by Ultem™. This polymer was introduced commercially by the General Electric Company in 1982 after many years of research and process development. An excellent review of its development by one of the inventors is available (*69*). The introduction of aromatic ether groups into polyimides has been driven by the desire to make polyimides that are melt processable and crystalline. This has been driven in large part by the search for matrix resins for use in advanced, thermoplastic, carbon fiber composites.

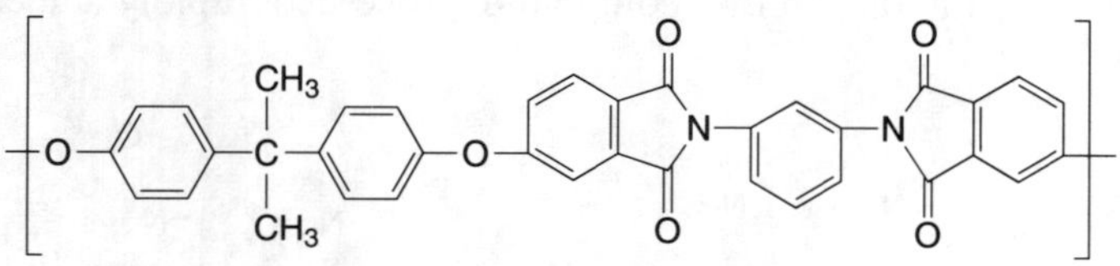

(I-33)

Ultem 1000™

1. Polyetherification

N-14. Diphenols and Nitro-Substituted Phthalimides

Fundamental studies on the nucleophilic displacement of aromatic nitro groups at the GE Company provided the basis for the development of commercial, thermoplastic polyetherimides. It was found that phenoxides reacted easily with both 3- and 4-nitro-N-phenylphthalimides to produce the corresponding phenyletherimides in excellent yields. This is shown in the accompanying equation.

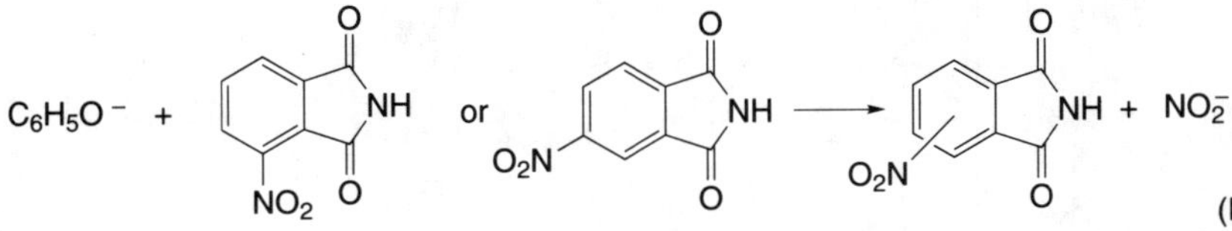

(I-34)

The analogous chloro and fluoro derivatives were studied in this system and the reactivity order was found to be nitro > fluoro > chloro. The 3-substituted isomers were also more reactive than the 4-substituted isomers. As expected, electron withdrawing groups on the imide nitrogen increased the the rate of nitro displacement. Phenoxides with electron-donating groups gave increased reaction rates. Ether exchange reactions occur between the etherimides and other phenoxides to give equilibrium mixtures determined by the nucleophilicities of the phenoxides (*69*).

Application of this basic chemistry was followed by the synthesis of bis-nitro-bis-imide monomers and their polymerization with bis-phenoxides (*70*, *71*). The chemistry outlined in the accompanying equations depicts the nitro displacement route to Ultem™ 1000. The bis-imide from m-phenylene diamine and 4-nitrophthalic anhydride was prepared in refluxing acetic acid in 92% yield (*71*). Preparation of the bisphenol A salt in an anhydrous form was essential for successful polymerization. Several methods for accomplishing this were studied and have been published (*71*). Polymerization of the bis-phenoxide with the dinitro compound proceeded rapidly under mild

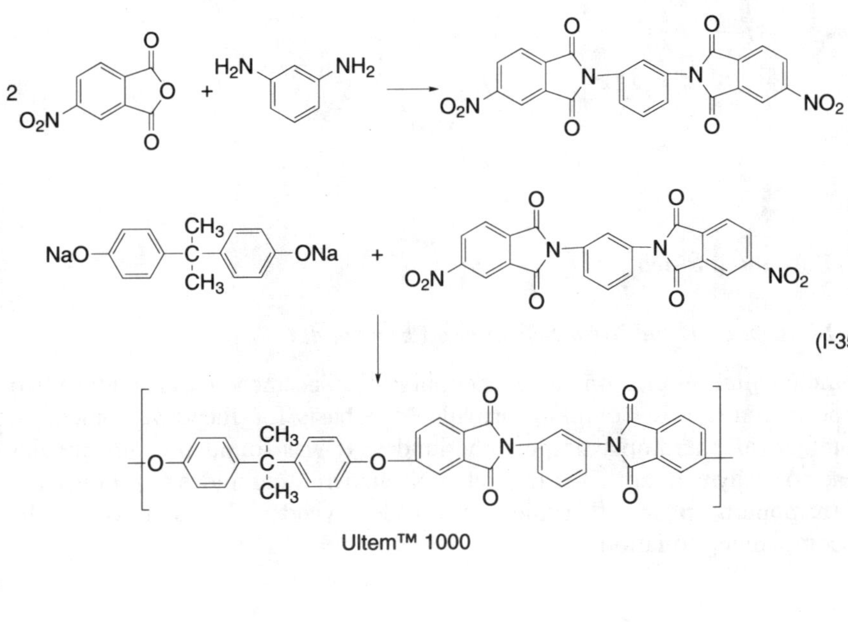

(I-35)

Ultem™ 1000

conditions in dipolar aprotic solvents in admixture with toluene. DMSO and DMF are the solvents of choice. Reaction temperatures of 40–60°C for a few hours gave high molecular weight polymers. As in the model compound systems, the 4-nitro monomers required longer reaction times and gave somewhat lower molecular weights (*71*).

Although this method gives high molecular weight polymer, it is not favored for commercial production for several reasons. Polymers formed by this method will have phenolic and nitro end groups, both of which can adversely affect the thermo-oxidative stability of the polymer. The need to manage and control the end groups in thermoplastic engineering polymers because of the temperatures encountered in their melt fabrication has been stressed earlier. Also, the polymerization system must be maintained in an essentially anhydrous state to prevent hydrolysis of the nitro-imide to a nitro-amic acid. The latter specie will not undergo the nitro-displacement reaction, thereby upsetting the stoichiometric balance and attainment of useful molecular weights (*69*).

EXAMPLE I: POLYMERIZATION OF THE DISODIUM SALT OF BISPHENOL A WITH BIS(4-NITROPHTHALIMIDO)-1,3-BENZENE (*71*)

A 300-ml, three-necked, round-bottomed flask fitted with a mechanical stirrer, thermometer, Dean-Stark trap, reflux condenser, and nitrogen bypass was charged with BPA (9.1943 g, 0.04027 mol), 60 ml DMSO, 50% aqueous sodium hydroxide (6.4434 g, 0.08055 mole of NaOH), and 40 ml toluene. The flask was heated to reflux with stirring under a nitrogen atmosphere and water was collected in the Dean-Stark trap and removed. After 2.5 hr. the temperature was 130°C and the Dean-Stark trap was replaced with a CaH_2 trap which allowed condensed solvent to pass over CaH_2 before returning to the flask and remove most of the last traces of water. Distillation of ca. 20 ml toluene raised the temperature to 145°C (oil bath at 175°C). The mixture was cooled to 80°C and then 17.9217 g (0.0391 mol) of the bis-nitro-bis-imide was added. After stirring vigorously 5.5 hr. at 80°C, sodium 4-methylphenoxide (1.2 g, 0.0092 m) was added to displace any remaining nitro end groups. After an additional 10 min, 2 ml of acetic acid was added which discharged the red-brown color and the polymer was precipitated by pouring the mixture into methanol in a Waring Blender. The precipitate was filtered, washed with methanol and water, and dried in vacuo at 108°C (10 Torr). The dried polymer was dissolved in ca. 200 ml of chloroform and reprecipitated with methanol. Filtration, washing, and drying yielded 20.3 g (88%) of a light yellow-tan product. Intrinsic viscosity (chloroform, 25°C) 0.39 dl/g.

2. Polyimidization

N-15. Polymerization of Dianhydrides and Diamines

The synthesis of polyetherimides by imidization polymerization utilizes the reaction of a diamine with a dianhydride. It is preferred over the polyetherification route using a bisphenoxide and a bis-nitro-bis-imide(*N*-14). However the nitro displacement chemistry that was elucidated for this latter route plays a role in the synthesis of the requisite bis(ether anhydride) monomers (*72*). The dianhydride monomer that is used for the production of Ultem™ 1000 is synthesized by the steps shown in the accompanying equations. 4-Nitro-*N*-substituted phthalimides are reacted with the disodium

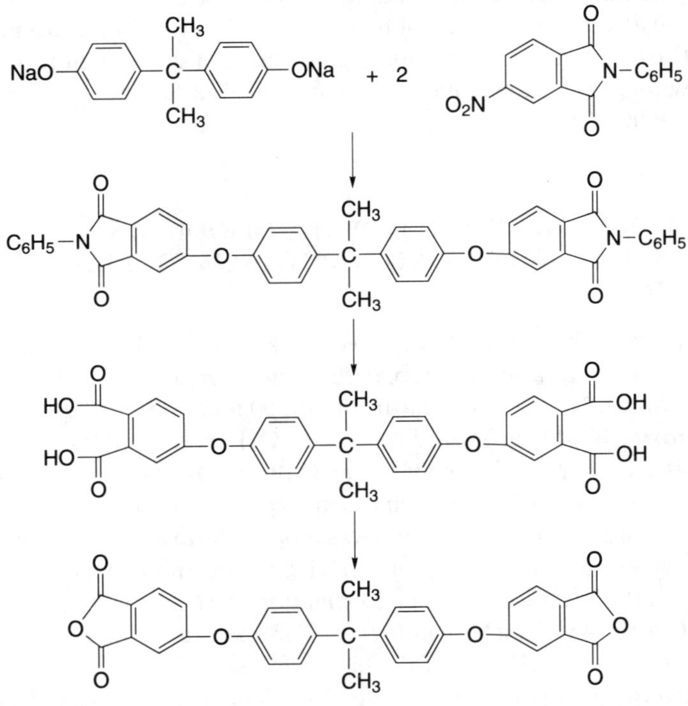

(I-36)

salt of Bisphenol A in a dipolar aprotic solvent to produce the bis(ether imide). The bis(ether imide) is hydrolyzed in aqueous sodium hydroxide to the tetra-acid which is then dehydrated by heating in acetic anhydride to produce the bis(ether anhydride).

The preparation of polyetherimides by imidization can be effected by a high temperature solution process (*73*) or by a continuous melt polymerization process (*74*). The former process is conducted in one step in a homogeneous solution of m-cresol. The dianhydride, the diamine and the solvents are heated to reflux at 160–180°C and the water of reaction removed azeotropically by toluene or chlorobenzene. Since m-cresol is acidic and a radical inhibitor, oxidation of air sensitive diamines under these conditions is minimal. The amorphous polyetherimides that were prepared by this route were all soluble in m-cresol at room temperature. An example is shown in the equation below.

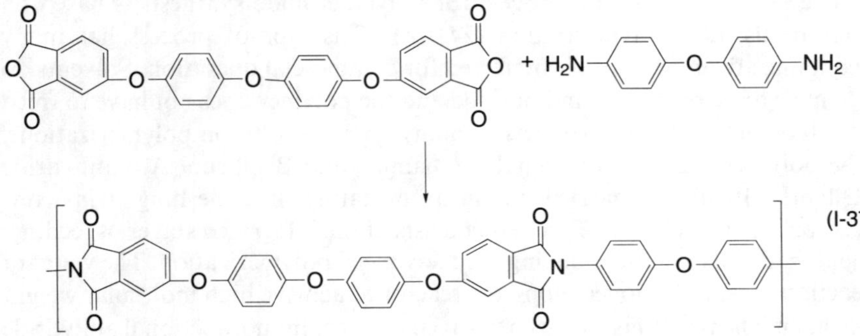

(I-37)

An advantage of this method is that the end groups can be readily converted to thermo-oxidatively stable imides. Amine end groups are convertible to imide end groups by reaction with phthalic anhydride and anhydride end groups are converted to imide end groups by reaction with aniline. Monofunctional anhydrides and amines can be used as molecular weight controllers in the polymerizing solution. This one step process has obvious advantages over the well known, two step amide-acid solution process. In this latter, two-step process, the dianhydride and diamine react to form a polyamic acid which is then imidized either in solution or in the solid state. The kinetics and mechanism of the solution imidization of polyamic acids has been elucidated (*75*).

EXAMPLE I: POLYMERIZATION OF 4,4′-OXYDIANILINE WITH 4,4′-BIS(2,3-DICARBOXYPHENOXY)DIPHENYL-ETHER ANHYDRIDE (*73*)

In a 100-ml, three-necked flask fitted with a Dean-Stark condenser were placed the dianhydride (4.4495 g, 9.000 mmol), ODA (1.8021g, 9.000 mmol), m-cresol (32 ml), and toluene (8 ml). The mixture was stirred with a magnetic stirrer under nitrogen and heated to reflux. Some of the refluxing toluene was distilled off so that the temperature of the reaction mixture was maintained at 160°C. Heating was continued for 1 hour, during which time most of the water was removed. The D-S condenser was then replaced with a recirculation trap filled with four A molecular sieves. The solution was heated to reflux for an additional 2.5 hour, allowing the toluene to recirculate through the bed of molecular sieves The solution was diluted with 30 ml of *m*-cresol and cooled. The polymer was precipitated by pouring the *m*-cresol solution into methanol, which was vigorously stirred in a blender. The polymer was filtered, washed with methanol, and dried under vacuum at 100°C. The yield of the polymer was 6.03 g. The intrinsic viscosity measured at 25°C in m-cresol was 1.92 dl/g.

The second imidization process for polyetherimide synthesis is based on continuous melt polymerization (*74*, *76*). This type of process has many economic advantages and is preferred for commercial operation. Solvents do not have to be recovered and purified and the polymer does not have to freed of salts that are the by-products of many types of solution polymerizations. The polymerization of *m*-phenylene diamine and Bisphenol A dianhydride (Ultem™ 1000) was carried out in a corotating, intermeshing twin-screw reactor/extruder (*74*). The process consists of the integrated stages of feeding, melting, devolatilization, mixing, and advanced polymerization. The water of reaction must be removed along the reactor to achieve high molecular weight polymer. The reactions were run with 3 mole % of monofunctional anhydride to cap the polymer at molecular weights that give the most desirable balance of mechanical and physical properties. The mechanical property profile of the polymer produced by this process compared favorably with the properties of commercially acceptable product.

N-16. Polymerization by Transimidization

Another method that has been developed by the GE group for the synthesis of polyetherimides is via transimidization (*77*). A bis-imide is exchanged or reacted with a diamine, as shown in the accompanying equation. The final stage of this route is conducted in the melt and, as such, has inherent

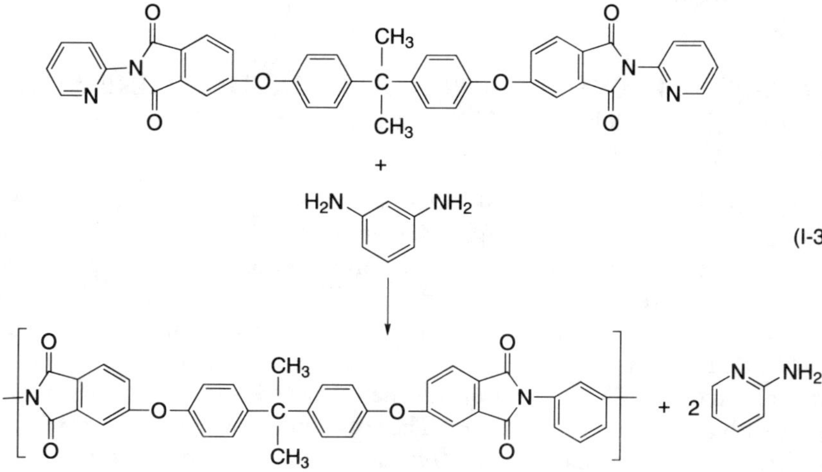

(I-38)

economic advantages. Two key factors necessary for the development of a useful process are the nature of the amine leaving group and the discovery of effective catalysts. This technology was developed by studying model compounds and polymerizations in great detail. The fastest leaving group was found to be 2-aminopyridine, followed closely by *o*-nitroaniline and other amines with electron withdrawing groups. Even aniline gave high molecular weight polymer when it was used as a leaving group. The most effective and preferred catalyst was found to be zinc acetate. Because 2-aminopyridine and m-phenylene diamine have similar boiling points, a prepolymer is formed initially in *o*-dichlorobenzene solution. This was carried out using 50 ppm of zinc acetate as catalyst and 3 mole % of phthalic anhydride as a chain stopper at 180°C for 2–3 hours for the system shown in the above equation. The solvent was then evaporated and the final molecular weight advancement was carried out in bulk.

A strong case was made for commercial use of this process over the nitro displacement route and the diamine-dianhydride route. Unattractive features of the nitro displacement process included the fact that (*1*) the bisphenoxide causes electron transfer side reactions (*2*) a solvent is needed (3) the sodium nitrite that is formed causes side reactions and (*4*) the polymer needs to be cleaned up extensively. Problems that surround the dianhydride preparation and purification that is used in the diamine-dianhydride process are eliminated in the transimidization process because no dianhydride is used. However, for small scale lab samples any of these three methods may have merit depending on the specific structure being sought.

D. POLY(PHENYLENE ETHERS)

1. Polyetherification

N-17. Diphenols and Activated Aromatic Dihalides: Carbonate Process

Bis(4-fluorophenyl)acetylene has been polymerized with bisphenols in a dipolar solvent using sodium carbonate as the base (*78*). High molecular weight, amorphous polymers were obtained. An example is shown in the accompanying equation. The activation of the aromatic fluoro groups by the acetylene group is considerably less than that of a carbonyl or sulfone group, but sufficient to give high polymer in several hours. High molecular weight polymers were not obtainable from hydroquinone or biphenol because they precipitated from the reaction mixture due to their crystallinity. The acetylene link in these polymers reacts at about 400°C to form crosslinked, brittle, insoluble products.

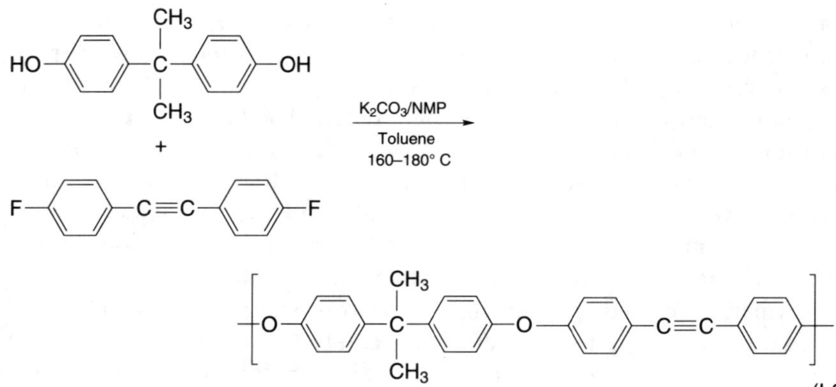

(I-39)

Heterocyclic 1,2,4-triazole units can activate para fluoro groups sufficiently to give high molecular weight polymers via the carbonate process (*79*). Perfluoroalkylene groups can activate the displacement of *para* fluoro groups to form polyaryl ethers from difunctional monomers(*80*).

N-18. Phase Transfer Catalyzed Systems

The polymerization of certain perfluorinated aromatic compounds with bis-nucleophiles has been studied under solid–liquid phase transfer conditions (*81*, *82*). In certain cases, increased yields and molecular weights were obtained when 18-crown-6 ether was used at a concentration of 27.7 mole %. Hexafluorobenzene (HFB) was shown to only undergo 1,4-disubstitution with model mono nucleophiles, and gave linear soluble polymers with certain bisphenols and aromatic dimercaptans. The polymerization of HFB with Bisphenol A is shown in the accompanying equation. High yields of high molecular weight polymer were obtained in the presence of the crown ether catalyst, whereas no polymer was obtained in its absence. The catalytic effect

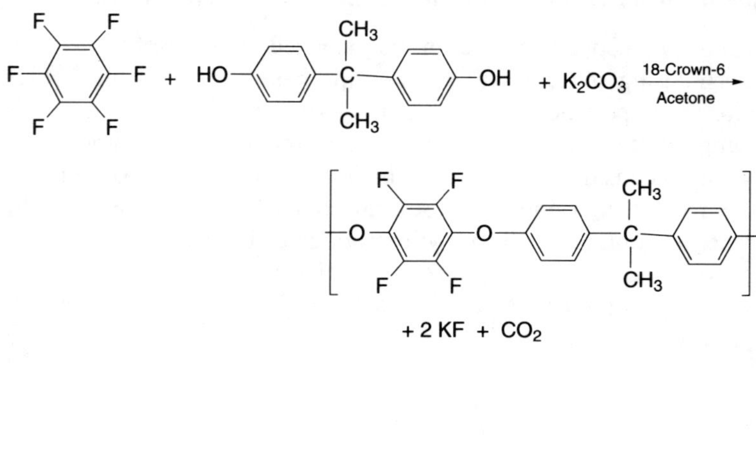

(I-40)

was most pronounced in solvents of intermediate polarity such as acetone; a small effect was noted in DMAC and little or no polymer was formed in nonpolar solvents like chlorobenzene. Reactivity is also favorably affected by the presence of trace amounts of water; optimum amounts were determined for several systems.

The fluorinated bisaryl reactants that were studied included perfluorobiphenyl (PFB), perfluorobenzophenone, and perfluorophenylsulfide. PFB polymerized with Bisphenol A and 4,4′-dimercaptodiphenyl ether to give soluble, high molecular weight polymers. Most of the other examples using the fluorinated bisaryl reactants gave insoluble or low molecular weight polymer.

Based on their work with these perfluorinated aromatics under solid-liquid phase transfer conditions, the authors have proposed that a free radical mechanism could be energetically competitive with direct, anionic substitution via Meisenheimer complexes (*82,83*). Initially, formation of a charge transfer complex between the electron deficient polyfluoroaryl

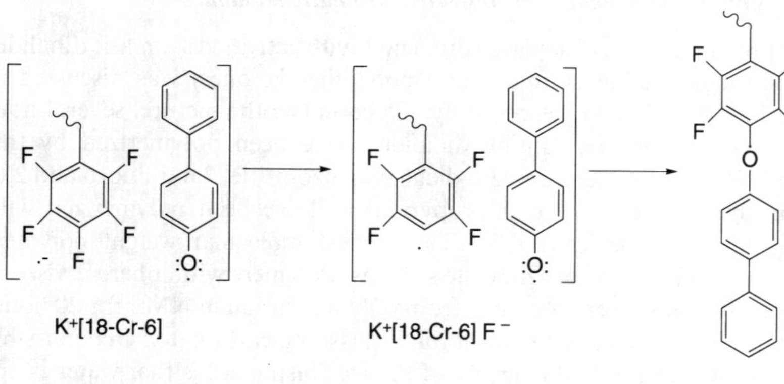

(I-41)

nucleus and the ion pair complex of catalyst and nucleophile occurs. Subsequent electron transfer from nucleophile to substrate leads to a transition state for aromatic substitution with a large degree of free radical character. Loss of fluoride ion from the complex and rapid collapse of the caged radical leads to product. Evidence for this very interesting proposal has been noted in other research programs (*50*).

EXAMPLE I: POLYMERIZATION OF HEXAFLUOROBENZENE AND BISPHENOL A IN ACETONE (*81*)

To a solution of 0.886 g (4.76 mmole) of HFB and 1.087 g (4.76 mmole) of Bisphenol A in 20 ml of dried acetone was added an appropriate amount of distilled water by microliter syringe to give a mole ratio of water/catalyst of 2.40. 2.84 g (20.6 mmole) of K_2CO_3 and 0.348 g (1.32 mmole) of 18-crown-6 ether were added to the reaction mixture. The magnetically stirred heterogeneous mixture was heated to 55°C and maintained under nitrogen for 48 hr. Upon cooling to room temperature the reaction mixture was slowly poured into ca. 150 ml of isopropanol and vigorously stirred. The filtered solids were washed three times with 300 ml portions of distilled water. The solids were air dried and subsequently placed in a vacuum oven (80°C) for 24 hour. The white fibrous polymer was obtained in 88.2% yield. The product was dissolved in chloroform and was filtered to remove insoluble material. The soluble polymer was reprecipitated in isopropanol and dried. The product was a linear, para-substituted polymer as determined by ^{19}F-NMR. Coherent films were cast from the chloroform solution. An inherent viscosity of 0.58 was determined in chloroform at 30°C. Tg = 153°C.

N-19. Silylated Diphenols and Activated Aromatic Dihalides

The polymerization of silylated diphenols with activated aromatic dihalides to produce poly(ether sulfones) and poly(ether ketones) was discussed in Methods N-6 and N-11, respectively. To complete the picture, several non-ketone, non-sulfone containing dihalides have been polymerized by this method (*38*). 2,6-Difluoro and 2,6-dichlorobenzonitrile, 2,6-difluoro and 2,6-dichloropyridine, and 3,6-dichloropyridazine have been polymerized with bis(*O*-trimethylsilyl)bisphenol A. The highest molecular weight polymers were obtained from the benzonitrile systems. Polymers with inherent viscosities in the 0.7 range were obtained from polymerization in NMP for 20 hours in the presence of an equivalent of potassium carbonate. One possible advantage of this method over use of the free bisphenol is that water is not formed or present. This could be beneficial for the polymerization of monomers with hydrolyzable groups.

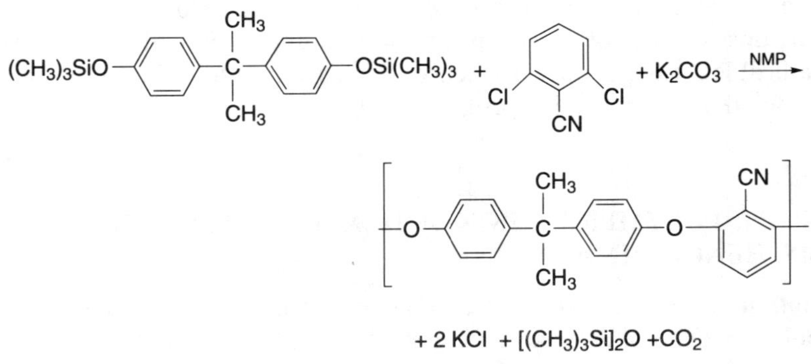

(I-42)

EXAMPLE I: POLYMERIZATION OF SILYLATED DIPHENOLS WITH DIHALO AROMATIC COMPOUNDS (*38*)

A silylated diphenol (30 mmol.), a dihaloaromatic compound (30.3 mmol) and dry K_2CO_3 (31 mmol) were weighed into a 150-ml three-necked flask, and 70 ml of freshly distilled NMP was added. The reaction mixture was heated in an oil bath with mechanical stirring. Carbon dioxide and hexamethyldisiloxane were gradually removed with a slow stream of nitrogen. Finally, the reaction mixture was poured into water. The precipitated polyether was dried in vacuum, dissolved in dichloromethane, filtered and precipitated with methanol. Viscosities were measured in methylene chloride.

N-20. Diphenols and Activated Aromatic Dinitro Compounds

The polymerization of nitro compounds activated by nitrile groups was studied in the original GE research aimed at the synthesis of polyetherimides from nitro-substituted monomers (*70*). High molecular weight polymers were obtained for several homo- and copolymers.

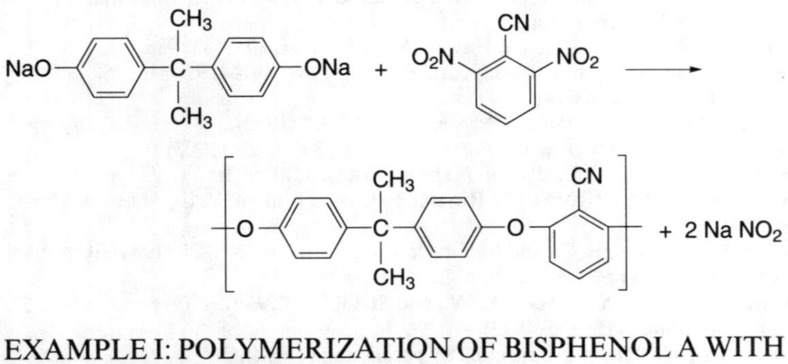

(I-43)

EXAMPLE I: POLYMERIZATION OF BISPHENOL A WITH 2,6-DINITROBENZONITRILE (*70*)

The disodium salt of Bisphenol A was prepared by heating the bisphenol with an equivalent amount of 50% aqueous sodium hydroxide in a toluene-DMSO mixture, and the water was removed by azeotropic distillation. The equivalent amount of 2,6-dinitrobenzonitrile was added and polymerization conducted at 145°C. The polymer was obtained in 75 % yield and possessed an intrinsic viscosity of 0.51.

References

1. Johnson, R. N., Farnham, A. G., Clendinning, R. A., Hale, W. F., and Merriam, C. N., *J. Polymer Sci.* A-1, **5**, 2375 (1967). Reprinted by permission of John Wiley & Sons. Inc.
2. Johnson, R. N., and Farnham, A. G., U.S. Pat. 4108837 (1978). Union Carbide Corporation. Also Brit. Pat.1078234 (1963).

3. Clendinning, R. A., Farnham, A. G., and Johnson, R. N., "The Development of Polysulfone and Other Polyarylethers" in "High Performance Polymers: Their Origin and Development", p. 149. Elsevier 1986.
4. Harris, J. E., and Johnson, R. N., "*Polysulfone, Encyclopedia of Polymer Science and Engineering*", **13**, 196 (1988).
5. Maiti, S., and Mandal, B. K., *Prog. Polym. Sci.*, **12**, 111 (1986).
6. Steiner, U. A., and Reichle, W. T., U.S. Pat. 4937387 (1990). Amoco Corp.
7. Jonsson, H., Hedrick, J. L., and Labadie, J. W., *Polym. Preprints*, **33**(1), 394 (1992).
8. Rose, J. B., "Discovery and Development of the Victrex™ Polyarylethersulfones", in "High Performance Polymers: Their Origin and Development", p. 169. Elsevier 1986.
9. Barr, D. A., and Rose, J. B., Brit. Pat. 1153035 (1965). ICI.
10. Newton, A. B., and Rose, J. B., *Polymer*, **13**, 465, (1972).
11. Attwood, T. E., Newton, A. B., and Rose, J. B., *Br. Polymer* J., **4**, 391 (1972).
12. Leslie, V. J., Rose, J. B., Rudkin, G. O., and Feltzin, J., "New Industrial Polymers," ACS Symposium Series 4. Washington, D. C. 1972. Pp 63-82.
13. "Textbook of Polymer Science", Billmeyer, F. W., Jr., John Wiley & Sons, Inc., New York, Third Edition, Chapter 2.
14. Verma, A. K., *Prog. Polym. Sci.*, **12**, 219 (1986).
15. U.S. Pat. 4307222 (1981). Process for Preparing Polyarylene Polyethers and a Novel Polyarylene Polyether. Union Carbide Corporation. Schwab, T. H., Ark, W. F., Clendinning, R. A. Example designated 'Control A'.
16. Sorenson, W. R., and Campbell, T. W., 2nd Edition. Preparative Methods of Polymer Chemistry. Interscience Publishers, Inc., New York., p. 181.
17. Clendinning, R. A., Farnham, A. G., Zutty, N. L., and D. C. Priest, Can. Pat. 847,963 (1970). Union Carbide Corporation.
18. Hergenrother, P. M., Jensen, B. J., and Havens, S. J., *Polymer*, **29**, 358 (1988).
19. Cotter, R. J., Rimsa, S. B., and Barclay, Jr., R., U.S. Pats. 4814419, 4816505, 4849503, 4939199 (1989, 1990). Amoco Corp.
20. Lucas, M., and Hedrick, J. L., *Polym. Preprints*, **33**(1), 1014 (1992).
21. Maier, G., Hecht, R., and Nuyken, O., *Polym. Preprints*, **34**(1), 429 (1993).
22. Carter, K. R., and Hedrick, J. L., *Polym. Preprints*, **34**(1), 521 (1993).
23. Percec, V., Clough, R. S., Grigoras, M., Rinaldi, P. L., and Litman, V. E., *Macromolecules*, **26**, 3650 (1993).
24. Viswanathan, R., Johnson, B. C., and McGrath, J. E., *Polymer*, **25**, 1827 (1984). Reprinted by permission of Butterworth–Heinemann, Ltd.
25. Miller, T. M., Neenan, T. X., Kwock, E. W., and Stein, S. M., *Polym. Preprints*, 34(1), 58, 1993. *J. Am. Chem. Soc.*, **115**, 356 (1993).
26. British Pat. 1558671 (1980). ICI. Reprinted by permission of HMSO.
27. Hedrick, J. L., Mohanty, D. K., Johnson, B. C., Viswanathan, R., and McGrath, J. E., *Polym. Preprints*, **24(1)**, 164 (1983). Reprinted by permission of authors.
28. Hedrick, J. L., Mohanty, D. K., Johnson, B. C., Viswanathan, R., Hinkley, J. A., and McGrath, J. E., *J. Polym. Sci., Polym. Chem. Ed.*, **24**, 287 (1986). Reprinted by permission of John Wiley & Sons, Inc.
29. Atwood, T. E., Barr, D. A., Feasey, G. G., Leslie, V. J., Newton, A. B., and Rose, J. B., *Polymer*, **18(4)**, 354 (1977).
30. Atwood, T. E., Barr, D. A., King, T., Newton, A. B., and Rose, J. B., *Polymer*, **18(4)**, 359 (1977). Reprinted by permission of Butterworth–Heinemann, Ltd.
31. Yu, T., Fu, S., Li, S., and Cheng, W., *Polymer*, **25**, 1363 (1984).
32. Jap. Pat.1087628 (1989). Toyo Soda Mfg. KK.
33. Carter, K. R., Kim, S. Y., and Labadie, J. W., *Polym. Preprints*, **34(1)**, 415 1993.
34. Radlmann, E., Schmidt, W., and Nischk, G. E., *Makromolekulare Chemie*, **130**, 45 (1969). Reprinted by permission of Hüthig & Wepf Verlag.
35. Gerbi, D. J., Dimotsis, G., Morgan, J. L., Williams, R. F., and Kellman, R., *J. Polym. Sci., Polymer Lett. Ed.*, **23**, 551 (1985). Reprinted by permission of John Wiley & Sons, Inc.
36. Savariar, R., and Cotter, R. J., Unpublished results, Amoco Performance Products, Inc. 1990.

37. Kricheldorf, H. R., and Bier, G., J., *Polym. Sci., Polym. Chem. Ed.*, **21**, 2283 (1983). Reprinted by permission of John Wiley & Sons, Inc.
38. Kricheldorf, H. R., and Jahnke, P., *Makromol. Chem.*, **191**, 2027 (1990). Reprinted by permission of Hüthig & Wepf Verlag.
39. Kricheldorf, H. R., and Bier, G., *Polymer*, **25**, 1151 (1984). Reprinted by permission of Butterworth–Heinemann, Ltd.
40. Kricheldorf, H. R., Chapter 9 in Handbook of Polymer Synthesis, Part A, Marcel Dekker, New York, 1992.
41. Imai, Y., and Oishi, Y., *Prog. Polym. Sci.*, **14**, 173 (1989).
42. Wan, I-Y., Priddy, D. B., Lyle, G. D., and McGrath, J. E., *Polym. Preprints*, **34**(1), 806 (1993).
43. Wang, Z. Y., de Carvalho, H.N., and Hay, A. S., *J. Chem. Soc., Chem Comm.*, **1991**, 1221.
44. Sato, M., Kondo, H., and Yokoyama, M., *Makromol. Chem., Rapid Commun.* **3**, 821 (1982). Reprinted by permission of Hüthig & Wepf Verlag.
45. Mulvaney, J. E., and Wu, S. J., J. *Polymer Sci.*, Part A: *Polym. Chem.*, **24**, 2373 (1986).
46. Rose, J. B., in "Recent Advances in Mechanistic and Synthetic Aspects of Polymerization", Fontanille, M., and Guyot, A., Eds., D. Reidel Publishing Co. 1987.
47. Mullins, M. J., and Woo, E. P., *JMS-Rev. Macromol. Chem. Phys.*, **C27(2)**, 313 (1987).
48. Atwood, T. E., Dawson, P. C., Freeman, J. L., Hoy, L. R. J., Rose, J. B., and Staniland, P.A., *Polymer*, **22**, 1096 (1981). Reprinted by permission of Butterworth–Heinemann, Ltd.
49. Lovering, J. B., Ph.D. Thesis, University College, London,1986.
50. Kelsey, D.R., Private communication. Amoco Performance Products, Inc.
51. Percec, V., Clough, R. S., Fanjul, J., and Grigoras, M., *Polym. Preprints*, **34**(1), 162 (1993).
52. Rose, J. B., Discovery and Development of the "Victrex" Polyaryletherketone PEEK" in "High Performance Polymers: Their Origin and Development", p. 187. Elsevier. 1986.
53. Rose, J. B., and Staniland, P. A., U.S. Pat. 4320224 (1982). ICI.
54. Cinderey, M. B., and Rose, J. B., U.S. Pat. 4176222 (1979). ICI.
55. Kelsey, D. R., U.S. Pat. 4774311 (1988). Amoco Corporation. See also U.S. Pat. 4638044 (1987). Amoco Corporation.
56. Harris, J. E., Winslow, P. A., Botkin, J. H., Maresca, L. M., Clendinning, R. A., Cotter, R. J., Matzner, M., and Kwiatkowski, G. T., *Macromolecules*, 26, 2366 (1993).
57. Singh, R., and Hay, A. S., *Macromolecules*, **25**, 1017 (1992). Reprinted by permission of authors.
58. Jap. Pats. 63,258,923 (1988), 1259029 (1989). Asahi Chemical Ind.
59. Kricheldorf, H. R., Delius, U., and Tonnes, K. U., *New Polymeric Mater.*, **1(2)**, 127 (1988).
60. Woo, E. P., Tuinstra, H. E., and Mullins, M. J., U.S. Pat. 4794155 (1988). Dow Chemical Co. *Chem Abstr.*, **111**:116256c (1989).
61. Kelsey, D. R., Robeson, L. M., Clendinning, R. A., and Blackwell, C. S., *Macromolecules*, **20**, 1204 (1987). D. R. Kelsey, U.S. Pat. 4774344 (1988), 4882397 (1989), 4906786 (1990), 5093460 (1992). Amoco Corporation. Reprinted by permission of authors.
62. Brink, A. E., Gutzeit, S., Lin, T., Marand, H., Lyon, K., McGrath, J. E., and Riffle, J. S., *Polym. Preprints*, **33**(1), 402 (1992).
63. Gibson, H. W., Wang, J., and Pandya, A., *Polym. Preprints*, **34**(1), 570 (1993).
64. Cotter, R. J., and Matzner, M., Chapter I in "Ring-Forming Polymerizations", Part B, 2, Academic Press, New York, 1972.
65. Sroog, C. E., in "Encyclopedia of Polymer Science and Technology", vol. 11. Ed. N. M. Bikales, H. F. Mark, and Gaylord, N. G.. Interscience, New York, 1969.
66. Mittal, K. L., "Polyimides, Synthesis, Characterization, and Applications", vol. 1 and 2. Plenum Press, New York, 1982.
67. Lee, C. J., *Rev. Macromol. Chem. Phys.* **C29**(4), 431 (1989).
68. King, J. J., and Lee, B. H., in High Performance Polymers: Their Origin and Development. p. 317. Elsevier. 1986.
69. Wirth, J. G., in High Performance Polymers: Their Origin and Development, p. 195. Elsevier. 1986.
70. Takekoshi, T., Wirth, J. G., Heath, D. R., Kochanowski, J. E., Manello, J. S., Webber, M. J., *J. Polym. Sci., Polym. Chem. Ed.*, **18**, 3069 (1980). Reprinted by permission of John Wiley & Sons. Inc.

71. White, D. M., Takekoshi, T., Williams, F.J., Relles, H. M., Donahue, P. E., Klopfer, H. J. Loucks, G. R., Manello, J. S., Matthews, R. O., and Schluenz, R. W., *J. Polym. Sci.: Polym. Chem. Ed.*, **19**, 1635 (1981). Reprinted by permission of John Wiley & Sons, Inc.
72. Takekoshi, T., Kochanowski, J. E., Manello, J. S., and Webber, M. J., *J. Polym. Sci.: Polym. Chem. Ed.*, **23**, 1759 (1985).
73. Takekoshi, T., and Kochanowski, J. E., and Manello, J. S., and Webber, M. J., *J. Polym. Sci.: Polym. Symp.*, **74**, 93 (1986). Reprinted by permission of John Wiley & Sons, Inc.
74. Schmidt, L. R., Lovgren, E. M., and Meissner, P. G., *Int. Polym. Processing*, **IV** (1989) 4, p. 270.
75. Kim, Y. J., Glass, T. E., Lyle, G. D., and McGrath, J. E., *Macromolecules*, **26**, 1344 (1993).
76. Takekoshi, T., and Kochanowski, J. E., U.S. Pat. 3803085 (1974). General Electric Co..
77. Takekoshi, T., Proceedings of the Symposium on "Recent Advances in Polyimides and other High Performance Polymers", P. Hergenrother, Chairperson, San Diego, CA, 1/22/90. p. E-1
78. Strukelj, M., Paventi, M., and Hay, A. S., *Macromolecules*, **26**, 1777 (1993).
79. Carter, K., R., Miller, R. D., and Hedrick, J. L., *Polym. Preprints*, **34**(1), 481 (1993).
80. Labadie, J. W., and Hedrick, J. L., *Macromolecules*, **23**, 5371 (1990).
81. Gerbi, D. J., Williams, R. F., Kellman, R., and Morgan, J. L., *Polymer Preprints*, **22(2)**, 385 (1981). Reprinted by permission of authors.
82. Kellman, R., Henry, M. T., Williams, R. F., Dimotsis, G., Gerbi, D. J., and Williams, J. C., "Phase Transfer Catalysis" Starks, C. M., Ed., ACS Symposium Series 326: *Am. Chem. Soc.*: Washington, DC, 1987, pp 128-142; also *Preprints-ACS, Div. Pet. Chem.*, **30(3)**, 408 (1985).
83. Gerbi, D. J., Williams, R. F., Kellman, R., and McPheeters, J. C., *Polymer Preprints*, **22(2)**, 387 (1981).

CHAPTER 2

Electrophilic Synthetic Methods

Introduction

Electrophilic methods for the synthesis of polyarylethers were developed and used for the production of some of the first commercial products. All of the electrophilic methods are based on the attack of an electron deficient specie on an aromatic ring. The electron deficient species include sulfonylium ($—SO_2^+$), acylium ($—CO^+$), and carbenium (R_3C^+) ions. The sources of these ions vary from method to method, but they are generally derived from the corresponding acid halide or the free acid. The aromatic ring that reacts with these species is activated most frequently by an ether group. If electron withdrawing groups are also on this ring or on neighboring rings that allow transmission of their effect to the reactive site, polymerization will be difficult or unsuccessful. Thus, examination of a proposed polymerization sequence on paper to determine whether some unreactive intermediate specie will thwart attainment of high molecular weight polymer is advisable.

The polyarylethersulfone, Astrel™ 360, that was introduced by the 3M Company in the 1960s was prepared by the electrophilic polysulfonylation route. Its structure is shown below.

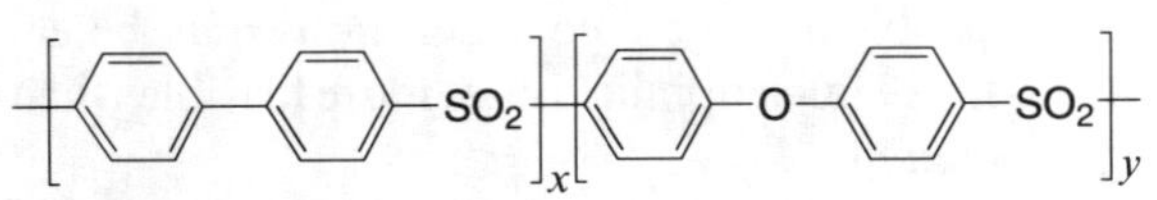

(II-1)

Astrel™ 360–x:y = 60–70: 30–40

The polymerization of sulfonyl chlorides with aromatic compounds can be effected with only catalytic amounts of Friedel-Crafts catalysts, which makes this route attractive. However, electrophilic routes to polyethers tend to give more defects in the polymer structure. These can adversely affect the melt processability and mechanical properties of the polymers. These defects include (1) formation of *ortho* structures during the chain growth processes and (2) chain branches by post reaction on polymer repeat units. Astrel™ 360 was withdrawn from the market in 1976 due at least in part to these factors.

In the early 1970's, Stilan™ was introduced commercially by the Raychem Corporation. It was produced by the electrophilic polymerization of *p*-

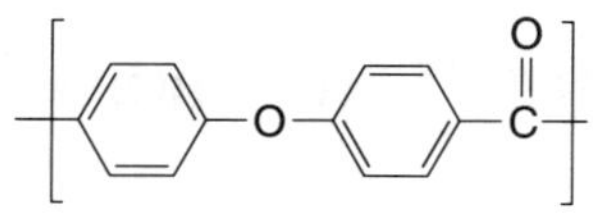

(II-2)

Stilan™

phenoxybenzoyl chloride in liquid HF, catalyzed by boron trifluoride. The development of this process technology was necessitated by the failure of the more conventional aluminum chloride based processes. The insolubility of this poly(ether ketone) in conventional solvents led to only low molecular weight polymer. Difficulties in managing the hazards associated with handling large amounts of liquid HF on a commercial scale led to the discontinuance of Stilan™ manufacture by this route. The polyaryletherketone, PEK, which was the basis of Stilan™, is commercially available today from other manufacturers. Many people feel that PEK possesses the best overall property profile of any poly(arylether ketone), a fact that should insure its continued availability.

The recent innovations on aluminum chloride based electrophilic processes described in this Chapter appear to have made this process commercially viable. The use of Lewis bases in conjunction with the Lewis acid catalyst (Raychem) and the development of ordered poly(ether ketones) (Declar™ by DuPont)) are two such advances. However, these electrophilic methods still require stoichiometric or larger amounts of Lewis acid, which makes for handling and disposal problems. These considerations may be enough of a hurdle in some cases to make nucleophilic routes more feasible when commercialization is being considered.

Essentially all polyarylethers can be synthesized by either electrophilic or nucleophilic routes. This capability has been utilized from time to time by investigators who are trying to track down the source of property differences that could be due to microstructural differences in polymer samples. In general, the nucleophilic routes that involve direct displacement of a group on an aromatic ring will yield the cleaner polymer. Electrophilic routes are inherently susceptible to formation of ortho structures in the chain and branch formation on the backbone. Neither of these structural features are desirable from property and processability viewpoints. These and other factors need to be weighed before selection of an electrophilic route for the synthesis of a polyarylether

A. POLY(ETHER SULFONES)

1. Polysulfonylation

E-1. Condensation of Sulfonyl Chlorides with Aromatic Compounds

One of the first aromatic poly(ether sulfones) to be commercialized was prepared by polysulfonylation technology. The 3M Company introduced Astrel™ 360 in 1967 as one of the first, very high Tg engineering polymers (*1*, *2*). This route to polysulfones takes advantage of the fact that the Friedel-Crafts reaction of sulfonyl chlorides with aromatic compounds is efficiently catalyzed by small amounts of Lewis acids. This is in contrast to the analogous reaction of acyl chlorides with aromatics which requires molar or greater amounts of catalysts. Astrel™ 360 was prepared by the route shown in the following equation. It was a copolymer containing the indicated repeat units in a ratio of x:y of about 60 – 70 to 30 – 40.

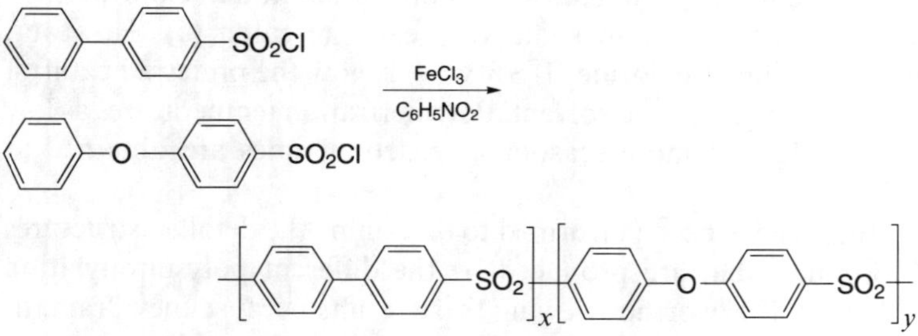

(II-3)

Commercial availability of Astrel™ 360 was discontinued in 1976 because of difficulties in melt fabricating the polymer and the availability of other polysulfones prepared by nucleophilic polyetherification methods. One needs to employ polymerization routes for high Tg or Tm polymers that produce linear, melt stable products, if commercial viability is being sought. Many times this cannot be assured by consideration of the polymerization route on paper only. It results from exhaustive, ongoing research on the process parameters, termination technologies and clean up procedures.

At about the same time that Astrel™ was developed, the ICI group developed polysulfonylation technology for the synthesis of Victrex™ PES, or Polyethersulfone (*2*, *3*). Their research focussed on the one monomer and two monomer routes to PES and the structure of the products obtained therefrom. They are shown in the following equations. Their preference for commercialization was the two monomer route because of monomer availability considerations. Unfortunately, it was found that the polymer from this route lacked toughness as compared to that from the one monomer route.

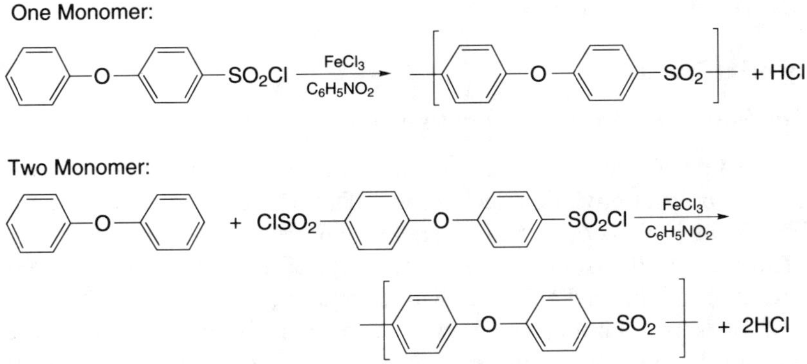

(II-4)

PES is a tough, rigid, high Tg (230°C) thermoplastic. Polysulfonylations can be conducted in bulk followed by a sintering operation, but polymerization in solution is preferred (*4*). Stoichiometric amounts of aluminum chloride were used to effect polymerization in some very early research (*5*). However, catalytic amounts of ferric chloride (1–3% w/w) is now the preferred catalyst and nitrobenzene is the preferred solvent. Polymerization temperatures as low as 80°C are feasible, but more reasonable reaction times are obtained at 120°–140°C.

Extensive studies have been performed to determine the detailed structures of the PES products that are produced by the different polysulfonylation routes (*3*, *6*, *7*, *8*, *9*). It is worth reviewing these results because they dramatically illustrate the importance of linear structures in achieving tough polymers. As mentioned above, the polymer prepared from the self-polymerization of *p*-phenoxybenzenesulfonyl chloride was tough, whereas that from the two monomer route was brittle. NMR studies (*6*, *7*) of the polymer from *p*-phenoxybenzene sulfonyl chloride showed that it consisted essentially (ca.

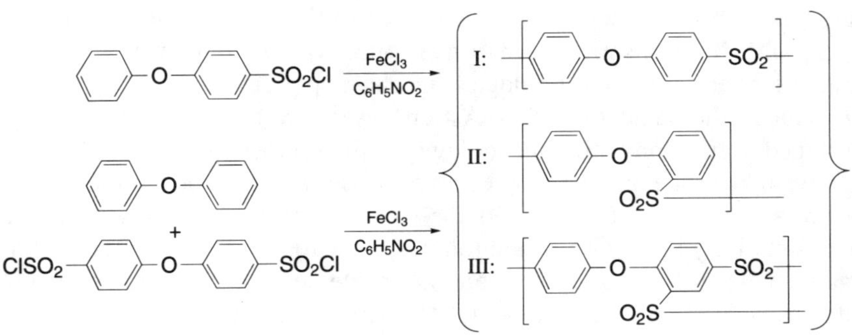

(II-5)

99%) of unit I, with less than 0.7% of the branched units, III. However, similar studies on polymer from the two monomer route showed that it contained units I:II in the percentage ratio of 80:20. It also had significantly more branching than polymer from the one monomer route. These structural differences were confirmed not only by direct examination of polymers from these two routes, but also by comparative studies with model linear polymers prepared by polyetherification. These model polymer studies showed that 20% of *o*, *p*-repeat units were enough to change a tough polymer into a brittle material, especially with the added presence of branching (*10*).

These structural differences can be explained by considering the reaction steps that occur during the polymerization. It is clear that polymerization of *p*-phenoxybenzenesulfonyl chloride occurs solely by substitution on rings that are linked via oxygen to rings containing sulfone groups. Sulfonylation of rings of this type occurs entirely in the para position due to the moderating influence of the sulfonyl group. In the two monomer route, half of the sulfone links are formed by the sulfonylation of diphenyl ether directly. This reaction, unmoderated by sulfone groups, can give a substantial quantity of *ortho*-substituted product based on model compound studies. Branching occurs as the reaction proceeds and the concentration of phenoxy end groups decreases. Sulfonylation of the less reactive phenylene groups in the polymer chains can then begin to compete with chain growth reactions (*2*). The direct displacement reactions that are the basis for nucleophilic routes to polyarylethers minimize the formation of isomeric repeat units that can adversely affect properties. This is the most fundamental reason why nucleophilic routes are preferred over electrophilic routes for aromatic polyether synthesis.

EXAMPLE I: POLYMERIZATION OF 4-PHENOXYBENZENESULFONYL CHLORIDE (*6*)

4-Phenoxybenzenesulfonyl chloride, 5.376 g, 0.02 m, (m. p. 45–47°C) was polymerized in stirred nitrobenzene (15 ml) solution at 120°C under dry nitrogen for 22 hours, catalyzed by anhydrous ferric chloride. One hundred per cent of the theoretical quantity of hydrogen chloride was liberated. The nitrobenzene solution was washed four times with 5% hydrochloric acid and with water (twice). The nitrobenzene solution was then dripped into a large volume of stirred methanol and the polymer, which precipitated, was collected, washed thoroughly with hot methanol and then dried in vacuo. The polymer had a reduced viscosity of 0.81(1% w/v solution in DMF at 25°C).

EXAMPLE II: POLYMERIZATION OF BIPHENYL WITH DIPHENYL ETHER 4,4′-DISULFONYL CHLORIDE (*1*)

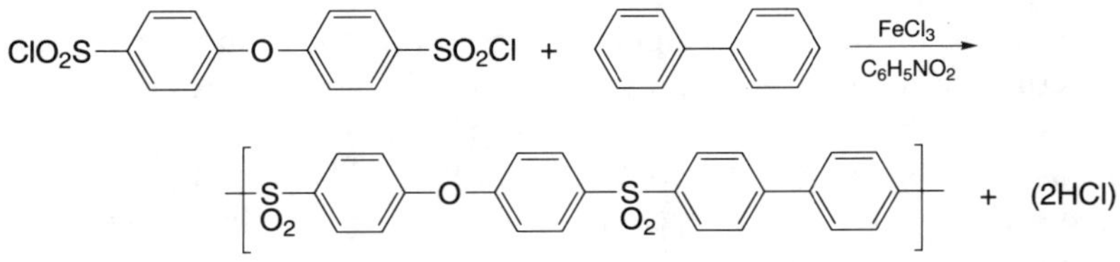

(II-6)

A dry, round-bottomed, three-necked flask equipped with a stirrer, nitrogen inlet, and temperature-control system was charged with stoichiometric quantities of the disulfonyl chloride and hydrocarbon monomers. Nitrobenzene was added to obtain a solids concentration of about 30–50%. The reactor contents were heated to 120°C and 0.3–1.0% of anhydrous ferric chloride (w/w based on polymer) was added. The reactor temperature was then raised to 130°C over a period of 1 hour and maintained at this temperature for about 20 hour or until a sample taken from the reactor showed an inherent viscosity of at least 0.3 (dimethylacetamide, 1%, 25°C). The polymer was then diluted further with dimethylformamide or N-methylpyrrolidinone to 20% solids content and used as such for the casting of films or directly precipitated into a bath of methanol or acetone with high shear agitation. The precipitated polymer crumb was extracted in boiling methanol and dried at temperature up to 500°F to remove residual solvent. The polymer depicted in the above equation possessed inherent viscosities of 0.3–0.7 and was soluble in DMF, DMAC and gamma-butyrolactone. It was slightly crystalline and exhibited a Tg of 275°C.

E-2. Condensation of Arenesulfonic Acids with Aromatic Compounds

Early research on polysulfonylation employed the free sulfonic acids instead of the sulfonyl chlorides (*5*). Formation of a disulfonic acid from diphenyl ether *in situ*, and subsequent polymerization with more diphenyl ether gave low molecular weight polymer. These prepolymers could be advanced to higher molecular weight by further dehydrative polymerization at 240°C in polyphosphoric acid. The polymers were unstable in the melt due to the presence of free sulfonic acid groups that induced crosslinking. Only low molecular weight polymers were obtained using phosphorous oxychloride as the dehydrating agent for polymerization of diphenyl ether with 4,4′-oxybenzenedisulfonic acid (*11*). More recently, a patent described the one step sulfonation/polymerization of diphenyl ether using tungstic or molybdic acids as catalysts. High molecular weight, soluble polymer was reportedly obtained

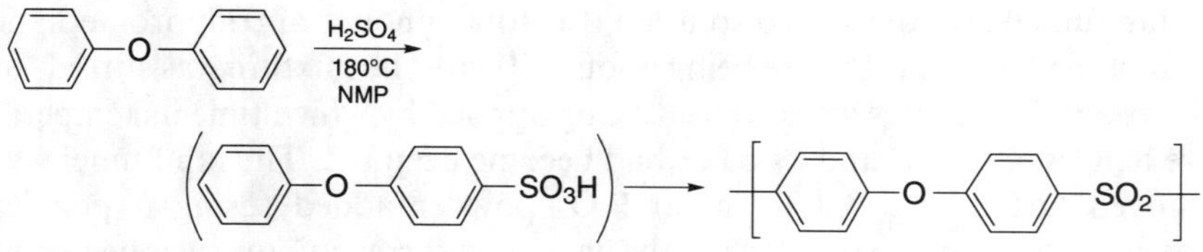

(II-7)

(*12*). This sulfonation-polysulfonylation process has also been carried out in trifluoromethane sulfonic acid (*13*).

Several other dehydrative media have been employed to successfully polymerize sulfonic acids or their alkali metal salts. Self-polymerization of sodium 4-phenoxybenzenesulfonate in a 10:1 molar mixture of a monosulfonic acid and phosphorous pentoxide yielded a polyethersulfone with an inherent viscosity of 0.85 (*14*). Trifluoromethane sulfonic acid/phosphorous pentoxide mixtures have produced polymers with reduced viscosities up to 0.6 from the polymerization of diphenyl ether with 4,4′-oxybenzenedisulfonic acid (*15*). This latter type of dehydrating media produces the best quality of polymer from sulfonic acid polymerizations. Even though this type of route to PES possesses excellent economics, it has never been commercialized. One can assume that the polymer quality does not measure up to that produced by nucleophilic pathways.

EXAMPLE I: POLYMERIZATION OF SODIUM DIPHENYL ETHER-4-SULFONATE (*14*)

Five ml of a mixture of methanesulfonic acid and phosphorus pentoxide (10:1) was placed in a reactor and 2.5 mmols. of sodium diphenylether-4-sulfonate was added. Polymerization was allowed to proceed for 72 hours at 100°C. The polymer was obtained in 99% yield and possessed an inherent viscosity of 0.84. By infrared spectroscopy, it was seen that the polymer had the structure of Polyethersulfone.

EXAMPLE II: SULFONATION/POLYSULFONYLATION OF DIPHENYL ETHER IN TRIFLUOROMETHANE SULFONIC ACID/P_2O_5 (*15*).

Diphenyl ether (34.0 g, 0.2 mole) was weighed into a 3-necked flask (capacity 250 ml) fitted with a stirrer, internal thermometer and reflux condenser (protected with a drying tube), under a nitrogen blanket. 150 ml of trifluoromethane sulfonic acid were added and the mixture cooled to – 25°C with solid carbon dioxide. A solution of Analar 98% w/w sulfuric acid (20.0 g, 0.2 mole) in 100 ml trifluoromethane sulfonic acid was added dropwise from the tap funnel into the stirred solution (the total amount of trifluoromethane sulfonic acid added therefore being about 250 ml) The mixture was stirred for 1 hour and 20 minutes without heat being applied by which time its temperature had risen to 8°C and its color had become deep red. The tap funnel was removed and 12.5 g (0.087 mole) P_2O_5 powder added (using a powder dispenser) through the free neck of the flask. The stirred mixture was heated to 80°C (using an oil bath) and maintained substantially at this temperature for 26.5 hours. Samples of about 17 g of the mixture were periodically removed with a clean syringe, each solution being injected into water to yield a white lace-like precipitate which was filtered off. The precipitate was washed with water and methanol and dried in a vacuum oven. The RV of the polymer was 0.60 in DMF after 26.5 hours. NMR spectroscopy (200 M Hz) showed the solid to be an aromatic polysulfone (PES) consisting predominantly of all *para* units.

B. POLY(ETHER KETONES)

1. Polyacylation

E-3. Condensation of Carboxylic Acid Chlorides with Aromatic Compounds Using Aluminum Chloride

The earliest reported synthesis of all-aromatic poly(ether ketones) used aluminum chloride as the reagent/catalyst and nitrobenzene as the solvent (*16*). The polymers obtained from polymerization of iso- and terephthaloyl chlorides with diphenyl ether had very low molecular weights because of their insolubility during polymerization. Similar results were obtained when methylene chloride was used for the polymerization of *p*-phenoxybenzoyl chloride (*17*). Solutions of aluminum chloride in polyphosphoric acid have yielded polymers of moderately high molecular weight (*18*). As has been noted earlier, the crystalline nature of poly(ether ketones) limits their solubility in the typical organic solvents. Greater than stoichiometric amounts of aluminum chloride aid solubilization of these polymers by complexing with the carbonyl groups. However, side reactions can lead to incorporation of aluminum in the polymer that adversely affects melt processability. The need to remove and dispose of large amounts of aluminum chloride, which is inherent to $AlCl_3$-based processes, is an unattractive feature.

Two recent developments have renewed interest in the use of aluminum chloride for the preparation of poly(ether ketones). The use of a Lewis base in conjunction with aluminum chloride and a two step synthesis have been reported by groups at Raychem and DuPont, respectively.

High molecular weight, substantially linear, melt processable polymers are obtained when "a controlling agent" such as a Lewis base is used in the reaction medium with aluminum chloride (*19*, *20*). Organic Lewis bases such as dimethyl formamide and inorganic Lewis bases such as sodium or lithium chloride are used. The presence of these bases reduces the amount of branching in the polymer and the amount of alkylation by solvents such as methylene chloride. Also, complexes between the Lewis bases and the $AlCl_3$ enhance the solubility of the polymers in the reaction media thereby allowing polymerization to proceed to high molecular weight. The Lewis acid (aluminum chloride) is used in an amount equal to one equivalent per equivalent of carbonyl groups present, plus a catalytic amount of about 0.3 equivalents per equivalent of acid halide group. About 0.5 equivalents of Lewis base per equivalent of acid halide is preferably used. This route has commercial potential over the liquid hydrogen fluoride route described below because of the less hazardous nature of the Lewis acid/base medium. Several other Raychem patents have expanded the scope of this technology (*21*).

EXAMPLE I: POLYMERIZATION OF 1,4-DIPHENOXYBENZENE WITH TEREPHTHALOYL CHLORIDE (*19*)

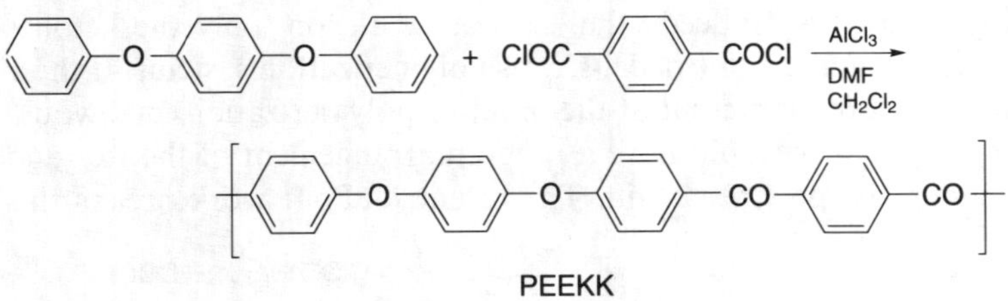

(II-8)

To a 250 ml three neck round bottom flask equipped with a mechanical stirrer, a nitrogen inlet, and a thermometer well fitted with a J-type thermocouple was added methylene chloride (30 ml., freshly distilled from phosphorus pentoxide). The flask and its contents were purged with and kept under nitrogen and cooled to −27°C with a dry ice-acetone bath. Anhydrous aluminum chloride (17.68 g, 132.6 mmole) was then added via Gooch tubing. To the resulting cold solution, *N,N*-dimethylformamide (DMF, 5.54 g, 75.8 mmole, freshly vacuum distilled from calcium hydride) in methylene chloride (15 ml) was added slowly (to control the vigorous exotherm). With continued cooling, a solution of 1,4-diphenoxybenzene (6.6811 g, 25.5 mmole), terephthaloyl chloride (5.1302 g, 25.3 mmole), and benzoyl chloride (0.571 g, 0.40 mmole) in methylene chloride (25 ml) was added. The transferring flask was rinsed with an additional volume of methylene chloride (10 ml) to ensure complete transfer. The stirred reaction mixture was allowed to warm up to room temperature over a 6.3 hour period, during which time it became a viscous, orange-red suspension. It was next cooled to 5–6°C with an ice bath and the reaction was quenched with DMF (50 ml). At this time the polymerization mixture became a highly viscous, white slurry. The polymer was isolated by filtration, transferred into a 1 quart glass blender containing cold (0°C) DMF (100 ml), and blended at high speed until room temperature was reached. This process was repeated. The polymer was then digested twice in DMF (250 ml, 50°C, 24 hours.), once in water (300 ml, room temperature, 24 hours) and once in hydrochloric acid (0.5 M, 500 ml, room temperature, 24 hours). After washing with more water and vacuum drying, polymer (8.9 g, 90% yield) of inherent viscosity 1.2 (0.1% in sulfuric acid) was obtained. It's melt stability and processability were demonstrated by the fact that slabs pressed for 30 min. at 400°C showed minimal change in inherent viscosity.

Ordered polyaryletherketones are the subject of some recent DuPont patents (*22*). The processes described therein produce polyetherketones that exhibit a higher level of crystallization and more rapid crystallization behavior than the corresponding random polyetherketones. This technology is believed to be the basis of the PEKK product line that DuPont introduced in 1988 as extruded sheet (Declar™) and as a composite matrix resin (*23*).

PEKK's are formed by polymerization of diphenyl ether with iso- and/or terephthaloyl chlorides. Indeed, the earliest work on polyetherketones employed these monomers in Friedel-Crafts polymerization systems. If these three monomers were all present at the onset of polymerization, one would expect that the polymer would have a random arrangement of the iso- and tere- units along the polymer chain. The recent DuPont work breaks this

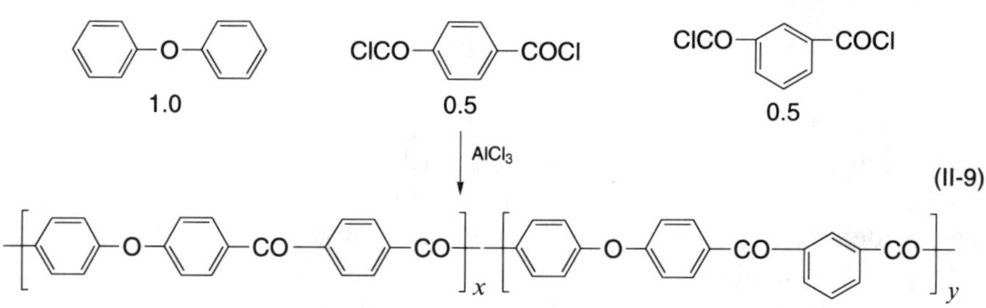

(II-9)

polymerization down into two steps by preforming a monomer from one of the acid chlorides and diphenyl ether. This monomer is subsequently polymerized with the other acid chloride to form the ordered PEKK polymer. This chemistry is shown in the equations below for the synthesis of 1,4-*bis*(4-phenoxybenzoyl)benzene and its polymerization with isophthaloyl chloride.

The ordered copolymers were characterized by three parameters to distinguish them from the prior art random copolymers. These parameters are the

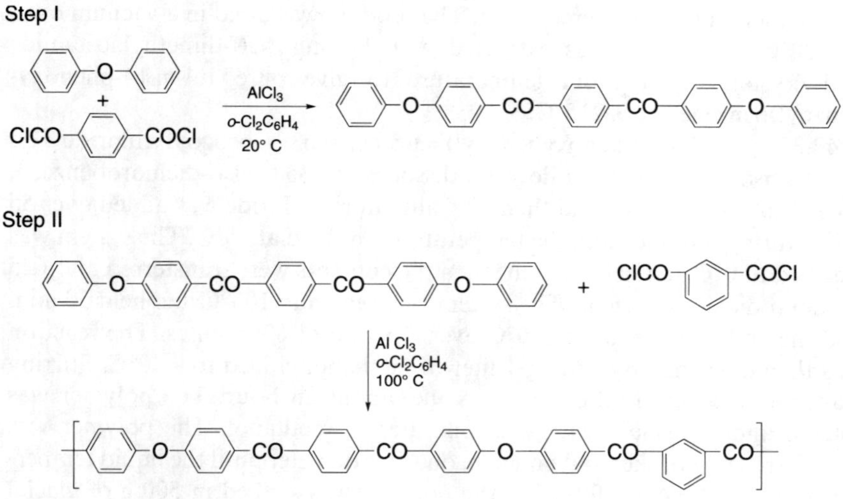

(II-10)

heats of fusion, crystallization rates (measured by the number of degrees difference between the peak melting temperature and the temperature at the onset of crystallization) and the polymer melting points. A random 50/50 tere/iso copolymer possessed a heat of fusion of 20 J/g, a crystallization rate of 50.4°C, and a melting point of 301°C. The perfectly alternating copolymer of the same composition exhibited values of 52.4 J/g, 60.5°C, and 330°C for these same parameters (*22*). The commercial product may be a 40/60 iso/terephthaloyl PEKK composition (*24*, *25*). These structural differences and attendant property improvements, especially the increased rate of crystallization, are sufficient to produce a commercially useful material from a well known, useless prior art composition.

EXAMPLE II: SYNTHESIS AND POLYMERIZATION OF 1,4-BIS(4-PHENOXYBENZOYL)BENZENE WITH ISOPHTHALOYL CHLORIDE (*22*)

The 1,4-*bis*(4-phenoxybenzoyl)benzene was prepared as follows. 6530 g terephthaloyl chloride and 13740 g diphenyl ether were dissolved in about 65 l. *o*-dichlorobenzene in a 44 gal glass lined reactor. The system was cooled to 0–5°C with stirring and 19714 g aluminum chloride were added with the temperature kept below 5°C. The mixture was warmed to 20°C and held 15 min. Then, 30 gal of methanol, precooled to –50°C, were added over a 15–20 min period so as to keep the pot temperature less than 70°C. The system was cooled to 20–30°C and stirred for 30 min. The slurry was filtered, reslurried in

30 gal. methanol and filtered again. The product was dried in a vacuum oven at 120°C. A portion was extracted with boiling *N,N*-dimethylacetamide and crystallized at room temperature to give pure 1,4-*bis*(4-phenoxybenzoyl)benzene, m. p. 215°C.

42.35 g 1,4-*bis*(4-phenoxybenzoyl)benzene, 0.63 g benzoyl chloride and 18.12 g isophthaloyl chloride were dissolved in 350 ml *o*-dichlorobenzene, cooled to –5° to –8°C, and then 72 g aluminum chloride was added over 55 min. During this addition, the temperature was held at –5°C. The system was warmed to 0°C and held 20 min when the contents were transferred into 625 cc. hot *o*-dichlorobenzene. The system was heated to 100°C and held 50 min. and then allowed to cool to 30°C over a period of 60 minutes. The reaction was then quenched by adding 1 liter of methanol chilled to –40°C. Stirring was continued until all color was gone, about an hour. The polymer was filtered and rewashed twice with one liter of methanol. The polymer was transferred to a beaker and boiled in one liter of water until the liquid temperature was 98°C. After filtration the polymer was soaked in 500 g of glacial formic acid for 30 minutes, filtered and dried overnight in a vacuum oven at 170°C with a light nitrogen bleed. The polymer had an inherent viscosity of 0.96.

E-4. Condensation of Carboxylic Acid Chlorides with Aromatic Compounds Using Boron Trifluoride/Hydrogen Fluoride

The insolubility of poly(ether ketones) in the common Friedel-Crafts solvents was overcome by the discovery that liquid hydrogen fluoride was an excellent solvent for this family of polymers (*26*). When used in conjunction with boron trifluoride as catalyst/reagent, very high molecular weight, tough poly(ether ketone) (PEK) was obtained from p-phenoxybenzoyl chloride (*26*). This chemistry is shown in the accompanying equation and was the process

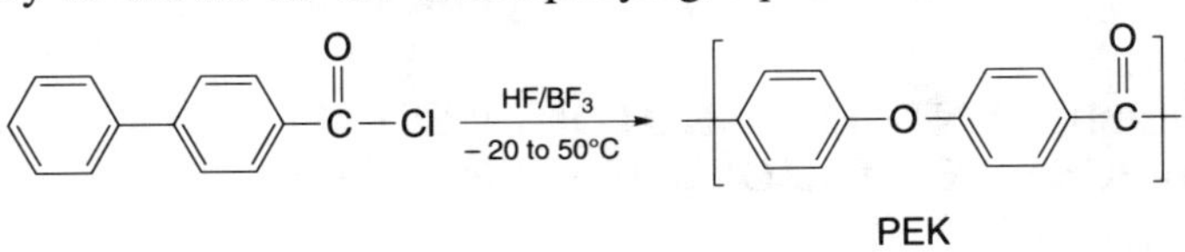

(II-11)

technology used for commercialization of PEK under the trade name Stilan™ by Raychem during the 1972–1976 period (*27*, *28*, *29*, *30*). Its major initial application was as an extruded wire coating for use in severe environments, e.g., nuclear reactors. The manufacture of PEK by this route was discontinued because of the hazards involved in handling liquid hydrogen fluoride.

The polymerization can be conducted in stainless steel reactors although product with improved color and melt stability is obtained by using reactors coated with poly(tetrafluoroethylene) (*31*). Two to three moles of boron trifluoride per mole of acid chloride group is preferred. The optimum amount of hydrogen fluoride is 2–10 moles per mole of boron trifluoride (*26*). These larger than catalytic amounts of both HF and BF3 play a role in solubilizing the polymer by protonation and complexation with the ether and carbonyl groups in the polymer. Carboxylic acids and their esters are also claimed to polymerize in this process. End capping agents such as biphenyl and diphenyl ether can be used to control molecular weight and improve melt stability.

Mechanistically, it may be assumed that acylation occurs by attack of aryl acylium ions ($ArCO^+$) on activated phenyl groups. Suitable comonomers for reaction with bis-acyl halides are diphenyl, diphenyl ether and 1,4-diphenoxybenzene, but not benzophenone or 1,4-dibenzoylbenzene. Electron withdrawal by the carbonyl groups deactivates the phenyl rings to electrophilic attack in the latter compounds. Benzene is unreactive because after monoacylation further reaction is prevented by the deactivating effect of the first carbonyl group that is introduced. It is interesting that the central phenylene ring in 1,4-diphenoxybenzene is not activated to electrophilic attack, because linear polymers are obtained from it. This is probably because the monomer is completely protonated in the highly acidic medium, which deactivates the central and one of the terminal phenylene groups. When one phenyl group has reacted, protonation shifts to the carbonyl group and the other terminal phenoxyl group becomes reactive. Delocalization of the positive charge continues deactivation of the phenylene ring so that a linear polymer is obtained.

This method is also applicable to aromatic compounds containing sulfonyl halide and phosphoryl halide groups (*26*). Several other Raychem patents have expanded the scope of this technology (*32*).

EXAMPLE I: POLYMERIZATION OF P-PHENOXYBENZOYL CHLORIDE TO PEK (*26*)

Twelve grams of p-phenoxybenzoyl chloride, purified by fractionation (B.P. 92–94°C/0.05 mm), was charged to a 100 ml. stainless steel shaker tube which had been flushed with nitrogen. After the tube was closed and cooled to – 78°C, 10 g of anhydrous hydrogen fluoride was distilled into the tube and 25 g of boron trifluoride was added under pressure. The reaction tube was warmed to – 20°C and kept between –20° and – 5°C with mechanical shaking for two hours, then heated at 50°C for 4.5 hours. The polymer product was discharged from the tube by inverting it and allowing the gas pressure to blow the polymer out into methanol. The yellow polymer was cut up in an Osterizer under methanol and washed thoroughly with methanol. Thus, 9.8 g (97% conversion) of polymer was obtained. The inherent viscosity of the product in concentrated sulfuric acid at 30°C was 2.76. The polymer was purified by dissolving in dichlorotetrafluoroacetone monohydrate, filtering the solution and reprecipitating in methanol as a white powder. The crystalline melting point of the polymer was 361°C, as determined by differential thermal analysis, while the glass transition temperature was 163°C.

E-5. Condensation of Carboxylic Acid Chlorides with Aromatic Compounds Using Other Very Strong Acids

Trifluoromethane sulfonic acid (TFMSA) and other strongly acidic media have been used as solvent/catalysts to polymerize carboxylic acid chlorides with aromatic compounds. Polymers with RV's of 0.22 were obtained from terephthaloyl chloride and diphenyl ether or 4,4′-diphenoxybenzophenone using TFMSA as the solvent/catalyst (*33*, *34*). Self polymerization of 4-phenylbenzoyl chloride gave a polymer with an RV of 0.86 (*33*). Another patent (*35*) describes the use of catalytic amounts of TFMSA and other very strong acids to effect polymerizations in strongly acidic media. one medium was a polyphosphoric acid/phosphorous pentoxide mixture that contained catalytic amounts of TFMSA, methanesulfonic acid, trifluoroacetic acid or trifluoroacetic acid. Polyphosphoric acid containing aluminum chloride has also been used successfully for these polymerizations (*18*, *36*).

EXAMPLE I: POLYMERIZATION OF 4-PHENYLBENZOYL CHLORIDE IN TRIFLUOROMETHANE SULFONIC ACID (*33*)

Thirty ml of trifluoromethane sulfonic acid was added, under nitrogen, to 21.65 g of 4-phenylbenzoyl chloride at approximately 20°C, heated to 40°C and stirred at this temperature overnight, raised to 70°C over 1.5 hr. and held at this temperature for 5 hours. The product was worked up to give 17.57 g of homopolymer, m. p. 365°C, RV 0.86.

E-6. Condensation of Carboxylic Acids with Aromatic Ethers

The direct polymerization of carboxylic acids with aromatic ethers is advantageous in that it obviates the need to prepare an acid chloride. On the other hand, the acids are less reactive than acid chlorides and in general require more vigorous reaction conditions. The super acid, trifluoromethane sulfonic acid, was found to be a very effective solvent/dehydrating agent for this polymerization (*33*, *37*, *38*). However, since the monomers react as protonated species in this medium, some limitations were encountered. For example, 4-phenoxybenzoic acid does not give polymer, but the homologous monomer shown in the accompanying equation polymerizes to PEEK of very high

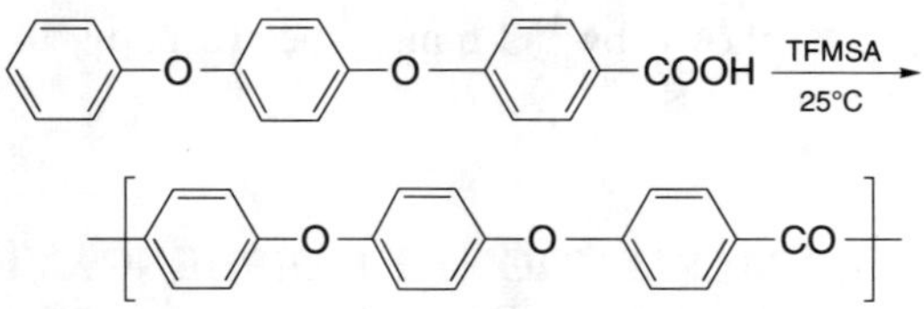

(II-12)

molecular weight. Additionally, terephthalic acid does not react with aryl ethers under these conditions. It was found that electron withdrawing groups *para* to a carboxyl group inhibit acylation in TFMSA and that the aromatic ether reactant must have at least three benzene rings and no carbonyl groups. For example, diphenoxybenzophenone did not polymerize. However, some of these limitations can be overcome by adding phosphorous pentachloride to the TFMSA (*34*). Polymers with inherent viscosities of 1.19–1.93 were obtained from terephthaloyl chloride and diphenyl ether or 4,4′-diphenoxy-benzophenone under these conditions. Perhaps acid chlorides were formed *in situ* as reactive intermediates from the phosphorous pentachloride.

The use of methanesulfonic acid containing phosphorous pentoxide (10%) as the dehydrating agent can give very high molecular weight polymers from the phenoxybenzoic acids (*39*, *40*). NMR studies on the products did not show the presence of branching. A cluster of recent patents has expanded on the use of this medium to prepare polyaryletherketones (*41*), ordered polyarylether-ketones (*42*), and poly(ether ketone sulfones) (*43*). Poly(phenylene etherim-ideketones) have been prepared by the condensation of carboxylic acids with aromatic ethers in a 1:10 P_2O_5—CH_3SO_3H mixture (*44*). Aliphatic and cycloaliphatic dicarboxylic acids were also polymerized successfully with aromatic ethers in the TFMSA/P_2O_5 medium (*41*).

Polymers of moderate molecular weight were obtained using polyphosphor-ic acid as the medium (*36*). Polymer inherent viscosities up to 0.53 were obtained from 4- and 3-phenoxybenzoic acids. Better results can be obtained sometimes when a Lewis acid is used in conjunction with polyphosphoric acid.

EXAMPLE I: POLYMERIZATION OF 4-(4-PHENOXYPHENOXY)BENZOIC ACID IN TFMSA (*37*)

The monomer was dissolved in a tenfold molar excess of trifluoromethane sulfonic acid and the clear orange-red solution allowed to stand at room temperature for several hours under a nitrogen atmosphere. The solution's viscosity normally increased rapidly, and the resulting polymer was isolated by precipitation in dilute sodium hydroxide, and purified by repeated extrac-tion with boiling water. Extraction with hot DMF was occasionally necessary to remove the last traces of CF_3SO_3Na. The polymer had an RV of 2.31 (in

TFMSA at 25°C). As shown by ^{13}C n.m.r. spectroscopy, the polymer had essentially all-para structures.

EXAMPLE II: POLYMERIZATION OF P-DIPHENOXYBENZENE WITH ADIPIC ACID IN TFMSA/P_2O_5 (*41*)

A 50 ml. three-necked flask is equipped with a glass stirrer, a nitrogen inlet, and an outlet for the system, and the flask is then flushed well with nitrogen and 1.10 g (0.0075 mole) of adipic acid, 1.97 g (0.0075 mole) of *p*-diphenoxybenzene, 2.56 g (0.018 mole) of phosphorus pentoxide, and 15 ml of trifluoromethanesulfonic acid are added. The contents of the flask are stirred at 23°C for 7 hours to obtain a high viscosity, very dark amber solution which is precipitated in 10% aqueous sodium hydroxide in a Waring blender and allowed to stand overnight. The mixture is then filtered, boiled 2X in dilute aqueous Na_2CO_3 for 1–2 hour each and filtered, boiled 2X in deionized water for 1–2 hour each and filtered and vacuum dried overnight at 110°C. The cream-colored polyketone has an I.V. of 1.64, a Tg of 92°C, and a Tm of 279°C.

E-7. Condensation of Diphenyl Ether with Phosgene

PEK, the parent poly(etherketone), is viewed by many to possess the best balance of properties of any members of the polyaryletherketone family. The most direct and economical route to this polymer that can be visualized is via the polymerization of diphenyl ether with phosgene.

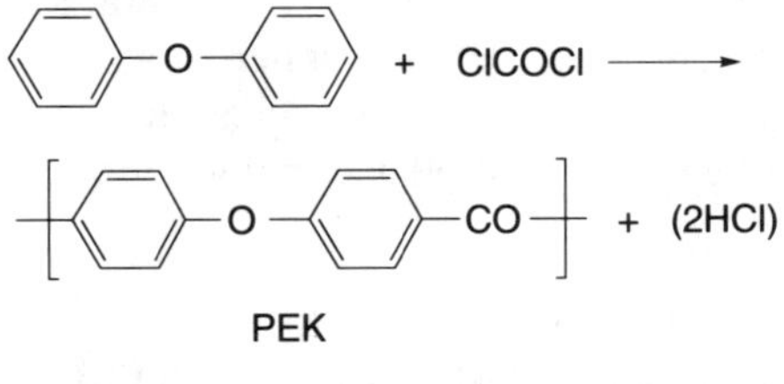

(II-13)

Several groups have attempted to develop this route, but without commercial success. The polymers produced by this method have either been of low molecular weight or not thermally processable. The HF/BF_3 system developed by Raychem produced low molecular weight polymer. PEK with IV's of 0.6 and 0.73 were obtained from reactions catalyzed by $AlCl_3/LiCl$ in 1,2-dichloroethane (*45*) and carbon disulfide (*46*), respectively. Polymerization in trifluoromethane sulfonic acid/$AlCl_3$ gave PEK with an IV of 0.95 (*46*).

E-8. Condensation of Aromatic Ethers with Alkyl Thiochloroformates and Dialkyl Dithiocarbonates

An alternate to the use of phosgene for polymerization with aromatic ethers is the use of derivatives of thio- or dithiocarbonic acids (*47*). Specifically, alkyl thiochloroformates and dialkyl dithiocarbonates are reported to be effective substitutes for phosgene. High molecular weight PEK and PEEK have been produced in the HF/BF_3 reaction/catalyst system developed by earlier work at Raychem.

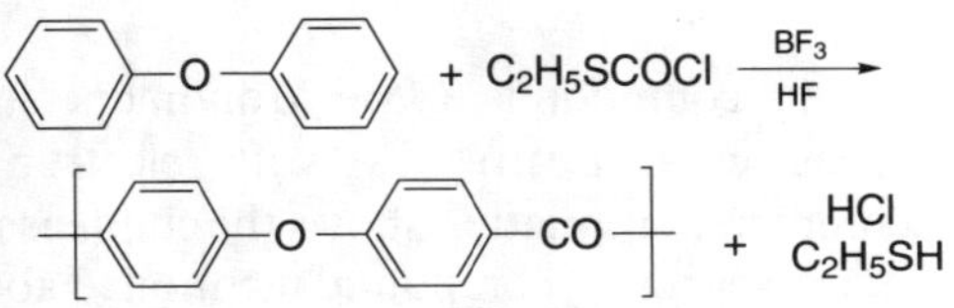

(II-14)

EXAMPLE I: POLYMERIZATION OF DIPHENYL ETHER WITH S-ETHYLCHLOROTHIOFORMATE (*47*)

A 50 ml poly(chlorotrifluoroethylene) tube was charged with 1.7021 g (10.0 mmoles) of diphenyl ether and 1.2457 g (10.0 mmoles) of S-ethyl chlorothioformate. The reaction mixture was cooled to –70°C, 10 milliliters of anhydrous hydrogen fluoride was added, and the reaction tube was connected to a PCTFE vacuum line. Boron trifluoride, at a pressure of 30 psi, was introduced while the tube was allowed to come to room temperature. The stirred reaction mixture was vented at room temperature to expel hydrogen chloride.The boron trifluoride pressure was reapplied (30 psi) and the reaction was allowed to proceed at room temperature for 20 hours, which produced a light orange colored, highly viscous solution. The reaction system was vented and the viscous solution was diluted with 30 ml of anhydrous hydrogen fluoride followed by precipitation into cold (– 20°C) methanol agitated in a Waring blender. The resulting fibrous polymeric precipitate was washed with water and methanol and then dried at 150°C. under a pressure of 20 mm Hg to give 1.90 g (97%) of a colorless fluffy material having an inherent viscosity of 1.24 (0.1 g/100 ml conc. H_2SO_4).

C. POLY(PHENYLENE ETHERS)

1. Polyalkylation

E-9. Polyalkylation of Aromatic Ethers

Very high molecular weight aromatic polymers have been prepared by adaptation of the well known Friedel-Crafts condensation of isopropyl

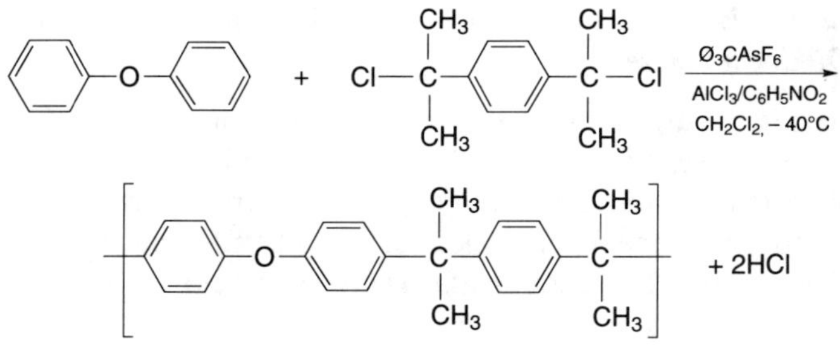

(II-15)

chlorides with aromatic compounds (*48*). Monomers such as *p-bis*-(2-chloroisopropyl)benzene condense readily with electron rich aromatic compounds. The accompanying equation shows the condensation of diphenyl ether with the bis-(chloroisopropyl) compound mentioned above. This type of polymerization proceeds well with a variety of bis(chloroisopropyl) compounds but is limited to electron rich aromatics such as napthalene, aromatic ethers, sulfides and amines. Aromatics which do not undergo alkylation successfully include diphenyl sulfone and benzophenone. High yields of the polyalkylation polymers are obtained using catalyst concentrations of 1–15 mole %. A catalyst recipe like that shown in the above equation was found to be necessary in order to get high molecular weight polymers. It was found advantageous to use metathetic catalysts like trityl hexafluoroarsenate to prevent reversibility and indane formation in these polymerizations. Many of the more conventional Friedel-Crafts catalysts gave polyindanes. Solvent choice was limited to aliphatic and aromatic chlorocarbons that are solvents for the polymers and inert toward the catalyst components.

EXAMPLE I: POLYARYLENEISOPROPYLIDENE (*49*)

A 1 l three-necked flask fitted with a mechanical stirrer, a thermometer, a gas inlet, and a drying tube is flame-dried and charged with 16.8 ml of distilled nitrobenzene, 51.0 ml of distilled phenyl ether, 76.5 g of *p*-di(chloroiso-propyl)benzene (recrystallized from *n*-hexane), 350 mg of triphenylmethyl-hexafluoroarsenate, and 400 ml. of distilled reagent grade methylene chloride. This reaction mixture is then cooled to −42°C. Sublimed, anhydrous aluminum chloride, 3.750 g, weighed in a dry box is added to the reaction mixture. Instantly, a brown-red complex forms and after an induction period of about 2 min, HCl begins to evolve. During the 2.5 hours reaction time, 300 ml of CH_2Cl_2 is added slowly as a diluent. At the end of the polymerization, 5 ml of phenyl ether is added to cap the end groups. After a total of 2 hours and 35 min., 5 ml. of methanol and 100 ml of methylene chloride are added to stop the reaction. This polymer solution is then diluted with 300 ml of methylene chloride, poured into 5 l. of acetone, stirred in a blender. and filtered. The polymer is washed twice with methanol and dried for 24 hours in a vacuum oven at 100°C. The yield of polymer is 97%. The inherent viscosity of the polyaryleneisopropylidene determined in o-dichlorobenzene at 30° is 1.3 dl./g; the melting range is 150–152°C.

References

1. Vogel, H. A., *J. Polymer Sci*: A-1, **8**, 2035 (1970). Reprinted by permission of John Wiley & Sons, Inc.
2. Rose, J. B., in "High Performance Polymers: Their Origin and Development". Elsevier. 1986. Page 169.
3. Rose, J. B., *Polymer*, **15**, 456 (1974).
4. Parodi, F., in "Comprehensive Polymer Science", Volume 5, Chapter 33, p. 561. Allen, G., Editor, Pergamon Press, 1989.
5. Cohen, S. M., and Young, R. H., *J. Polymer Sci.*: A-1, **4**, 722 (1966). also U.S. Pat. 3418277 (1968). Monsanto Co.
6. Cudby, M. E. A., Feasey, R. G., Gaskin, S., Kendall, V., and Rose, J. B., *Polymer*, **9(5)**, 265 (1968). Reprinted by permission of Butterworth–Heinemann, Ltd.
7. Cudby, M. E. A., Feasey, R. G., Jennings, B. E., Jones, M. E. B., and Rose, J. B., *Polymer*, **6**, 589 (1965).
8. Cudby, M. E. A., Feasey, R. G., Gaskin, S., Jones, M. E. B., and Rose, J. B., *J. Polym. Sci.*: C, No. 22, 747 (1969).
9. Ivin, K. J. and Rose, J. B., *Advances Macromol. Chem.*, **1**, 335 (1968).
10. Rose, J. B., et. al., *Polymer*, **18**, 369 (1977).
11. Jones, M. E. B., Brit. Pat. 1166624 (1969). ICI, Ltd.
12. Jap. Pat. 2011634 (1990). New Japan Chem.
13. Rose, J.B., Eur. Pat. Appl., 49070 (1982). ICI, Ltd. *CA.*, **97**, 6999 (1982).
14. Ueda, M., *Makromol. Chem., Rapid Commun.*, **6**, 271 (1985). also Jap. Pat. 60-228541 (1985). Idemitsu Ind., KK. Reprinted by permission of Hüthig & Wepf Verlag.
15. Rose, J. B., Brit. Pat., 2099006 (1982). ICI, Ltd. Reprinted by permission of HMSO.
16. Bonner, W. H., U.S. Pat. 3065205 (1962). DuPont Co.
17. Goodman, I., McIntyre, J. E., and Russell, W., Brit. Pat. 971227 (1964).
18. Niume, K., Toda, F., Uno, K., Hasegawa, M., and Iwakura, Y., *J. Polym. Sci., Polym. Chem. Ed.*, **20**, 1965 (1982).
19. Jansons, V., and Gors, H. C., U.S. Pat. 4709007 (1987). Raychem Corp.
20. Jansons, V., and Gors, H. C., U.S. Pat. 4593086 (1986). Raychem Corp.
21. Reamey, R. H., Dahl, K. J., Jansons, V., Moore, S., Gors, H. C., Horner, P. J., Mazzanti, J. B., and Towle, I. D. H., U.S. Pats. 4665151 (1987); 4808693 (1989); 4814508 (1989); 4826947 (1989); 4841013 (1989).
22. Gay, F. P., and Brunette, C. M., U.S. Pat. 4816556 (1989); also Eur. Pat. Appl. 192260 (1986). DuPont Co.
23. Chang, I. Y., SAMPE Quarterly, July, p. 29 (1988).
24. Sauer, B. B., and Hsiao, B. S., *Polym.*, **34**, 3315 (1993).
25. Gardner, K. H., Hsiao, B. S., Matheson, R. R., Jr., and Wood, B. A., *Polym.*, **33**, 2483 (1992).
26. Marks, B. M., U.S. Pat. 3441538 (1969). Du Pont Co.
27. Rose, J. B., in "High Performance Polymers: Their Origin and Development". Elsevier. 1986. Page 187.
28. Rose, J. B., in "Recent Advances in Mechanistic and Synthetic Aspects of Polymerization," p. 207, D. Reidel Publishing Co., 1987.
29. Staniland, P. A., in "Comprehensive Polymer Science," Vol. 5, Chapter 29, p. 483. Edited by G. Allen. Pergamon Press.

30. Mullins, M. J., and Woo, E. P., *JMS-Rev. Macromol. Chem. Phys.*, **C27(2)**, 313 (1987).
31. Dahl, K. J., U.S. Pat. 3953400 (1976). Raychem Corp.
32. Dahl, K. J., and Jansons, V., U.S. Pats. 3914298 (1975); 3956240 (1976); 3965146 (1976); 4111908 (1978); 4229564 (1980); 4247682 (1981).
33. Rose, J. B., Eur. Pat. Appl. 63874 (1982). ICI, Ltd. *C. A.*, **98**, 180081 (1983).
34. Rose. J. B., U.S. Pat. 4396755 (1983). ICI, Ltd.
35. Towle, I. D. H., U.S. Pat. 4820792 (1989). Raychem, Ltd.
36. Iwakura, Y., Uno, K., and Takiguchi, T., *J. Polym. Sci., Polym. Lett. Ed.*, **15**, 283 (1977); J. Polym. Sci., A-1, 6, 3345 (1968).
37. Colquhoun, H., *Polym. Preprints*, **25**, 17 (1984); and Lewis, D.F., Brit. Pat. 2116990 (1985). ICI, Ltd. C. A., **100**, 7426 (1984). Reprinted by permission of author.
38. Colquhoun H. M., and Lewis, D. F., *Polymer*, **29**, 1902 (1988).
39. Ueda, M., and Kano, T., *Makromol. Chem., Rapid Commun.*, **5**, 833 (1986).
40. Ueda, M., and Sato, M., *Macromolecules*, **20**, 2675 (1987).
41. Darnell, W. R., and Jackson, W. J. Jr., U.S. Pat. 4820794 (1989); WO 8808010 (1988). Eastman Kodak Co.
42. U. S. Pat. 4839459 (1989). Eastman Kodak Co.
43. Darnell, W. R., and Jackson, W. J. Jr., U.S. Pat. 4861856(1989). Eastman Kodak Co.
44. Parthiban, A., and Sundaram, N., *J. Polym. Sci.: Pt. A*: *Polym. Chem.* **31**, 1233 (1993).
45. Litter, M. I., and Marvel, C. S., *J. Polym. Sci., Polym. Chem. Ed.*, **23**, 2205 (1985).
46. Nozawa, S., and Nakata, M., Eur. Pat. 135938 (1985). Mitsubishi Chem. Ind.
47. Jansons, V., U.S. Pat. 4361693 (1982). Raychem Corp.
48. Fritz, A., and Rees, R. W., *J. Polym. Sci.*: A-1, **10**, 2365 (1972). also Fritz, A., *Polym. Preprints*, **12(1)**, 232, 1971.
49. Fritz, A., *Macromolecular Syntheses*, **9**, 69 (1985). Reprinted by permission of John Wiley & Sons, Inc.

CHAPTER 3

Additional Synthetic Methods

Introduction

This Chapter discusses some additional synthetic methods that do not fit neatly into the nucleophilic and electrophilic classifications of Chapters I and II. This division is not meant to indicate that these methods are of lesser importance, especially since at least one of them is practiced commercially. Oxidative coupling technology is utilized to prepare the Poly(phenylene oxide) or PPO™ polymers, which are blended with polystyrene resins to produce the Noryl™ family of engineering resins.

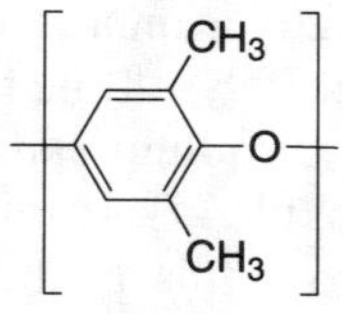

(III-1)

PPO

Ullman's early researches on the synthesis of aryl ethers (*1*, *2*, *3*) and the coupling of haloaromatics to form aryl-aryl bonds (*4*) laid the basis for several of the modern methods for the synthesis of aromatic polyethers. Many attempts to use the features of the Ullman aryl ether synthesis to prepare aromatic polyethers preceded the successful development of the now, well known nucleophilic and electrophilic process technologies. Diaryl ethers were synthesized by Ullman by reacting alkali metal phenoxides with aryl halides catalyzed by copper. The yields varied considerably with different copper-catalyst preparations. Ullman's original work on the coupling of haloaromatics by copper powders was carried out in the early 1900's (*4*). Two molecules of aryl halide are coupled by reaction in the presence of activated bronze or freshly precipitated copper at temperatures above 100°C. The coupling of halo- and other aromatics to produce high molecular weight polyethers in the presence of metals, such as nickel and palladium, is described in this chapter. This latter process technology is based on the fundamental advances in organometallic chemistry that have been made over the last several decades.

The development of the Scholl reaction for the synthesis of polyarylethers is a welcome addition to the arsenal of synthetic methods for polyarylethers.

It is, however, somewhat limited in scope because of the structural requirements of the monomers for successful polymerization. The polymerization methods based on the ring opening of macrocycles and the decomposition of polycarbonates have only seen limited utility to date.

A. POLY(ETHER SULFONES)

1. Ullman Polymerization

A-1. Ullman Polymerization of Bis(phenates) with Haloaromatic Sulfones

The Ullman polymerization of *bis*(phenates) with haloaromatic compounds has not been widely used for the preparation of poly(ether sulfones). The ease of halide displacement under Ullman conditions is I > Br > Cl > F. Since the more reactive halo compounds are also more expensive, there has been no incentive to develop this process technology for polyarylethers that are derived from activated halo aromatic compounds. One early report obtained low molecular weight Polysulfone from the reaction of 4,4′-dibromodiphenyl sulfone with the disodium salt of Bisphenol A in DMSO at 153°C catalyzed by cuprous oxide (*5*). It might be possible to obtain high molecular weight polymer from this route, but when one can use 4,4′-dichlorodiphenyl sulfone successfully in some of the nucleophilic methods described in Chapter I, there is no incentive to pursue it. The order of halogen reactivity for activated halo aromatics in nucleophilic displacement reactions is F > Cl > Br > I. For nonactivated halo aromatic compounds, the Ullman process is a valuable route to poly(phenylene ethers) as is discussed later in this chapter.

2. Cation-Radical Polymerization

A-2. Cation-Radical Polymerization (Scholl reaction)

The Scholl reaction (*6*) has been successfully used for the preparation of aromatic polyethersulfones (*7, 8, 9*). In the simplest terms, the Scholl reaction can be described as the elimination of two aryl hydrogen atoms by the formation of an aryl-aryl bond under the influence of Friedel-Crafts catalysts. A cation-radical mechanism is generally accepted as being operative in the Scholl reaction (*10*). An example is shown in the accompanying equation. Ferric chloride was found to be the preferred catalyst and is used at a 4/1 molar ratio to monomer. The polymerizations were run in nitrobenzene at room temperature.

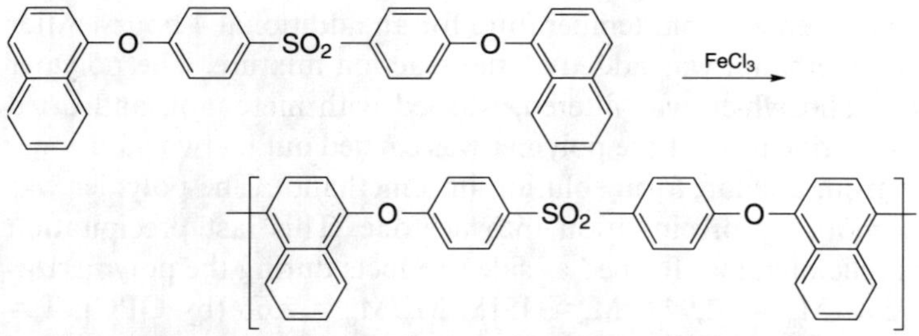

(III-2)

The mechanism of this cation-radical polymerization has been studied and discussed (*11*, *12*). Initiation consists of a one electron oxidation of the monomer by the ferric chloride to form a cation-radical at one of the naphthoxy groups. Most of the successful polymerizations using this chemistry that have been reported have been with monomers that contain naphthoxy groups, which may be due to their ease of oxidizability in the initiation step. Both electrophilic and radical propagation steps have been proposed with termination resulting from the cation radical combining with the counter-anion or a fragment of it to form a dead chain end. An extensive attempt was made to adapt this reaction for the preparation of some of the commercial polysulfones without success (*13*). For example, polymerization of 4,4′-diphenoxydiphenyl sulfone would yield Radel R™, which is normally

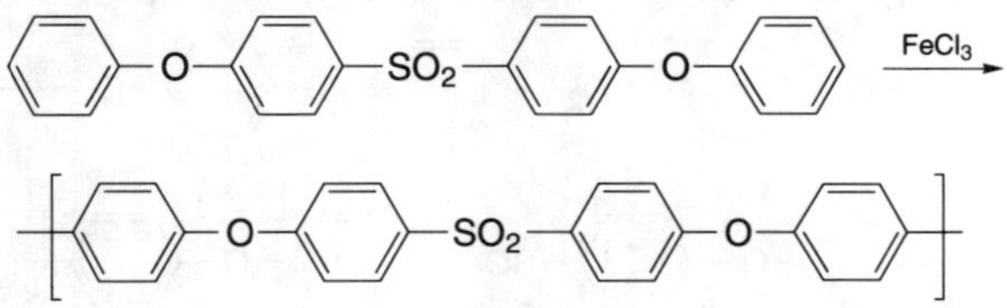

(III-3)

Radel R™

made by the polymerization of biphenol with 4,4′-dichlorodiphenyl sulfone. Only low molecular weight polymers were obtained from 4,4′-diphenoxydiphenyl sulfone and many other of its derivatives. The corresponding phenylthiodiphenyl sulfones gave slightly higher, but still not useful molecular weights.

EXAMPLE I: POLYMERIZATION OF 4,4′-DI-(1-NAPHTHOXY)DIPHENYL SULFONE (*7*)

1.0 g (1.98 mmol) of 4,4′-di-(1-naphthoxy)diphenyl sulfone was dissolved in 5 ml of dried nitrobenzene while a stream of nitrogen was passed through the solution. Separately 1.2883 g (7.942 mmol) of $FeCl_3$ was dissolved in 10 ml of dried nitrobenzene and added dropwise at room temperature to the stirred monomer solution. After the $FeCl_3$ solution was added, the reaction was stirred under nitrogen at room temperature for an additional 4 hours. After this time acidic methanol was added to the reaction mixture. The polymer separated as a solid which was filtered, washed with methanol, and dried under vacuum. Purification of the polymer was carried out by two additional precipitations from a chloroform solution into methanol. The polymer was purified once more by precipitation into acetone. This last precipitation removed the cyclic oligomer formed as side products during the polymerization. Yield: 75%. M_n = 12,942; M_w=34518; M_w/M_n = 2.66 (by GPC); T_g= 271.8.

3. Metal-catalyzed Polymerization of Aromatic Monomers with Functional Groups

A-3. Nickel-catalyzed Polymerization of Aryl Dichlorides

One of the newer and more novel methods for the preparation of polyarylethersulfones is the nickel(0) promoted coupling of aryl dihalides (*14*, *15*). This route is based on earlier researches on the coupling of aryl chlorides by nickel and reducing metals (*16*). An example is shown in the accompanying equation. This route offers an alternate pathway to polymers

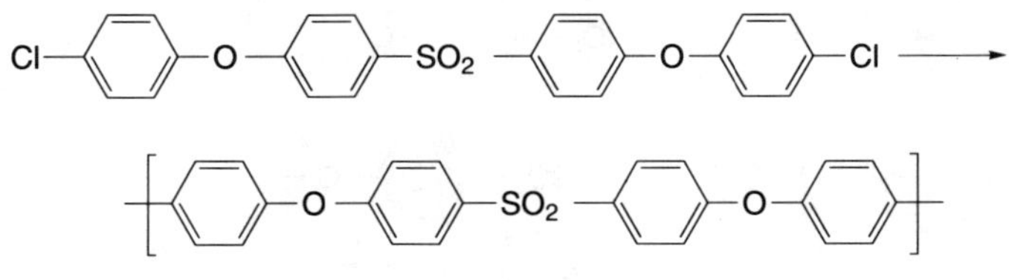

(III-4)

Radel R™

that can be prepared by other methods. The example in the above equation is for the synthesis of Radel R™, a product prepared commercially by nucleophilic aromatic substitution. For this reason, the parameters affecting the nickel coupling route have been studied in great detail (*14*, *15*, *17*). The polymerization is performed under an inert atmosphere in a dry, aprotic solvent like DMAC using catalytic amounts of zero valent nickel, triphenylphosphine ligand and zinc metal. Low amounts of nickel (< 2 mole % based on the amount of aryl dichloride), high triphenylphosphine/nickel ratios, excess zinc metal, and moderate temperatures (70°C) are used. Catalysts such as 2,2′-bipyridine or sodium bromide can be used to accelerate the reaction. Two additional prerequisites for the successful preparation of high molecular weight product are the solubility of the monomer, and more particularly of the obtained polymer, in the reaction medium. Also, any functional groups present in the monomer must be inert toward the zinc/nickel catalyst system. Nitro groups, for example, are not tolerated (*14*). Ester, ketone, amide, and imide groups can be present in the monomers. 3-Phenyl-2,5-dichlorothiophene has been polymerized successfully by this method (*18*). A disadvantage of this process on a commercial scale would be the necessity to insure the separation of the catalyst components from the polymer (*14*).

EXAMPLE I: GENERAL PROCEDURE FOR NICKEL COUPLING OF ARYL DICHLORIDES (*14*)

Into a three-neck 250 ml flask equipped with a magnetic stirring bar are placed 8.0 g zinc dust, 6.0 g triphenylphosphine, 0.15 g 2,2′-bipyridine, and 0.10 g nickel chloride. One neck of the flask is sealed with a serum cap. A sidearm flask containing 20 g of crystalline monomer(s) is attached to the second neck and a stopcock is attached to the third. The stopcock is connected to a double-manifold vacuum line by pressure tubing, and the reaction apparatus is then evacuated and filled with nitrogen several times. Sixty milliliters of dry, nitrogen-sparged DMAC (*N*,*N*-dimethylacetamide) are added by syringe to the reaction flask, and the reaction apparatus is placed in an oil bath at 70°C. Once the red-brown catalyst forms, the monomer is added from the sidearm flask into the catalyst solution. The polymerization is run at 70°C for 10–15 hours. At the end of this period 0.5 ml chlorobenzene are added to the polymer solution and reacted for 0.5 hour to remove any nickel attached to polymer, i.e., to terminate the polymer with phenyl groups. The reaction mixture is diluted with an equal volume of DMAC and filtered through a medium glass filter to remove any excess zinc. The polymer is coagulated in 1 l of methanol in a blender. The polymer is then slurried at 90°C for 1 hour in 1 l of deionized water. This is followed by filtration, washing the polymer cake several times with an equal volume of methanol, and drying in a vacuum oven.

A-4. Palladium-catalyzed Cross-coupling of Tin Reagents

The palladium-catalyzed cross-coupling of bifunctional tin reagents with dihalides has been investigated and yields moderately high molecular weight polysulfones (*19*). The analogous zinc and magnesium reagents were found to give only low molecular weight polymers. 4,4′-Bis(tri-n-butylstannyl)diphenyl ether was polymerized with 4,4′-dibromodiphenyl sulfone in DMAC at 165°C for four hours. The 2:1 triphenyl phosphine:palladium chloride complex was the best catalyst, although polymer yields were low. The polysulfone depicted in the equation below possessed a Tg of 168°C and a melting point of 320°C.

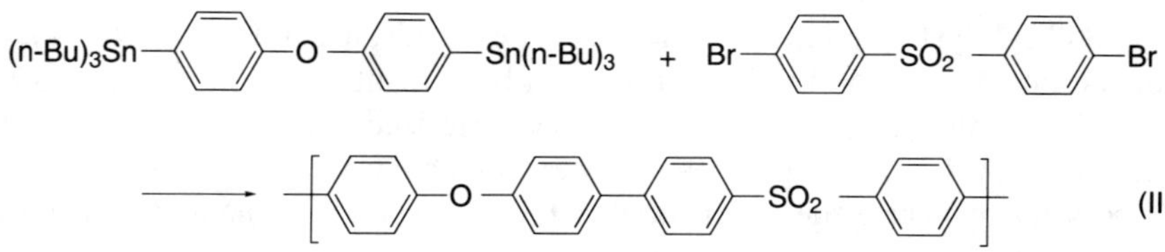

(III-5)

No other polysulfones were prepared by the route in this publication.

4. Ring Opening Polymerization

A-5. Ring Opening Polymerization of Sulfone-Containing Macrocyclic Oligomers

Recently, the synthesis and ring opening polymerization of several sulfone-containing macrocyclic oligomers has been reported (*20*, *21*, *22*). Ring opening polymerization processes have advantages over preformed high molecular weight polymers in several fabrication technologies because of their very low initial viscosities. This property facilitates the impregnation of the fibers used in making composites. The drive to using polyarylethers as matrix resins for the manufacture of thermoplastic composites focussed attention on this type of polymerization technology. Another advantage that this type of polymerization offers is that it is a bulk polymerization that does not have large amounts of salts and solvents present. While these advantages have been realized on a laboratory scale, the lack of economical syntheses of the required monomers has stymied commercial development.

Two of the macrocyclic oligomers that have been synthesized and polymerized are pictured below. The spiroindane based oligomer was prepared from the bisphenol and 4,4′-difluorodiphenyl sulfone by the carbonate process in 90% yield (*22*). The spiroindane bisphenol is unique in that it has a pronounced tendency to form cyclics. The other macrocycle pictured below

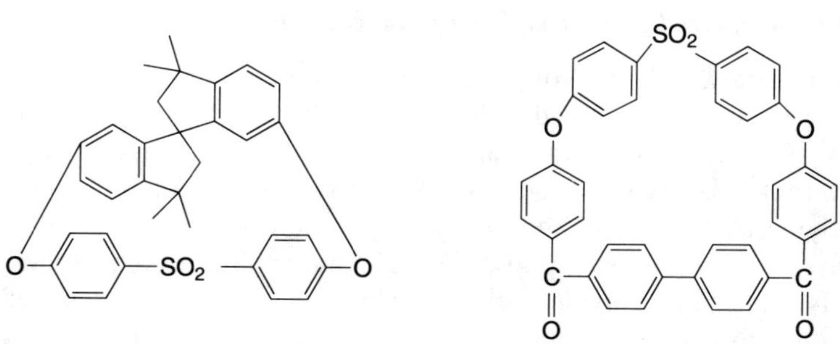

(III-6)

was prepared by the nickel-promoted cyclization at high dilution of the corresponding chlorobenzoyl terminated linear oligomer. Another report (*23*) describes synthetic conditions that produced mixtures of cyclic oligomers from conventional monomers in yields of about 50%. It is not necessary for the spiroindane moiety to be present. The cyclic dimer of Bisphenol A Polysulfone has been prepared in 10% yield by polyetherification (*24*). Polymerization of cyclics is rapid when conducted in the melt and catalyzed by nucleophilic species. The cyclic ether sulfone oligomer below gave polymer

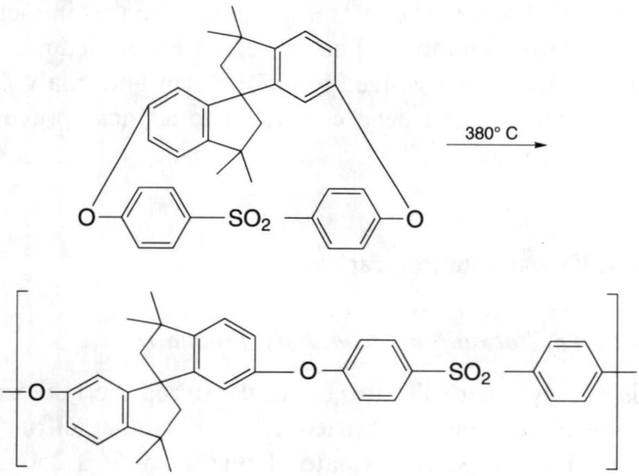

(III-7)

with a molecular weight of 80,000 in fifteen minutes when catalyzed by the disodium salt of Bisphenol A. High molecular weight Polyethersulfone was obtained from its cyclic oligomers in two hours at 300°C when catalyzed by cesium fluoride (*23*). Since these same polymers can be made by direct methods, this route only offers advantages when used in conjunction with a fabrication method.

EXAMPLE I: PREPARATION AND POLYMERIZATION OF A MACROCYCLIC ETHER SULFONE (*21*, *22*)

A 3-neck 250 ml round bottom flask was equipped with provisions for a magnetic stirrer and a nitrogen inlet. The flask was also fitted with a Dean-Stark trap, condenser and thermometer. The flask was charged with dimethyl sulfoxide (100 ml), toluene (50 ml.) and potassium carbonate (2.073 g, 20.9 mmol.). To this flask was added 2,2′,3,3′-tetrahydro-3,3,3′,3′-tetramethyl-1,1′-spiro[1H-indene]-6,6′-diol (SBI) (3.084 g, 10.0 mmol) and *bis*(4-fluorophenyl) sulfone (2.543 g, 10.0 mmol). The reaction mixture was heated and sufficient toluene was removed such that the pot temperature was 140°C.

After 4 hours, the reaction mixture was cooled, then poured into methanol (400 ml) and the white precipitate was collected by vacuum filtration. The resulting solid was oven dried (house vacuum) at 110°C for several hours, then triturated (or blended) twice with water. The solid was dried for 24 hours in a vacuum at 110°C. The yield was 4.73 g (90%).

A toluene solution of 1.0 g (1.9 mmol) of the macrocyclic polyethersulfone oligomer from 4,4′-difluorodiphenyl sulfone and SBI plus 5.0 mg (0.02 mmol.) of the disodium salt of Bisphenol A was distilled to remove toluene and dry the system. The vessel containing the residue was immersed in a salt bath at 380-400°C for 15 minutes. The contents were cooled to room temperature and the residue was dissolved in chloroform and analyzed by HPLC and GPC. The oligomer had been converted to a linear polymer having a molecular weight of about 80,000.

5. Poly(ether sulfones) from Polycarbonates

A-6. Preparation of Polysulfones from Polycarbonates

The synthesis of poly(ether sulfones) by the catalyzed thermal decomposition of polycarbonates has been reported (*25*). Polyethersulfone, PES, was obtained by heating the polycarbonate of Bisphenol S at 250–280°C in the

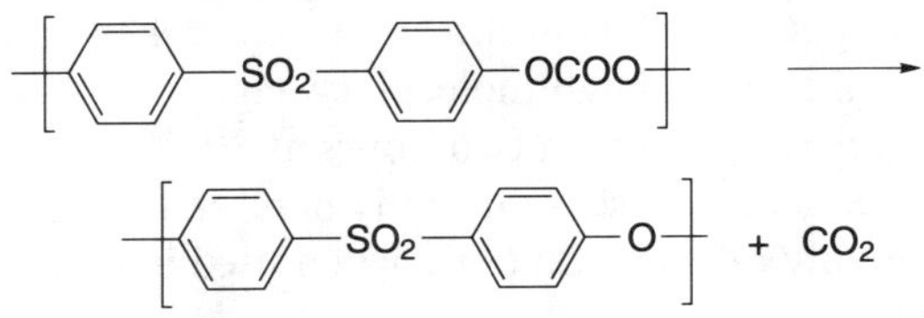

(III-8)

presence of catalytic amounts of the potassium salt of bisphenol S. This patent also gives examples for the preparation of PES in one step by reaction of Bisphenol S with diphenyl carbonate with simultaneous evolution of carbon dioxide. This methodology has also been used to prepare poly(aryl ether)-poly(aryl carbonate) block copolymers (26). A copolycarbonate derived from two bisphenols, one of which has an electron withdrawing group in the position ortho or para to the phenol group, is decomposed. Only the carbonate links derived from the more acidic bisphenol decompose to give aryl ether linkages. Bisphenol S and 4,4′-dihydroxybenzophenone copolycarbonates with Bisphenol A were the only compositions studied. Tough block copolymers were produced. Conceptually, the preparation of one polymer from another polymer when the desired polymer can be made directly, severely limits the practical utility of this method

The reaction of sulfone- and carbonyl-activated dihalo aromatics with Bisphenol A polycarbonate in NMP and potassium carbonate at 150–165°C also gave polyarylethers (*27*).

EXAMPLE I: PREPARATION OF POLYETHERSULFONE (*25*)

One hundred grams of 4,4′-dihydroxydiphenyl sulfone were mixed with 87.8 g of diphenyl carbonate and 0.13 g. of the dipotassium salt of 4,4′-dihydroxydiphenyl sulfone; the mixture was melted whereupon evolution of phenol began. Evolution of carbon dioxide started at 250–280°C. After 120 minutes, a polymer softening at 225°C was obtained. An infrared spectrum and elemental analysis indicated that the product was Polyethersulfone.

B. POLY(ETHER KETONES)

1. Ullman Polymerization

A-7. Ullman Polymerization of Bis(phenates) with Haloaromatic Ketones

The polymerization of bis(phenates) with haloaromatic compounds has not been widely used for the preparation of poly(ether ketones). The ease of halide displacement under Ullman conditions is I > Br > Cl > F. Since the more reactive halo compounds are also more expensive, there has been no incentive to develop this process technology for the synthesis of polyarylethers from activated halo aromatic compounds. One early report obtained low molecular weight polymer from the reaction of 4,4′-dichlorobenzophenone with the disodium salt of Bisphenol A in DMSO at 153°C catalyzed by cuprous oxide (*5*). For non-activated halo aromatic compounds, the Ullman process is a valuable route to poly(phenylene ethers) as is discussed later in this chapter.

2. Cation-Radical Polymerization

A-8. Cation-Radical Polymerization (Scholl reaction)

Dinaphthoxy monomers containing keto groups have been successfully polymerized to high molecular weight polymer by the Scholl reaction. The catalysts, solvents and reaction conditions are essentially the same as detailed above (Method A-2) for the synthesis of polyarylethersulfones. An example is shown in the accompanying equation (*7*).

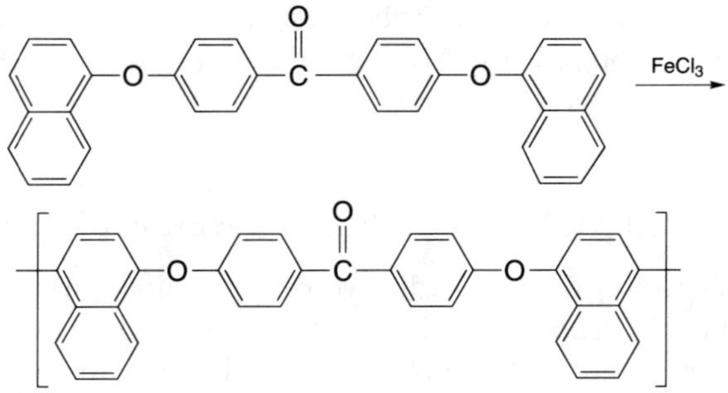

(III-9)

3. Metal-catalyzed Polymerization of Aromatic Monomers with Functional Groups

A-9. Nickel-catalyzed Polymerization of Aryl Dichlorides

Polyaryletherketones can be prepared by the nickel(0) promoted coupling of aryl dichlorides (*17*, *28*). The more crystalline nature of the polyaryletherketones imposes a limitation on the scope of this method. High molecular weights cannot be attained if the polymer precipitates from solution at the relatively moderate polymerization temperatures that are optimum for this method. The example shown in the accompanying equation has been studied

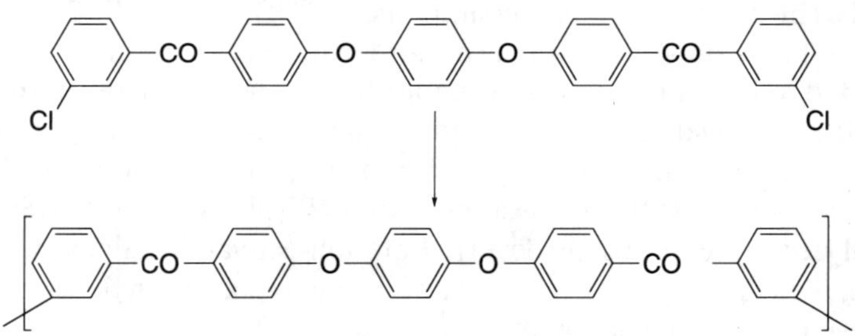

(III-10)

in great detail because it is a homogeneous polymerization (*28*). The inherent viscosity of the polymer was 0.87 (in conc. sulfuric acid) when polymerization was conducted at 90°C in DMAC. The optimum mole ratios of nickel chloride, triphenyl phosphine, zinc, and 2,2′-bipyridine to the monomer were 0.05, 0.1, 3.1, and 0.05, respectively.

EXAMPLE I: POLYMERIZATION OF 1,1′-(*p*-PHENYLENEDIOXY)*BIS*[4-(3-CHLOROBENZOYL)BENZENE (*28*)

In a 50-ml two-necked round-bottomed flask were placed $NiCl_2$ (16.2 mg, 0.125 mmol), PPh_3 (65.6 mg, 0.25 mmol), bipyridine (bpy) (19.5 mg, 0.125 mmol), zinc (506 mg, 7.75 mmol), and 1,1′-(*p*-phenylenedioxy)*bis*[4-(3-

chlorobenzoyl)benzene (1.35 g, 2.5 mmol). The flask was evacuated and filled with nitrogen three times. Then, dry DMAC (3.0 ml) was added via syringe through the serum cap. The mixture was stirred at 90°C for 2.5 hours. The mixture became red-brown in 20 min. The resulting viscous mixture was diluted with DMAC (25 ml) and poured into a large amount of HCl/methanol. The polymer was collected, washed with methanol, and dried in vacuo at 80°C for 10 hours. The yield was 1.17 g (100%). The inherent viscosity of the polymer in NMP was 0.87 dl-g^{-1}, measured at a concentration of 0.5 g-dl^{-1} at 30°C.

A-10. Palladium-catalyzed Cross-coupling Reactions

The palladium-catalyzed, cross-coupling polymerization of 4,4′-bis(tri-n-butylstannyl) diphenyl ether with 4,4′-dibromobenzophenone or the analogous benzil monomer yielded polymers of moderate molecular weight (*19*). As was the case with this route for polyarylethersulfones, the corresponding magnesium and zinc derivatives gave inferior results. The equation illustrates the synthesis of a benzil group-containing polymer by this method.

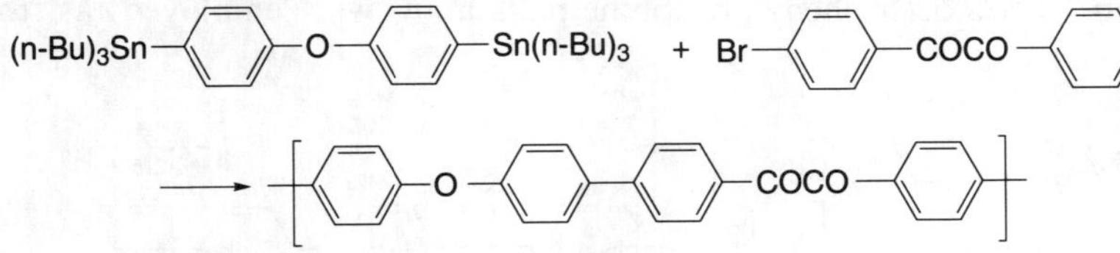

(III-11)

The 2:1 triphenyl phosphine:palladium chloride complex was the best catalyst, although polymer yields were low. The preferred polymerization solvent was DMAC at 165°C.

A recent study successfully explored the feasibility of using Stille's palladium-catalyzed ketone synthesis for synthesizing poly(aryl ketones) (*29*). The specific system studied is shown in the following equation. Polymers

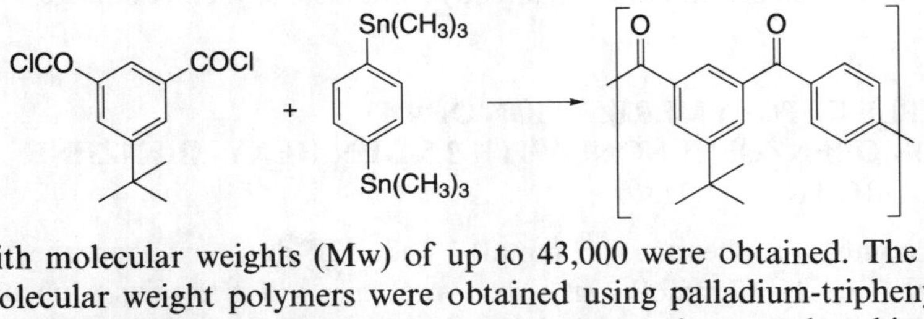

(III-12)

with molecular weights (Mw) of up to 43,000 were obtained. The highest molecular weight polymers were obtained using palladium-triphenylarsine catalysts ($Pd[As(C_6H_5)_3]_2Cl_2$) instead of the analogous phosphine based catalysts ($Pd[P(C_6H_5)_3]_2Cl_2$). The best catalysts were postulated to possess an optimum balance between reactivity and thermal stability.

EXAMPLE I: POLYMERIZATION OF 5-T-BUTYLISOPHTHALIC ACID CHLORIDE WITH 1,4-BIS(TRIMETHYLSTANNYL)BENZENE (*29*)

5-t-Butylisophthalic acid chloride (105.9 mg, 0.4087 mmol) and ($Pd[As(C_6H_5)_3]_2Cl_2$) (5.7 mg, 0.0072 mmol) were weighed into a polymerization tube in a dry box. The necessary amount of 1,4-*bis*(trimethylstannyl)benzene (167.9 mg, 0.4159 mmol) was weighed into a separate vial and quantitatively transferred to the polymerization tube using 1,2-dichloroethane (ca. 1.5 ml). The tube was sealed, removed from the dry box and heated for 14 hour at 50°C. At this point the tube was opened and the solution poured into rapidly stirred ethanol. Following centrifugation the solid was dried and analyzed using size exclusion chromatography. Clear amorphous films could be cast from a number of different solvents (e.g. tetrahyrofuran, benzene, chloroform, methylene dichloride, 1,2-dichloroethane).

Moderately high molecular weight polyarylethers have been obtained from the Pd-catalyzed condensation of dibromo aromatic compounds with aromatic diboronic acids (*30*). An example is shown in the accompanying equation. Tetrakis(triphenylphosphine)palladium was employed as the

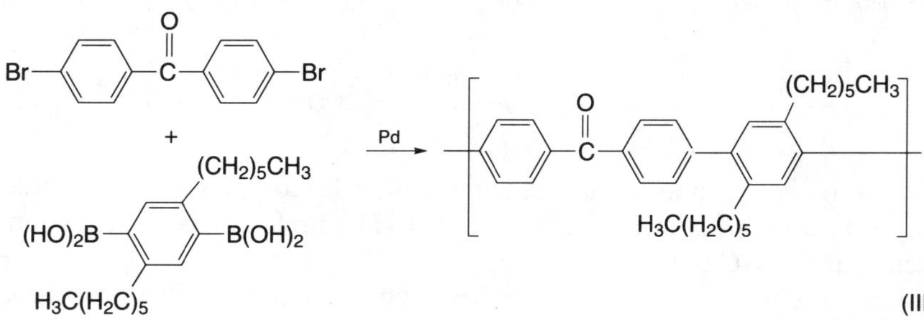

(III-13)

catalyst. This method forms an aryl-aryl bond. Thus, polyarylethers are formed only when the monomer(s) already contain an aryl ether linkage.

EXAMPLE II: POLYMERIZATION OF 4,4′-DIBROMOBENZOPHENONE WITH 2,5-DI(N-HEXYL)BENZENE-1,4-DIBORONIC ACID (*30*)

4,4′-Dibromobenzophenone (3 mmol) and 2,5-di(n-hexyl)benzene-1,4-diboronic acid (3 mmol) in the presence of tetrakis(triphenylphosphine)palladium (0), are refluxed in the heterogeneous system toluene/water (with 2 mol Na_2CO_3/l water) (30 ml each) for 2 days. After the mixture has cooled to room temperature, the layers are separated and the organic one is washed with water (2 × 20 ml). The polymer is precipitated from this solution with methanol, recovered and dried in vacuum over P_2O_5. Further purification is achieved with a Soxhlet extraction using toluene or chloroform as solvent, followed by repeated precipitation with methanol, The polymer is obtained as a yellowish material in yields of 97–100%.

A-11. Polymerization of Organometallic-aromatic Complexes with Bisphenates

A ruthenium(II) complex of p-dichlorobenzene has been polymerized with the disodium salt of dihydroxybenzophenone to produce PEEK, or poly(ether ether ketone) (*31*). This is illustrated in the accompanying equation. The initial condensation of the cyclopentadienyl complex is

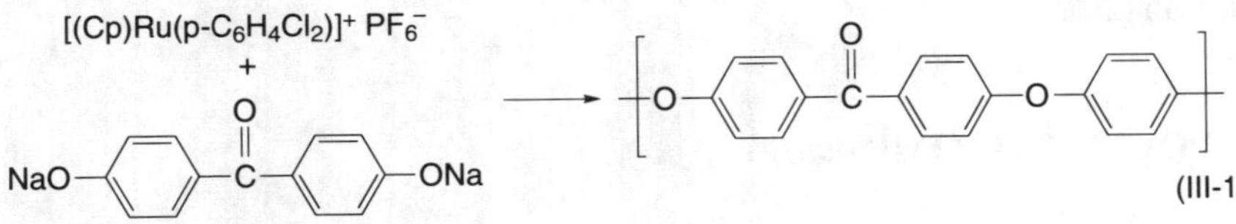

(III-14)

conducted in DMSO at 85-90°C. The intermediate metalated polymer was then demetalated by heating to 160°C for 1.5 hours in the same solvent. Extensions of this approach for the synthesis of poly(arylene sulfides) and other poly(arylene ethers) have been reported (*32*). The reaction of 1,4-dichlorobenzene chromium tricarbonyl with phenoxides has been studied as a route to monomers that are useful for the synthesis of high temperature polymers (*33*). Methods for using simple chloroaromatics like p-dichlorobenzene for making high performance polyarylethers is a continuing desirable, but elusive goal.

4. Ring Opening Polymerization

A-12. Ring Opening Polymerization of Ketone-Containing Macrocyclic Oligomers

This method is formally similar to that described in Section A (Method A-5) above for poly(ether sulfones). The oligomers were made both by the carbonate process (*22*) and by nickel promoted coupling of bis(chlorobenzoyl) terminated oligomers (*20*). The spiroindane bisphenol gave about 50% yields of cyclic oligomers from several keto-containing monomers, including 4,4′-difluorobenzophenone and 1,5-dichloroanthroquinone. The macrocycle depic-

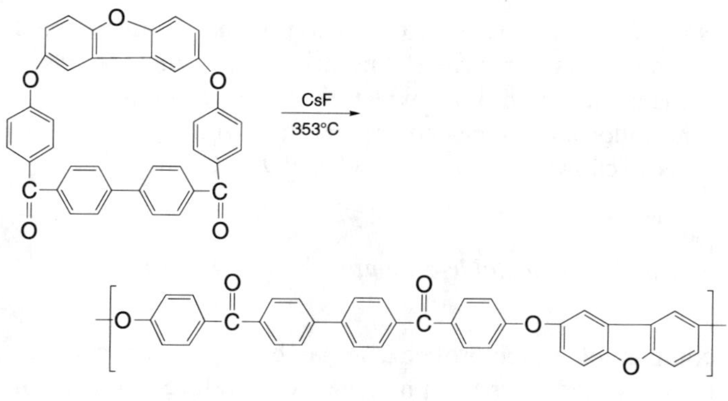

(III-15)

ted above was obtained in 40% yield by running the nickel(0)-promoted coupling reaction of the bis(chlorobenzoyl) precursor monomer at high dilution (*20*).

C. POLY(ETHER IMIDES)

1. Ring Opening Polymerization

A-13. Ring Opening Polymerization of Imide-Containing Macrocyclic Oligomers

A large number of macrocyclic etherimides have been synthesized from that spiroindane structure that has a pronounced tendency toward cyclization. The bisphenol, SBI (2,2′,3,3′-tetrahydro-3,3,3′,3′-tetramethyl-1,1′-spiro[1H-indene]-6,6′-diol), has been converted to a dianhydride by reaction of its disodium salt with 4-nitro-N-methylphthalimide and subsequent conversion of this bis(imide) to the dianhydride. The dianhydride was reacted with a number of diamines to give cyclics in 25–77% yields. A similar displacement reaction with 4-chloronitrobenzene, followed by reduction of the nitro groups gave the diamine derivative. This spiroindane-based diamine was reacted with a number of dianhydrides to give cyclics in 30–75% yields.

Many of these cyclic oligomers were very high melting, sparingly soluble solids that did not allow their polymerization in solution or the melt to be studied. A study of some mixed cyclics did show that they could be polymerized to high molecular weight polymers by ether exchange when catalyzed by sodium sulfide in DMAC at 200°C. Polymerization by imide exchange was too slow to be an operative mechanism (*21*).

EXAMPLE I: PREPARATION OF A CYCLIC SILICONE ETHER IMIDE (*22*)

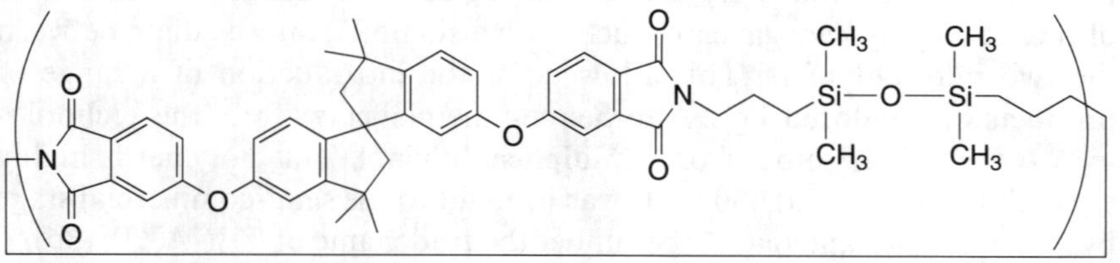

(III-16)

A solution of 1.274 g (5.0 mmol) of 1,9-diamino-4,4,6,6-tetramethyl-4,6-disila-5-oxanonane in 50 ml of *o*-dichlorobenzene (ODCB) was added over 1 hour to a solution of 3 g (5.0 mmol) of the dianhydride prepared from reaction of the bisphenol SBI and 4-nitro-N-methylphthalimide as described above, and 2 mg of sodium phenylphosphonate in 250 ml of ODCB at 140°C. When the addition was completed, the temperature was raised to 225°C and ODCB and water were removed by distillation until the distillate was no longer cloudy; a total of about 100 ml of ODCB was thus removed. The residual solution was heated under reflux for 2 hours and then reduced to about 10% of its original volume by distillation. Upon cooling and pouring into 5 volumes of methanol, a solid precipitated which was collected by filtration and dried in a vacuum oven at 110°C. It was shown by field desorption mass spectrometry to comprise the desired macrocyclic siloxane polyetherimide monomer and dimer. A further portion of macrocyclic monomer was obtained by evaporation of the methanol from the filtrate. The total yield of macrocyclic oligomers was 3.34 g or 82% of theoretical.

D. POLY(PHENYLENE ETHERS)

One of the milestones in the scientific and commercial development of polyarylethers was the discovery of the oxidative coupling route in the late 1950s (*34*, *35*). This research led to the commercialization of poly(2,6-dimethyl-1,4-phenylene) oxide under the tradename of PPO™ by the General Electric Co. This polymer is commonly referred to as polyphenylene oxide. Because of the polymer's high Tg (208°C), melt fabrication of the polymer required high temperatures. It was found that oxidative reactions involving the methyl groups caused degradation in the melt. These reactions adversely affected processability and finished part properties to the point where the neat polymer was withdrawn from the market. Fortunately, at about the same time it was discovered that PPO™ was thermodynamically miscible over the entire composition range with polystyrene. This discovery was the basis for the Noryl™ family of engineering resins that were then developed and introduced by GE. Blends of these two polymers gave products with a single Tg intermediate between the two pure polymers. This allowed for the introduction of a range of products with tailored Tg's. Another commercial entry from this technology was the polyarylether from 2,6-diphenylphenol. This polymer is highly crystalline and melts at 480°C. It was brought to the semi-commercial stage by Akzo (Netherlands) as a fiber under the tradename of TENAX™ (*36*).

1. Oxidative Coupling

A-14. Oxidative Coupling of 2,6-Disubstituted Phenols

The oxidative coupling of 2,6-dimethylphenol to polyphenylene oxide is shown in the accompanying equation. An excellent account of the development of this chemistry by the inventor is available (*34*). The failure of phenol

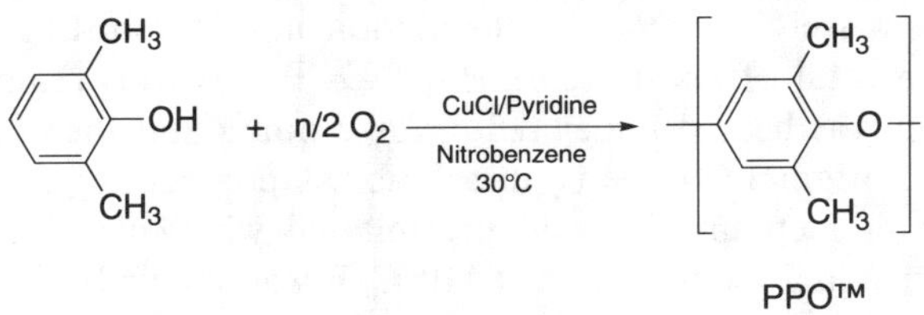

(III-17)

itself to give a useful product under these conditions led to the study of substituted phenols and the subsequent development of PPO™.

Phenol Structure

The yields and molecular weights of polymers obtained from this polymerization depend strongly on the ring substituents and their locations. 2,6-Disubstituted phenols with hydrogen atoms in the 4 position polymerize the most easily. Some 2,6-disubstituted phenols with methyl or aryloxy groups in the 4 position can also be polymerized. 2,3,6-Trisubstituted phenols can be homo- and copolymerized. Bulky groups (t-butyl and isopropyl groups) in the 2- and 6-positions favor formation of the coupled diphenoquinone product, viz.,

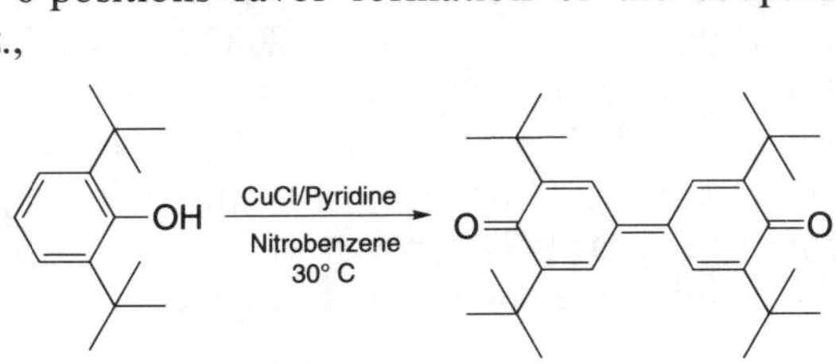

(III-18)

Methoxy groups in the 2 and 6 positions also favor formation of the diphenoquinone type of coupled product (*37*). 2,6-Dinitrophenol does not polymerize and only low molecular weights are obtained when two halogen groups are present.

Phenols with one *ortho* substituent and an open *para* position can be oxidatively coupled, but the product is highly branched and colored. Presumably, coupling at the open *ortho* position leads to branching while other types of oxidation produce quinones and other colored species. Nevertheless, small amounts of phenol and o-cresol are incorporated into PPO when they are copolymerized with 2,6-dimethylphenol. The resultant polymers are not highly branched.

Catalysts

A stoichiometric amount of molecular oxygen is used as the oxidant for the polymerization of phenols with a transition-metal catalyst such as a copper or manganese compound combined with ligands. The type of oxidizing system, its composition, concentration, and other variables can have a large effect on reaction rate, selectivity, and product quality.

Copper halide salts are used with ligands such as diethylamine, dibutylamine and morpholine. Diamines, oligomeric amines and even polymeric amines have been reported to be useful. The amine solubilizes the metal salt in the organic medium and increases the pH, thereby lowering the oxidation potential of the phenol. Amines that form insoluble complexes with copper, do not function as catalysts and can be used to stop the reaction. An excess of amine relative to copper is used. Copper-amine catalysts are susceptible to hydrolysis and deactivation by the water generated during the polymerization. To avoid hydrolysis, a drying agent is added to the reaction mixture. Some catalysts are stable enough to operate without a desiccant. Ligands used for manganese complex catalysts include benzoin oxime, amino alcohols, and mercaptans. Some of these systems require a strong base, such as aqueous sodium hydroxide. Excellent summaries of the literature on catalysts exist (*37*, *38*).

Solvent Selection

The major criteria for solvent selection are that they be inert to the oxidant and catalyst system and that they are solvents for the polymer being formed. Toluene and benzene are suitable solvents. PPO™ is also soluble in halogenated hydrocarbons, but only slightly soluble in aliphatic hydrocarbons, acetone, most alcohols and tetrahydrofuran.

Polymerization Mechanism

The mechanism for the oxidative polymerization of 2,6-dimethylphenol has been studied extensively and is now accepted as involving free radical processes (*37*, *38*). Aryloxy radicals are formed first by the oxidation of the phenol with the oxidized form of the copper-amine complex. The aryloxy radicals couple to form a quinol ether which enolizes to form a dimer as shown in the accompanying equation.

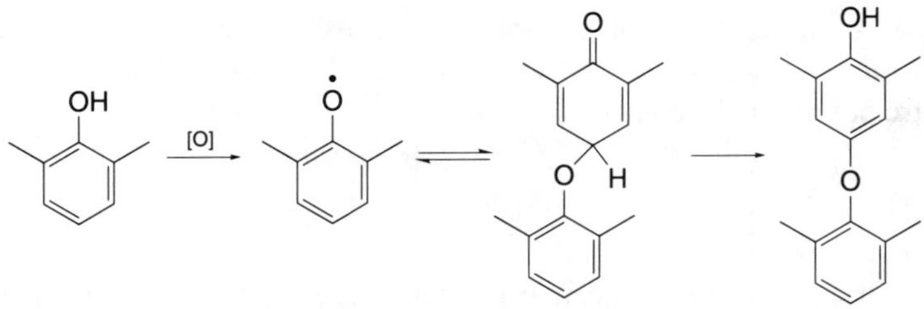

(III-19)

Oxidation of the dimer and its coupling with the 2,6-dimethylphenoxy radical, followed by enolization produces the analogous trimer. Continuation of the processes of oxidation, radical coupling, dissociation and enolization results eventually in a high molecular weight polymer. Another process that occurs concurrently is an intramolecular rearrangement in which cyclohexadiene groups appear to move along a polymer chain, and are then converted to a phenol when they reach a terminal position by enolization. Overall, the polymerization resembles a step-growth polymerization more closely than a chain process, such as in a vinyl polymerization.

EXAMPLE I: POLYMERIZATION OF 2,6-DIMETHYLPHENOL TO PPO™ (38)

2,6-Dimethylphenol (DMP) is purified by recrystallization from hexane until at least 99.5% pure. Reagent-grade solvents and catalyst components are used. A vessel is equipped with a mechanical stirrer (paddle-type or a vibromixer), an oxygen inlet tube arranged to impinge oxygen on the rotating impeller, a thermometer, and a water bath to provide cooling. Oxygen is introduced into a vigorously stirred solution of cuprous bromide and dibutylamine in toluene. A ratio of CuBr:dibutylamine:DMP of 1:22:75 and a final DMP concentration of 15% gives an acceptable reaction rate. A 25% solution of DMP in toluene is added over a 20-min period. The temperature rises during the addition and should be maintained at about 40°C with the help of the water bath. After the addition is complete and the exotherm subsides, warming of the water bath is necessary to maintain the temperature near

40°C. The time required for a high molecular weight product is approximately 75 min. with the system described here, but depends on the specific conditions (stirring, purity, temperature, and so on). The reaction mixture is transferred to a large beaker using additional toluene, giving about a 10% solution. Methanol (3–5 volumes) containing 0.5–1% acetic acid is added gradually to the stirred solution. As the polymer precipitates it often appears to agglomerate, but as more methanol is added, it separates into a granular powder. The polymer is filtered, washed with methanol, and dried in a vacuum oven at 60°C. The yield is over 95%; the intrinsic viscosity is ca 0.5 dL/g in chloroform at 25°C. If diethylamine or another short-chain secondary amine is used instead of dibutylamine, anhydrous magnesium sulfate or molecular sieves (1 g per 2 g DMP) are added with the catalyst and filtered off before precipitating the polymer.

Two other detailed preparations of PPO™ are also available (*39*, *40*).

EXAMPLE II: GENERAL PROCEDURE FOR THE POLYMERIZATION OF 2,6-DIARYLPHENOLS (*41*)

To a test tube there was added 8.2 mg of CuCl, 6.2 mg of *N*,*N*,*N'*,*N'*-tetramethylethylene diamine, 0.4 g of anhydrous magnesium sulfate and 7 ml. of *o*-dichlorobenzene. The test tube was put in an oil bath at 65°C and the mixture was agitated with a magnetic stirrer. Oxygen was bubbled into the mixture for 10 min. When the solution became green colored, a solution of 1 g of the substituted 2,6-diphenylphenol in 8 ml of *o*-dichlorobenzene was added slowly over 20 min. A dark red color developed instantly. The reaction was continued for between 2 and 20 hour. When the reaction was finished, the mixture was diluted with 10 ml of chlorobenzene and several drops of anhydrous hydrazine were added to the reaction mixture to reduce the diphenoquinone by-product whereupon the deep red-orange color of the diphenoquinone disappeared. The inorganic solids were removed by filtration through a layer of Celite (1 cm) and the solution was added dropwise to 200 ml of methanol containing several drops of hydrazine. After stirring for several hours, the polymer was collected by filtration. The polymer was dissolved in about 20 ml of chloroform and precipitated in 200 ml of methanol. The polymer was collected again by filtration and dried in a vacuum oven at 80°C for 24 hours.

A-15. Polymerization of Para-Substituted Phenols

Another general method for preparing polyarylethers is by displacement of halo and other groups from the 4 position of 2,6-disubstituted phenols. This displacement polymerization requires only catalytic amounts of an oxidizing agent to serve as an initiator, unlike oxidative coupling polymerizations which require stoichiometric amounts. High molecular weight PPO™ is produced rapidly under the conditions shown in the accompanying equation (*42, 43*).

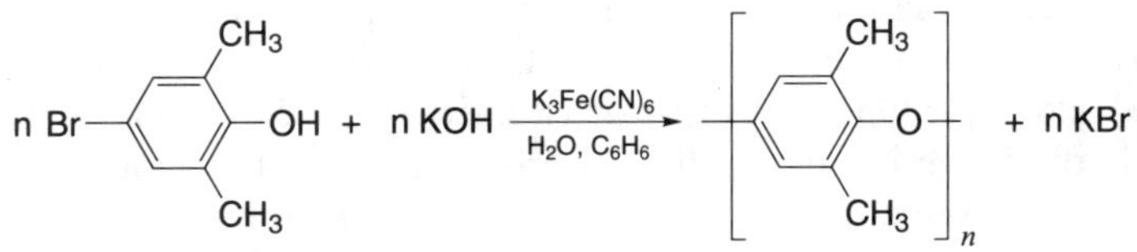

(III-20)

Suitable initiators include potassium ferricyanide, lead dioxide, iodine and sodium hypochlorite.

Substituents in the *ortho* positions are required for these reactions. Apparently, oxidation at an open *ortho* position consumes initiator and may interfere with the propagation step in the chain process. Moderately hindered phenols can be polymerized. 3,4-Dibromo-2,6-dimethylphenol is converted to high molecular weight poly(3-bromo-2,6-dimethyl-1,4-phenylene oxide) (*44*). 2,4,6-Trihalophenols are polymerized under similar conditions to give highly branched aromatic polymers (*45*). The sodium salt of 4-bromo-2,6-dichlorophenol in dimethyl sulfoxide with an initiator is converted to poly(2,6-dichloro-1,4-phenylene oxide) (*46*).

The mechanism of these polymerizations involves a coupling reaction between an aryloxy radical (generated by the initiator) and a phenoxide anion (generated by an excess of base) to form a radical-anion intermediate that loses bromide ion to form a new aryloxy radical (*47*). Evidence has recently been presented that shows that this reaction has the characteristics of a single electron transfer mechanism (*48*). For completeness sake, it should be mentioned that some very early work on the synthesis of poly(phenylene ethers) utilized benzene-1,4-diazo-oxides. Studies on their solid state photodecomposition (*49*) and their thermal decomposition (*50*) met with only limited success.

EXAMPLE I: POLYMERIZATION OF 4-BROMO-2,6-DIMETHYLPHENOL (*42*)

4-Bromo-2,6-dimethylphenol, m.p. 78.5–79°C, 2.01 g (0.01 mole), was dissolved in 100 ml of water containing 0.01 mole of potassium hydroxide, 100 ml. of benzene was added, and the flask was stirred for three days. The aqueous layer was separated and shown by titration with silver nitrate to contain 75% of the theoretical possible bromide ion. By adding petroleum ether to the benzene layer and cooling to 0°C, a total of 0.63 g (52%) of polymer was obtained, [n] 0.205, m.p. 232–244°C. Another polymerization of 2.01 g of 2,6-dimethyl-4-bromophenol initiated by 0.001 mole of potassium ferricyanide and 0.03 mole of base in 100 ml each of water and benzene under prepurified nitrogen gave 1.12 g of polymer in five minutes. The polymer had an intrinsic viscosity of 0.495.

2. Ullman Polymerization

A-16. Ullman Polymerization of Bis(phenates) with Haloaromatic Compounds

Polyarylethers derived from haloaromatics that are not activated by electron withdrawing groups can be prepared successfully by a modification of the basic Ullman reaction (*2*, *51*, *52*). Although Ullman showed that copper powder is a catalyst for making aromatic ethers by this reaction, later workers showed that cuprous ion is the active catalytic specie (*53*, *54*, *55*). These mechanism studies concluded that Cu+ coordinates with the pi system of the aromatic halide thus facilitating carbon-halogen cleavage. This scheme is essentially a nucleophilic aromatic substitution. The order of ease of halide replacement is I > Br > Cl >> F which is the reverse of the order observed for polyether formation from activated halides. The Ullman reaction utilizes high temperatures (e.g. 180–230°C) in inert media that are solvents for the polyether being formed. A cuprous chloride-pyridine solution is a preferred catalyst (*51*). An example is shown in the accompanying equation. Poly(1,4-

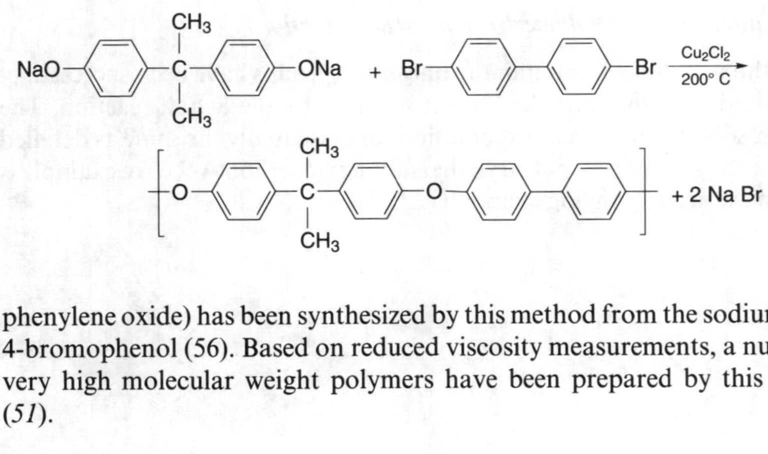

(III-21)

phenylene oxide) has been synthesized by this method from the sodium salt of 4-bromophenol (56). Based on reduced viscosity measurements, a number of very high molecular weight polymers have been prepared by this method (*51*).

EXAMPLE I: POLYMERIZATION OF 4,4′-DIBROMODIPHENYL WITH THE DISODIUM SALT OF BISPHENOL A (*51*)

Into a flask was placed 45.64 g (0.12 mole) of the hexahydrate of the disodium salt of 2,2-*bis*(4-hydroxyphenyl)-propane and 100 ml. of toluene. The flask was fitted with a moisture trap and the mixture refluxed for 2.5 hours to remove substantially all of the water present. One hundred forty grams of benzophenone as the reaction diluent was added and the remaining toluene distilled off to a pot temperature of 210°C. The mixture was cooled, the moisture trap removed, and 37.44 g (0.12 mole) of 4,4′-dibromodiphenyl was added. The flask was sparged with nitrogen and 9 ml of a cuprous chloride-pyridine solution (0.09 gram cuprous chloride, 0.9 mmol) was added. The reaction mass was heated for 6 hours at 185–215°C after which 5 mole-percent of bromobenzene was added and heating continued for an additional hour. The reaction mass was diluted with 200 ml of toluene and the polymer product precipitated by adding ethanol containing acetic acid. The fibrous precipitate was washed with alcohol, hot water containing small amounts of acetic and hydrochloric acids, and again with hot alcohol-acetone. The product was then vacuum dried overnight at 120–5°C. The yield of polymer was 44.6 g (45.4 calculated) and the RV was 1.11. The polymer was dissolved in chlorobenzene, treated with oxalic acid and filtered through a bed of filter aid. The filtered solution was precipitated in ethanol, washed and vacuum dried. A film was molded at 300°C and 3000 p.s.i The film had a Tg of 175°C, a tensile modulus of 240,000 p.s.i., a tensile strength of 11,000, an elongation of 150% and a pendulum impact strength of greater than 700 ft lb/in^3. The film was crystallized by orientation and was found to have a melting point of 240°C.

3. Cation-Radical Polymerization

A-17. Cation-Radical Polymerization (Scholl reaction)

Dinaphthoxy monomers containing only ether groups have been successfully polymerized to high molecular weight polymer by the Scholl reaction. The catalysts, solvents and reaction conditions are essentially the same as detailed above for the synthesis of polyarylethersulfones (Method A-2). An example is shown in the accompanying equation (*12*).

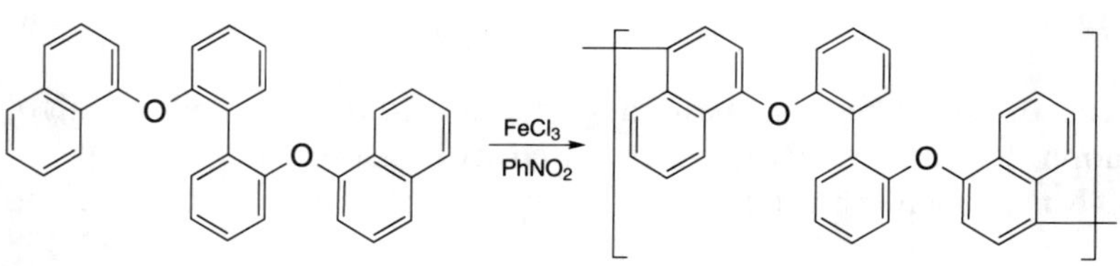

(III-22)

EXAMPLE I: POLYMERIZATION OF 2,2′-BIS(1-NAPHTHOXY)BIPHENYL (*12*)

2,2′-*Bis*(1-naphthoxy)biphenyl (0.224 g, 0.51 mmol) was dissolved in 0.5 ml of dry nitrobenzene in a 25-ml round-bottom flask equipped with a nitrogen inlet/outlet. To the magnetically stirred solution of monomer, a solution of $FeCl_3$ (0.203 g, 1.25 mmol) in 1.0 ml of dry nitrobenzene was added dropwise during 20 min. via a syringe with the needle penetrating through a rubber septum. The reaction mixture was stirred at room temperature for 5 hours. The content was precipitated into 150 ml of methanol acidified with 2% hydrochloric acid. The precipitate was filtered, washed with methanol and dried under vacuum. The yield was 0.204 g (92%). $M_n = 13{,}900$ g/mol, $M_w/M_n = 5.3$. The polymer was dissolved in 5 ml of $CHCl_3$ and precipitated into 150 ml. of acetone acidified with 2% HCl. The precipitate was filtered, washed with acetone, and dried in vacuum. The yield of the acetone-insoluble fraction was 0.130 g (64%). M_n (GPC) = 30,300 g/mol, $M_w/M_n=3.2$.

4. Metal-catalyzed Polymerization of Aromatic Monomers with Functional Groups

A-18. Palladium-catalyzed Cross-coupling Reactions

4,4′-Dibromodiphenyl ether has been polymerized with 2,5-di(n-hexyl)benzene-1,4-diboronic acid to yield moderately high molecular weight polymer. The reaction is catalyzed by Pd(0) complexes (*30*). This is an example of the synthesis of a poly(phenylene ether) by formation of an aryl–aryl bond. See Method A-10 in this Chapter.

A-19. Nickel-catalyzed Polymerization of Aryl Dichlorides

The nickel (0)-catalyzed polymerization of 2,2′-*bis*[4-(4-chlorophenoxy)phenyl propane has been reported (*33*). Relatively low molecular weight polymer was obtained.

References

1. Ullman, F., and Sponagel, P., *Ann.*, **350**, 83 (1906).
2. Ullman, F., and Sponagel, P., *Ber.*, **38**, 2211 (1905).
3. Ungnade, and Orwoll, *Org. Syntheses*, **26**, 50 (1946).
4. Fanta, *Chem. Revs.*, **38**, 139 (1946).
5. Jennings, B. E., Jones, M. E. B., and Rose, J. B., *J. Polym. Sci.*: C, No. 16, 715 (1967).
6. Scholl, R., and Seer, C., *Ber.*, **55**, 330 (1922).

7. Percec, V., and Nava, H., J. *Polym. Sci., Polym. Chem.*, 26, 783 (1988). Reprinted by permission of John Wiley & Sons, Inc.
8. Percec, V., and Nicholas, P. P., *Eur. Pat. Appl.* 282096 (1988). B. F. Goodrich Co.
9. Percec, V., Wang, J. H., Oishi, Y., Feiring, A. E., *J. Polym. Sci.: A: Polym. Chem.* **29**, 965 (1991).
10. Clowes, G. A., *J. Chem. Soc.*, C, 2517 (1968).
11. Percec, V., and Wang, J.H., *Polym. Preprints*, **32**(1), 391 (1991). Reprinted by permission of authors.
12. Percec, V., Okita, S., and Wang, J. H., *Macromolecules*, **25**, 64 (1992).
13. Percec, V., Wang, J. H., and Oishi, Y., *J. Polym. Sci.: A: Polym. Chem.*, **29**, 949 (1991).
14. Colon, I., and Kwiatkowski, G. T., *J. Polym. Sci.: A: Polym. Chem.*, **28**, 367 (1990). Reprinted by permission of John Wiley & Sons, Inc.
15. Ueda, M., and Ito, T., *Polymer* J., **23**, 297 (1991).
16. Colon, I., and Kelsey, D. R., *J. Org. Chem.*, **51**, 2627 (1986).
17. Kwiatkowski, G. T., and Colon, I., *Polym. Preprints*, **32**(1), 342 (1991).
18. Ueda, M., Miyaji, Y., and Ito, T., *Macromolecules*, **24**, 2694 (1991).
19. Bochmann, M., Kelly, K., and Lu, J., *J. Polym. Sci.: A: Polym. Chem.* **30**, 2511 (1992).
20. Colquhoun, H. M., Dudman, C. C., Thomas, M., O'Mahoney, C. A., and Williams, D. J., *J. Chem. Soc., Chem. Commun.*, 336 (1990).
21. Cella, J. A., Fukuyama, J., and Guggenheim, T. L., *Polym. Preprints*, 30(2), 142 (1989). Reprinted by permission of authors.
22. Cella, J. A., Talley, J. J., and Fukuyama, J., *Polym. Preprints*, 30(2), 581 (1989). Reprinted by permission of authors.
23. Mullins, M.J., Woo, E.P., Chen, C. C., Murray, D. J., Bishop, M. T., Balon, K. E., *Polym. Preprints*, **32**(2), 174 (1991); Mullins, M.J., Woo, E.P., Bishop, M. T., Galvan, R., Gorman, D.B., Chamberlin, T. A., *ibid.* **33**(1), 414 (1992).
24. Gibson, H.W., Ganguly, S., Yamaguchi, N., Xie, D., Chen, M., Bheda, M., and Miller, P., *Polym. Preprints*, **34**(1), 576 (1993).
25. Mayske, P. J., and Tresper, E., *Eur. Pat. Appl.* 218064 (1987). Bayer, A. G.
26. Mullins, M. J., Crain, S. P., Woo, E. P., and Murray, D. J., U.S. Pat. 4880884 (1989). Dow Chemical Co.
27. Wang, Z. Y., de Carvalho, H.N., and Hay, A. S., *J. Chem. Soc., Chem Comm.*, 1991, 1221.
28. Ueda, M., and Ichikawa, F., *Macromolecules* **23**, 926 (1990). Reprinted by permission of authors.
29. Moore, J., *Makromol. Chem., Rapid Commun.* **13**, 91 (1992). Deeter, G. A., and Moore, J. S., *Macromolecules*, **26**, 2535 (1993). Reprinted by permission of Hüthig & Wepf Verlag.
30. Rehahn, M., Schluter, A., and Wegner, G., *Makromol. Chem., Rapid Commun.* **11**, 535 (1990). Reprinted by permission of Hüthig & Wepf Verlag.
31. Segal, J. A., *J. Chem. Soc., Chem. Commun.*, (19), 1338 (1985).
32. Dembek, A. A., Fagan, P. J., and Marsi, M., *Macromolecules*, **26**, 2992 (1993).
33. Percec, V., and Okita, S., *J. Polym. Sci.: Part A: Polym. Chem*, **31**, 923 (1993).
34. Hay, A. S., "Discovery and Development of Noryl™ Resins," in "High Performance Polymers: Their Origin and Development," p. 209, Elsevier 1986.
35. Anon., *Chem. & Eng. News*, p. 57, 12/7/64.
36. Elias, H.-G., Neue Polymere Werkstoffe 1969-1974, p. 69 (1975). Carl Hanser, Munchen.
37. White, D. W., "Poly(phenylene oxide)s" in *Comp. Polym. Sci.*, **5**, 473 (1989). Pergamon Press.
38. Aycock, D., Abolins, V., White, D. W., "Polyphenylene Ether" in *Encycl. Polym. Sci. & Eng.*, **13**, 18 (1988). Reprinted by permission of Hüthig & Wepf Verlag.
39. Hay, A. S., et al., *Macromolecular Syntheses*, Collective Volume I, p. 81, 1977.
40. Sorenson, W. R., and Campbell, T. W., Preparative Methods of Polymer Chemistry, 2nd Edition. Interscience Publishers, Inc., New York, p. 429.
41. Yang, H., and Hay, A. S., *J. Polym. Sci.: Polym. Chem.*, **31**, 1261 (1993). Reprinted by permission of John Wiley & Sons, Inc.
42. Price, C. C., and Chu, N. S., *J. Polym. Sci.*, **61**, 135, 1962. Reprinted by permission of John Wiley & Sons, Inc.

43. Staffin, G. D., and Price, C. C., *J. Am. Chem.* Soc., **82**, 3632 (1960).
44. White, D. M., and Klopfer, H. J., *ACS Symp. Ser.*, **6**,169 (1975).
45. Kim, Y. H., *Polym. Preprints*, **34**(1), 56 (1993).
46. Stamatoff, G. S., U S Pat. 3236807 (1966). *C.A.*, **64**, 14386 (1966).
47. Price, C. C., *J. Paint Tech.*, **38**, 705 (1966).
48. Percec, V., and Shaffer, T. D., *J. Polym. Sci., Polym. Lett. Ed.*, **24**, 439 (1986).
49. Sus, O., Moller, K., and Heiss, H., *Annals,* **598**, 123 (1956).
50. Dewar, M. J. S., and James, A. N., *J. Chem Soc.*, 917 (1958).
51. Farnham, A. G., and Johnson, R. N., U.S. Pat. 3332909 (1967). Union Carbide Corp.
52. Robeson, L. M., Farnham, A. G., and McGrath, J. E., *Appl. Polym. Symp.* No. 26, **373** (1975).
53. Bunnett, J. F., and Zahler, R. E., *Chem. Rev.*, **49**, 394 (1951).
54. Bacon, R. G. R., and Hill, H. A. O., *Proc. Chem. Soc.*, 113 (1962).
55. Weingarten H., *J. Org. Chem.*, **29**, 977 (1964); *ibid.*, **29**, 3624 (1964).
56. van Dort, H. M., Hoefs, C. A. M., Magre, E. P., Schopf, A. J., and Yntema, K., *Eur. Polym. J.*, **4**, 275 (1968).

Tables

Guide to the Tables

Tables 1–17 provide a wealth of information about aromatic polyethers. Although many ways can be conceived to present these data, the protocols described below were felt to be the most straightforward and useful. A little time spent studying this section will provide the reader with easy access to the data. The tables (18–22) in Chapters IV–VII were derived from Tables 1–17, so that they follow the same protocols.

The polyarylethers are arranged in the tables according to the following hierarchy.

Polyarylethersulfones	Tables 1–5
Polyaryletherketones	Tables 6–10
Polyaryletherimides	Tables 11–13
Polyarylethers	Tables 14–17

The tables within each polymer family are based on the number of rings in the backbone of the repeat unit. For example, Tables 1–5 contain polyarylethersulfones possessing < = 3, 4, 5, 6, and > = 7 rings in the backbone of the repeat unit. A similar arrangement was used for the other polymer classes. All polymers with a sulfone group in the backbone are listed in Tables 1–5 irrespective of what other groups are present. If a polymer has a sulfone and a ketone group in the backbone, it will be listed both in a polyarylethersulfone and a polyaryletherketone table. Polymers with a ketone or an imide group will be listed in a polyaryletherketone or a polyetherimide table, respectively. Finally, the polyarylether tables contain all the polymers that do not contain a sulfone, ketone or an imide group. Although many of the entries in the polyarylether tables possess only ether and aromatic groups in the backbone, others possess additional structures. Many of the heterocyclic (non-imide) ring-containing polymers are listed in the polyarylether tables.

The following column by column description provides a guide and explanation of the data that are in the tables.

Polymer Structure - Column 1

The structure of the polymer repeat unit is presented in this column. The individual entries are arranged in order of increasing structural complexity. The three main structural features that were used to determine the order of the individual entries are (*1*) the number of rings in the backbone of the repeat unit, (*2*) the number of backbone atoms (of any kind) that connect the rings in the backbone, and (*3*) the number of carbon atoms that are pendant to the rings or atoms in the polymer backbone. The entries were then sorted and arranged by parameters (*1*), (*2*) and (*3*) in that order.

A simple example follows. The second column in the table below lists the number of rings that are in one repeat unit of the polymer backbone. The third column lists the number of non-ring atoms that are in the backbone of the repeat unit, but does not count the non-carbon atoms that are attached to these backbone atoms. Note that the sulfone group in the backbone is counted as only one backbone atom, namely the sulfur atom. The "pendant" oxygens on the sulfur are not counted. The fourth column lists the number of carbon atoms that are pendant to the backbone elements. Note that the second table entry has two pendant carbon atoms on the backbone carbon atom that forms the isopropylidene group.

Polymer Structure	No. of Backbone Rings	No. of Backbone Atoms	No. of Pendant Carbon Atoms
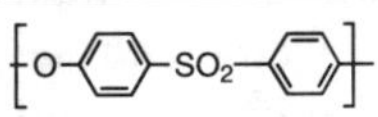	2	2	0
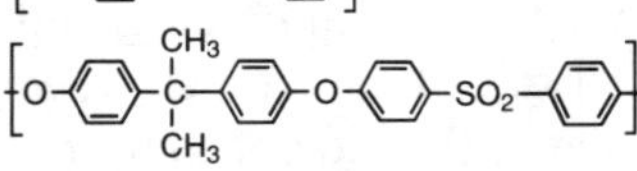	4	4	2

Carbon atoms substituted on rings would be counted in a similar manner. The entries were then sorted sequentially by the number of backbone rings, the number of backbone atoms and then by the number of pendant carbon atoms. These sorts were done in increasing order for each of these parameters. An understanding of these sorting considerations allows one to rapidly determine whether a specific polymer has been reported. After assigning the above three parameters to a new or old repeat unit of interest, one can then directly enter the area of a table where it should be listed.

The parameters selected for arranging the individual table entries also give a qualitative picture of the thermal and mechanical properties that one would expect from a given structure. For example, a polymer with only backbone rings would be expected to possess high transition temperatures and/or be intractable. As the number of backbone or pivot atoms increase in a polymer backbone, a trend toward lower transition temperatures and improved toughness would ensue. Similarly, increasing the number of carbon atoms and/or rings that are pendant to the main backbone elements would lead to increased transition temperatures and decreased toughness.

For polyarylethersulfones, the monomer from which the diaryl sulfone moiety was derived is drawn on the right. The left hand side of the repeat unit is derived from

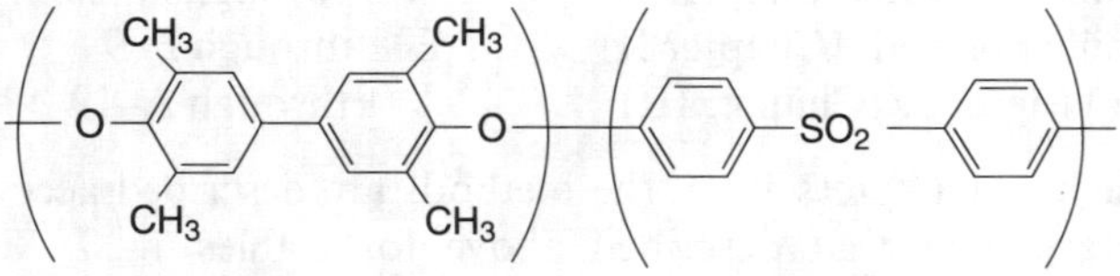

Bisphenol-derived Diarylsulfone-derived

the other reactant, which in many cases is a bisphenol. Certainly, in polyarylethersulfones that contain four aromatic rings in the backbone, the structural element on the left side of the repeat unit was derived from a bisphenol. Thus, these tables can serve as a key to references where "polymerization-grade" bisphenols have been reported and used successfully in a condensation polymerization. The wide diversity of bisphenol structures that have been employed is readily apparent from scanning the entries that are listed in the table for polyarylethersulfones that contain four rings in the backbone (Table 2). Similarly, the more limited number of diaryl sulfone monomers that have been used can also be seen by inspecting the structures in this table. This convention was also followed for the other classes of polyarylethers, where feasible.

Tg - Glass Transition Temperature - Column 2

Tm - Crystalline Melting Temperature - Column 3

The glass transition temperature and crystalline melting temperature of the polymers are reported in columns 2 and 3 , respectively. Of all the parameters that allow one to describe and compare polymers, these are the most useful and universal. The presence or absence of crystallinity in a polymer immediately places it in the amorphous or crystalline category. Members of the amorphous polymer group can then be ranked and compared based on their glass transition temperatures. Similarly, the utility of crystalline polymers will be initially dependent on their Tg, Tm and degree of crystallinity. Cross comparisons with non-polyarylethers are also facilitated by these two parameters.

Synthesis - Column 4

This column describes the method by which the polymer was synthesized. The designation before the first semicolon in the column denotes the method that was employed. These designations refer to the methods that are described in Chapters I–III. A summary of the methods follows.

Nucleophilic Methods (Chapter I)	N–1 through N–20
Electrophilic Methods (Chapter II)	E–1 through E–9
Additional Methods (Chapter III)	A–1 through A–19

Within each of Chapters I–III, the methods are described according to the same hierarchy as that described above for Tables 1–17. The order is polyarylethersulfones, polyaryletherketones, polyaryletherimides, followed by polyarylethers. Although a specific method may have been applied to synthesize polymers from all of these classes, the detailed procedures can differ. For this reason, different method numbers have been assigned to the same overall method when used with different polymer classes. For example, the synthesis of a crystalline polyaryletherketone by the carbonate process (N–9) requires markedly different conditions than the synthesis of an amorphous polyarylethersulfone (N–2). The table on the following page summarizes the methods that are described in Chapters I–III and subsequently used in Tables 1–17.

Method Numbers Classified by Polymer Type

Method Name	Poly (ether sulfones)	Poly (ether ketones)	Poly (ether imides)	Poly (aryl ethers)
Arenesulfonic acids with Aromatic Compounds	E-2			
Aromatic Ethers with Alkyl Thiochloroformates and Dialkyl Dithiocarbonates		E-8		
Carboxylic Acid Chlorides with Aromatic Compounds Using Aluminum Chloride		E-3		
Carboxylic Acid Chlorides with Aromatic Compounds Using Boron Trifluoride/Hydrogen Fluoride		E-4		
Carboxylic Acid Chlorides with Aromatic Compounds Using Other Very Strong Acids		E-5		
Carboxylic Acids with Aromatic **Ethers**		E-6		
Cation Radical Polymerization	A-2	A-8		A-17
Diphenols and Activated Aromatic Dihalides: Carbonate Process	N-2	N-9	N-2; N-9	N-17
Diphenols and Activated Aromatic Dihalides: Caustic Process	N-1	N-8	N-1	N-1
Diphenols and Activated Aromatic Dinitro Compounds	N-4	N-12		N-20
Diphenols and Nitro-substituted Phthalimides			N-14	
Diphenyl Ether with Phosgene		E-7		
Disulfinates and Activated Aromatic Dihalides	N-7			
Nickel-catalyzed Polymerization of Aryl Dihalides	A-3	A-9		A-19
Organometallic-aromatic Complexes with Bisphenates		A-11		
Oxidative Coupling of 2,6-Disubstituted Phenols				A-14
Palladium-catalyzed Cross-coupling Reactions	A-4	A-10		A-18
Phase Transfer Catalyzed Systems	N-5			N-18
Polyalkylation of Aromatic Ethers				E-9
Polymerization by Transimidization			N-16	
Polymerization of Dianhydrides and Diamines			N-15	
Polymerization of Para-substituted Phenols				A-15
Preparation of Poly(ether sulfones) from Polycarbonates	A-6			
Ring Opening Polymerization of Macrocyclic Oligomers	A-5	A-12	A-13	
Self-Polymerization of Halophenylsulfonyl Phenoxides	N-3	N-10		
Silylated Diphenols and Activated Aromatic Dihalides	N-6	N-11		N-19
Sulfonyl Chlorides with Aromatic Compounds	E-1			
Synthesis of Poly(ether ketones) via Amorphous Precursors		N-13		
Ullman Polymerization	A-1	A-7		A-16

The data after the first semicolon in column 4 lists the inorganic reagents and solvents that were used for the synthesis in the cited reference. Melt or bulk is used to indicate that no solvent was used. Abbreviations are explained in the accompanying list.

Ac_2O	Acetic Anhydride
CHP	*N*-Cyclohexylpyrollidone
CsPhenate	Cesium Phenate
Cu-TMEDA	Copper Tetramethylethylene Diamine
DCB	*o*-Dichlorobenzene
DCE	1,2-Dichloroethane
DMAC	*N,N'*-Dimethylacetamide
DMF	*N,N'*-Dimethylformamide
DMPU	*N,N'*-Dimethylpropylene urea
DMSO	Dimethyl sulfoxide
DPS	Diphenyl sulfone
Et_3N	Triethyl Amine
HMPA	Hexamethyl Phosphoramide
KPhenate	Potassium Phenate
NMP	*N*-Methylpyrollidone
PAEK	Polyaryletherketone
PEEK	Poly(etheretherketone)
PEEKK	Poly(etheretherketoneketone)
PEKK	Poly(etherketoneketone)
PES	Polyethersulfone
PPA	Polyphosphoric Acid
PPO	Poly(phenylene oxide)
PSF	Polysulfone
SiBP	Silylated Bisphenol
TBAC	Tetrabutylammonium Chloride
TFMSA	Trifluoromethanesulfonic Acid
THF	Tetrahydrofuran

After the second semicolon, the atom(s) that was (were) displaced in the synthesis, particularly when the nucleophilic displacement route was used is listed. If no atoms are shown, this type of data is not applicable to the synthetic route that was used. This information is quite valuable since it lets one know at a glance which activated monomer was used in the synthesis. For example, an entry of F would indicate that fluoride was displaced in a nucleophilic displacement polymerization. A simple examination of the repeat unit structure would tell which specific fluoro monomer was used. and allows one to quickly identify a reference to a specific activated, fluoro monomer.

Finally, when a reference has focussed on a particular non-synthetic property, this information is shown in this column. Some examples are steam sterilization, kinetics, thermal stability, and permeability.

Reference - Column 5; Year - Column 6

A reference for the polymer is listed in column 5. It is not necessarily the first or only synthesis of a given polymer. Standard journal abbreviations were used. The publication year for the reference is in column 6. This presentation mode was chosen to show at a glance when the work was done, and whether it appeared in a patent or a journal article. These considerations are especially important for industrial personnel. This arrangement obviates the need to refer to a separate reference list to find this information. The authors of the publications cited in the tables appear in an index at the end of the book.

Organization - Column 7

The organization from which the publication emanated is shown in this column.

Reference Lists

The complete references that are cited in the tables are indexed at the end of the book p. 337. They are arranged alphabetically by the name of the publication in conformity with the listing in the tables.

TABLE 1
Polyarylethersulfones containing three or fewer rings in the backbone of the repeat unit

Polymer structure	Tg (°C)	Tm (°C)	Synthesis	Reference	Year	Organization
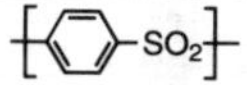	350	520		*Chimia*, **28**(9),561	1974	ICI
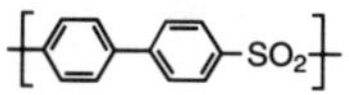		375, 410	E-1; Chlorobiphenyl, Bulk, $FeCl_3$	*Chem. Ind.* (15), 461; *Polym.*, **6**, 589	1968 1968	ICI
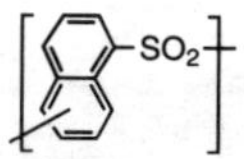			E-1; Bulk, $FeCl_3$	*Polym.*, **6**, 589	1965	ICI
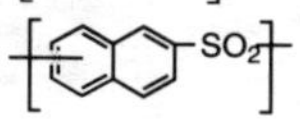			E-1; Bulk, $FeCl_3$	*Polym.*, **6**, 589	1965	ICI
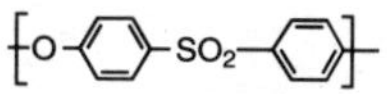 Victrex™ PES	230		N-1; DMSO; F, Cl, Br; Kinetics	*Br. Polym. J.*, **4**, 391; *Polym.*, **13**, 465; *ACS Symp. Ser.*, **4**, 63;	1972 1972 1974	ICI
			Steam Sterilization	*Polym. News*, **11**, 359	1986	Union Carbide
		250–265	E-1; $AlCl_3$, $C_6H_5NO_2$ E-2; PPA	*J.Polym. Sci.*, A-1, **4**, 722;	1966	Monsanto
			E-1; Bulk, $FeCl_3$	*Polym.*, **6**, 589	1965	ICI
	250		E-1; Dimethyl Sulfone, $FeCl_3$	*Chem. Ind.* (15), 461	1968	ICI
			E-1; BF_3, HF	US Pat. 3441538	1969	DuPont
			E-2; NMP	Jap. Pat. 2011634	1990	New Japan Chem.
			E-2; CH_3SO_3H, P_2O_5	Jap. Pat. 60-228541	1985	Idemitsu Inds.
			A-6; K Phenate, Bulk	Eur. Pat. 218064	1987	Bayer, A. G.
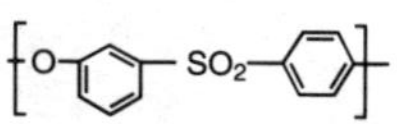	165–175		N-1; Melt, Sulfolane; F, Cl	*Polym.*, **18**, 359; **18**, 370	1977	ICI

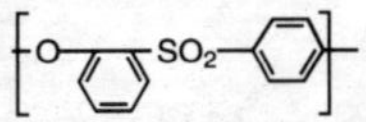	221	368	N-1; Melt, Sulfolane; F, Cl	*Polym.*, **15**, 456; **18**, 359; **18**, 370	1974 1977	ICI
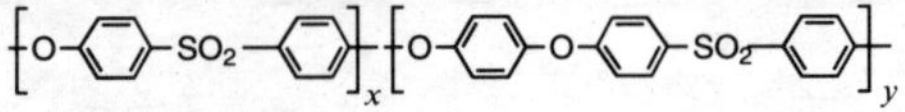 Radel ™ A-100-x/y=4.7	220		N-2; NMP, Sulfolane; Cl	*High Perf. Polym.*, Elsevier, p.157;	1986	Union Carbide
				J. Polym. Sci., B; *Polym. Phys*, **31**, 891;	1993	
			Gas Permeability	*J. Memb. Sci.*, **78**, 123	1993	
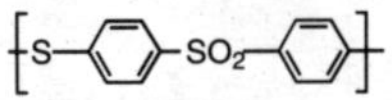	240		N-1; Sulfolane; Cl	*Polym.*, **18**, 359	1977	ICI
			E-1; Bulk, $FeCl_3$	*Polym.*, **6**, 589	1965	
				Chem. Ind. (15), 461	1968	
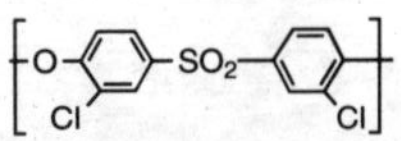			N-1; Melt, Sulfolane; F, Cl	*Polym.*, **18**, 359	1977	ICI
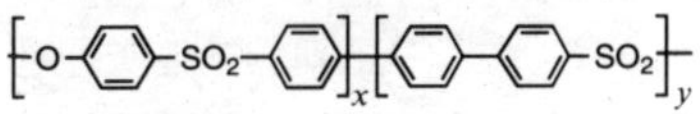 Victrex™ 720P- x>>y	250		E-1;	*Chimia*, **28**(9), 561; *High Perf. Polym.*, Elsevier, p. 171	1974 1986	ICI
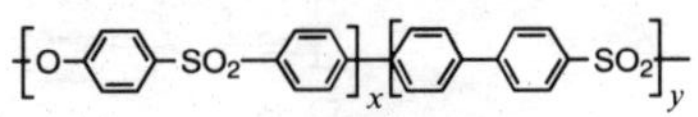 Astrel™ 360- y > > x	285		E-1;	*Chimia*, **28**(9), 561	1974	ICI
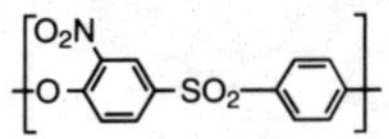	234		Post reaction on PES. 50% nitration	J. *Memb. Sci.*, **74**, 193	1992	U. of Twente
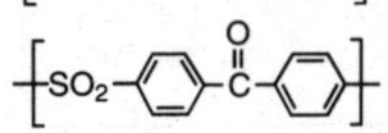		375	Oxidation of sulfide polymer	*Brit. Polym. J.*, **4**, 391	1972	ICI
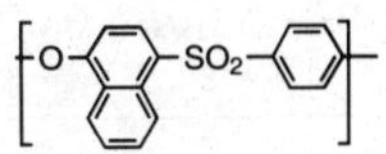	238	367	N-1; Melt, Sulfolane; F, Cl	*Polym.*, **15**, 456; **18**, 359; **18**, 370	1974 1977	ICI

TABLE 1—*continued*

Polymer structure	Tg (°C)	Tm (°C)	Synthesis	Reference	Year	Organization
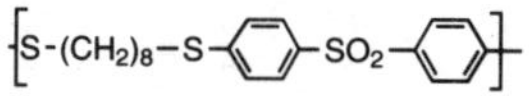		120-125		*J. Org. Chem.*, **29**, 207	1964	
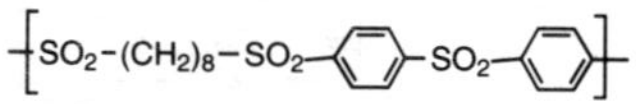		250-255		*J. Org. Chem.* **29**, 207	1964	
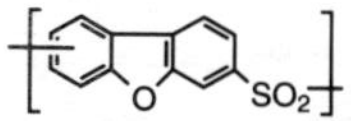		410-440	E-1;	*Adv. Macromol. Chem.*, **1**, 335	1968	3M
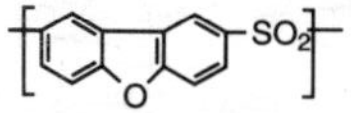	360		E-1; Dimethyl sulfone	*Polym.*, **15**, 456	1974	ICI
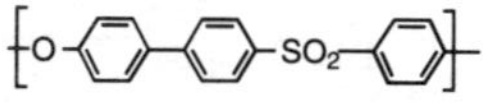	265		N-1; Melt, Sulfolane; F, Cl	*Polym.*, **15**, 456; **18**, 370	1974 1977	ICI
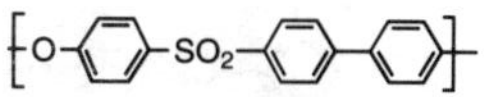	216	310	N-1; NaOH, DMSO; Cl	*Polym.*, **22**, 1096	1981	ICI
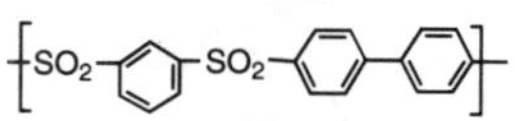			E-1; Bulk, $FeCl_3$	*J. Polym. Sci.*, C, No. **16**, 715	1967	ICI
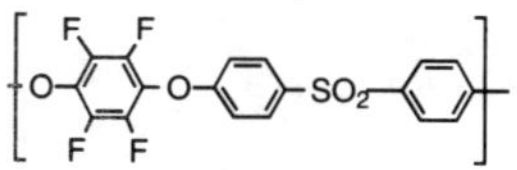			N-5; DMAC; F	*Polym. Prep.*, **21**(2),164	1980	
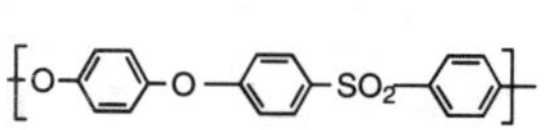	210	310	N-1; NaOH, DMSO, Sulfolane; Cl	*J. Polym. Sci.*, A-1, **5**, 2375; *Ency. Polym. Sci. & Eng.*, **13**, 196	1967 1988	Union Carbide
	217	310	N-2; K_2CO_3,NMP, DMAC; Cl; Kinetics	*J. Polym. Sci., Polym. Chem.*, **23**, 287; *Polym.*, **25**, 1827	1986 1984	Va. P. I. & State U.

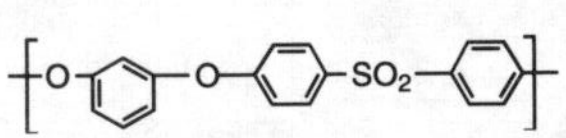			N-1; NaOH, DMSO	Eur. Pat. 0161576	1986	Sumitomo Chemical Co.
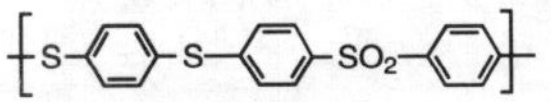			N-6; Sulfolane; Cl	*Prog. Polym. Sci.*, **14**,173	1989	Tokyo Inst. Tech.
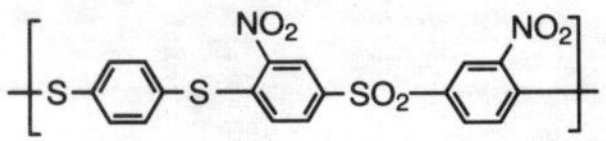			N-6; HMPA; Cl	*Prog. Polym. Sci.*, **14**,173	1989	Tokyo Inst. Tech.
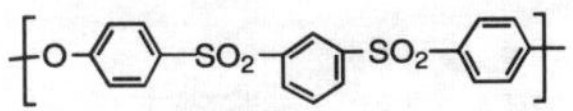			E-1; Bulk, $FeCl_3$	*J. Polym. Sci., C,No.* **16**, 715	1967	ICI
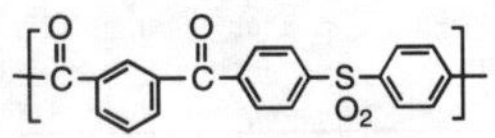	195	257	N-13; NaH, DMF; F	*Polym. Prep.*, **34**(1), 570.	1993	Va. P. I. & State U.
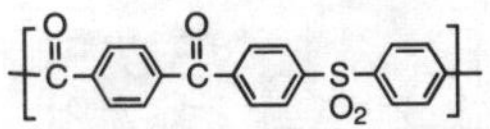	228	414	N-13; NaH, DMF; F	*Polym. Prep.*, **34**(1), 570	1993	Va. P. I. & State U.
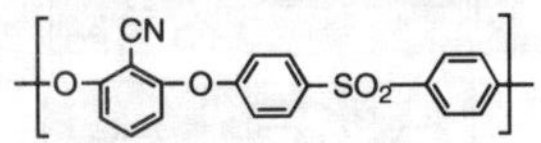	222		N-2; K_2CO_3, NMP; Cl	*Polym. Prep.*, **23**(1), 284	1982	Va. P. I. & State U.
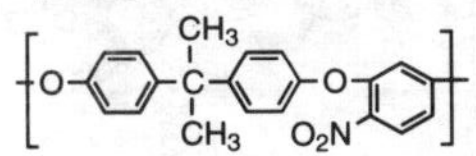	150		N-1; KOH, DMSO; Cl	GB Pat. 1078234	1967	Union Carbide
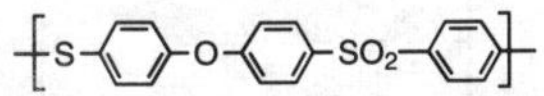	188		N-2; K_2CO_3, Sulfolane; Cl	Can. Pat. 946091	1974	Uniroyal
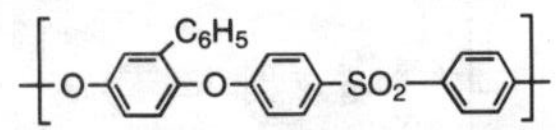			N-6; K_2CO_3, NMP; Cl	*Makromol. Chem.*, **191**, 2027	1990	Hamburg U.
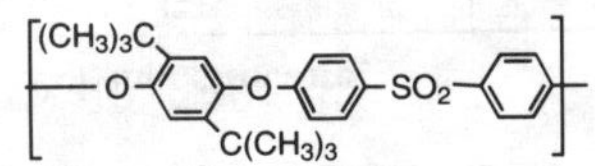	225-230	387	N-2; K_2CO_3, Diphenyl sulfone; Cl	US Pat. 4849503	1989	Amoco

TABLE 1—*continued*

Polymer structure	Tg (°C)	Tm (°C)	Synthesis	Reference	Year	Organization
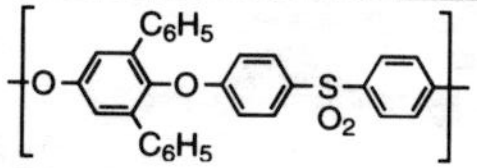	227		N-2; K_2CO_3, NMP; F	*J.M.S.-Pure Appl. Chem.*, **A29**(12), 1141	1992	McGill U.
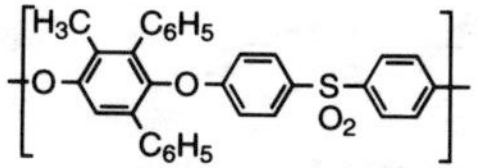	235		N-2; K_2CO_3, NMP; F	*J.M.S.-Pure Appl. Chem.*, **A29**(12), 1141	1992	McGill U.
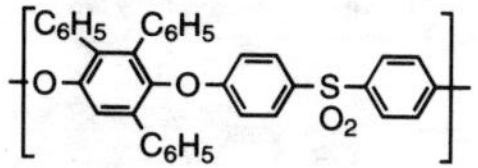	240		N-2; K_2CO_3, NMP; F	*J.M.S.-Pure Appl. Chem.*, **A29**(12), 1141	1992	McGill U.
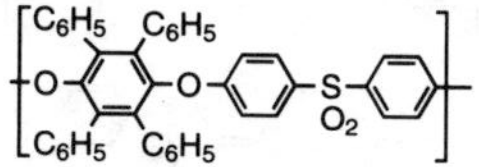	265	411	N-2; K_2CO_3, NMP; F	*J.M.S.-Pure Appl. Chem.*, **A29**(12), 1141	1992	McGill U.

TABLE 2
Polyarylethersulfones containing four rings in the backbone of the repeat unit

Polymer structure	Tg (°C)	Tm (°C)	Synthesis	Reference	Year	Organization
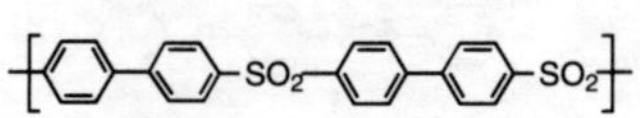		>400	E-1; $FeCl_3$, Nitrobenzene, Dimethyl sulfone	*J. Polym. Sci.*, A-1, **8**, 2035	1970	3M
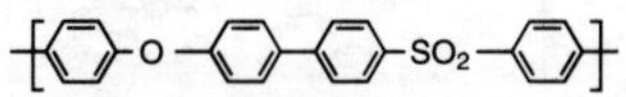	168	320	A-4;Sn/Pd, DMAC; Br	*J. Polym. Sci.*, A, **30**, 2511	1992	U. of East Anglia
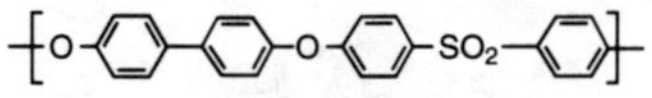 Radel™ R	220		N-1; NaOH, DMSO, Sulfolane; Cl	*Ency. Polym. Sci & Eng.*, **13**,196	1988	Union Carbide
	220		N-2; K_2CO_3, Sulfolane, NMP; Cl	*High Perf. Polym.*, Elsevier, p. 157;	1986	Amoco/Union Carbide
				Polym. Prep., **32**(2),166;	1991	
	232			*J. Polym. Sci., Polym. Chem.*, **23**, 287	1986	Va. P. I. & State U.
			N-6; CsPhenate, Bulk, Phenol	US Pat. 4794155	1988	Dow
	215		A-3; DMAC; Cl	*J. Polym. Sci., Polym. Chem.*, **28**, 367;	1990	Amoco/Union Carbide;
		190		*Polym. J.*, **23**, 297	1991	U. Yamagata
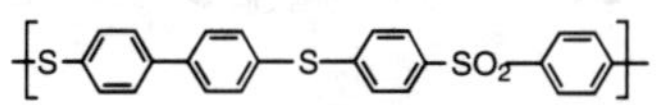		286	N-2; K_2CO_3, DMF; Cl	*Makromol. Chem.*, **140**, 83	1970	Celanese
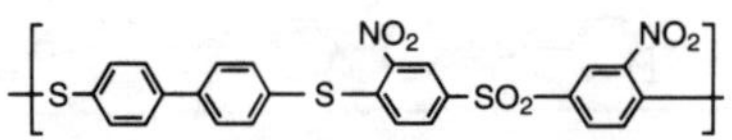			N-6; Sulfolane; Cl	*Prog. Polym. Sci.*, **14**,173	1989	Tokyo Inst. Tech.
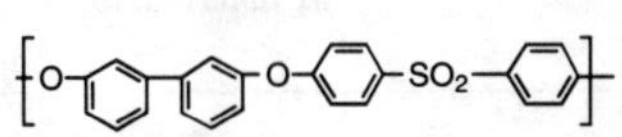	142		A-3; DMAC; Cl	*Polym. J.*, **23**, 297	1991	U. Yamagata

TABLE 2—*continued*

Polymer structure	Tg (°C)	Tm (°C)	Synthesis	Reference	Year	Organization
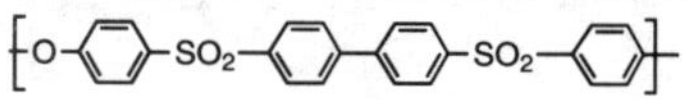	276–289	395	N-1; Melt, Sulfolane; F, Cl	*Polym.*, **18**(4), 359; **18**(4), 370	1977	ICI
	269		A-3; DMAC; Cl	*Polym. J.*, **23**, 297	1991	U. Yamagata
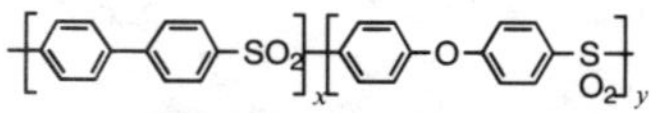 Astrel™ 360- x:y = 60-70: 30-40	285		E-1	Anon.		3M
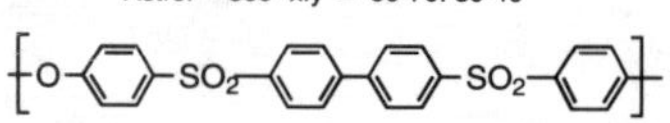	276–285		N-2; K_2CO_3,Sulfolane; Cl	*J. Polym. Sci.*, *Polym. Chem.*, **30**, 221	1992	Amoco
			E-1; $FeCl_3$, Bulk;	*J. Polym. Sci.*, C, No.**16**, 715	1967	ICI
	275	Sl. Cryst.	E-1; $FeCl_3$ Nitrobenzene, Dimethyl sulfone	*J.Polym.Sci.*, A-1, **8**, 2035	1970	3M
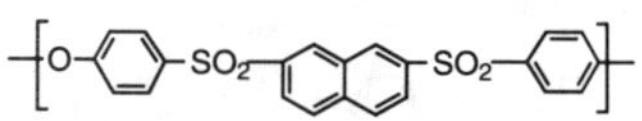	250		E-1; $FeCl_3$ Nitrobenzene, Dimethyl sulfone	*J.Polym.Sci.*, A-1, **8**, 2035	1970	3M
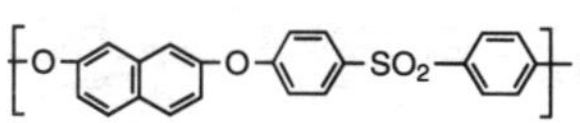	215		N-2; K_2CO_3, Sulfolane; Cl	Jap. Pat. 61-275326	1986	Idemitsu Kosan
	225			*J. Polym. Sci.*, *B*, *Polym Phys.* **31**, 1061	1993	U.Texas
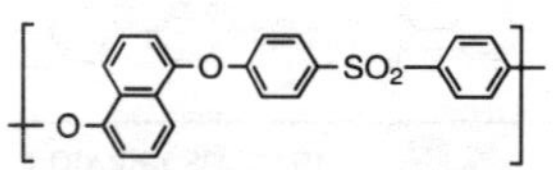	238		N-2; K_2CO_3, Sulfolane; Cl	Jap. Pat. 61-275326	1986	Idemitsu Kosan
	240			*J. Polym. Sci.*, *B*, *Polym Phys.* **31**, 1061	1993	U.Texas
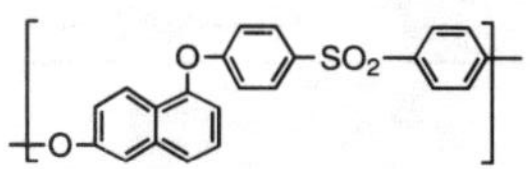	221		N-2; K_2CO_3, Sulfolane; Cl	Jap. Pat. 61-275326	1986	Idemitsu Kosan

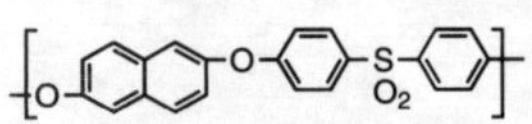	227	N-2; Permeability	*J. Polym. Sci.*: B, *Polym. Phys.*, **31**, 1061	1993	U. Texas
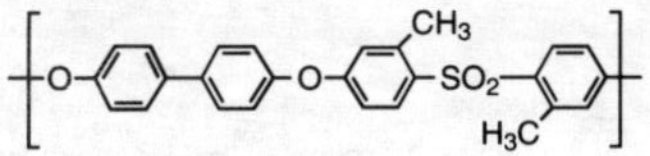		Post Reaction on Radel	US Pat. 4797457; US Pat. 4833219	1989	Nat. Res. Council, Canada
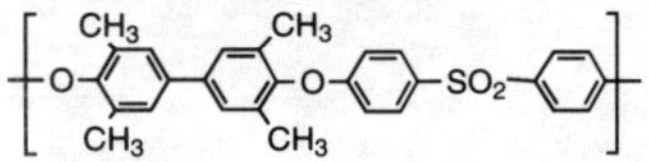	262	N-1; NaOH, DMSO; Cl	*Appl. Polym. Symp.*, No. **26**, 373	1975	Union Carbide
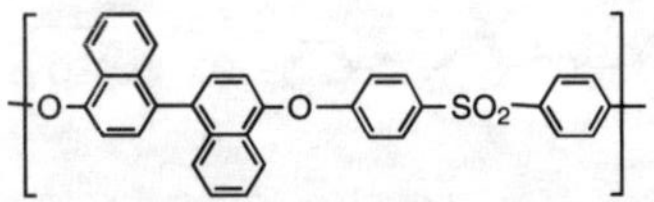	281	A-2; $FeCl_3$, Nitrobenzene	*J. Polym. Sci.*, A, **29**, 965; Eur. Pat. 282096	1991 1988	Case West. Res. U
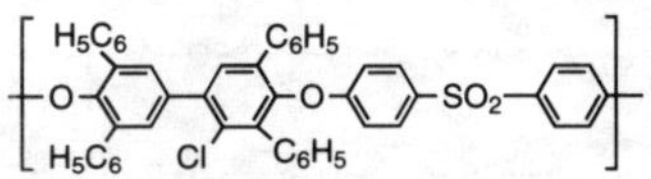	280	N-2; K_2CO_3, NMP; F	*Polym. Prep.*, **32**(1), 389	1991	McGill U.
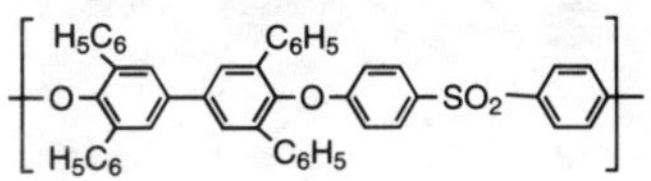	270	N-2; K_2CO_3, NMP; F	*Polym. Prep.*, **32**(1), 389	1991	McGill U.
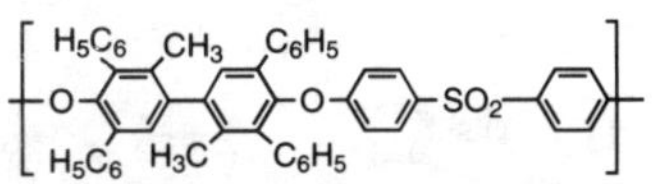	285	N-2; K_2CO_3, NMP; F	*Polym. Prep.*, **32**(1), 389	1991	McGill U.
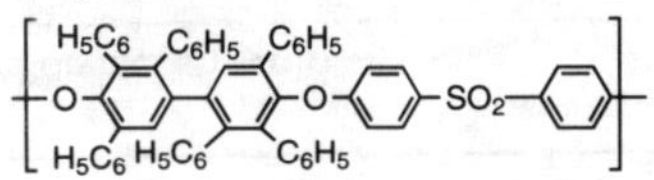	265	N-2; K_2CO_3, NMP; F	*Polym. Prep.*, **32**(1), 389	1991	McGill U.

TABLE 2—*continued*

Polymer structure	Tg (°C)	Tm (°C)	Synthesis	Reference	Year	Organization
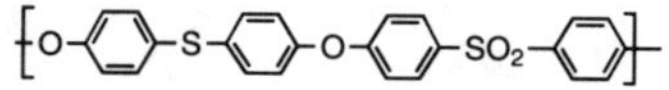	175		N-1; NaOH, DMSO, Sulfolane; Cl, F	*J.Polym. Sci.*, A-1, **5**, 2375;	1967	Union Carbide
				Ency.Polym.Sci. & Eng., **13**,196;	1988	
			N-2; K_2CO_3, NMP; Cl	*Polym. Prep.*, **23**(1), 284	1982	Va. P. I. & State U.
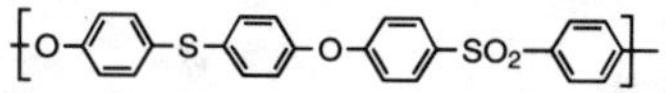	180		N-2; K_2CO_3, DMAC; Cl; Kinetics	*Polym.*, **25**, 1827	1984	Va. P. I. & State. U.
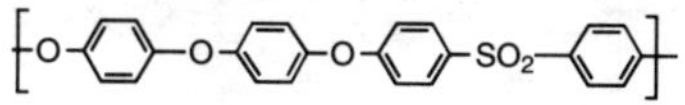	180		N-1; NaOH, DMSO, Sulfolane; Cl, F	*J.Polym.Sci.*, A-1, **5**, 2375;	1967	Union Carbide
				Ency. Polym. Sci. & Eng.,**13**,196	1988	
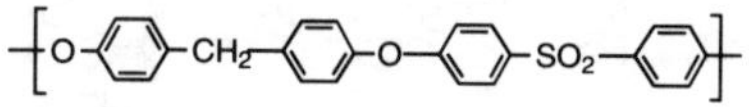	180		N-1; NaOH, DMSO, Sulfolane; Cl, F	*J. Polym. Sci.*, A-1, **5**, 2375	1967	Union Carbide
	180		N-1; NaOH, DMSO; Cl	*Ency. Polym. Sci. & Eng.*, **13**,196	1988	Union Carbide
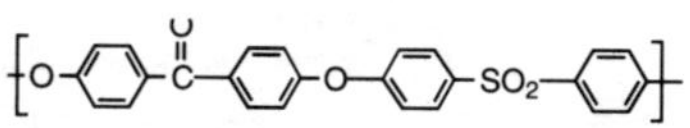	205		N-1; NaOH, DMSO, Sulfolane; F	*J. Polym. Sci.*, A-1, **5**, 2375;	1967	Union Carbide
				Ency. Polym. Sci. & Eng.,**13**,196	1988	
			N-1; DMSO; NO_2	*Makromol. Chem.*, **130**, 45	1969	Bayer, A-G
	185		N-6; Bulk; F	*Polym.*, **25**(8),1151	1984	Hamburg U.
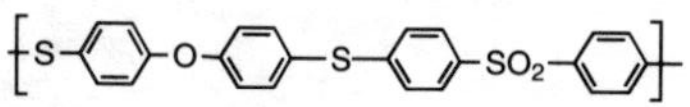		225-245	N-2; K_2CO_3, DMF; Cl	*Makromol. Chem.*, **140**, 83	1970	Celanese

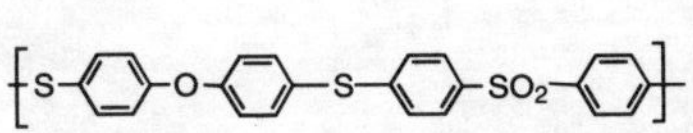		N-6; Sulfolane; Cl	*Prog. Polym. Sci.*, **14**,173	1989	Tokyo Inst. Tech.
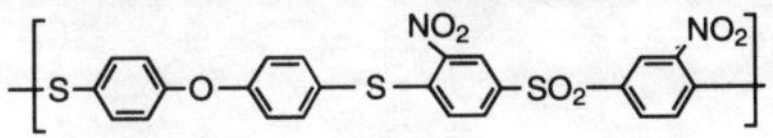		N-6; HMPA; Cl	*Prog. Polym. Sci.*, **14**,173	1989	Tokyo Inst. Tech.
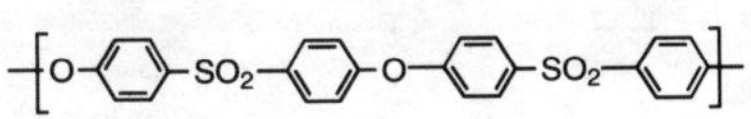	245	N-1; NaOH, DMSO, Sulfolane; Cl, F	*J. Polym. Sci.*, A-1, **5**, 2375;	1967	Union Carbide
			Ency. Polym. Sci. & Eng., **13**,196;	1988	
		N-2; K_2CO_3, NMP; Cl	*Polym. Prep.*, **23**(1), 284	1982	Va. P. I. & State U.
	220	E-1; $FeCl_3$, Nitrobenzene, Dimethyl sulfone	*J. Polym. Sci.*, A-1, **8**, 2035	1970	3M
	250	E-1; $FeCl_3$, Bulk	*J. Polym. Sci.*, C, No.**16**, 715	1967	ICI
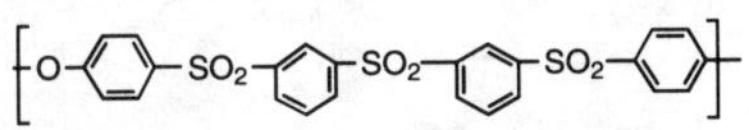		E-1; $FeCl_3$, Bulk	*J. Polym. Sci.*, C, No.**16**, 715	1967	ICI
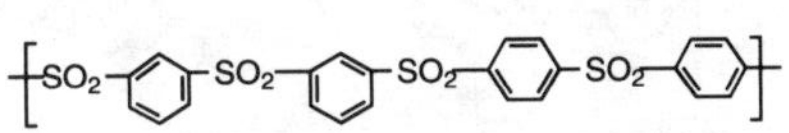		E-1; $FeCl_3$, Bulk	*J. Polym. Sci.* , C, No.**16**, 715	1967	ICI
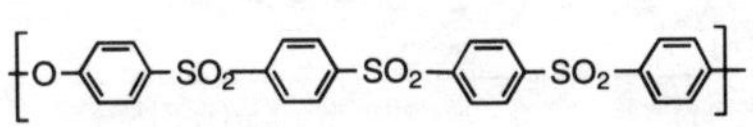		E-1; $FeCl_3$, Bulk	*J. Polym. Sci.*, C, No.**16**, 715	1967	ICI
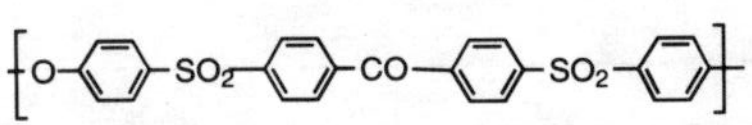		E-1; $FeCl_3$, Bulk	*J. Polym. Sci.*, C, No.**16**, 715	1967	ICI

TABLE 2—*continued*

Polymer structure	Tg (°C)	Tm (°C)	Synthesis	Reference	Year	Organization
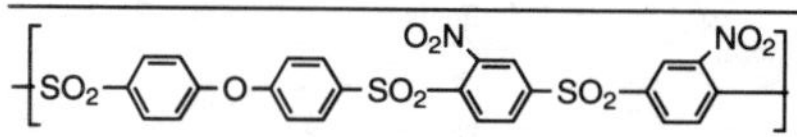			N-5; NaOH, TBAC, H_2O-Nitrobenzene; Cl	*Makromol. Chem., Rapid Comm.*, **3**, 821	1982	Kogakuin U., Tokyo
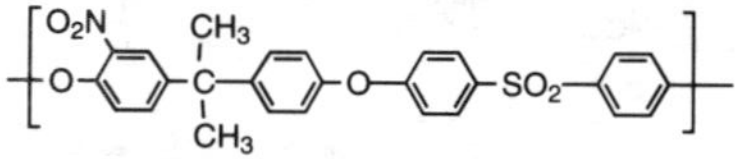	190		Post Reaction on PSF. Permeation Data	*J. Memb. Sci.*, **72**, 91	1992	N. C. State U.
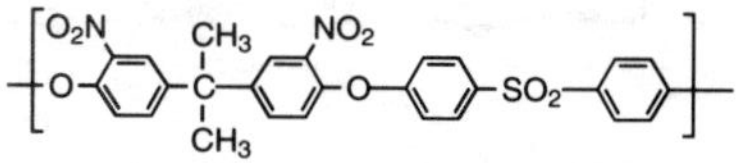	195		Post Reaction on PSF. Permeation Data	*J. Memb. Sci.*, **74**, 193	1992	U. of Twente
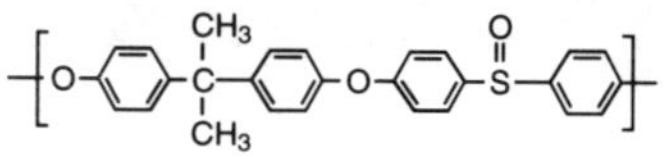	165		N-1; NaOH, DMSO, Sulfolane; Cl, F	*J. Polym. Sci.*, A-1, **5**, 2375;	1967	Union Carbide
				Ency. Polym. Sci. & Eng., **13**, 196	1988	Union Carbide
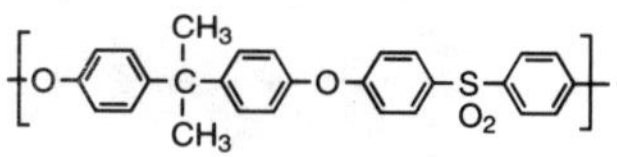 Udel™ Polysulfone	184, 195		N-1; NaOH, DMSO, Sulfolane; Cl, F	*J. Polym. Sci.*, A-1, **5**, 2375, 2399;	1967	Union Carbide
			Thermal Stability	*Appl. Polym. Symp.*, No. **26**, 373	1975	
			N-2; K_2CO_3, DMSO; NO_2	Jap. Pat. 1087628	1989	Toyo Soda
			N-2; K_2CO_3, NMP; Cl	*Polym. Prep.*, **23**(1), 284	1982	Va. P. I. & State U.
			N-6; K_2CO_3, NMP; Cl	*Makromol. Chem.*, **191**, 2027	1990	Hamburg U.
	186		Steam Sterilization	*Polym. News*, **11**, 359;	1986	Union Carbide
			Permeability Data	*J. Appl. Polym. Sci.*, **28**, 2925;	1983	U. Texas
				Polym., **32**, 840	1991	

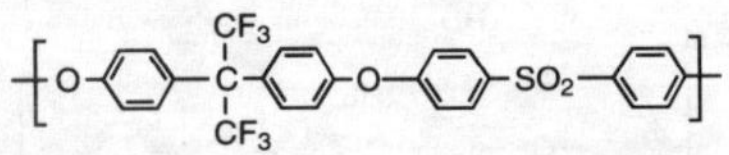	205	255	N-1; NaOH, DMSO, Sulfolane; Cl,F	*J.Polym. Sci.*, A-1, **5**, 2375;	1967	Union Carbide
				Ency. Polym. Sci. & *Eng.*, **13**,196	1988	
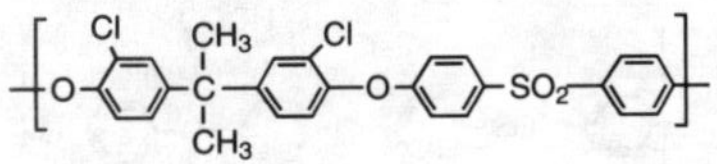	205		N-1; NaOH, DMSO, Sulfolane; Cl	*J. Polym. Sci.*, A-1, **5**, 2375;	1967	Union Carbide
				Polym. Prep., **16**(1), 476	1975	
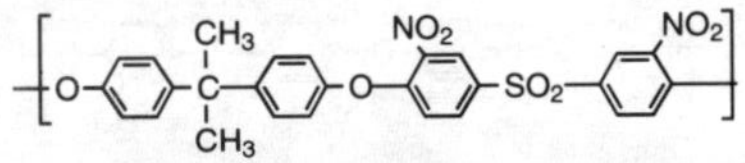	138		N-2; K_2CO_3, Sulfolane; Cl	*J. Polym. Sci., Polym. Lett. Ed.*, **17**, 85	1979	
			N-2; K_2CO_3, Sulfolane; Cl	*Makromol. Chem.*, **179**, 2989	1978	
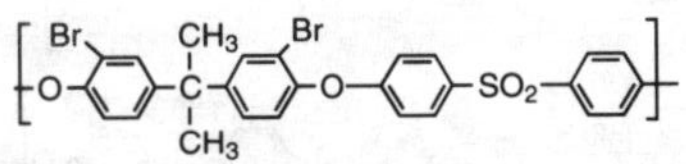			Bromination of Udel 3500	*Polym.*, **30**, 1137	1989	Nat. Res. Council, Canada
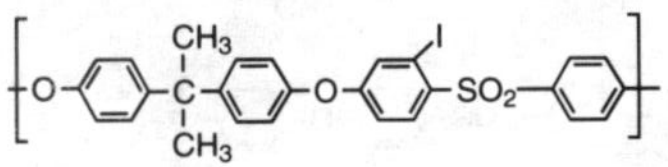			Post Reaction on PSF	US Pat. 4797457; US Pat. 4833219	1989	Nat. Res. Council, Canada
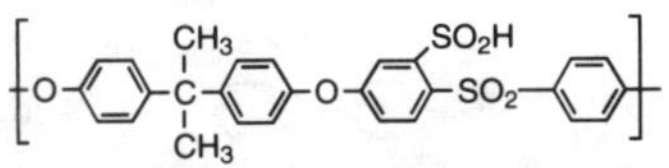			Post Reaction on PSF	US Pat. 4797457; US Pat. 4833219	1989	Nat. Res. Council, Canada
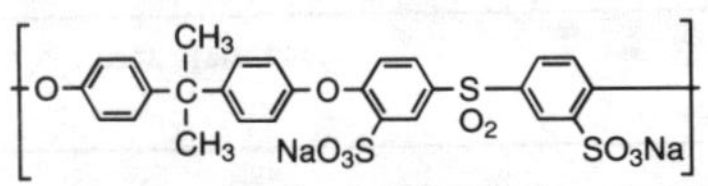	192		N-2; K_2CO_3, DMAC; Cl	*J. Polym. Sci., Polym. Chem.*, **31**, 853	1993	Nippon Steel

TABLE 2—*continued*

Polymer structure	Tg (°C)	Tm (°C)	Synthesis	Reference	Year	Organization
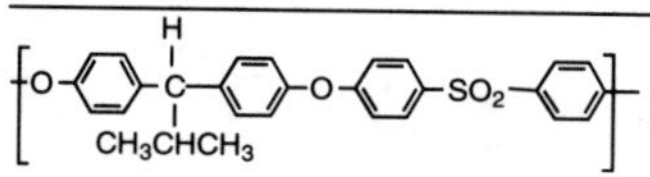	200		N-1; NaOH, DMSO, Sulfolane; Cl, F	*J. Polym. Sci.*, A-1, **5**, 2375	1967	Union Carbide
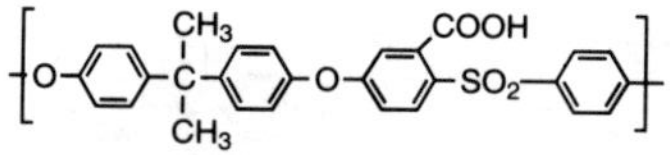	210		Post reaction on PSF	*Polym. Bull.*, **29**, 521	1992	U. of Singapore
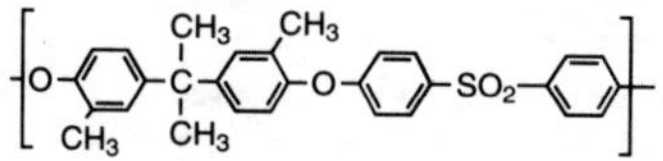	178		N-1; NaOH, DMSO; Cl	*Appl. Polym. Symp.*, No. **26**, 373	1975	Union Carbide
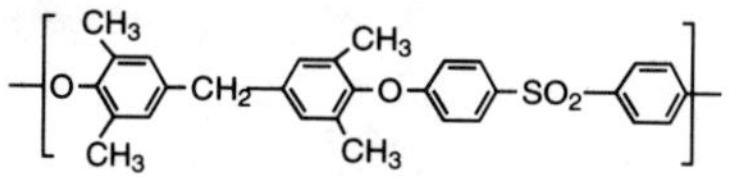	228		N-1; NaOH, DMSO; Cl	US Pat. 4156068	1979	
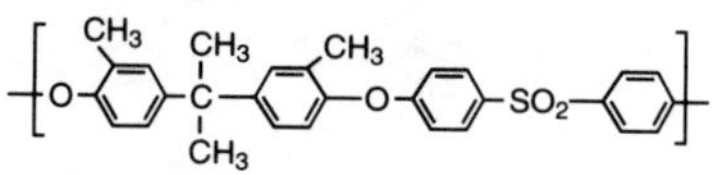	178		N-1; NaOH, DMSO; Cl	*Polym. Prep.*, **16**(1), 476	1975	Union Carbide
	180		Permeability Data	*Polymer*, **32**, 840	1991	U. Texas
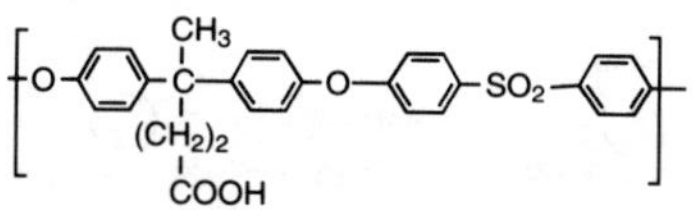			N-1; NaOH, DMSO; Cl	*Adv. Chem. Ser.*, **176**, 559	1979	Union Carbide
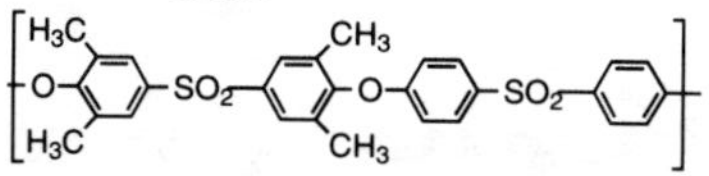	265		N-2; K_2CO_3, Sulfolane; Cl	US Pat. 4473684	1984	Amoco
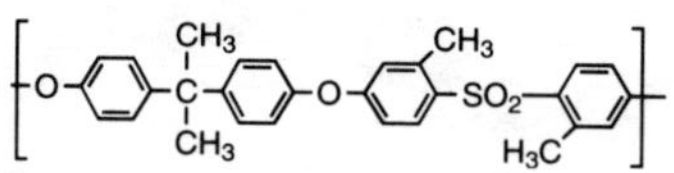			Post Reaction on PSF	US Pat. 4797457; US Pat. 4833219	1989	Nat. Res. Council, Canada
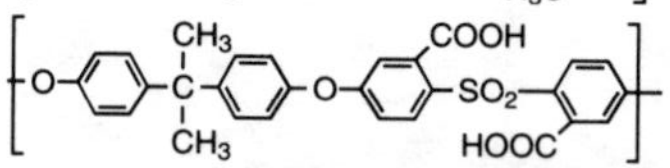			Post Reaction on PSF	US Pat. 4797457; US Pat. 4833219;	1989	Nat. Res. Council, Canada
	220			*Polym. Bull.* **29**, 521	1992	U. of Singapore

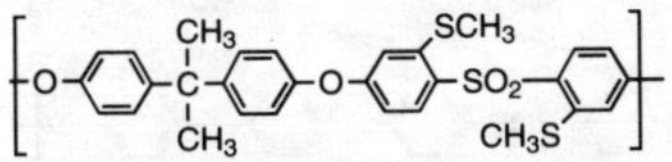		Post Reaction on PSF	US Pat. 4797457; US Pat. 4833219	1989	Nat. Res. Council, Canada
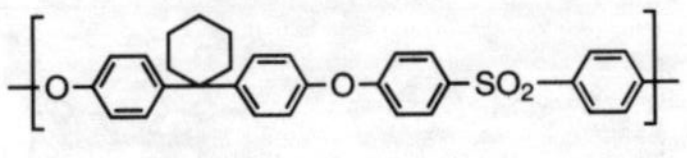	205	N-1; NaOH, DMSO, Sulfolane; Cl, F	*J. Polym. Sci.*, A-1, **5**, 2375; *Ency. Polym. Sci. & Eng.*, **13**,196	1967 1988	Union Carbide
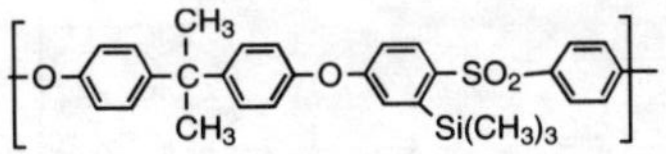	171	Post Reaction on PSF	*Polym. Prep.*, **32**(2), 258	1991	Amoco
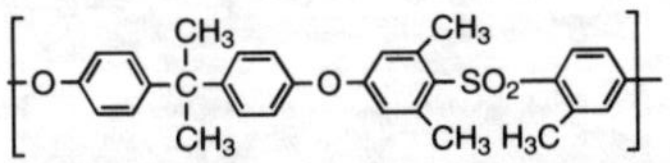		Post Reaction on PSF	US Pat. 4797457; US Pat. 4833219	1989	Nat. Res. Council, Canada
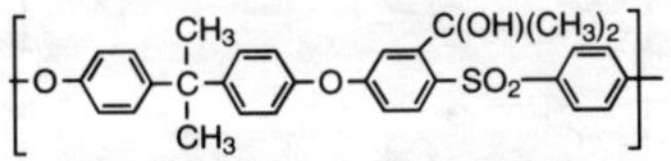		Post Reaction on PSF	US Pat. 4797457; US Pat. 4833219	1989	Nat. Res. Council, Canada
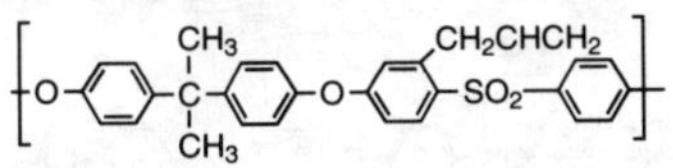		Post Reaction on PSF	US Pat. 4797457; US Pat. 4833219	1989	Nat. Res. Council, Canada
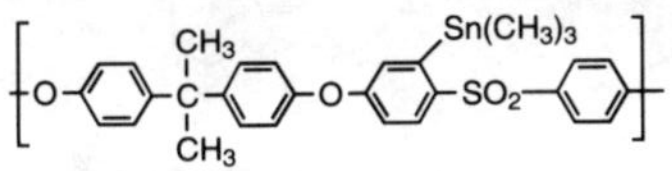		Post Reaction on PSF	US Pat. 4797457; US Pat. 4833219	1989	Nat. Res. Council, Canada
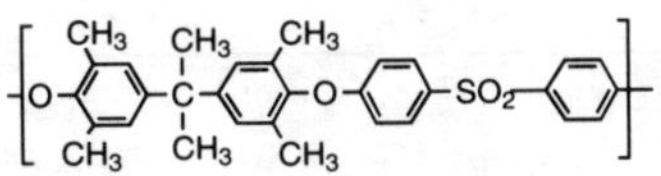	235	N-1; NaOH, DMSO, Sulfolane; Cl	*J. Polym. Sci.*, A-1, **5**, 2375; *Appl. Polym. Symp.*, No. **26**, 373;	1967 1975	Union Carbide

TABLE 2—*continued*

Polymer structure	Tg (°C)	Tm (°C)	Synthesis	Reference	Year	Organization
	242 235		Permeability Data N-2; K_2CO_3, NMP; F	*Polymer*, **32**, 840 *Polym. Prep.*, **25**(2), 19	1991 1984	U. Texas Va. P. I. & State, U.
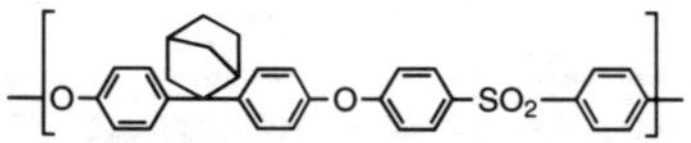	250		N-1; NaOH, DMSO, Sulfolane; Cl, F	*J. Polym. Sci.*, A-1, **5**, 2375	1967	Union Carbide
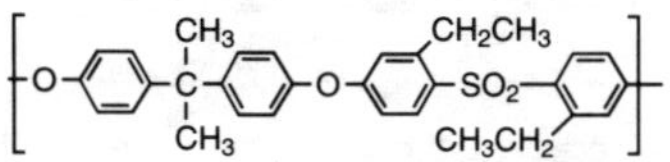			Post Reaction on PSF	US Pat. 4797457; US Pat. 4833219	1989	Nat. Res. Council, Canada
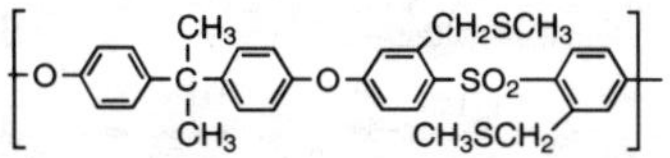			Post Reaction on PSF	US Pat. 4797457; US Pat. 4833219	1989	Nat. Res. Council, Canada
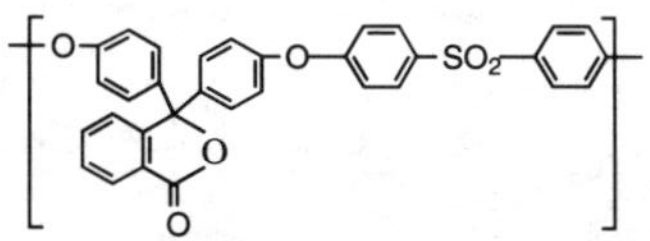			N-6; K_2CO_3, NMP; Cl	*Makromol. Chem.*, **191**, 2027	1990	Hamburg U.
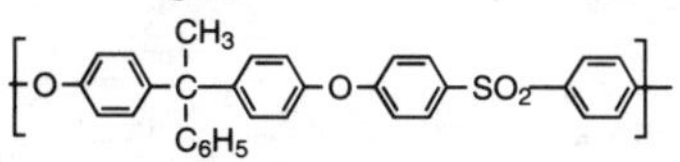	200		N-1; NaOH, DMSO, Sulfolane; Cl,F	*J. Polym. Sci.*, A-1, **5**, 2375	1967	Union Carbide
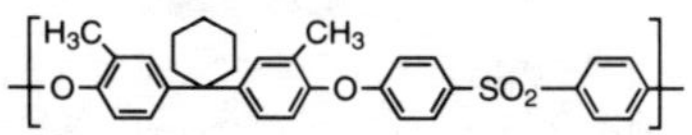	197		N-1,N-2; DMSO, NMP; Cl Permeabilty Data	*Polym.*, **32**, 840	1991	U. Texas
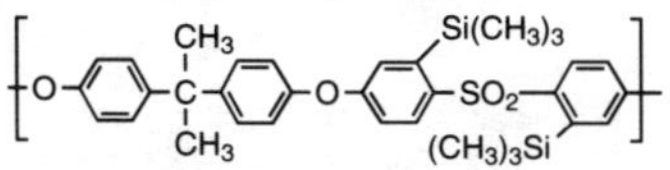			Post Reaction on PSF	US Pat. 4797457; US Pat. 4833219	1989	Nat. Res. Council, Canada

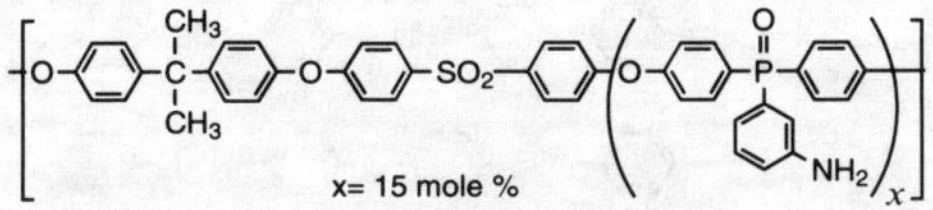	198	N-2; K_2CO_3, NMP; F, Cl	*Polym. Bull.*, **29**, 477	1992	Va. P. I. & State, U
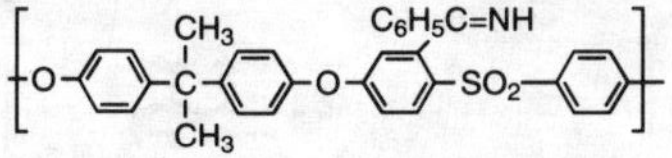		Post Reaction on PSF	US Pat. 4797457; US Pat. 4833219	1989	Nat. Res. Council, Canada
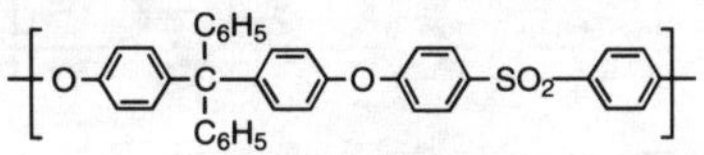	230	N-1; NaOH, DMSO, Sulfolane; Cl, F	*J. Polym. Sci.* A-1, **5**, 2375; *Ency. Polym. Sci. & Eng.*,**13**,196	1967 1988	Union Carbide
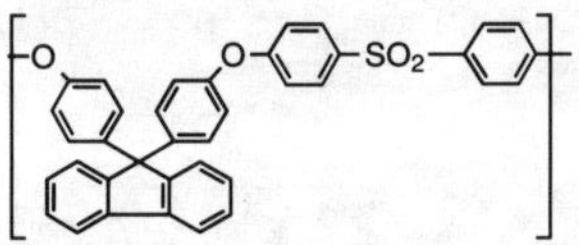	280 280	N-2; K_2CO_3, DMAC; Cl	*Polym.*, **29**, 358; US Pat. 4806618	1988 1989	NASA, Langley; Central Glass (Jap.)
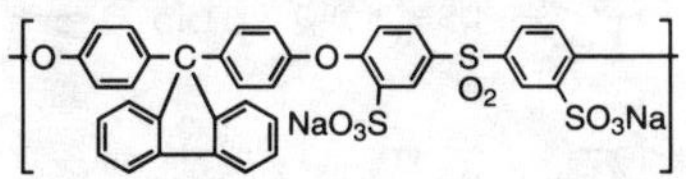	274	N-2; K_2CO_3, DMAC; Cl	*J. Polym. Sci.*: Pt. A: *Polym. Chem.*, **31**, 853	1993	Nippon Steel
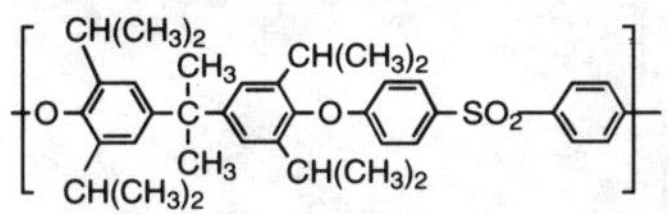	175	N-1; NaOH, DMSO; Cl	*Appl. Polym. Symp.*, No. **26**, 373	1975	Union Carbide
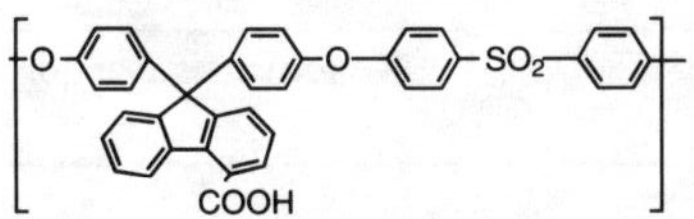		N-1; KOH, DMSO; F, Cl	Proc. Int. Membrane Conf., p. 187	1986	Nat. Res. Council, Canada

TABLE 2—*continued*

Polymer structure	Tg (°C)	Tm (°C)	Synthesis	Reference	Year	Organization
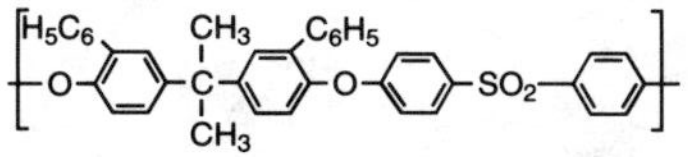	191		N-1; NaOH, NMP; Cl	Jap. Pat. 1098629	1989	Idemitsu Kosan
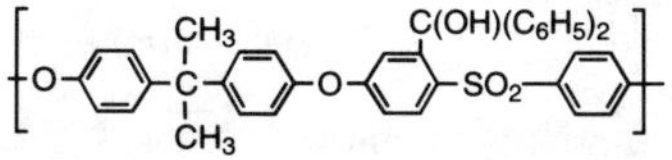			Post Reaction on PSF	US Pat. 4797457; US Pat. 4833219	1989	Nat. Res. Council, Canada
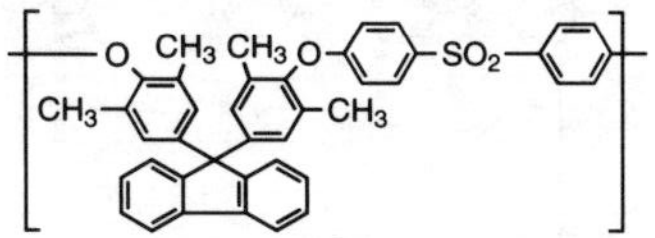	310		N-2; K_2CO_3, DMAC; Cl, F	*Polym.*, **29**, 358	1988	NASA/ Langley
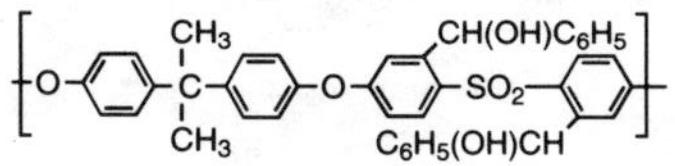			Post Reaction on PSF	US Pat. 4797457; US Pat. 4833219	1989	Nat. Res. Council, Canada
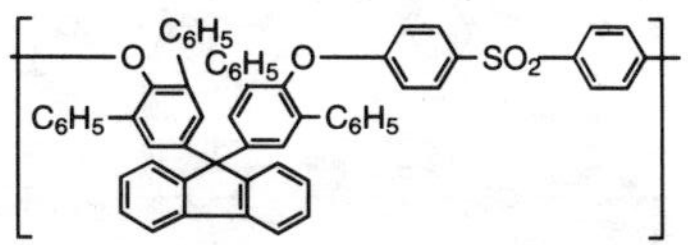	262		N-2; K_2CO_3, DMAC; F	*J. Polym. Sci., Polym. Chem.*, **29** , 1045	1991	McGill U.
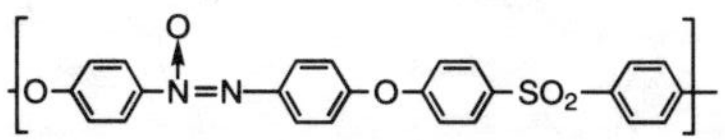			N-1; NaOH, DMSO; Cl	*Prog. Polym. Sci.*, **12**, 111	1986	Ind. Inst. Tech.
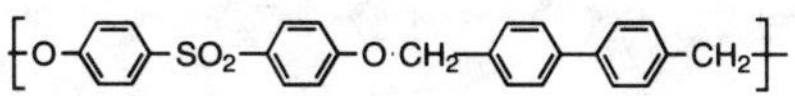	176		N-2; K_2CO_3, DMAC; Cl	*Angew. Makromol. Chemie*, **188**,143, 155	1991	U. Lublin
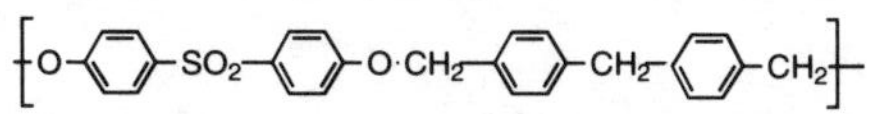	146		N-2; K_2CO_3, DMAC; Cl	*Angew. Makromol. Chemie*, **188**,143, 155	1991	U. Lublin

Structure		Conditions	Reference	Year	Source
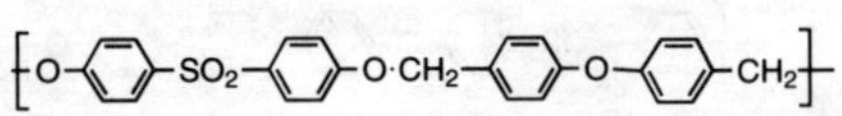	145	N-2; K_2CO_3, DMAC; Cl	*Angew. Makromol. Chemie*, **188**, 143, 155	1991	U. Lublin
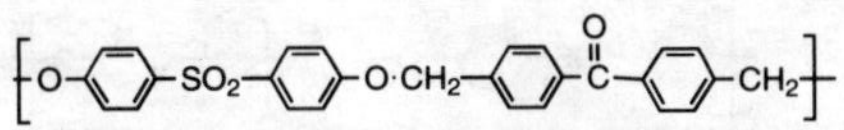	176	N-2; K_2CO_3, DMAC; Cl	*Angew. Makromol. Chemie*, **188**, 143, 155	1991	U. Lublin
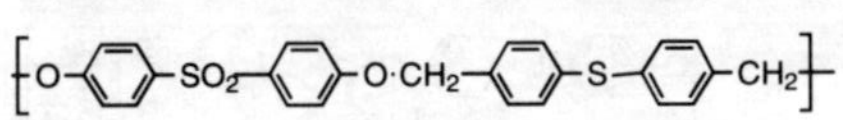	146	N-2; K_2CO_3, DMAC; Cl	*Angew. Makromol. Chemie*, **188**, 143, 155	1991	U. Lublin
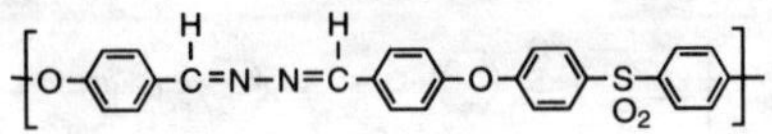	211	N-2; K_2CO_3, DMAC; Cl	*Angew. Makromol. Chemie*, **188**, 143, 155	1991	U. Lublin
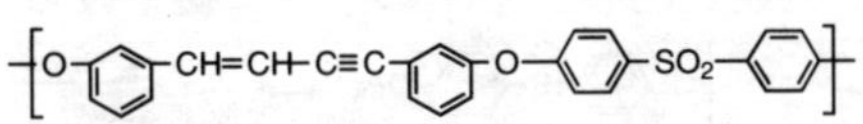	179	N-2; K_2CO_3, DMSO, Sulfolane; F	*Polym. Prep.*, **20**(1), 211	1979	
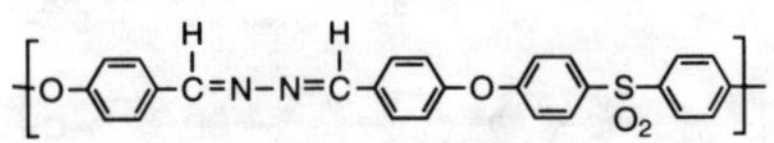	151	N-2; K_2CO_3, DMPU; F	*Polym. Prep.*, **34**(1), 521	1993	IBM
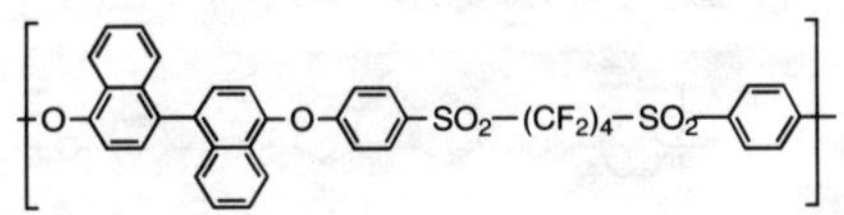	206	A-2; $FeCl_3$, Nitrobenzene	*J. Polym. Sci., Polym. Chem.*, **29**, 965	1991	Case West. Res. U.
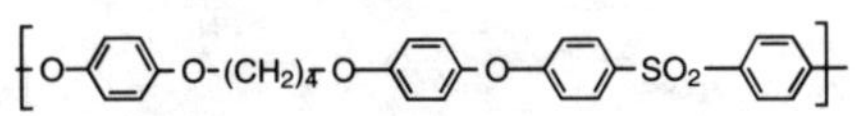	116	N; NaOH, DMSO; Cl	Eur. Pat. 327501	1989	Ciba-Geigy
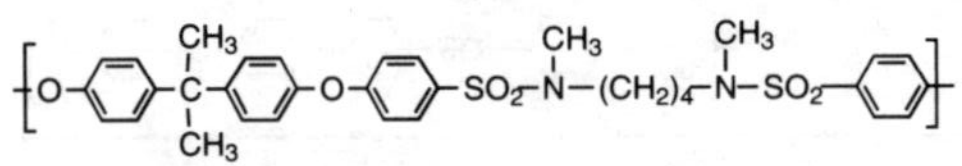	120	N-1; NaOH, DMSO, Sulfolane; Cl, F	*J. Polym. Sci.*, A-1, **5**, 2375	1967	Union Carbide
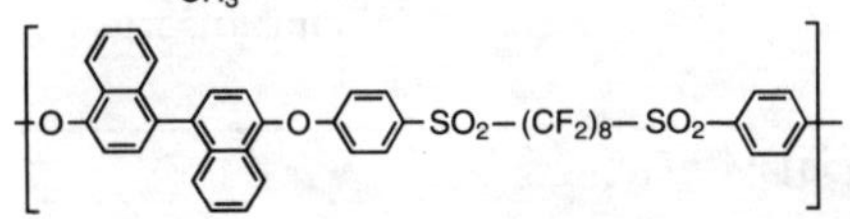	165	A-2; $FeCl_3$, Nitrobenzene	*J. Polym. Sci., Polym. Chem.*, **29**, 965	1991	Case West. Res. U.

TABLE 3
Polyarylethersulfones containing five rings in the backbone of the repeat unit

Polymer structure	Tg (°C)	Tm (°C)	Synthesis	Reference	Year	Organization
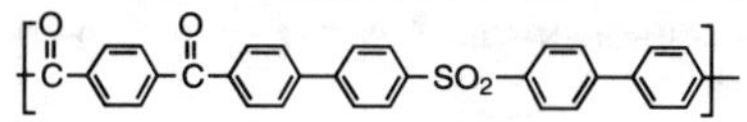			E-5; TFMSA, P_2O_5	US Pat. 4861856	1989	Eastman Kodak
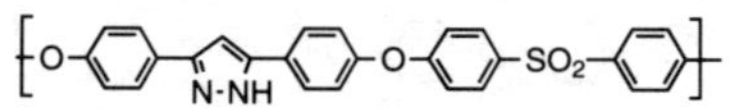	260		N-2; K_2CO_3,DMAC, NMP; Cl	*Polym. Prep.*, **32**(1), 619	1991	Va. Comm. U.
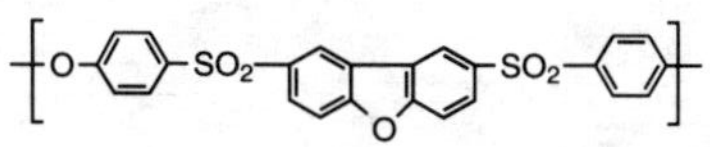	290		E-1; $FeCl_3$ Nitrobenzene, Dimethyl sulfone	*J.Polym. Sci.*, A-1, **8**, 2035	1970	3M
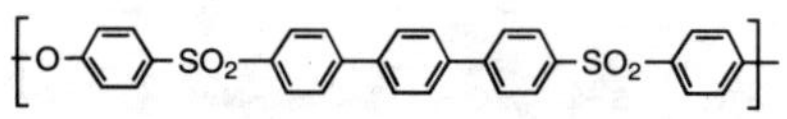	311	398	N-2; K_2CO_3, Diphenyl sulfone; Cl	*Bull. Soc. Chim. Belg.*, **98**, 667	1989	ICI
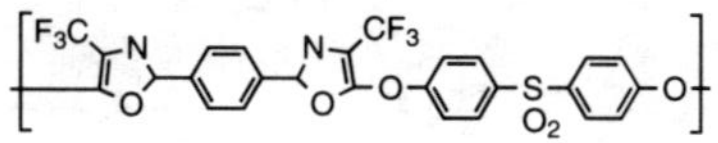	191		N-2; $(Et)_3N$, NMP; F	*Polym. Prep.*, **34**(1), 429	1993	U. Bayreuth
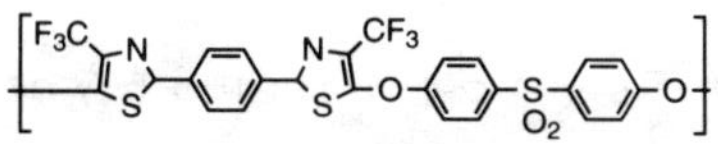	203		N-2; $(Et)_3N$, NMP; F	*Polym. Prep.*, **34**(1), 429	1993	U. Bayreuth
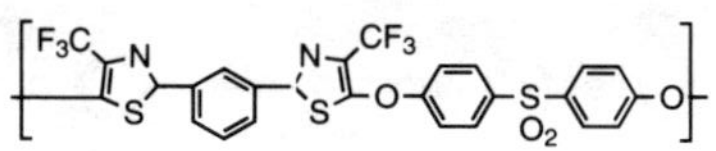			N-2; $(Et)_3N$, NMP; F	*Polym. Prep.*, **34**(1), 429	1993	U. Bayreuth
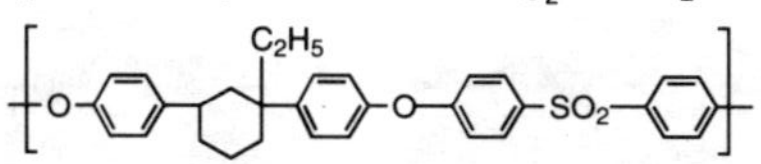	230		N-1; NaOH, DMSO; Cl	GB Pat. 1078234	1967	Union Carbide
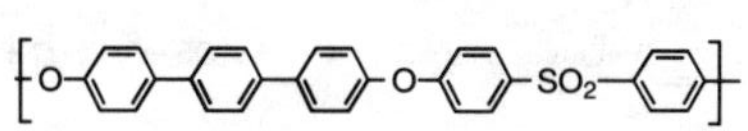	251	389	N-2; K_2CO_3, Diphenyl sulfone; Cl, F	*Bull. Soc. Chim. Belg.*, **98**, 667	1989	ICI

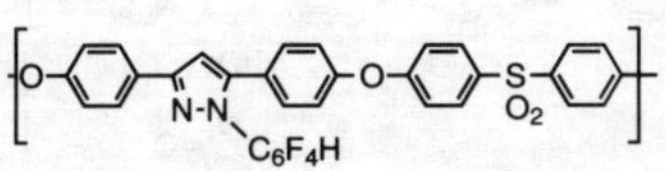	230	N-9; K_2CO_3, DMAC; F	*Polym. Prep.*, **34**(1), 441	1993	Va. Comm. U.
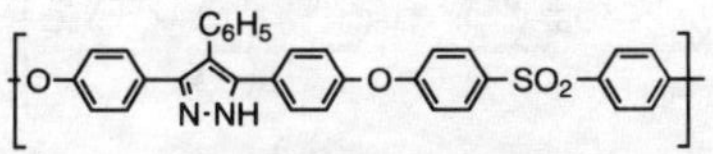	237	N-2; K_2CO_3, DMAC; Cl	*Polym. Prep.*, **32**(2), 160	1991	Va. Comm. U.; NASA/ Langley
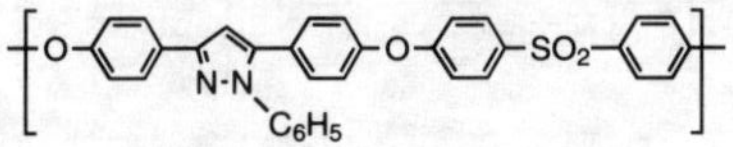	235	N-2; K_2CO_3, DMAC, NMP; F	*Polym. Prep.*, **32**(1), 619	1991	Va. Comm. U.
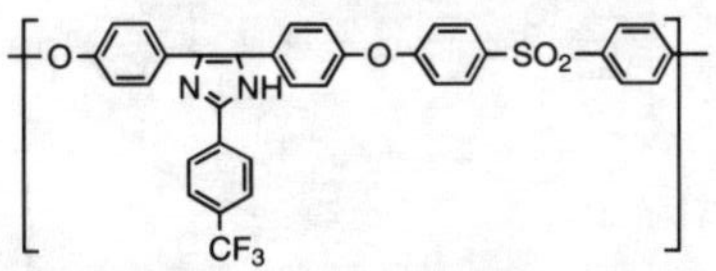	284	N-2; K_2CO_3, DMAC; F	*Polym. Prep.*, **32**(2), 162	1991	NASA/ Langley
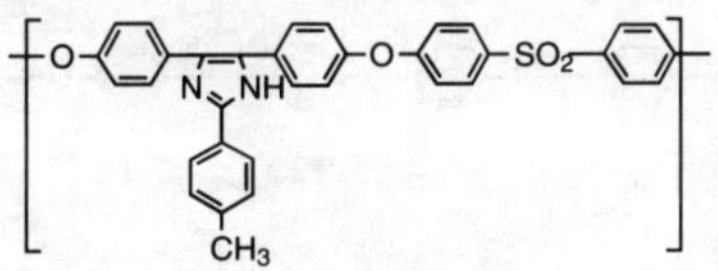	275	N-2; K_2CO_3, DMAC; F	*Polym. Prep.*, **32**(2), 162	1991	NASA/ Langley
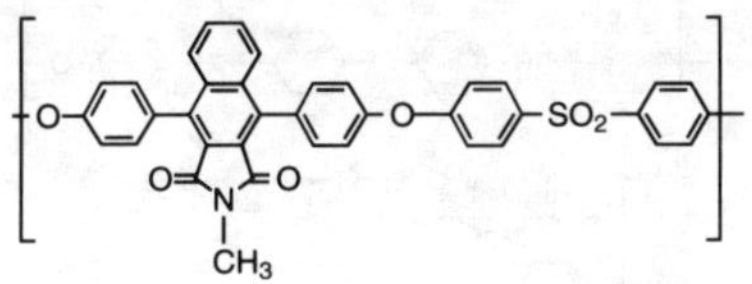	314	N-2; K_2CO_3, NMP, Sulfolane; F	*Polym. Prep.*, **32**(2), 164	1991	McGill U.
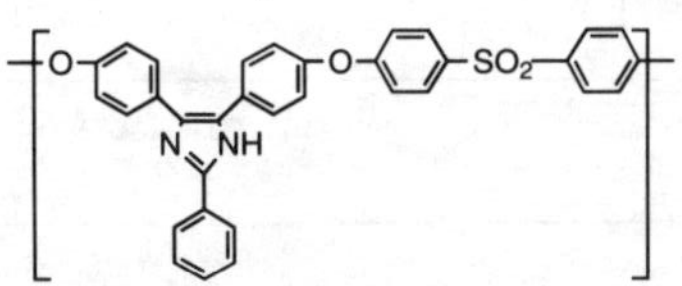	277	N-2; K_2CO_3, DMAC; F	NASA Tech. Briefs, Aug., p. 48	1990	NASA/ Langley

TABLE 3—*continued*

Polymer structure	Tg (°C)	Tm (°C)	Synthesis	Reference	Year	Organization
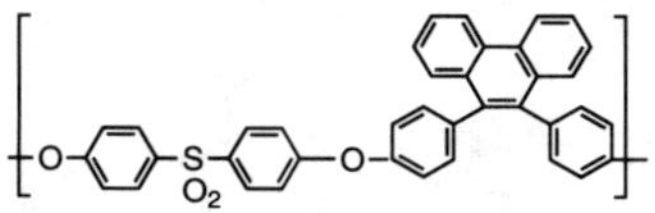	238		Post Reaction on PAEK	*Polym. Prep.*, **33**(2), 351	1992	Carleton U.
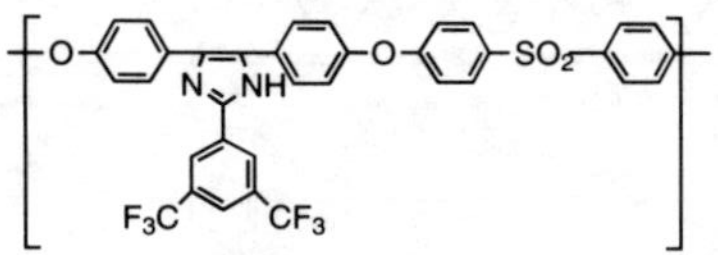	295		N-2; K_2CO_3, DMAC; F	*Polym. Prep.*, **32**(2), 162	1991	NASA/ Langley
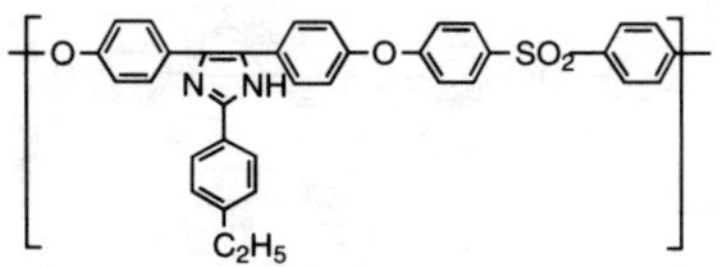	271		N-2; K_2CO_3, DMAC; F	*Polym. Prep.*, **32**(2), 162	1991	NASA/ Langley
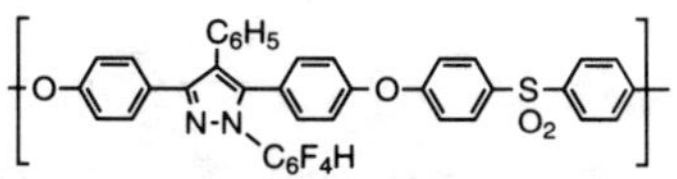	221		N-2; K_2CO_3, DMAC; F	*Polym. Prep.*, **34**(1), 441	1993	Va. Comm. U.
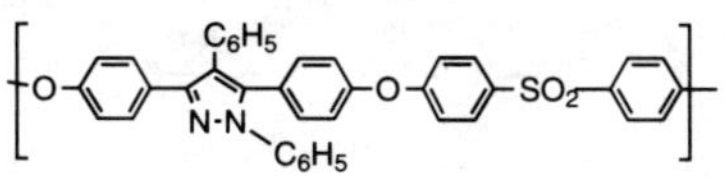	243		N-2; K_2CO_3, DMAC; F	*Polym. Prep.*, **32**(2), 160	1991	Va. Comm. U.; NASA/ Langley
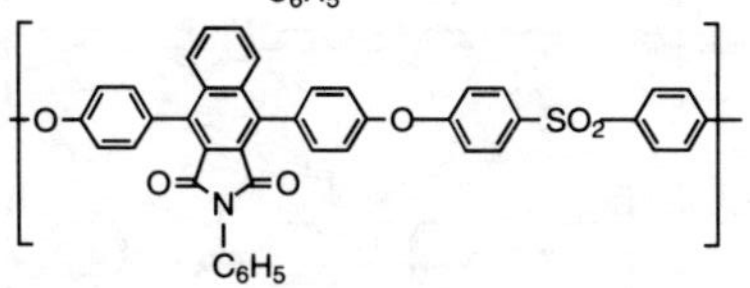	310		N-2; K_2CO_3, NMP, Sulfolane; F	*Polym. Prep.*, **32**(2), 164	1991	McGill U.
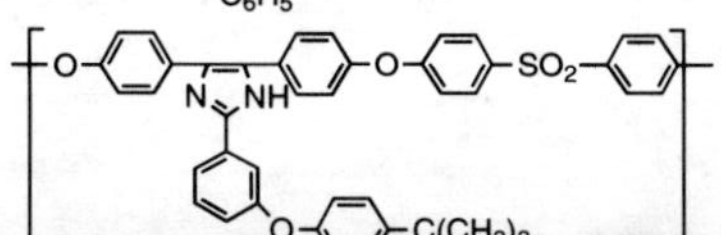	235		N-2; K_2CO_3, DMAC; F	*Polym. Prep.*, **32**(2), 162	1991	NASA/ Langley

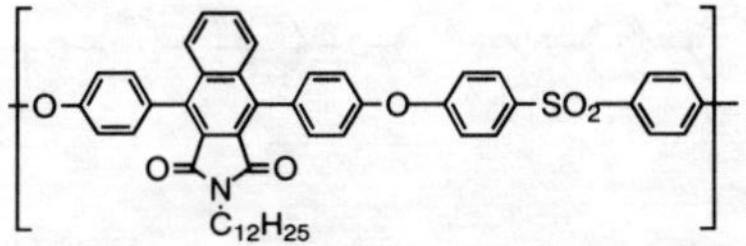	167		N-2; K_2CO_3, NMP, Sulfolane; F	*Polym. Prep.*, **32**(2), 164	1991	McGill U.
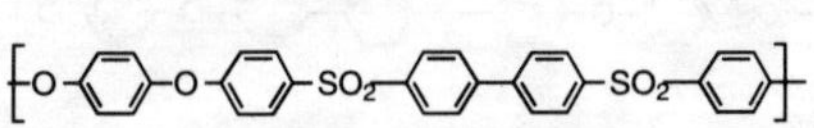	250		N-2; K_2CO_3, Sulfolane; F, Cl	*Polym. Prep.*, **32**(2), 166;	1991	Amoco
	261			*Polym.*, **18**(4), 370	1977	ICI
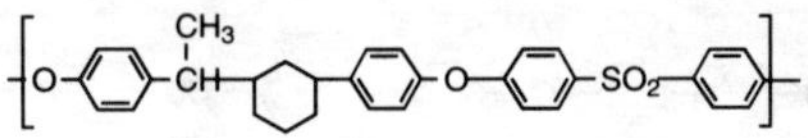	230		N-1; NaOH, DMSO, Sulfolane; Cl, F	*J. Polym.. Sci.*, A-1, **5**, 2375	1967	Union Carbide
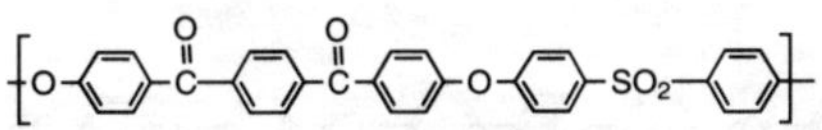	175		N-2; K_2CO_3, NMP; Cl	*CA*, **96**, 7266; Ger. Pat. 3014230	1982	BASF
	196	(214)	E-5; TFMSA, P_2O_5	US Pat. 4861856	1989	Eastman Kodak
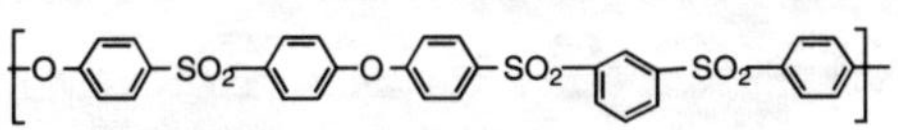			E-1; $FeCl_3$, Bulk	*J. Polym. Sci.*, C, No.**16,** 715	1967	ICI
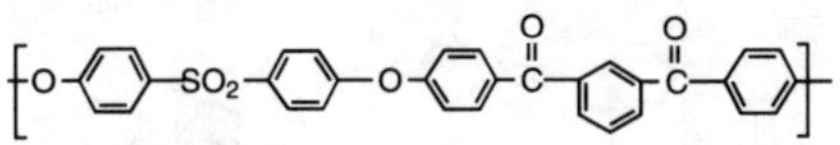	172		N-2; K_2CO_3, Diphenyl sulfone; F	*Bull. Soc. Chim. Belg.*, **98**, 667	1989	ICI
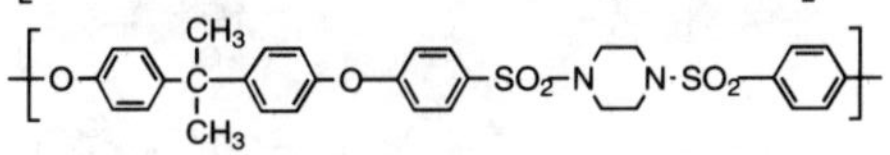		255–260	N-1; NaOH, DMSO; Cl	US Pat. 4108837	1978	Union Carbide
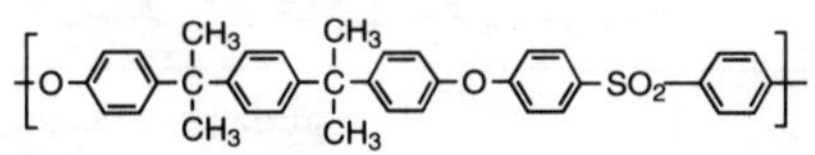	158 (HDT)	4.1%. Cryst	N-1; NaOH, Sulfolane; Cl	Polym. Rxns. & New Polymers, ACS No.129, 131	1973	Uniroyal
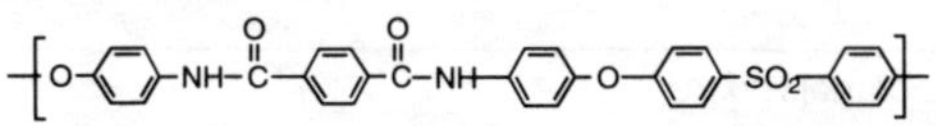			N-2; K_2CO_3, NMP; Cl	*CA*, **95**, 8035; Ger. Pat. 2940190	1981	BASF

TABLE 4
Polyarylethersulfones containing six rings in the backbone of the repeat unit.

Polymer structure	Tg (°C)	Tm (°C)	Synthesis	Reference	Year	Organization
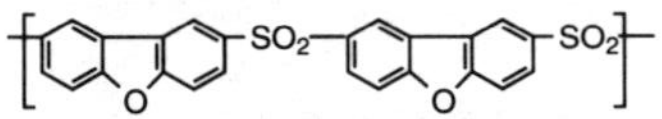	360		E-1; Nitrobenzene, Dimethyl sulfone; $FeCl_3$	*J. Polym. Sci.*, A-1, **8**, 2035	1970	3M
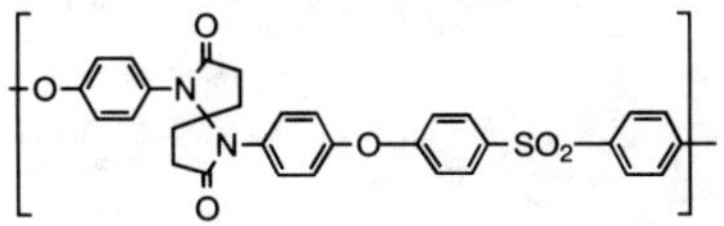	272		N-2; Cl	*Polym. Prep.*, **32**(2), 180	1991	Shell
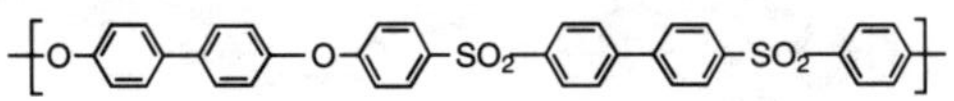	270 280	336	N-2; K_2CO_3, Sulfolane; Cl, F	*Polym. Prep.*, **32**(2), 166; *Polym.*, **18(4)**, 370	1991 1977	Amoco ICI
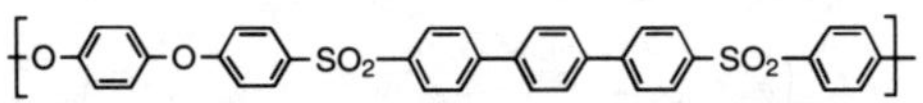	259	393	N-2; K_2CO_3, Diphenyl sulfone; Cl	*Bull. Soc. Chim. Belg.*, **98**, 667; Eur. Pat. 331305	1989	ICI
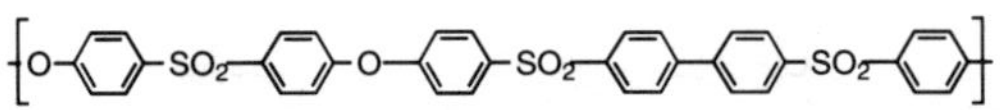 Victrex™ HTA	265 265		N-2; K_2CO_3, Sulfolane; F, Cl N-2; K_2CO_3, Sulfolane; Cl	*Polym.*, **18(4)**, 359; **18(4)**, 370 *Polym. Prep.*, **32**(2), 166	1977 1991	ICI Amoco
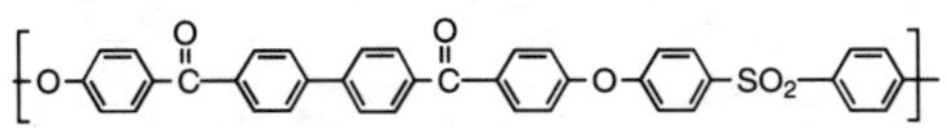	219		A-5; CsF, Bulk Permeability	*J. Chem. Soc., Chem. Commun.*, 336 *J. Polym. Sci.*, B, *Polym. Phys.*, **31**, 983	1990 1993	ICI U. Texas
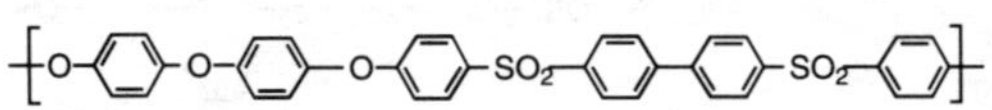	235		N-2; K_2CO_3, Sulfolane; Cl	*Polym. Prep.*, **32**(2), 166	1991	Amoco

Structure					
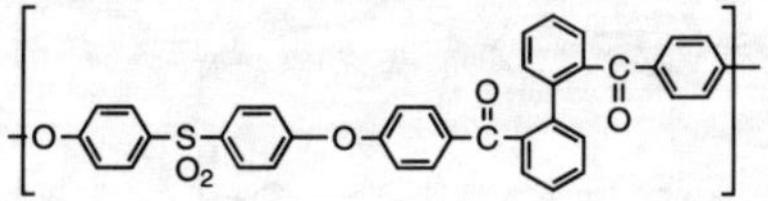	201	N-2; K_2CO_3, NMP; F	*Polym. Prep.*, **33**(2), 351	1992	Carleton U.
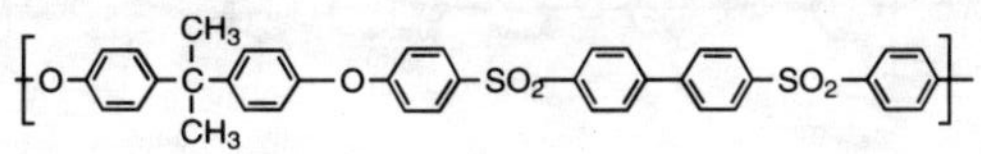	233–242	N-2; K_2CO_3, Sulfolane; F, Cl	*Polym.*, **18**(4), 370	1977	ICI
	240	N-2; K_2CO_3, Sulfolane; Cl;	*Polym. Prep.*, **32**(2), 166;	1991	Amoco
		Permeability	*J. Polym. Sci.*, B, *Polym. Phys.*, **31**, 983	1993	U. Texas
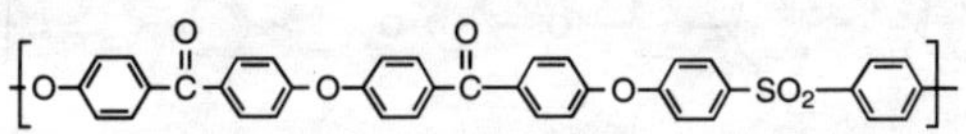		N-2; K_2CO_3, Sulfolane; Cl	*CA*, **87**, 136615; Ger. Pat. 2549529	1977	BASF
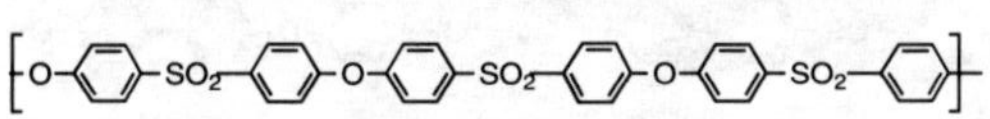		E-1; $FeCl_3$, Bulk	*J. Polym. Sci.*, C, No.**16**, 715	1967	ICI
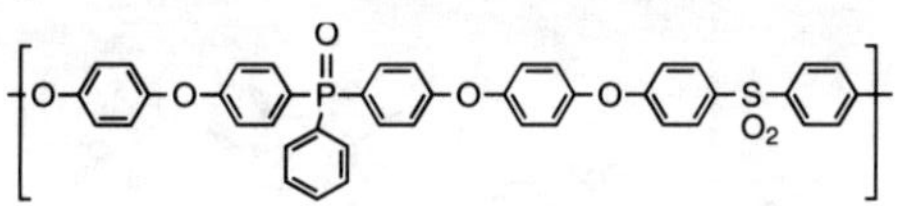	207	N-2; K_2CO_3, DMAC; F	*Polym. Prep.*, **33**(2), 241	1992	NASA/ Langley
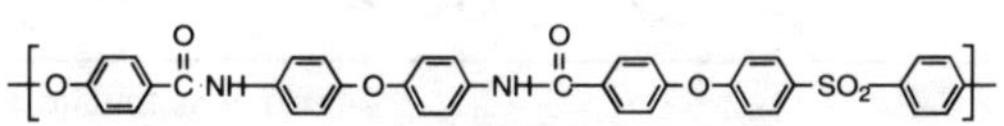		N-2; K_2CO_3, Sulfolane; Cl	*CA*, **87**, 102832; Ger. Pat. 2557652	1977	BASF
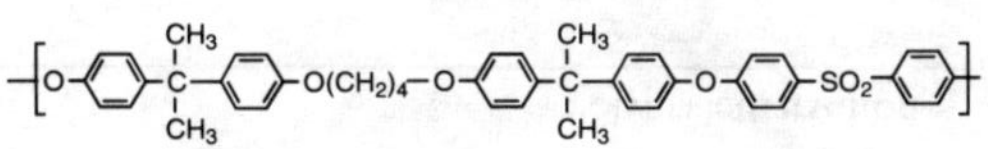	140	N-1; NaOH, DMSO, Sulfolane; Cl	*J. Polym. Sci.*, A-1, **5**, 2375	1967	Union Carbide

TABLE 5
Polyarylethersulfones containing seven or more rings in the backbone of the repeat unit

Polymer structure	Tg (°C)	Tm (°C)	Synthesis	Reference	Year	Organization
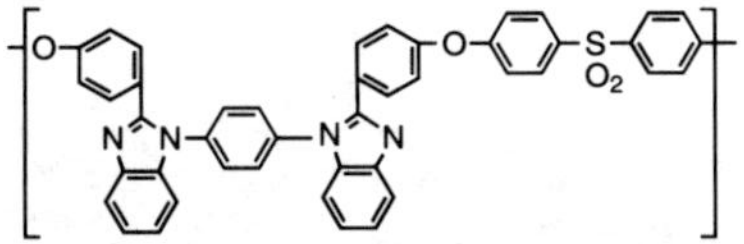	270		N-2; K_2CO_3, Sulfolane; F	*Polym. Prep.*, **33**(1), 1098	1992	NASA/ Langley
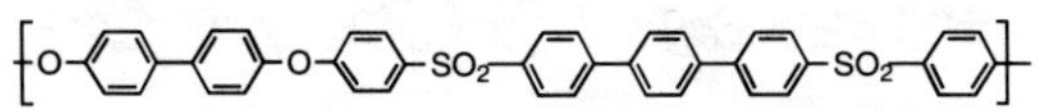	264	383	N-2; K_2CO_3, Diphenyl sulfone; Cl	*Bull. Soc. Chim. Belg.*, **98**, 667	1989	ICI
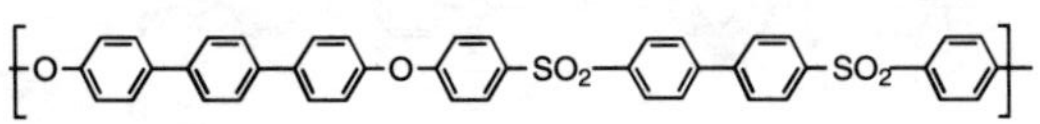	263	408	N-2; K_2CO_3, Diphenyl sulfone; Cl	*Bull. Soc. Chim. Belg.*, **98**, 667	1989	ICI
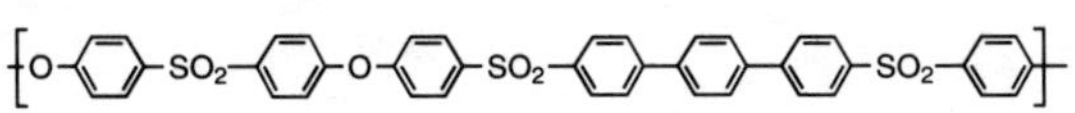	290		N-2; K_2CO_3, Sulfolane, Diphenyl sulfone; F, Cl.	*Polym.*, **18**(4), 370; *Bull. Soc. Chim. Belg.*, **98**, 667;	1977 1989	ICI
			E-1	*Polym.*, **15**, 456	1974	
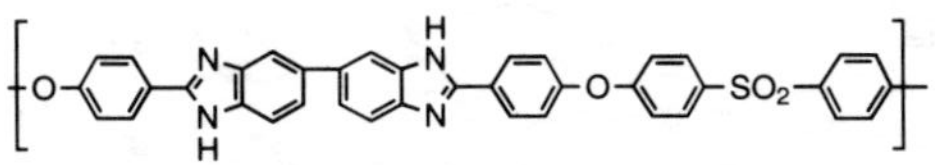	352		N-2; K_2CO_3, DMAC; F	*Polym. Prep.*, **32**(3), 195	1991	NASA/ Langley
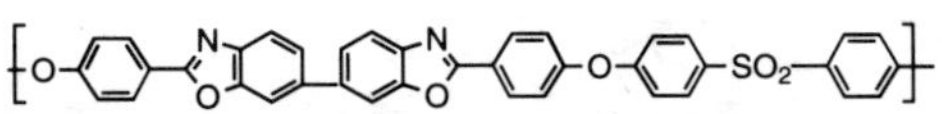	247		N-2; K_2CO_3, DMAC, Diphenyl sulfone; F	*Polym. Prep.*, **32**(1), 646	1991	NASA/ Langley
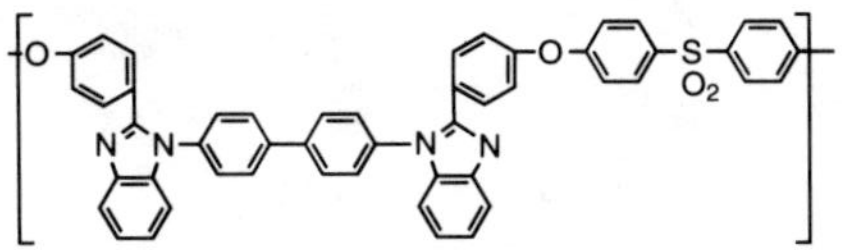	289		N-2; K_2CO_3, Sulfolane; F	*Polym. Prep.*, **33**(1), 1098	1992	NASA/ Langley

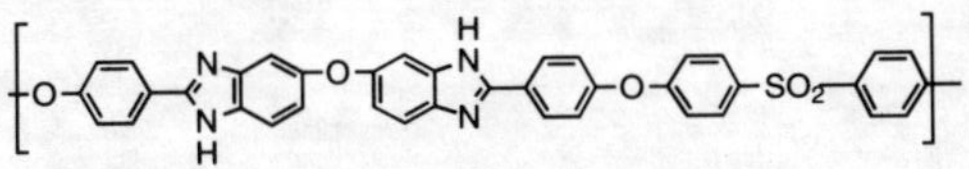	322	N-2; K_2CO_3, DMAC; F	*Polym. Prep.*, **32**(3), 195	1991	NASA/ Langley
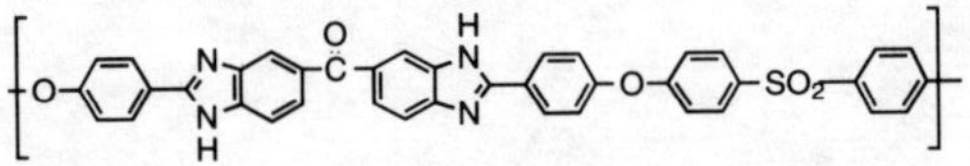		N-2; K_2CO_3, DMAC; F	*Polym. Prep.*, **32**(3), 195	1991	NASA/ Langley
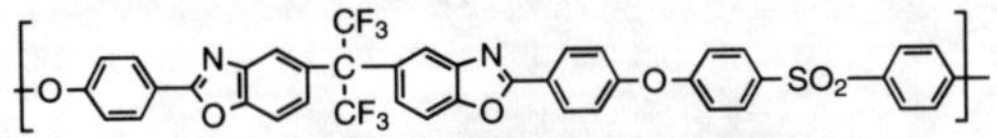	285	N-2; K_2CO_3, DMAC, Diphenyl sulfone; F	*Polym. Prep.*, **32**(1), 646	1991	NASA/ Langley
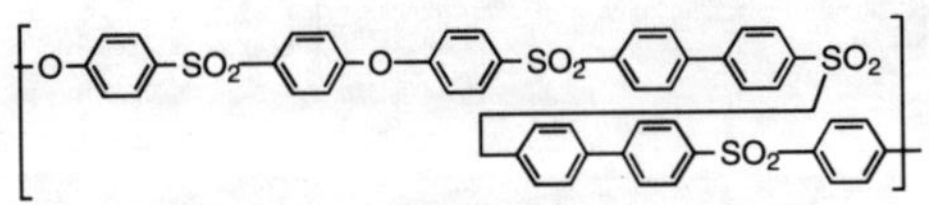	284–289	N-2; Sulfolane; F, Cl	*Polym.*, **18**(4), 359; **18**(4), 370	1977	ICI
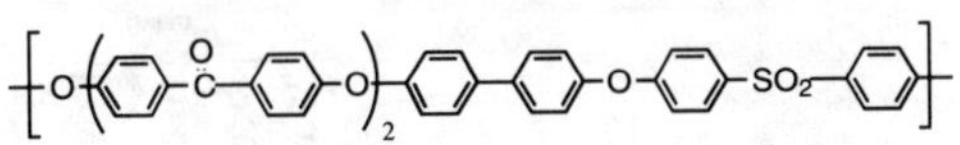	201		*SAMPE*, **32**, 382	1987	ICI
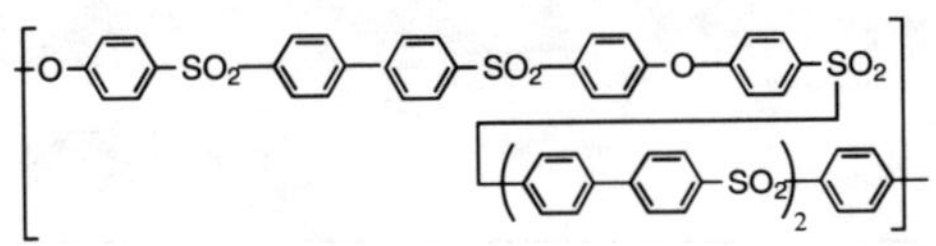	292–298	N-2; Sulfolane; F, Cl	*Polym.*, **18**(4), 370	1977	ICI

TABLE 6
Polyaryletherketones containing three or fewer rings in the backbone of the repeat unit

Polymer structure	Tg (°C)	Tm (°C)	Synthesis	Reference	Year	Organization
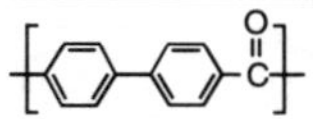		500	E-4; HF, BF_3	US Pat. 3441538	1969	DuPont
			E-5; TFMSA, P_2O_5	US Pat. 4820794	1989	Eastman Kodak
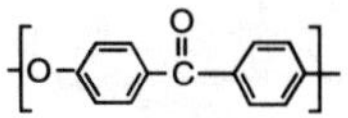 Stilan™	170	365	E-4; HF, BF_3	US Pats. 3953400; 4229564; 4247682	1976 1980 1981	Raychem
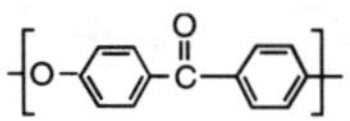	163	361	E-4; HF, BF_3	US Pat. 3441538	1969	DuPont
			E-5; Polyphosphoric acid	*J. Polym.Sci.*, A-1, **6**, 3345; *Macromol.*, **20**, 2675	1968 1987	U. Tokyo
	162	363	E-5; TFMSA, P_2O_5	US Pat. 4820794	1989	Eastman Kodak
			E-8; BF_3, HF (S-ethylchlorothioformate)	US Pat. 4361693	1982	Raychem
	154	367	N-9; K_2CO_3, Diphenyl sulfone; F	*Polym.*,**18**, 365; **21**, 577; **22**,1096	1977 1980 1981	ICI
			N-11;K_2CO_3, Phenol	US Pat. 4794155	1988	Dow
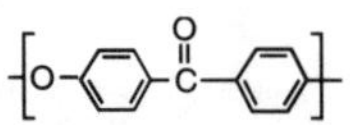	156	366	N-13; Heating the polyketal with acid	*Macromol.*, **20**,1204 US Pat. 4882397	1987 1989	Amoco

from 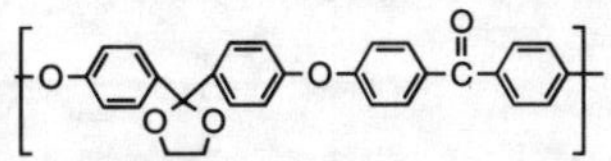						
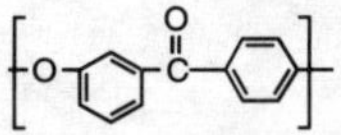			E-5; Polyphosphoric acid	*J. Polym.Sci.*, A-1, **6**, 3345	1968	U. Tokyo
	133		E-5;P_2O_5, CH_3SO_3H	*Macromol.*, **20**, 2675	1987	Yamagata U.
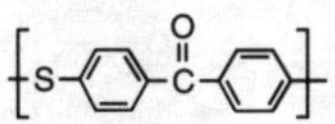	152	335	N; Na_2S, CHP; F	*J. Polym. Sci., Polym. Chem.*, **27**, 661	1989	Dow
		352	N	*Brit. Polym. J.*, **4**, 391	1972	ICI
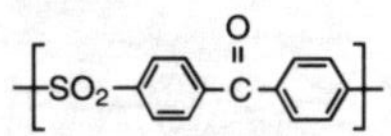		375	Oxidation of sulfide polymer	*Brit. Polym. J.*, **4**, 391	1972	ICI
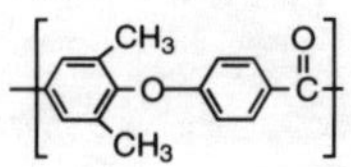			E-4; BF_3, HF	Eur. Pat. 254431	1988	DuPont
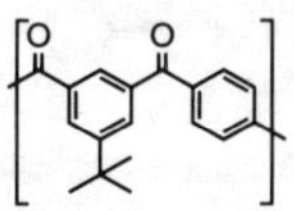			A-10; 1,2-DCE; Cl	*Makromol. Chem. Rapid Comm.*, **13**, 91	1992	U. Michigan
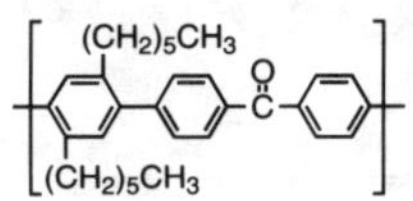			A-10; Na_2CO_3, Pd(0), Toluene, H_2O; Br	*Makromol. Chem., Rapid Comm.*, **11**, 535	1990	U. Karlsruhe
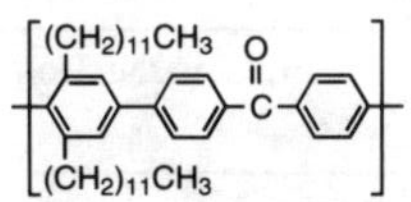			A-10; Na_2CO_3, Pd(0), Toluene, H_2O; Br	*Makromol. Chem., Rapid Comm.*, **11**, 535	1990	U. Karlsruhe

TABLE 6—*continued*

Polymer structure	Tg (°C)	Tm (°C)	Synthesis	Reference	Year	Organization
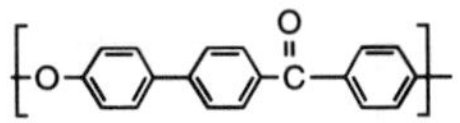	216	478	E-5 ; TFMSA	*Polym. Prep.*, **25**(2), 17; *Polym.*, **29**, 1902	1984 1988	ICI
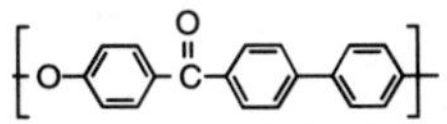		435	E-4; HF, BF_3	US Pats. 3914298; 3965146; 4111908; 4229564	1975 1976 1978 1980	Raychem
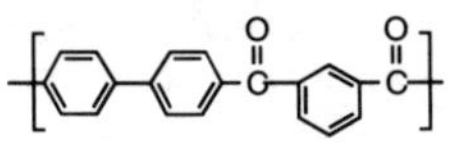			E-4; HF, BF_3	US Pats. 3441538; 3956240	1969 1976	DuPont
	201	332	E-5; TFMSA, P_2O_5	US Pat. 4820794	1989	Eastman Kodak
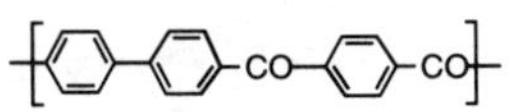		>450	E-5; TFMSA, P_2O_5	US Pat. 4820794	1989	Eastman Kodak
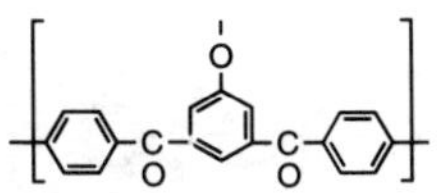	144		N-8; NaH, DMAC; F	*Polym. Prep.*, **34**(1), 58	1993	AT&T
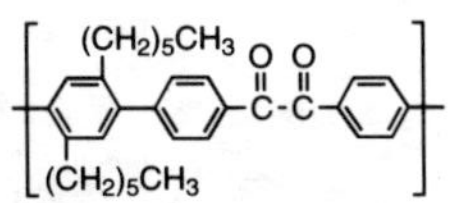			A-10; Na_2CO_3, Pd(0), Toluene, H_2O; Br	*Makromol. Chem., Rapid Comm.*, **11**, 535	1990	U. Karlsruhe
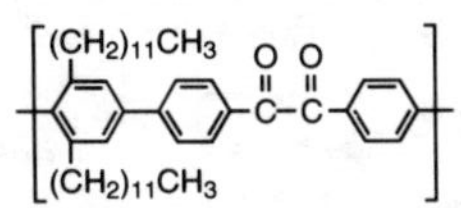			A-10; Na_2CO_3, Pd(0), Toluene, H_2O; Br	*Makromol. Chem., Rapid Comm.*, **11**, 535	1990	U. Karlsruhe

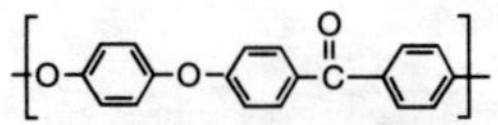 Victrex™ PEEK	144	335	N-9; K_2CO_3, Diphenyl sulfone; F	*Polym. Prep.*, **20**(1), 191; *Polym.*, **21**, 577; **22**,1096; Mod. *Plast.*, p. 86, 11/81;	1979 1980 1981	ICI
				Polym. Prep., **25**(2),17;		
	152	334	E-5 ; TFMSA Morphology	*Polym.*, **24**, 953; **29**,1902	1984 1983 1988	ICI
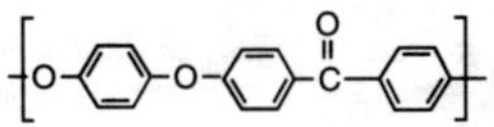	160	350	N-8; NaOH, DMSO, Sulfolane; F	*J. Polym. Sci.*, A-1, **5**, 2375	1967	Union Carbide
		345	N-11; CsF, Bulk	*Polym.*, **25**, 1151;	1984	Hamburg U.
		344		US Pat. 4331798	1982	ICI
			E-4; BF_3, HF (S-ethylchlorothioformate)	US Pat. 4361693	1982	Raychem
	144	335	A-11; Ru/C, DMF, DMSO; Cl	*J. Chem. Soc., Chem. Commun.*, (19), 1338	1985	ICI
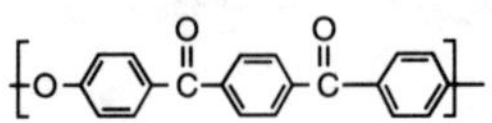	185	385	E-3; $AlCl_3$, o-Dichlorobenzene	Preprints, Japan-US Pol. Symp.	1985	DuPont
				Polym. Prep., **32**(2), 262, 264	1991	
	165	391	E-3 Blends	*J. Polym. Sci.*, B, *Polym. Phys.*, **25**, 311	1987	Union Carbide
		389	E-3; $AlCl_3$, $(CH_3)_2SO_2$, CH_2Cl_2	US Pat. 4826947	1989	Raychem
			E-4; BF_3, HF	US Pat. 4645819	1987	BASF

TABLE 6—*continued*

Polymer structure	Tg (°C)	Tm (°C)	Synthesis	Reference	Year	Organization
		384	N-9; K_2CO_3, Diphenyl sulfone; F	*Polym.*, **22**, 1096	1981	ICI
		384	N-9; K_2CO_3, Diphenyl sulfone; F	*Polym. Prep.*, **20**(1), 191	1979	ICI
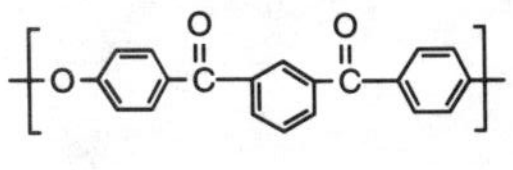			E-3; $AlCl_3$, $(CH_3)_2SO_2$, CH_2Cl_2, 1,2-DCE	US Pat. 4826947; *Polym.*, **34**, 1512	1989 1993	Raychem Toyobo Co.
	175	280	E-5; TFMSA	US Pat. 4361693	1982	
	139	285	N-9; Diphenyl sulfone; F. E-3; $AlCl_3$, Dichloroethane	*Bull. Soc. Chim. Belg.*, **98**, 667; *Gaofenzi Tongxun*, **6**, 442	1989 1984	ICI
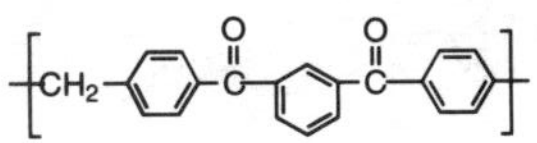			E-4; BF_3, HF	US Pat. 3441538	1969	DuPont
			E-5; TFMSA	US Pat. 3956240	1976	
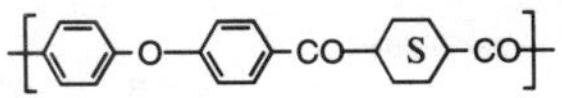	170	332	E-5; TFMSA, P_2O_5	US Pat. 4820794	1989	Eastman Kodak
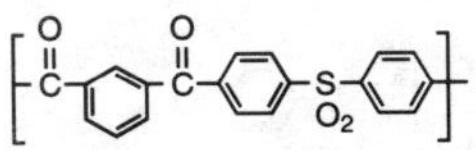	195	257	N-13; NaH, DMF; F	*Polym. Prep.*, **34**(1), 570	1993	Va. P. I. & State U.
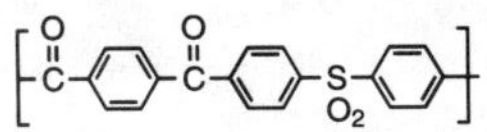	228	414	N-13; NaH, DMF; F	*Polym. Prep.*, **34**(1), 570	1993	Va. P. I. & U. State

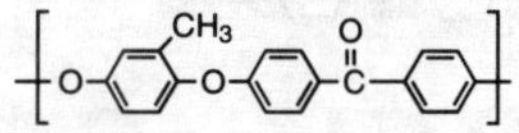	141	223	N-11; CsF, Bulk; F	*New Polym. Mater.*, **1**, 127	1988	Hamburg U.
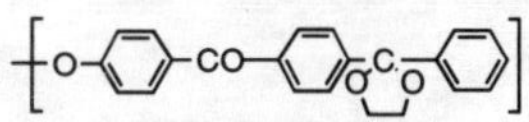 20% Ketalization		315	Post Ketalization on PEKK	US Pat. 3734888	1973	DuPont
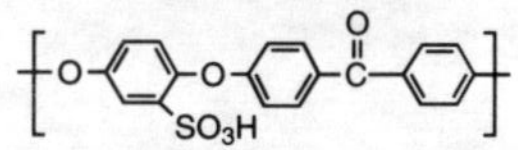	206		Sulfonation of PEEK	*Brit. Polym. J.*, **17**, 4	1985	U. Mass.
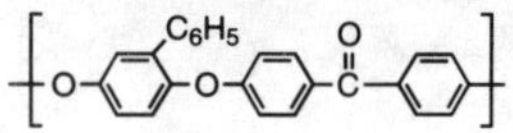	168		N-11; CsF, Bulk; F	*New Polym. Mater.*, **1**, 127	1988	Hamburg U.
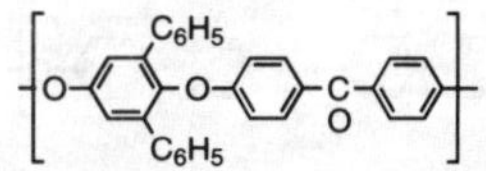	205		N-9; K_2CO_3, NMP; F	*J. M. S.-Pure Appl. Chem.*, **A29**(12), 1141	1992	McGill U.
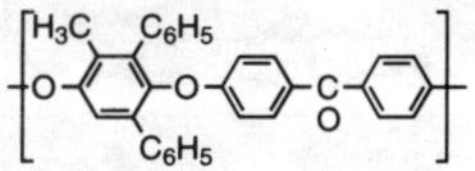	219		N-9; K_2CO_3, NMP; F	*J. M. S.-Pure Appl. Chem.*, **A29**(12), 1141	1992	McGill U.
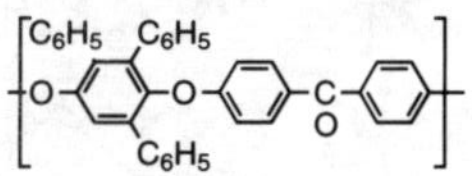	222		N-9; K_2CO_3, NMP; F	*J. M. S.-Pure Appl. Chem.*, **A29**(12), 1141	1992	McGill U.
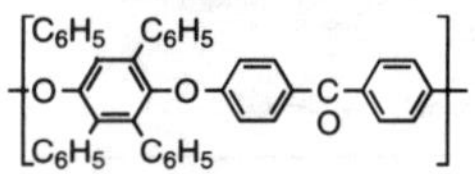	222	383	N-9; K_2CO_3, NMP; F	*J. M. S.-Pure Appl. Chem.*, **A29**(12), 1141	1992	McGill U.
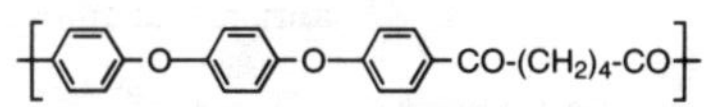	92	279	E-5; TFMSA, P_2O_5	US Pat. 4820794	1989	Eastman Kodak

TABLE 7
Polyaryletherketones containing four rings in the backbone of the repeat unit

Polymer structure	Tg (°C)	Tm (°C)	Synthesis	Reference	Year	Organization
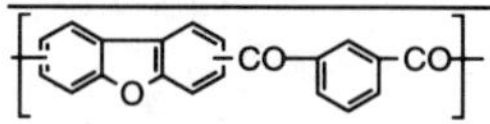			E-4; BF_3, HF	US Pat. 3441538	1969	DuPont
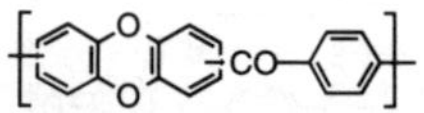	260	525	E-4; BF_3, HF	GB Pat. 1250251	1971	DuPont
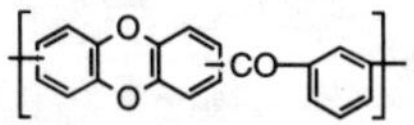	200	430	E-4; BF_3, HF	GB Pat. 1250251	1971	DuPont
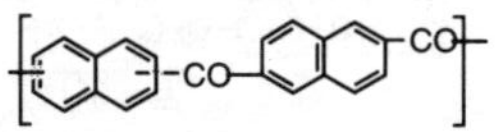			E-5; TFMSA, P_2O_5	US Pat. 4820794	1989	Eastman Kodak
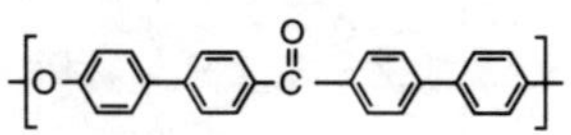	152	354	A-10; DMAC; Br	*J. Polym. Sci., Polym. Chem.*, **30**, 2511	1992	U. East Anglia
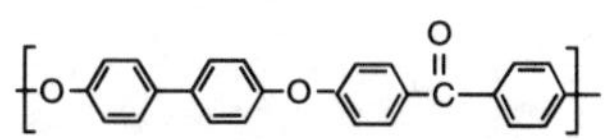	170	428	N-9; K_2CO_3, Diphenyl sulfone; F	*Polym. Prep.*, **32**(2), 166;	1991	Amoco
	167	416		*Polym.*, **18**, 365	1977	ICI
		421	N-11; Bulk; F	US Pat. 4331798;	1982	ICI
		422		*Polym.*, **25**(8), 1151	1984	ICI
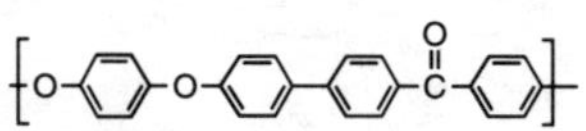		236–255	N-8; NaOH, DMSO; NO_2	*Makromol. Chem.*, **130**, 45	1969	Bayer A-G
	167	416	N-9; K_2CO_3, Diphenyl sulfone; F	*Polym.*, **22**, 1096	1981	ICI
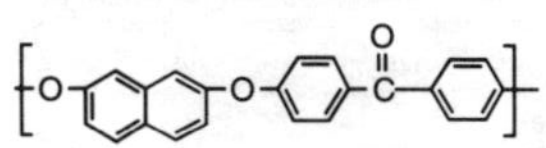	154	243	N-11; Bulk; F	*Polym.*, **25**(8),1151;	1984	Hamburg U.
	153	242		US Pat. 4331798	1982	ICI

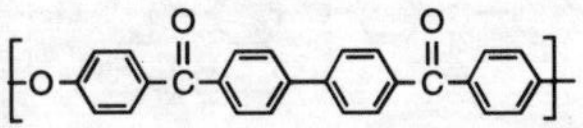			E-4; BF_3, HF E-5; TFMSA, P_2O_5 A-9; DMAC; Cl	US Pat. 3441538 US Pat. 4820794 *Macromol.*, **23**, 926	1969 1989 1990	DuPont Eastman Kodak Yamagata U.
		361				
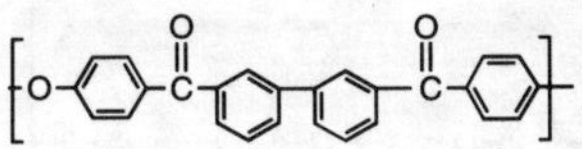	123		A-9; DMAC; Cl	*Macromol.*, **23**, 926	1990	Yamagata U.
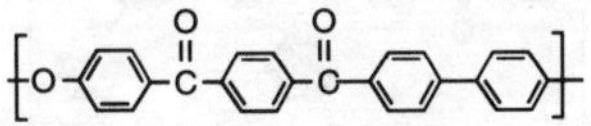		402	E-4; HF, BF_3	Rec. Adv. Mech. & Syn. Asp. of Polym., p. 207, Reidel.	1987	ICI
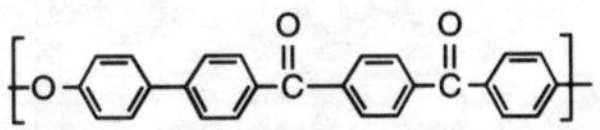		402	E-4; HF, BF_3	US Pat. 3956240	1976	Raychem
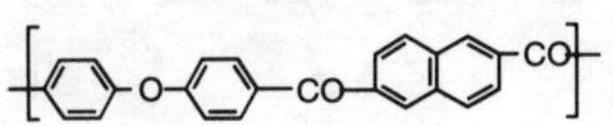	210	395	E-5; TFMSA, P_2O_5	US Pat. 4820794	1989	Eastman Kodak
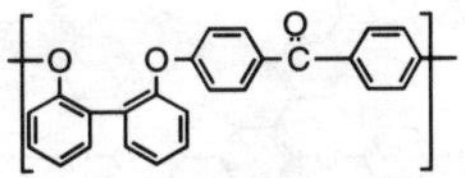	152		N-11; CsF, Bulk; F	*New Polym. Mater.*, **1**, 127	1988	Hamburg U.
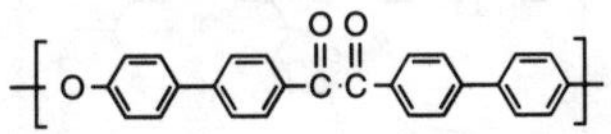	142	362–378	A-10; DMAC; Br	*J. Polym. Sci., Polym. Chem.*, **30**, 2511	1992	U. East Anglia
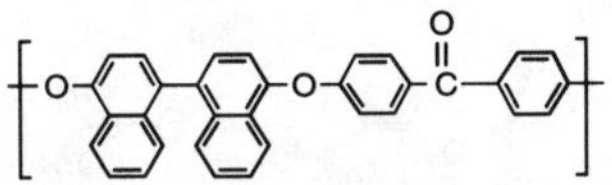	251		A-8; $FeCl_3$, Nitrobenzene	*Polym. Prep.*, **32**(1), 392; Eur. Pat. 282096	1991 1988	Case West. Reserve U.
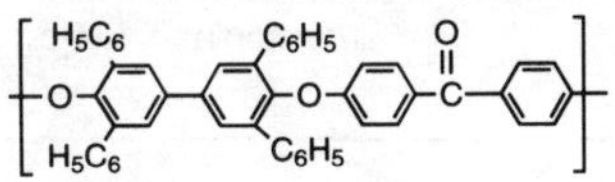	220	>400	N-9; K_2CO_3, NMP; F	*Polym. Prep.*, **32**(1), 389; *J. Polym. Sci.: Polym. Chem.*, **31**, 1907	1991 1993	McGill U.

TABLE 7—*continued*

Polymer structure	Tg (°C)	Tm (°C)	Synthesis	Reference	Year	Organization
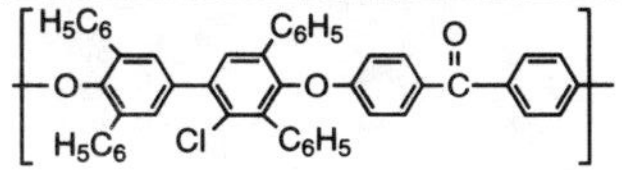	230		N-9; K_2CO_3, NMP; F	*Polym. Prep.*, **32**(1), 389	1991	McGill U.
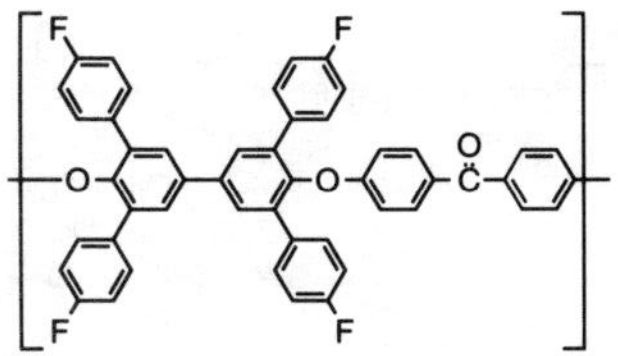	253		N-9; K_2CO_3, NMP; F	*J. Polym. Sci., Polym. Chem.*, **31**, 1907	1993	McGill
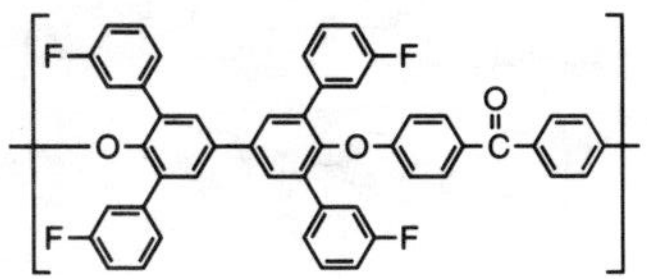	221		N-9; K_2CO_3, NMP; F	*J. Polym. Sci., Polym. Chem.*, **31**, 1907	1993	McGill
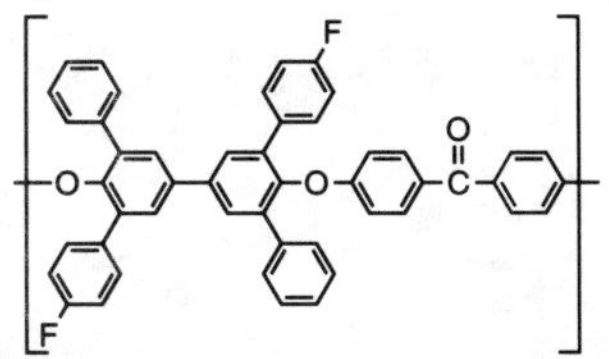		392	N-9; K_2CO_3, NMP; F	*J. Polym. Sci., Polym. Chem.*, **31**, 1907	1993	McGill
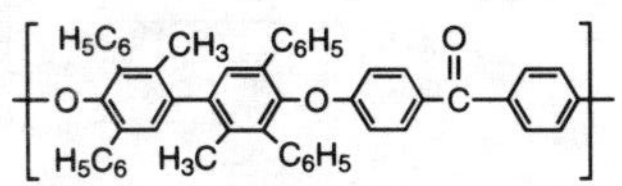	285		N-9; K_2CO_3, NMP; F	*Polym. Prep.*, **32**(1), 389	1991	McGill U.
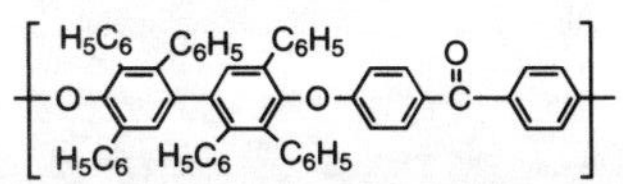	265		N-9; K_2CO_3, NMP; F	*Polym. Prep.*, **32**(1), 389	1991	McGill U.

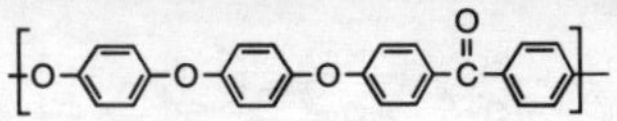	150	315	N-8; DMSO, Sulfolane; F.	*J.Polym. Sci.*, A-1, **5**, 2375;	1967	Union Carbide
	129	324	Blends	*J. Polym. Sci., B, Polym. Phys.*, **25**, 311	1987	
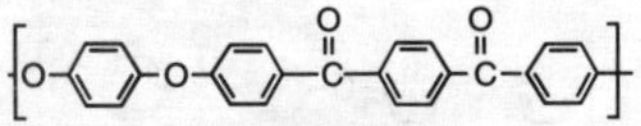	154	358	N-9; K_2CO_3, Diphenyl sulfone; F.	*Polym.*, **22**, 1096;	1981	ICI
Hostatec™ X915			Blends	*J. Polym. Sci., B, Polym. Phys.*, **25**, 311	1987	Union Carbide
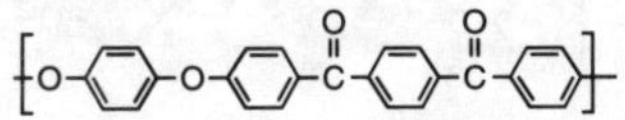		360	E-4; HF, BF_3	US Pat. 3956240	1976	Raychem
			E-5; TFMSA, P_2O_5	US Pat. 4820794	1989	Eastman Kodak
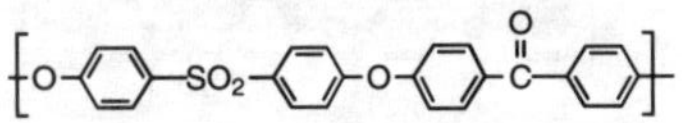	185		N-11, SiBP, Bulk; F	*Polym.*, **25**(8),1151	1984	Hamburg U.
	181			*Comp. Polym. Sci.* (Allen), **5**, 483	1989	ICI
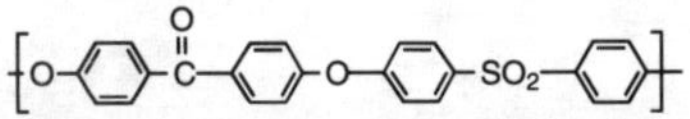	205		N-8; NaOH, DMSO, Sulfolane; F	*J.Polym.Sci.* ,A-1, **5**, 2375;	1967	Union Carbide
				Ency. Polym. Sci. & Eng., **13,**196	1988	
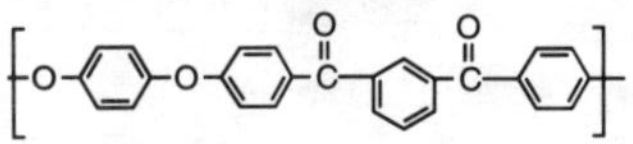	136	277	N-9; K_2CO_3, Diphenyl sulfone; F	*Bull. Soc. Chim. Belg.*, **98**, 667	1989	ICI
	153	271	E-4; HF, BF_3	US Pat. 3956240	1976	Raychem
	156	224	E-5; TFMSA, P_2O_5	US Pat. 4820794	1989	Eastman Kodak
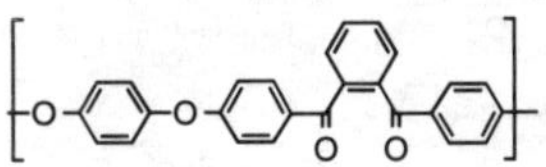	185		N-9; K_2CO_3, DMAC; F	*Macromol.*, **25**, 1017	1992	McGill U.
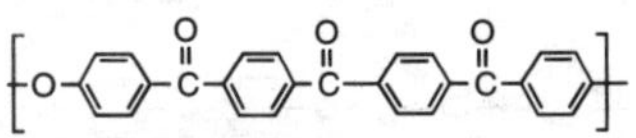		370-380	E-3; $AlCl_3$	GB Pat. 971227	1964	
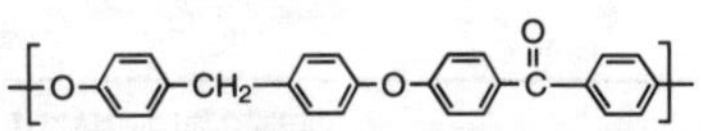	123			*Comp. Polym. Sci.* (Allen), **5**, 483	1989	ICI
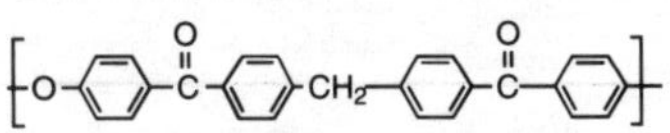		254–256	E-3; $AlCl_3$	GB Pat. 971227	1964	

TABLE 7—*continued*

Polymer structure	Tg (°C)	Tm (°C)	Synthesis	Reference	Year	Organization
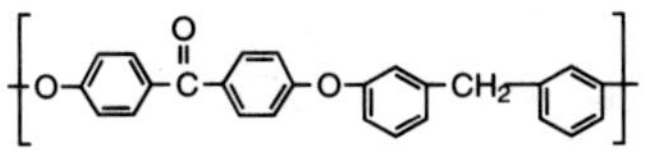	122		N-9; K_2CO_3, NMP; Cl	*Polym.*, **26**, 174	1985	NASA/ Langley
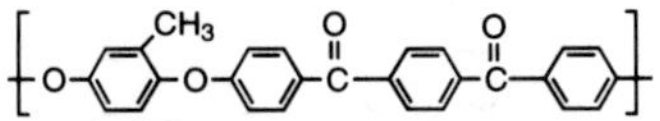	159	305	N-11; CsF, Bulk; F	*New Polym. Mater.*, **1**, 127	1988	Hamburg U.
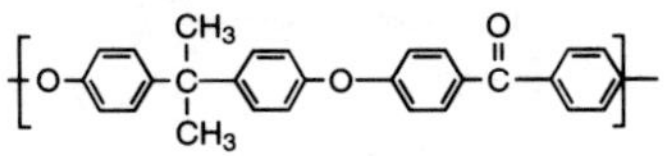	155	(248)	N-8; NaOH, DMSO, Sulfolane; F	*J. Polym. Sci.*, A-1, **5**, 2375;	1967	Union Carbide
				Ency. Polym. Sci. & Eng. **13**,196;	1988	
		190–220		*Makromol. Chem.*, **130**, 45	1969	Bayer A-G
	155		N-9; K_2CO_3, DMAC; Cl	*Polym.*, **29**, 358	1988	NASA/ Langley
	154		N-11; Bulk; F	*Polym.*, **25**(8),1151; US Pat. 4331798	1984 1982	Hamburg U.
	152					ICI
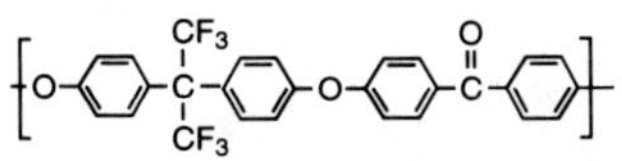	175		N-8; NaOH, DMSO, Sulfolane; F	*J.Polym.Sci.*, A-1, **5**, 2375	1967	Union Carbide
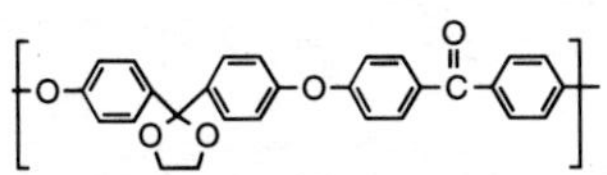	144		N-9; K_2CO_3, DMAC, Sulfolane; F	*Macromol.*, **20**, 1204; US Pat. 4882397	1987 1989	Amoco
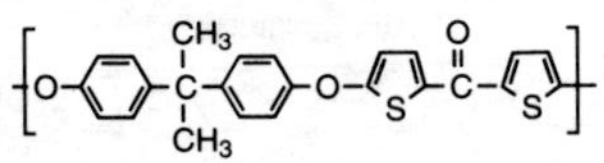	124		N-9; K_2CO_3, NMP; Cl	*Polym. Prep.*, 33(1), 418; **33**(2), 170	1992	U. No. Carolina
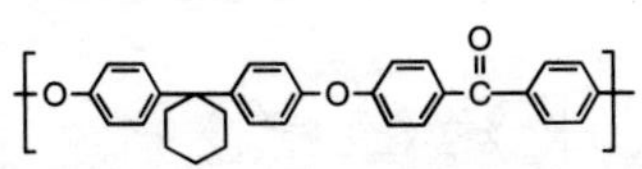	175		N-11; Bulk; F	*Polym.*, **25**(8),1151	1984	Hamburg U.

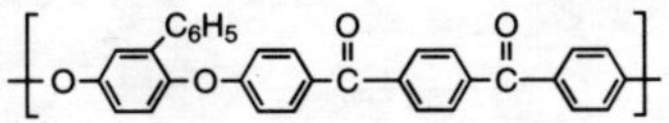	126	N-11; CsF, Bulk; F	*New Polym. Mater.*, **1**, 127	1988	Hamburg U.
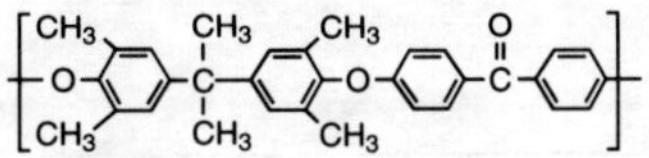	145 144	N-11; Bulk; F	*Polym.*, **25**(8),1151; US Pat. 4331798	1984 1982	Hamburg U. ICI
		N-9; K_2CO_3, NMP; F	*Polym. Prep.*, **25**(2), 19;	1984	Va. P. I. & State U.
	235		*J. Polym. Sci., Polym. Chem. Ed.*, **20**, 609	1982	U. Arizona
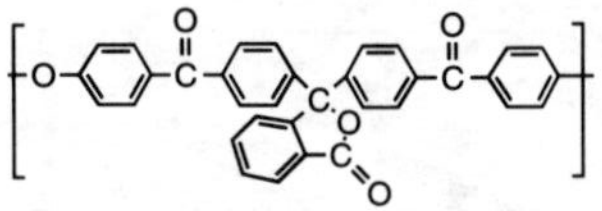		E-3; $AlCl_3$, Dichlorethane	*Acta Polymerica*, **39**, 452	1988	F. Schiller Univ., Jena
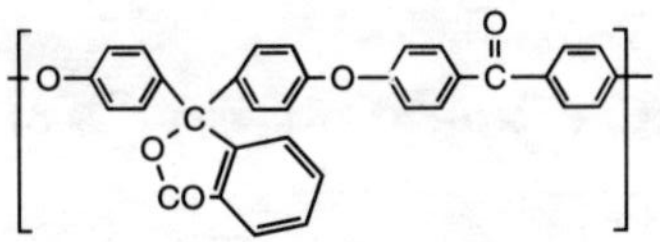	214	N-11; CsF, Bulk; F	*New Polym. Mater.*, **1**, 127	1988	Hamburg U.
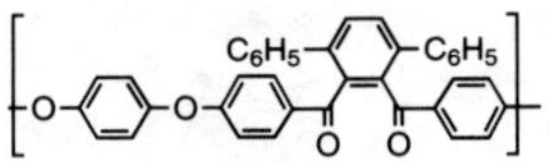	240	N-9; K_2CO_3, DMAC; F	*Macromol.*, **25**, 1017	1992	McGill U.
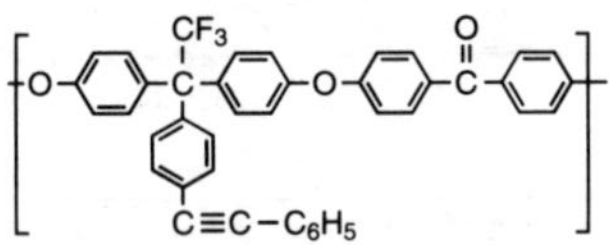		N-9; K_2CO_3, DMAC; F	*Polym. Prep.*, **34**(1), 461	1993	NASA/ Langley Hampton U.
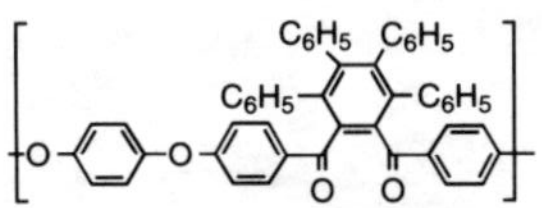	273	N-9; K_2CO_3, DMAC; F	*J. Polym. Sci., Polym. Chem.*, **29**, 1045	1991	McGill U.

TABLE 7—*continued*

Polymer structure	Tg (°C)	Tm (°C)	Synthesis	Reference	Year	Organization
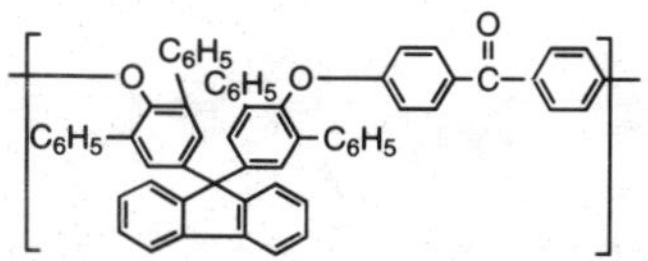	260		N-9; K_2CO_3, DMAC; F	*J. Polym. Sci., Polym. Chem.*, **29**, 1045	1991	McGill U.
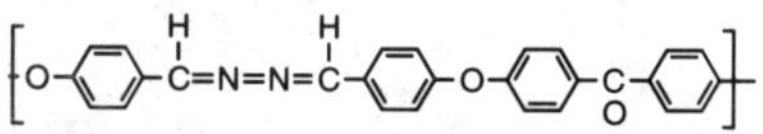	179		N-9; K_2CO_3, DMPU; F	*Polym. Prep.*, **34**(1), 521	1993	IBM
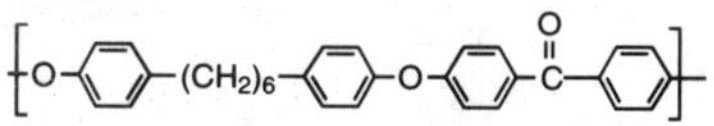	66	207	N-11; CsF, Bulk; F	*Macromol.*, **22**, 517	1989	Hamburg U.
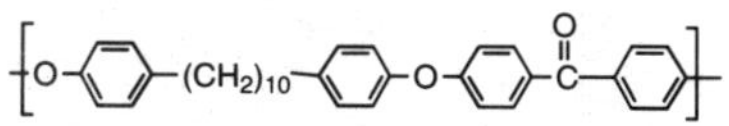	41	176	N-11; CsF, Bulk; F	*Macromol.*, **22**, 517	1989	Hamburg U.
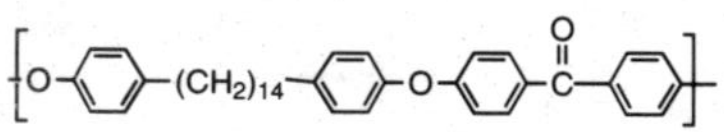	22	130	N-11; CsF, Bulk; F	*Macromol.*, **22**, 517	1989	Hamburg U.

TABLE 8

Polyaryletherketones containing five rings in the backbone of the repeat unit

Polymer structure	Tg (°C)	Tm (°C)	Synthesis	Reference	Year	Organization
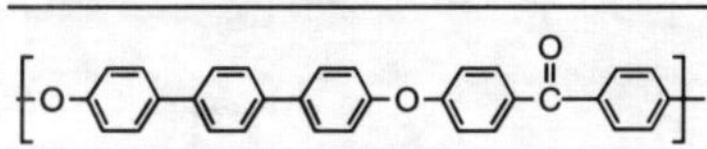		Highly crystalline	N-9; K_2CO_3, Diphenyl sulfone; F	*Bull. Soc. Chim. Belg.*, **98**, 667	1989	ICI
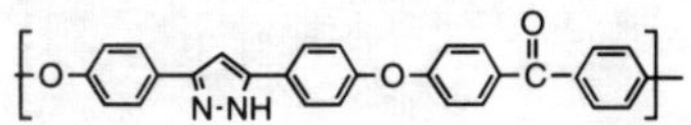	229		N-9; K_2CO_3, DMAC, NMP; Cl	*Polym. Prep.*, **32**(1), 619	1991	Va. Comm. U.
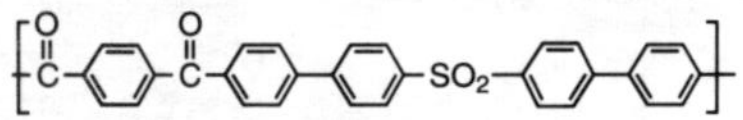			E-5; TFMSA, P_2O_5	US Pat. 4861856	1989	Eastman Kodak
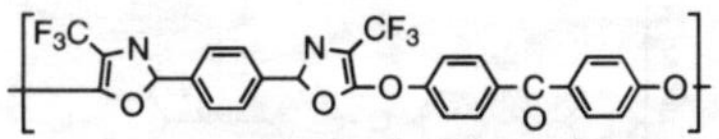	167	278	N-9; $(Et)_3N$, NMP; F	*Polym. Prep.*, **34**(1), 429	1993	U. Bayreuth
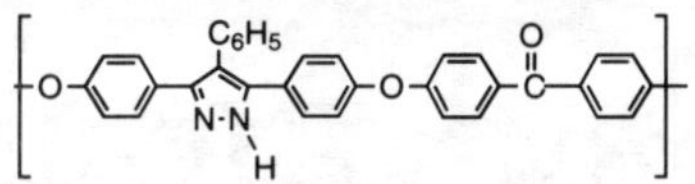	208		N-9; K_2CO_3, DMAC; Cl	*Polym. Prep.*, **32**(2), 160	1991	Va. Comm. U.; NASA/ Langley
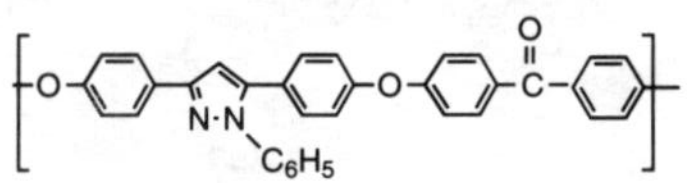	205		N-9; K_2CO_3, DMAC, NMP; F	*Polym. Prep.*, **32**(1), 619	1991	Va. Comm. U.
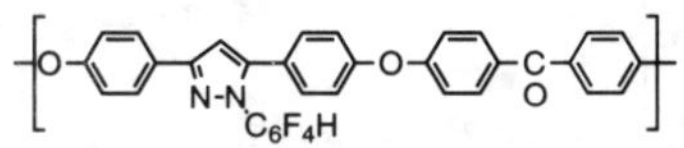	204		N-9; K_2CO_3, DMAC; F	*Polym. Prep.*, **34**(1), 441	1993	Va. Comm. U.
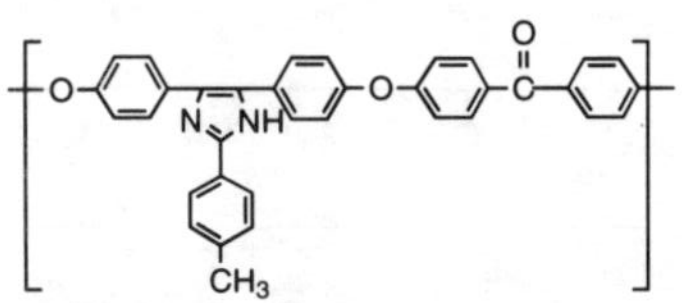	254		N-9; K_2CO_3, DMAC; F	*Polym. Prep.*, **32**(2), 162	1991	NASA/ Langley

TABLE 8—*continued*

Polymer structure	Tg (°C)	Tm (°C)	Synthesis	Reference	Year	Organization
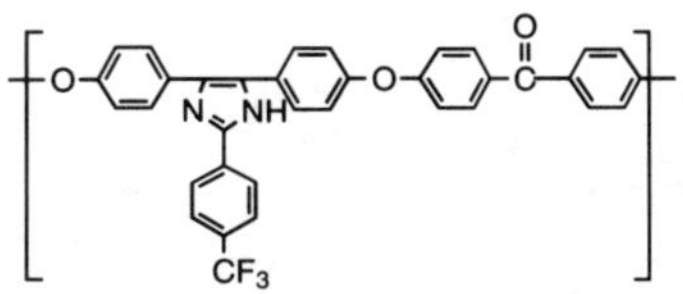	263		N-9; K_2CO_3, DMAC; F	*Polym. Prep.*, **32**(2), 162	1991	NASA/ Langley
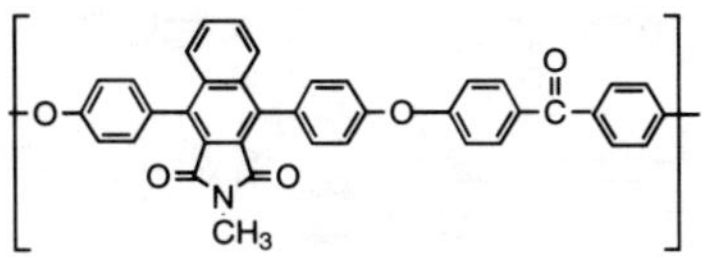	283		N-9; K_2CO_3, NMP, Sulfolane; F	*Polym. Prep.*, **32**(2), 164	1991	McGill U.
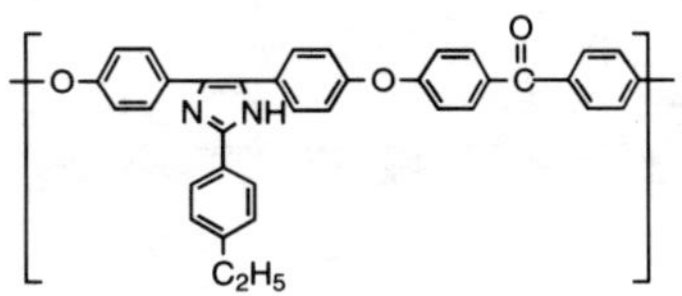	243		N-9; K_2CO_3, DMAC; F	*Polym. Prep.*, **32**(2), 162	1991	NASA/ Langley
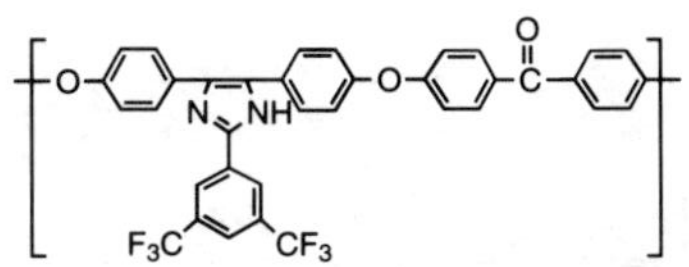	270		N-9; K_2CO_3, DMAC; F	*Polym. Prep.*, **32**(2), 162	1991	NASA/ Langley
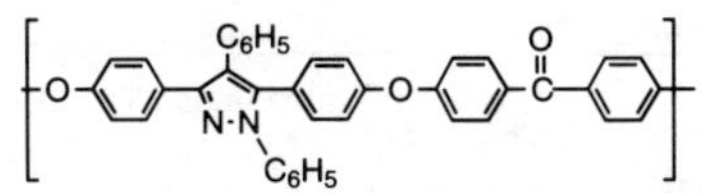	205		N-9; K_2CO_3, DMAC; F	*Polym. Prep.*, **32**(2), 160	1991	Va. Comm. U.; NASA/ Langley
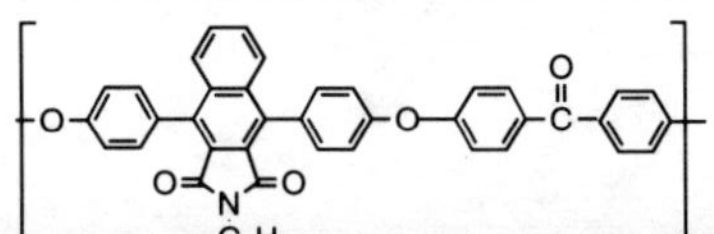	275		N-9; NMP, Sulfolane; F	*Polym. Prep.*, **32**(2), 164	1991	McGill U.

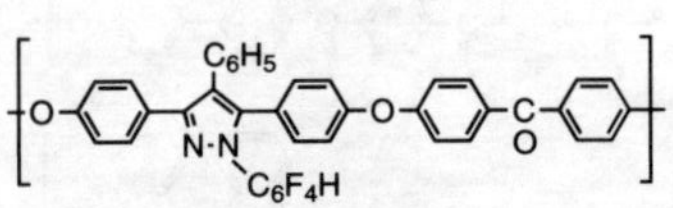	213		N-9; K_2CO_3, DMAC; F	*Polym. Prep.*, 34(1), 441	1993	Va. Comm. U.
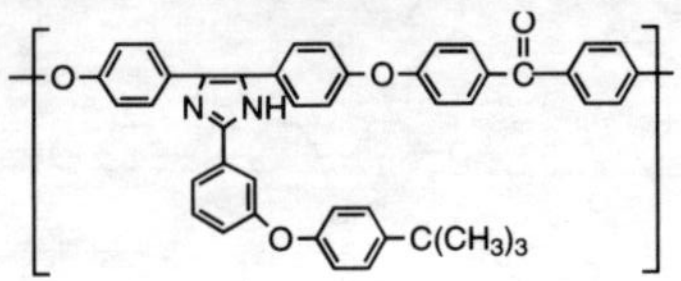	217		N-9; K_2CO_3, DMAC; F	*Polym. Prep.*, **32**(2), 162	1991	NASA/ Langley
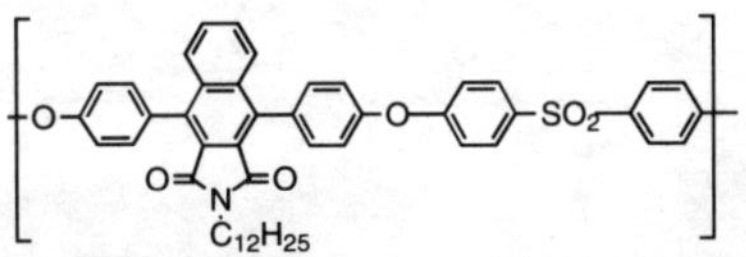	147		N-9; NMP, Sulfolane; F	*Polym. Prep.*, **32**(2), 164	1991	McGill U.
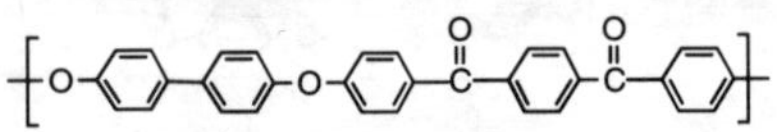	168	409	N-9	*SAMPE*, **32**, 382	1987	ICI
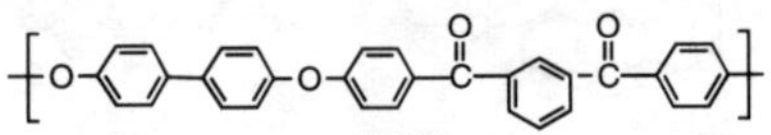 PK 99	160	305	N-9; K_2CO_3, Diphenyl sulfone; F	*Bull. Soc. Chim. Belg.*, **98**, 667	1989	ICI
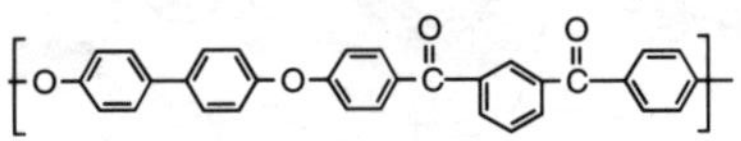			N-11; Cs Phenate, Bulk, Phenol	US Pat. 4794155	1988	Dow
			E-5; TFMSA, P_2O_5	US Pat. 4820794	1989	Eastman Kodak
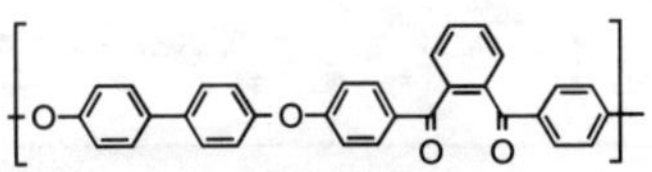	190		N-9; K_2CO_3, DMAC; F	*Macromol.*, **25**, 1017	1992	McGill U.
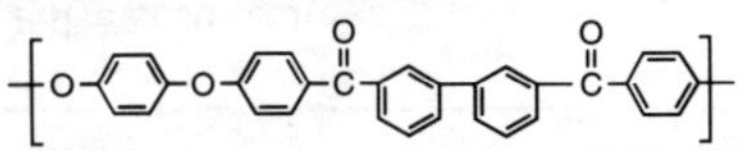	154		A-9; DMAC; Cl	*Macromol.*, **23**, 926	1990	Yamagata U.

TABLE 8—*continued*

Polymer structure	Tg (°C)	Tm (°C)	Synthesis	Reference	Year	Organization
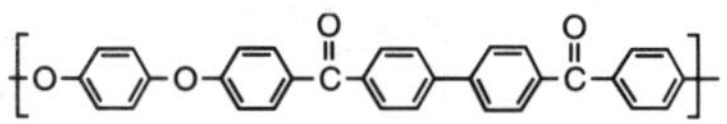	192 180	428 411 323	E-5 ; TFMSA E-5; TFMSA, P_2O_5 A-9; DMAC; Cl	*Polym. Prep.*, **25**(2),17; *Polym.*, **29**, 1902; US Pat. 4820794 *Macromol.*, **23**, 926	1984 1988 1989 1990	ICI Eastman Kodak Yamagata U.
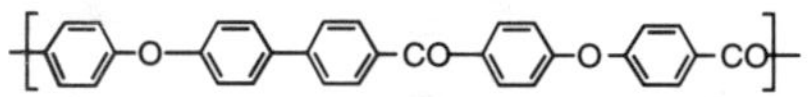		300–350	E-4; BF_3, HF	US Pat. 3956240	1976	Raychem
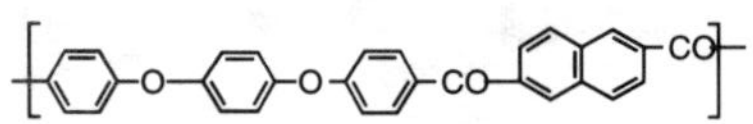	184	361	E-5; TFMSA, P_2O_5	US Pat. 4820794	1989	Eastman Kodak
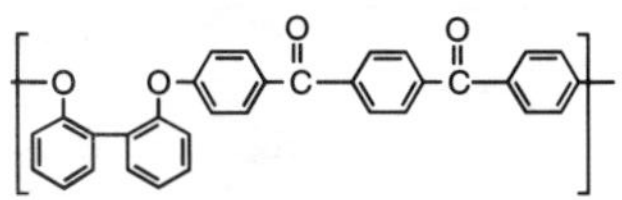	150		N-11; CsF, Bulk; F	*New Polym. Mater.*, **1**, 127	1988	Hamburg U.
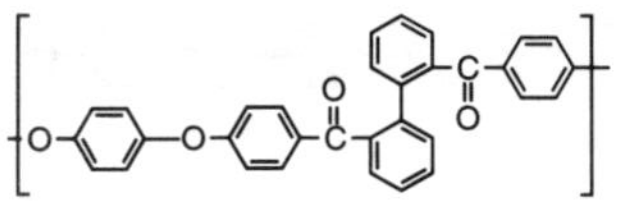	175		N-9; K_2CO_3, NMP; F	*Polym. Prep.*, **33**(2), 351	1992	Carleton U.
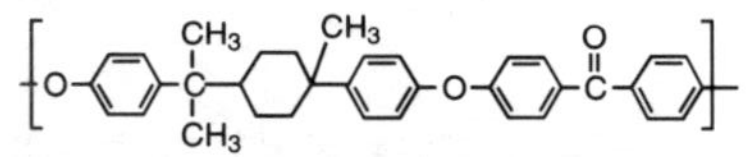	200		N-8; NaOH, DMSO, Sulfolane; F	*J. Polym. Sci.*, A-1, **5**, 2375	1967	Union Carbide
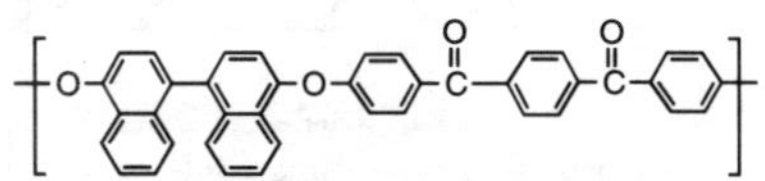			A-8; $FeCl_3$, Nitrobenzene	*Polym. Prep.*, **32**(1), 391	1991	Case West. Reserve U.
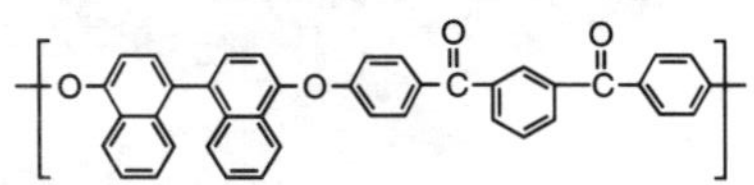			A-8; $FeCl_3$, Nitrobenzene	*Polym. Prep.*, **32**(1), 391	1991	Case West. Reserve U.

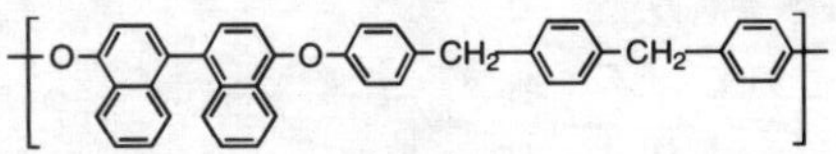			A-8; $FeCl_3$, Nitrobenzene	*Polym. Prep.*, **32**(1), 391	1991	Case West. Reserve U.
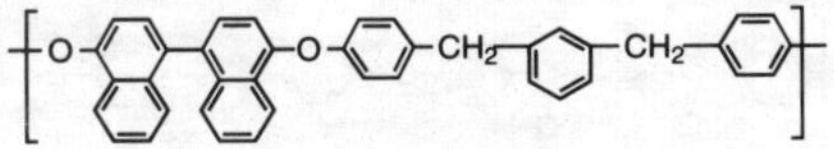			A-8; $FeCl_3$, Nitrobenzene	*Polym. Prep.*, **32**(1), 391	1991	Case West. Reserve U.
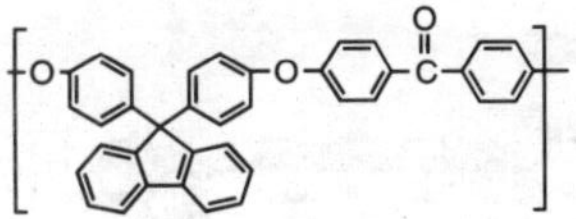	252 251		N-9; K_2CO_3, DMAC; Cl	*Polym.*, **29**, 358; US Pat. 4806618	1988 1989	NASA/ Langley Central Glass (Jap.)
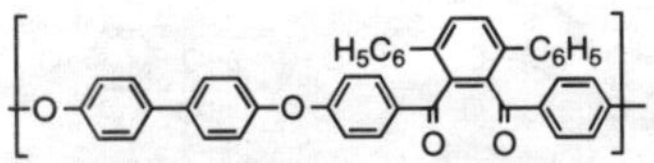	240		N-9; K_2CO_3, DMAC; F	*Macromol.*, **25**, 1017	1992	McGill U.
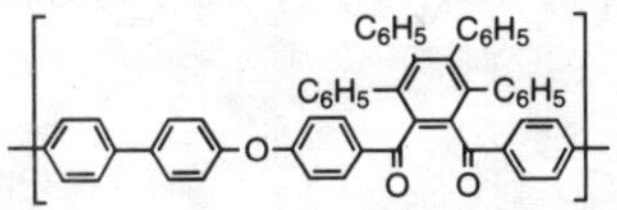	292		N-9; K_2CO_3, DMAC; F	*J. Polym. Sci., Polym. Chem.*, **29**,1045	1991	McGill U.
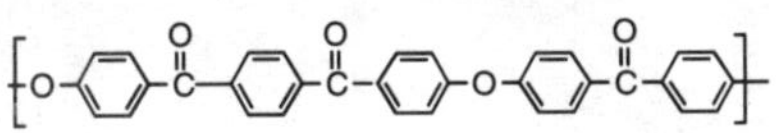		383	N-9; K_2CO_3, Diphenyl sulfone; F	*Polym.*, **22**, 1096;	1981	ICI
	160	384	Blends	*J. Polym. Sci.*, B, *Polym. Phys.*, **25**, 311	1987	Union Carbide
	175	375	E-3; $AlCl_3$, CH_2Cl_2, Dimethyl sulfone	Ger. Pat. 3829520	1990	BASF
		370	E-4; HF, BF_3 E-3; $AlCl_3$, $(CH_3)_2SO_2$, CH_2Cl_2; H E-5; TFMSA, P_2O_5	US Pat. 3956240; US Pats. 4843131; 4912181 US Pat. 4861856	1976 1989 1990 1989	Raychem Eastman Kodak

TABLE 8—*continued*

Polymer structure	Tg (°C)	Tm (°C)	Synthesis	Reference	Year	Organization
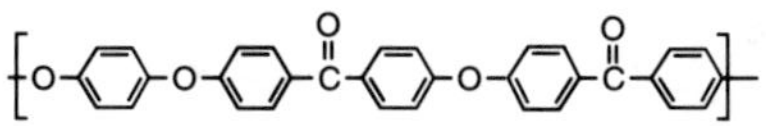	143	336	N-9; K_2CO_3, CHP; F	*J. Polym. Sci., Polym. Chem.*, **27**, 661	1989	Dow
		340	E-4; HF, BF_3	US Pat. 3956240; GB Pat. 3956240	1976 1976	Raychem
	151	345	E-5 ; TFMSA Blends	GB Pat. 2116990; *J. Polym. Sci.*, B, *Polym. Phys.*, **25**,311	1983 1987	ICI Union Carbide
	165	352	E-5; P_2O_5, CH_3SO_3H	*Macromol.*, **20**, 2675	1987	Yamagata U.
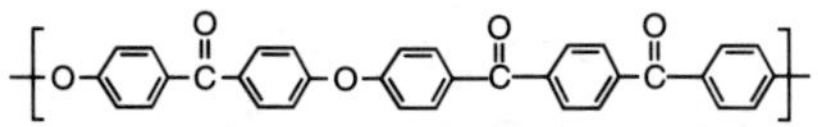		375		*Kunst. Ger. Plastics*, **79**, 75	1989	BASF
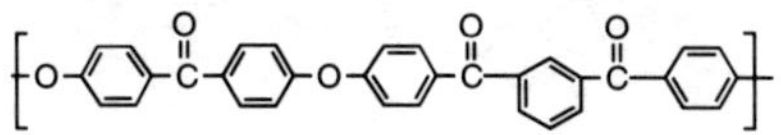	152	285	N-9; K_2CO_3, Diphenyl sulfone; F	*Bull. Soc. Chim. Belg.*, **98**, 667	1989	ICI
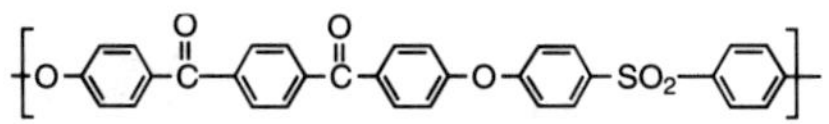	175		N-9; K_2CO_3, NMP; Cl	CA, **96**, 7266 Ger. Pat. 3014230	1982	BASF
	196	(214)	E-5; TFMSA, P_2O_5	US Pat. 4861856	1989	Eastman Kodak
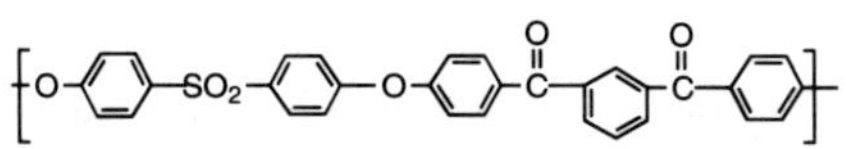	172		N-9; K_2CO_3, Diphenyl sulfone; F	*Bull. Soc. Chim. Belg.*, **98**, 667	1989	ICI
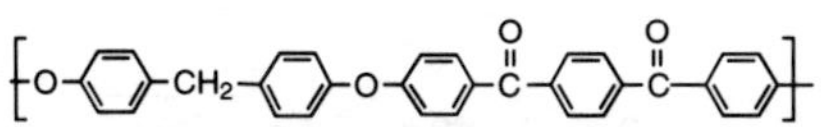	120	275	N-9; K_2CO_3, DMAC; Cl	*Polym.*, **29**, 358	1988	NASA/ Langley
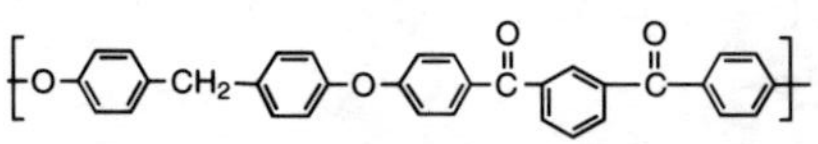	145		N-9; K_2CO_3, DMAC; Cl	*Polym.*, **29**, 358	1988	NASA/ Langley

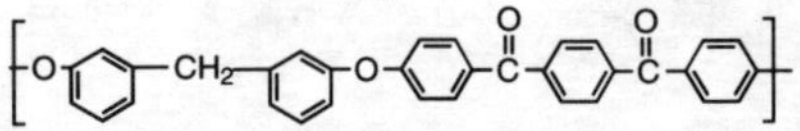	122		N-9; K_2CO_3, DMAC; Cl	*Polym.*, **29**, 358	1988	NASA/ Langley
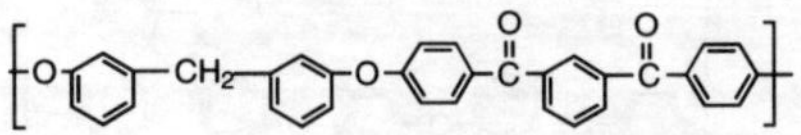	114		N-9; K_2CO_3, DMAC; Cl	*Polym.*, **29**, 358	1988	NASA/ Langley
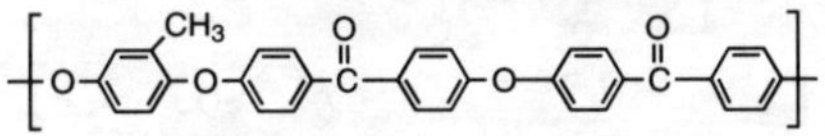	150	259	N-11; CsF, Bulk; F	*New Polym. Mater.*, **1**, 127	1988	Hamburg U.
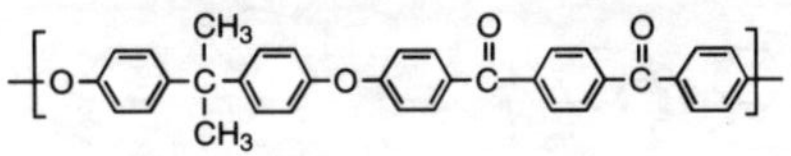	165		N-8; NaOH, DMSO, Sulfolane; Cl, F.	*J. Polym. Sci.*, A-1, **5**, 2375	1967	Union Carbide
	166		N-9; K_2CO_3, NMP; F	*Polym. Prep.*, **32**(2), 172	1991	U. No. Carolina
	170		N-9; K_2CO_3, NMP, DMAC; Cl	*Polym.*, **26**,174; **29**, 358	1985 1988	NASA/ Langley
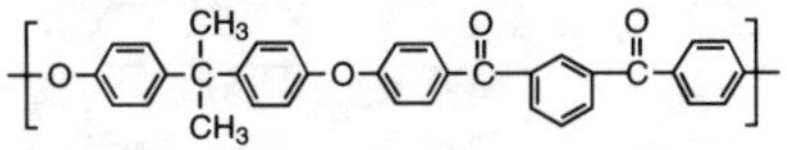	152		N-9; K_2CO_3 , NMP; F	*Polym. Prep.*, **32**(2) 172	1991	U. No. Carolina
	152		N-9; K_2CO_3, DMAC; Cl, F	*Polym.*, **29**, 358	1988	NASA/ Langley
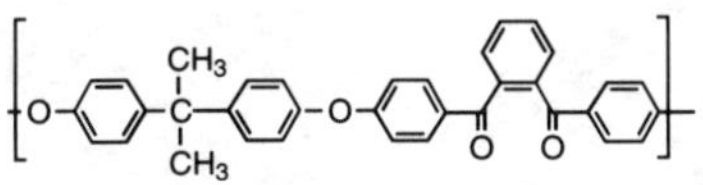	160, 182		N-9; K_2CO_3, DMAC; F	*Macromol*, **24**, 2637	1991	McGill U.
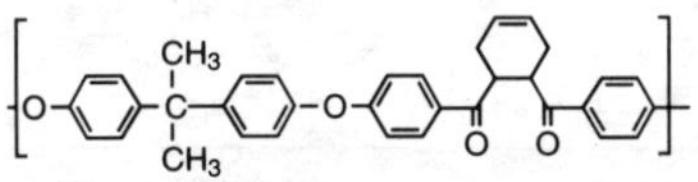	188		N-9; K_2CO_3, DMAC; F	*Macromol*, **24**, 2637	1991	McGill U.
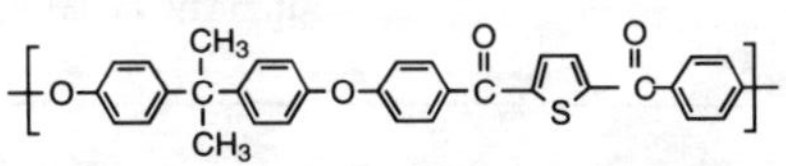	158		N-9; K_2CO_3, NMP; F	*Polym. Prep.*, **32**(2) 172	1991	U. No. Carolina

TABLE 8—*continued*

Polymer structure	Tg (°C)	Tm (°C)	Synthesis	Reference	Year	Organization
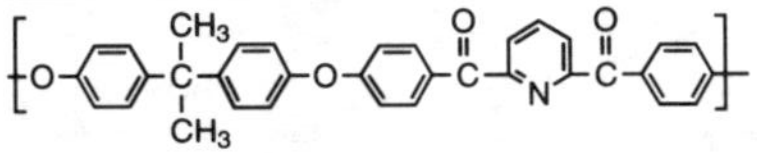	132		N-9; K_2CO_3, DMAC; F	*Polym.*, **29**, 358	1988	NASA/ Langley
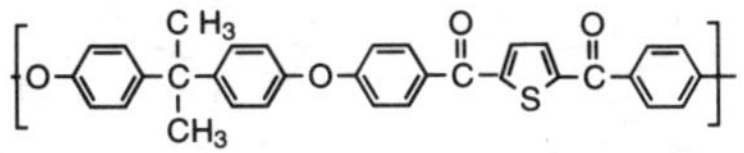	147		N-9; K_2CO_3, NMP; F	*Polym. Prep.*, **33**(1), 485	1992	U. Florida, U. No. Carolina
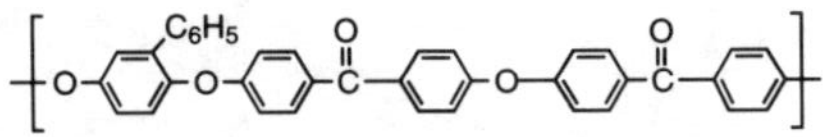	146		N-11; CsF, Bulk; F	*New Polym. Mater.*, **1**, 127	1988	Hamburg U.
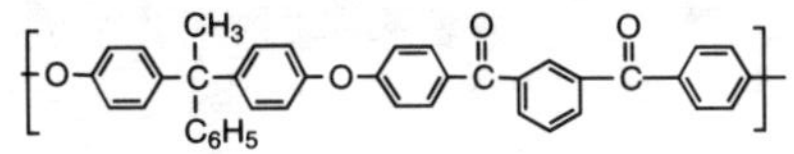	172		N-9; K_2CO_3, DMAC; Cl	*Polym.*, **29**, 358	1988	NASA/ Langley
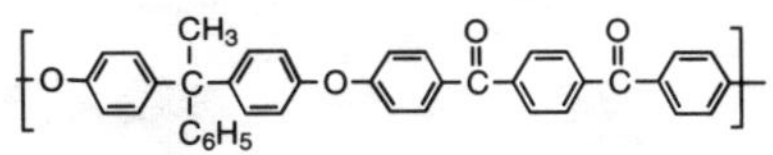	189		N-9; K_2CO_3, DMAC; Cl	*Polym.*, **29**, 358	1988	NASA/ Langley
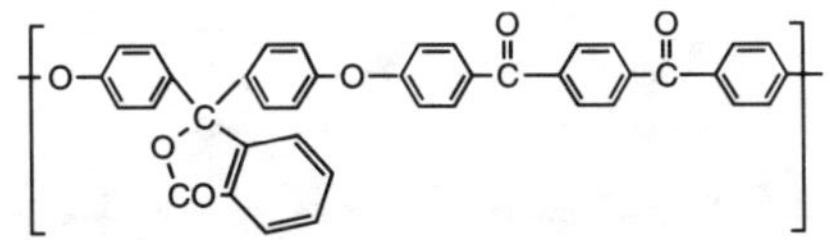	210		N-11; CsF, Bulk; F	*New Polym. Mater.*, **1**, 127	1988	Hamburg U.
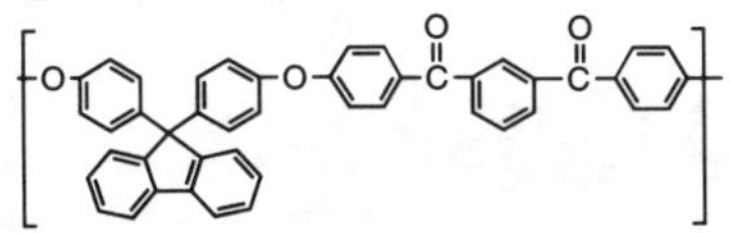	223		N-9; K_2CO_3, DMAC; Cl	*Polym.*, **29**, 358	1988	NASA/ Langley
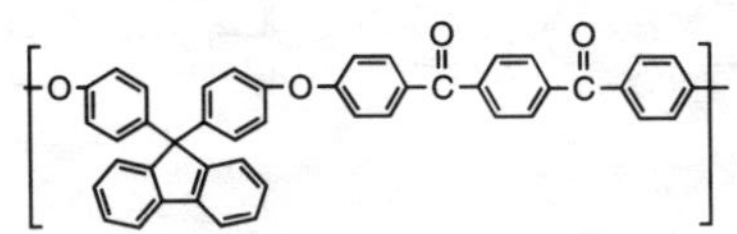	243		N-9; K_2CO_3, DMAC; Cl	*Polym.*, **29**, 358	1988	NASA/ Langley

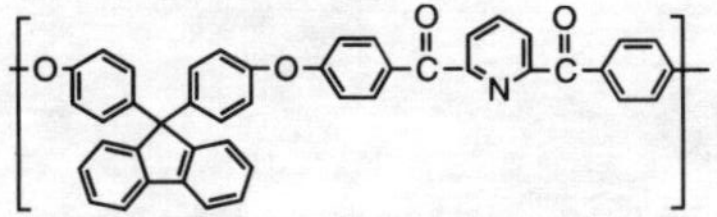	223	N-9; K_2CO_3, DMAC; F	*Polym.*, **29**, 358	1988	NASA/ Langley
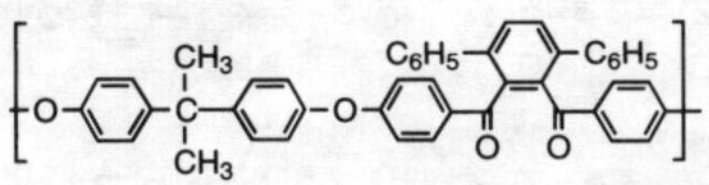	221	N-9; K_2CO_3, DMAC; F	*Macromol*, **24**, 2637	1991	McGill U.
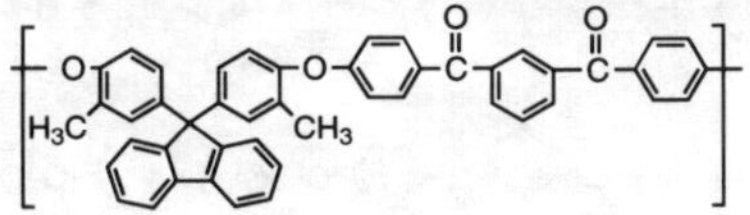	218	N-9; K_2CO_3, DMAC; Cl, F	*Polym.*, **29**, 358	1988	NASA/ Langley
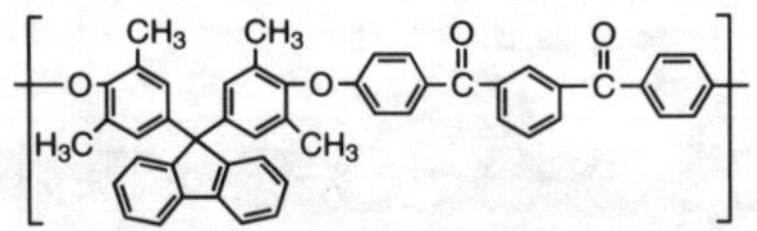	257	N-9; K_2CO_3, DMAC; Cl	*Polym.*, **29**, 358	1988	NASA/ Langley
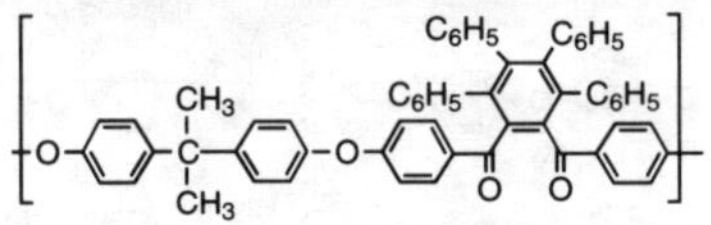	268	N-9; K_2CO_3, DMAC; F	*Macromol*, **24**, 2637	1991	McGill U.
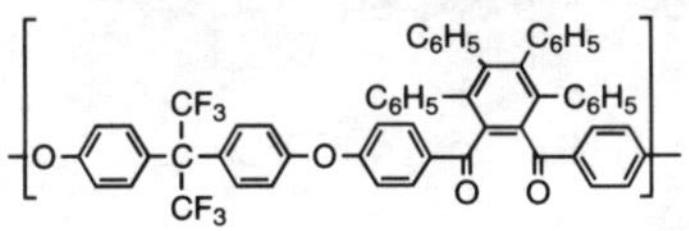	278	N-9; K_2CO_3, DMAC; F	*J. Polym. Sci., Polym. Chem.*, **29**, 1045	1991	McGill U.
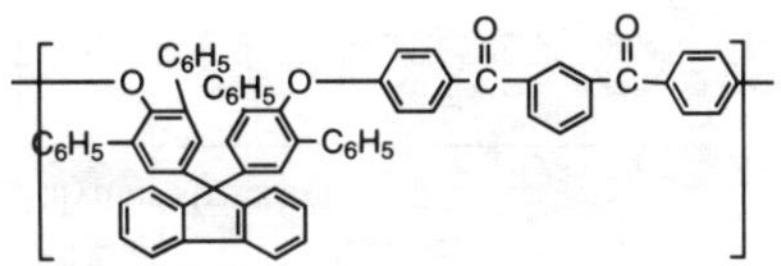	236	N-9; K_2CO_3, DMAC; F	*J. Polym. Sci., Polym. Chem.*, **29**,1045	1991	McGill U.
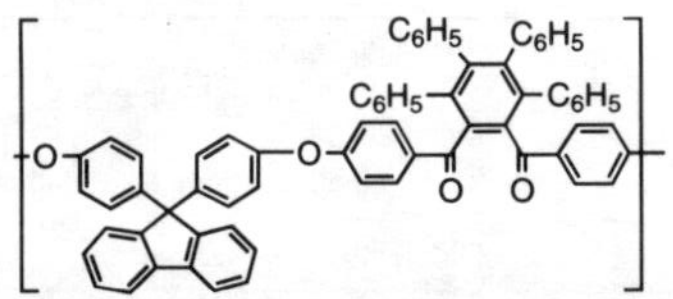	313	N-9; K_2CO_3, DMAC; F	*J. Polym. Sci., Polym. Chem.*, **29**,1045	1991	McGill U.

TABLE 9
Polyaryletherketones containing six rings in the backbone of the repeat unit

Polymer structure	Tg (°C)	Tm (°C)	Synthesis	Reference	Year	Organization
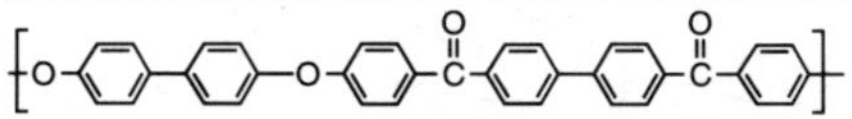	209	469	E-5; TFMSA	*Polym. Prep.*, **25**(2),17; *Polym.*, **29**,1902	1984 1988	ICI
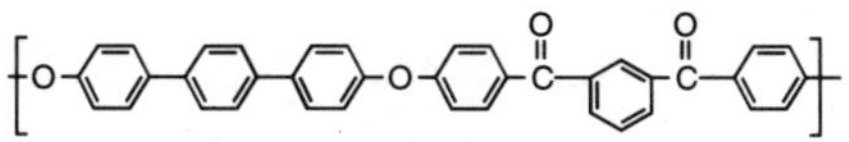	174	394	N-9; K_2CO_3, Diphenyl sulfone; F	*Bull. Soc. Chim. Belg.*, **98**, 667	1989	ICI
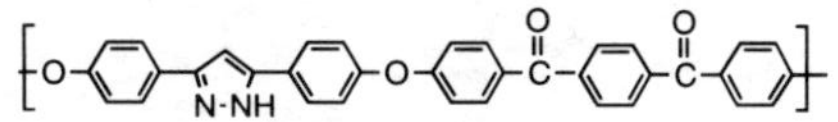	235		N-9; K_2CO_3, DMAC, NMP; F	*Polym. Prep.*, **32**(1), 619	1991	Va. Comm. U.
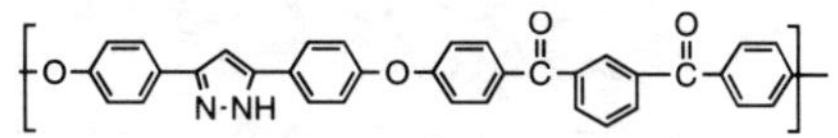	205		N-9; K_2CO_3, DMAC, NMP; Cl	*Polym. Prep.*, **32**(1), 619	1991	Va. Comm. U.
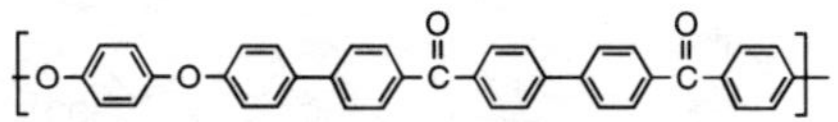	200	407	E-5; TFMSA	GB Pat. 2116990	1983	ICI
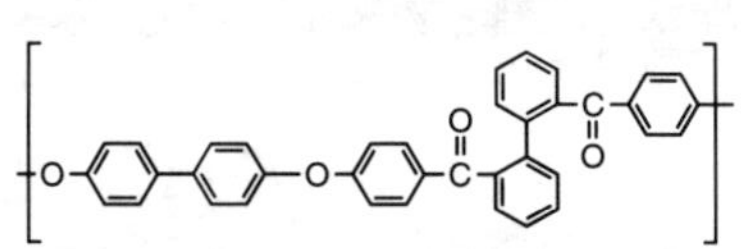	193		N-9; K_2CO_3, NMP; F	*Polym. Prep.*, **33**(2), 351	1992	Carleton U.
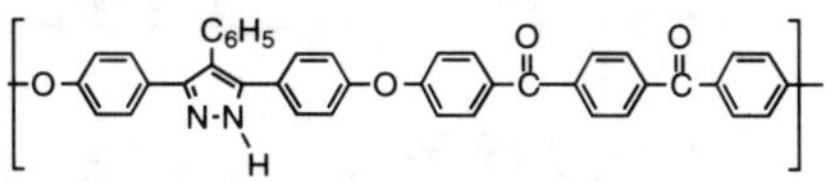	200		N-9; K_2CO_3, DMAC; Cl	*Polym. Prep.*, **32**(2), 160	1991	Va. Comm. U.; NASA/Langley
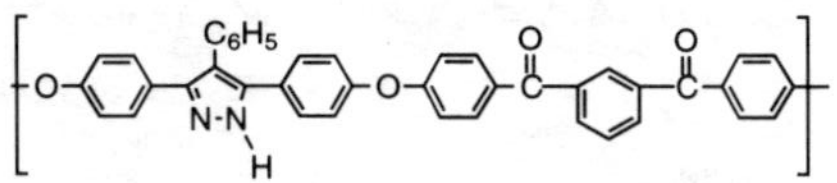	188		N-9; K_2CO_3, DMAC; Cl	*Polym. Prep.*, **32**(2), 160	1991	Va. Comm. U.; NASA/Langley

Structure					
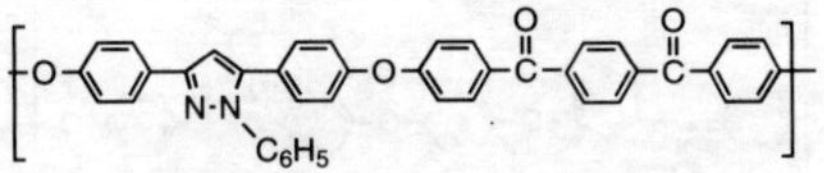	205	N-9; K_2CO_3, DMAC, NMP; F	*Polym. Prep.*, **32**(1), 619	1991	Va. Comm. U.
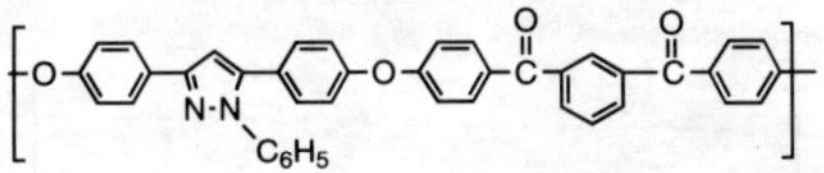	190	N-9; K_2CO_3, DMAC, NMP; F	*Polym. Prep.*, **32**(1), 619	1991	Va. Comm. U.
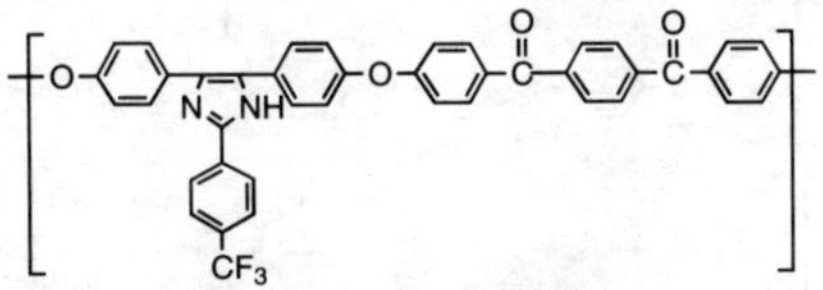	249	N-9; K_2CO_3, DMAC; F	*Polym. Prep.*, **32**(2), 162	1991	NASA/Langley
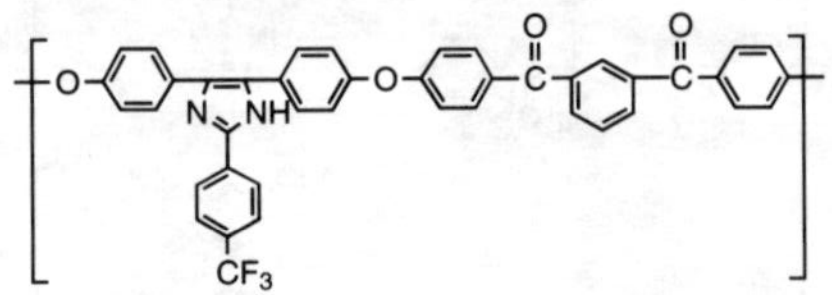	233	N-9; K_2CO_3, DMAC; F	*Polym. Prep.*, **32**(2), 162	1991	NASA/Langley
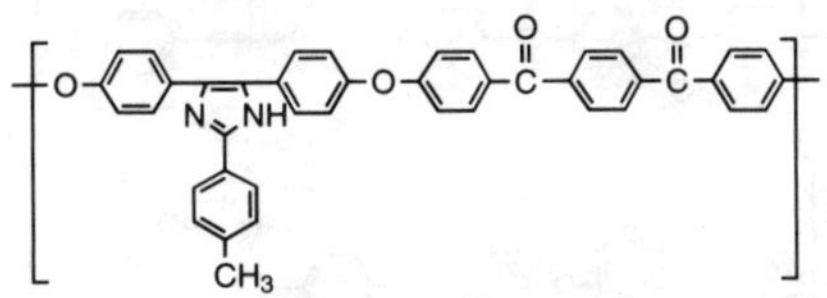	245	N-9; K_2CO_3, DMAC; F	*Polym. Prep.*, **32**(2), 162	1991	NASA/Langley
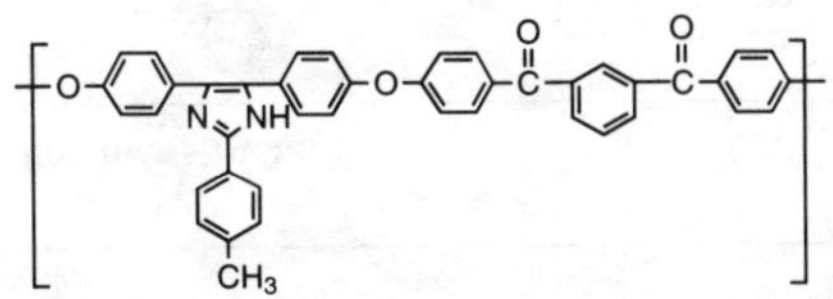	227	N-9; K_2CO_3, DMAC; F	*Polym. Prep.*, **32**(2), 162	1991	NASA/Langley

TABLE 9—*continued*

Polymer structure	Tg (°C)	Tm (°C)	Synthesis	Reference	Year	Organization
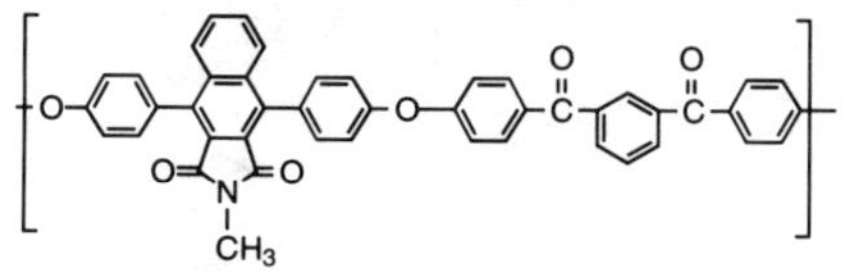	251		N-9; NMP, Sulfolane; F	*Polym. Prep.*, **32**(2), 164	1991	McGill U.
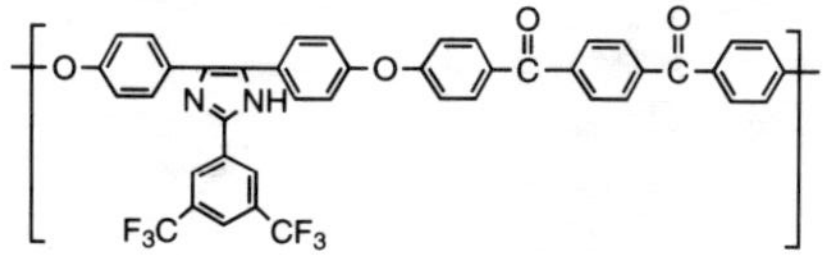	259		N-9; K_2CO_3, DMAC; F	*Polym. Prep.*, **32**(2), 162	1991	NASA/ Langley
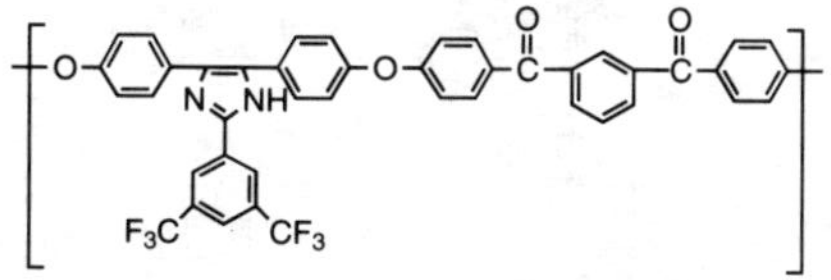	234		N-9; K_2CO_3, DMAC; F	*Polym. Prep.*, **32**(2), 162	1991	NASA/ Langley
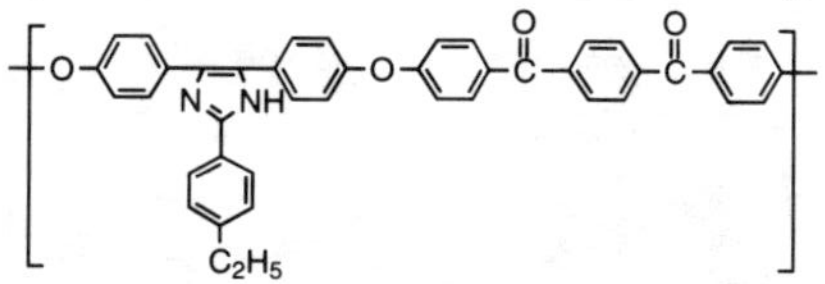	236		N-9; K_2CO_3, DMAC; F	*Polym. Prep.*, **32**(2), 162	1991	NASA/ Langley
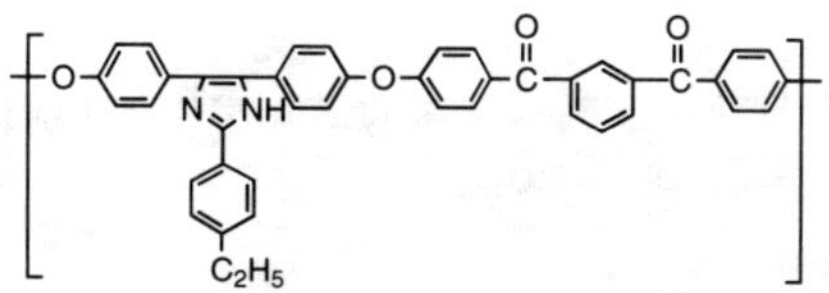	218		N-9; K_2CO_3, DMAC; F	*Polym. Prep.*, **32**(2), 162	1991	NASA/ Langley
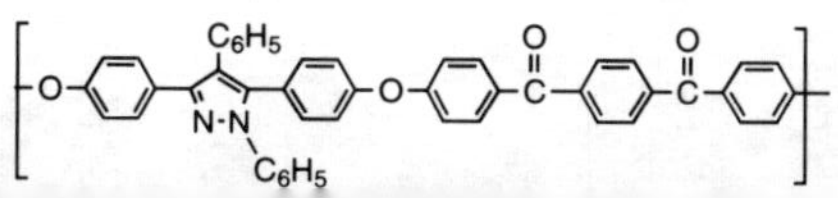	210		N-9; K_2CO_3, DMAC; F	*Polym. Prep.*, **32**(2), 160	1991	Va. Comm. U.; NASA/ Langley

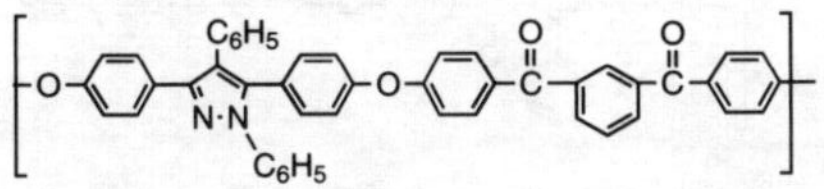	196	N-9; K_2CO_3, DMAC; F	*Polym. Prep.*, **32**(2), 160	1991	Va. Comm. U.; NASA/ Langley
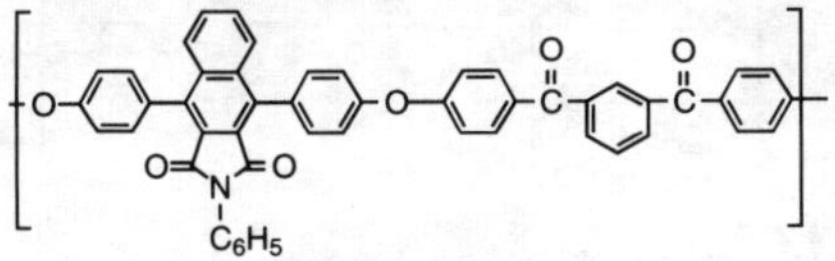	245	N-9; K_2CO_3, NMP, Sulfolane; F	*Polym. Prep.*, **32**(2), 164	1991	McGill U.
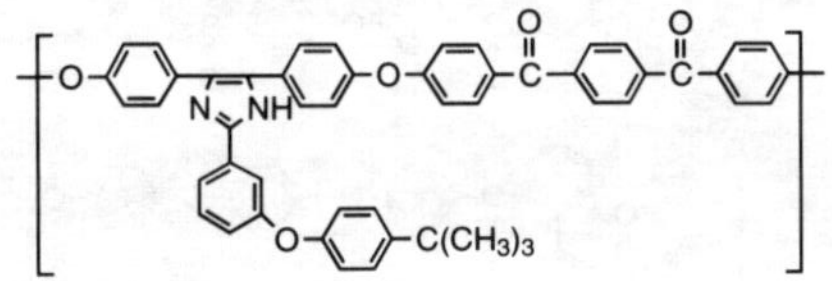	213	N-9; K_2CO_3, DMAC; F	*Polym. Prep.*, **32**(2), 162	1991	NASA/ Langley
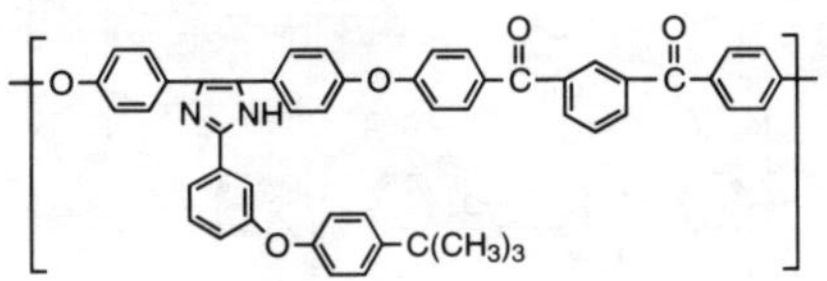	200	N-9; K_2CO_3, DMAC; F	*Polym. Prep.*, **32**(2), 162	1991	NASA/ Langley
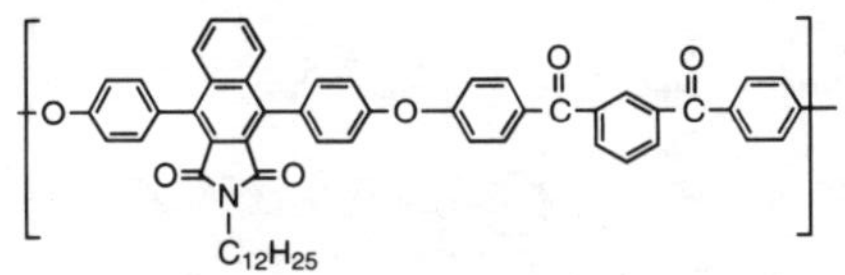	140	N-9; NMP, Sulfolane; F	*Polym. Prep.*, **32**(2), 164	1991	McGill U.
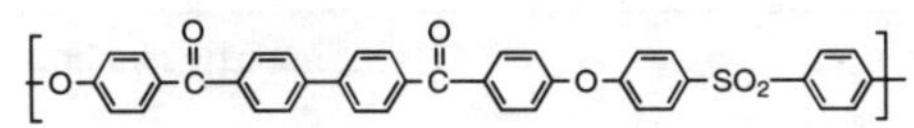	219	A-12; Bulk, CsF	*J. Chem. Soc., Chem. Commun.*, 336;	1990	ICI
		Permeability	*J. Polym. Sci.*, B, *Polym. Phys.*, **31**, 983	1993	U. Texas

TABLE 9—*continued*

Polymer structure	Tg (°C)	Tm (°C)	Synthesis	Reference	Year	Organization
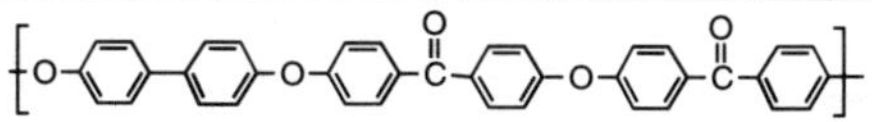	172	374	E-5; TFMSA	GB Pat. 2116990	1983	ICI
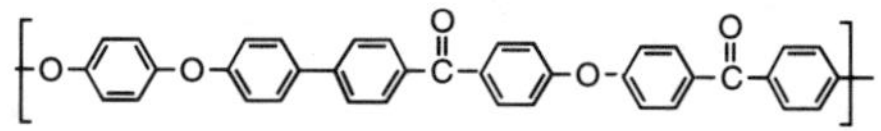	165	348	E-5; TFMSA	GB Pat. 2116990	1983	ICI
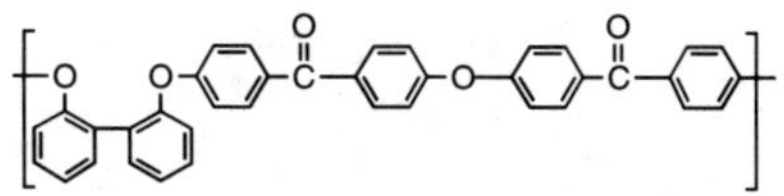			N-11; CsF, Bulk; F	*New Polym. Mater.*, **1**, 127	1988	Hamburg U.
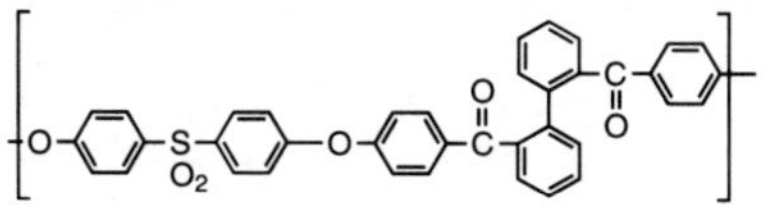	201		N-9; K_2CO_3, NMP; F	*Polym. Prep.*, **33**(2), 351	1992	Carleton U.
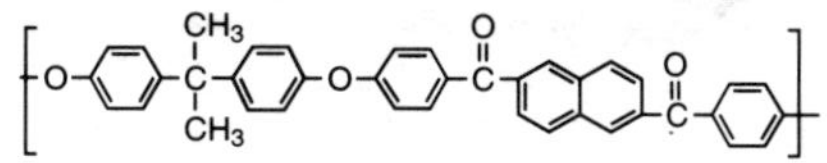	185	220	N-9; K_2CO_3, DMAC; F	*Polym.*, **29**, 358	1988	NASA/ Langley
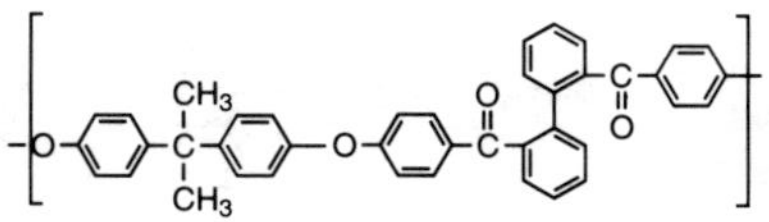	172		N-9; K_2CO_3, NMP; F	*Polym. Prep.*, **33**(2), 351	1992	Carleton U.
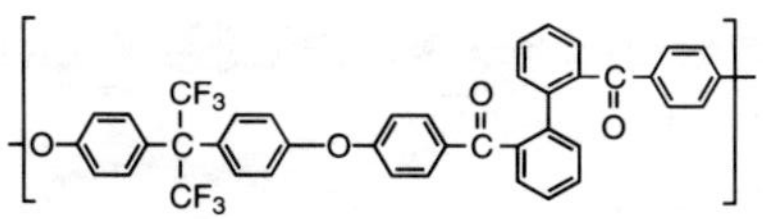	179		N-9; K_2CO_3, NMP; F	*Polym. Prep.*, **33**(2), 351	1992	Carleton U.
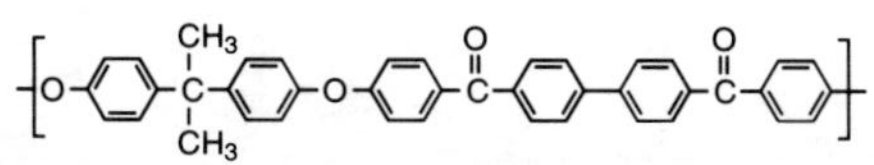	180		N-9 Permeability	*J. Polym. Sci.*, B, *Polym. Phys.*, **B**, **31**, 983	1993	U. Texas

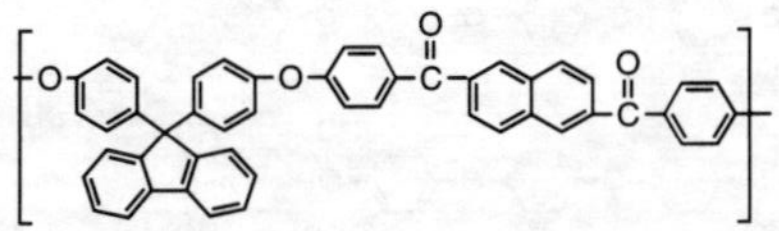	252		N-9; K_2CO_3, DMAC; F	*Polym.*, **29**, 358	1988	NASA/ Langley
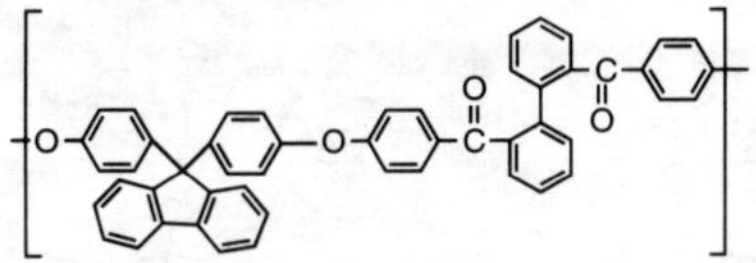	230		N-9; K_2CO_3, NMP; F	*Polym. Prep.*, **33**(2), 351	1992	Carleton U.
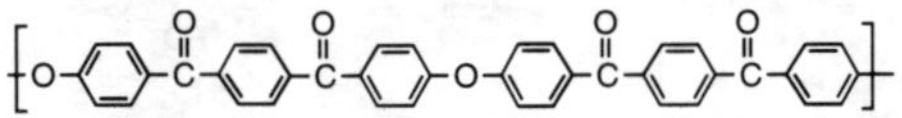		375	E-3; $AlCl_3$, *o*-Dichloro-benzene	*Polym. Prep.*, **32**(3), 255, 258	1991	DuPont
	180	389	E-3; $AlCl_3$, $(CH_3)_2SO_2$, CH_2Cl_2	US Pat. 4826947	1989	Raychem
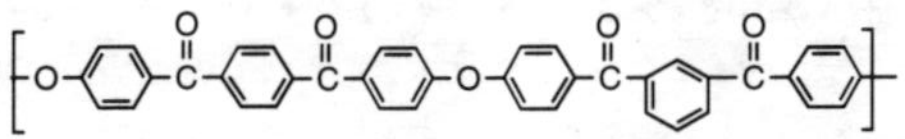		325	E-3; $AlCl_3$, *o*-Dichloro-benzene	*Polym. Prep.*, **32**(3), 255, 258; Eur. Pat. 192260; US Pat. 4816556	1991 1986 1989	DuPont
			E-3; $AlCl_3$, $(CH_3)_2SO_2$, CH_2Cl_2	US Pat. 4826947	1989	Raychem
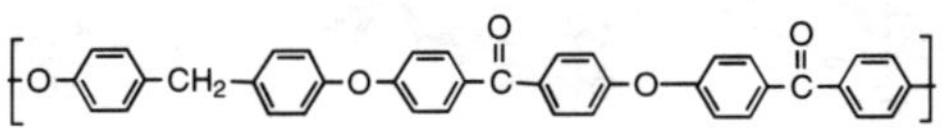	128	275	N-9; K_2CO_3, DMAC; F	*Polym.*, **29**, 358	1988	NASA/ Langley
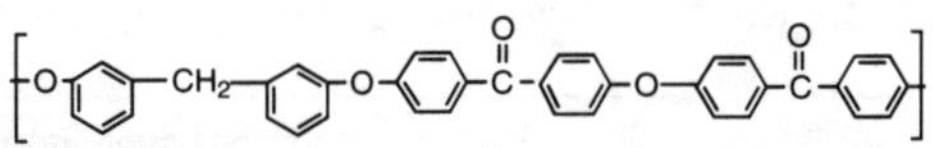	123		N-9; K_2CO_3, DMAC; F	*Polym.*, **29**, 358	1988	NASA/ Langley
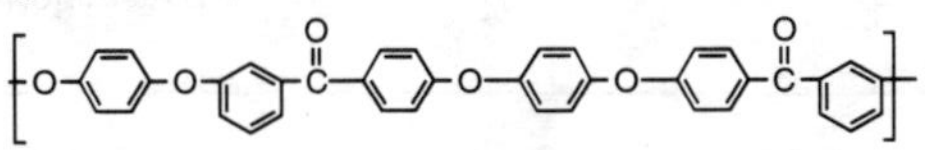	147	350	E-5; P_2O_5, CH_3SO_3H	*Macromol.*, **20**, 2675	1987	Yamagata U.

TABLE 9—*continued*

Polymer structure	Tg (°C)	Tm (°C)	Synthesis	Reference	Year	Organization
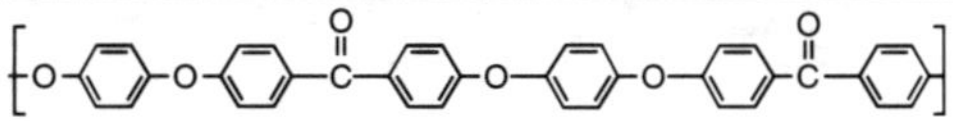			E-5; P_2O_5, CH_3SO_3H	*Macromol.*, **20**, 2675	1987	Yamagata U.
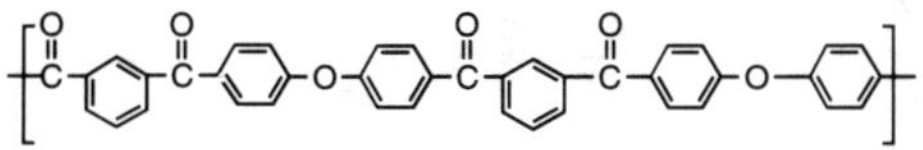	158	269	E-5; TFMSA, P_2O_5 E-3; $AlCl_3$, $(CH_3)_2SO_2$, CH_2Cl_2	US Pat. 4839459 US Pat. 4826947	1989 1989	Eastman Kodak Raychem
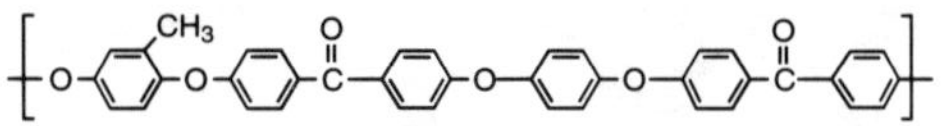	121	266	N-11; CsF, Bulk; F	*New Polym. Mater.*, **1**, 127	1988	Hamburg U.
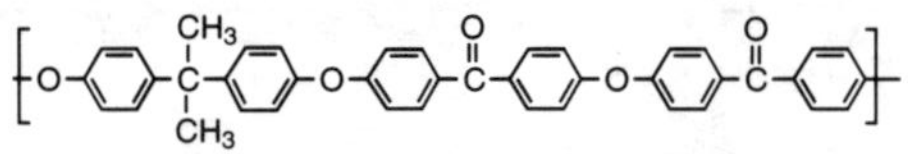	162		N-9; K_2CO_3, DMAC; F	*Polym.*, **29**, 358	1988	NASA/Langley
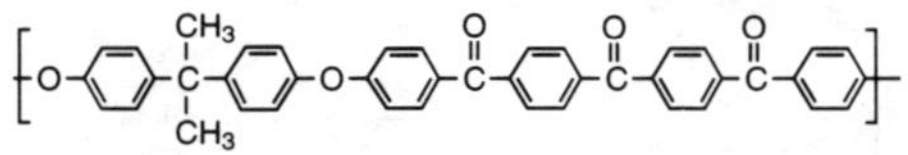	155	283	N-9; K_2CO_3, DMAC; F	*Polym.*, **29**, 358	1988	NASA/Langley
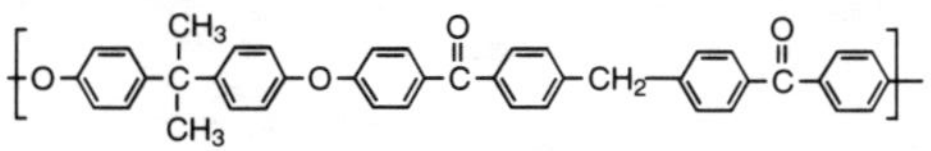	155		N-9; K_2CO_3, DMAC; F	*Polym.*,**29**, 358	1988	NASA/Langley
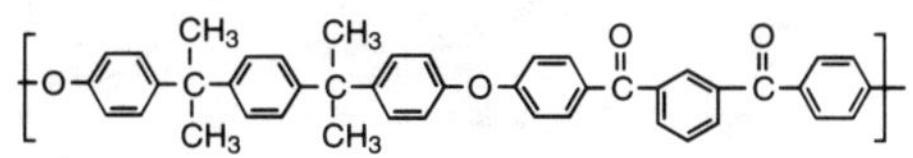	150		N-9; K_2CO_3, DMAC; Cl	*Polym.*, **29**, 358	1988	NASA/Langley
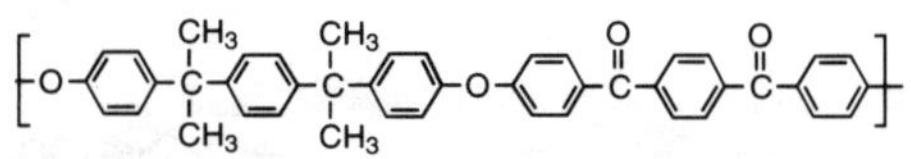	168	207	N-9; K_2CO_3, DMAC; Cl	*Polym.*, **29**, 358	1988	NASA/Langley
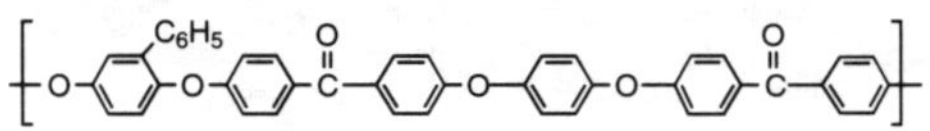	142		N-11; CsF, Bulk;	*New Polym. Mater.*, **1**, 127	1988	Hamburg U.

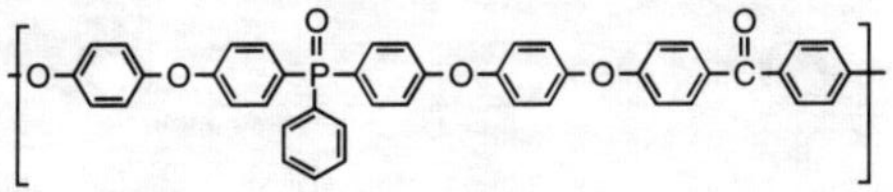	187		N-9; K_2CO_3, DMAC: F	*Polym. Prep.*, **33**(2), 241	1992	NASA/Langley
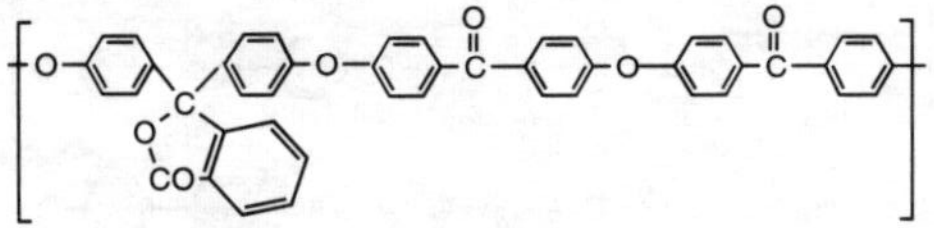	191		N-11; CsF, Bulk; F	*New Polym. Mater.*, **1**, 127	1988	Hamburg U.
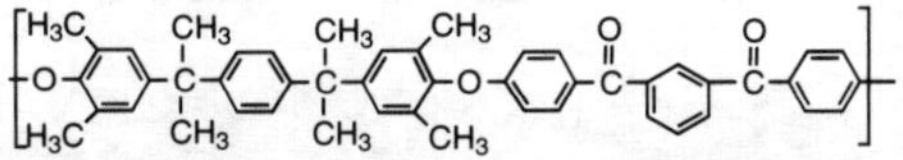	191		N-9; K_2CO_3, DMAC; F	*Polym.*, **29**, 358	1988	NASA/Langley
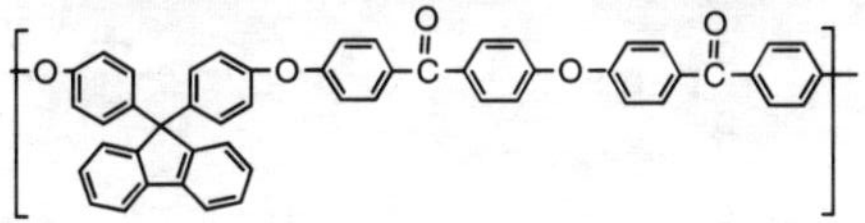	231	430	N-9; K_2CO_3, DMAC; F	*Polym.*, **29**, 358	1988	NASA/Langley
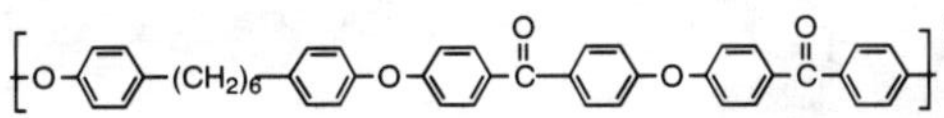	88	282	N-11; CsF, Bulk; F	*Macromol.*, **22**, 517	1989	Hamburg U.
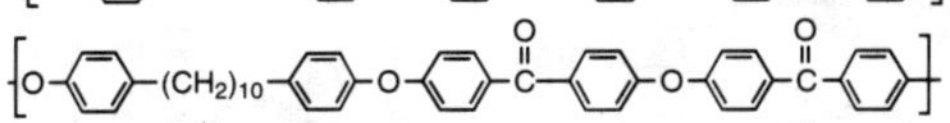	72	241	N-11; CsF, Bulk; F	*Macromol.*, **22**, 517	1989	Hamburg U.
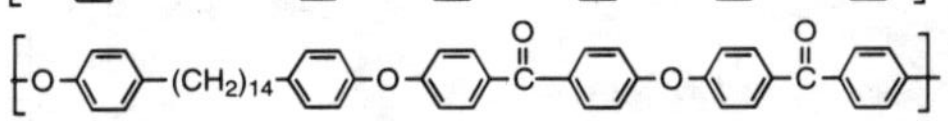	26	192	N-11; CsF, Bulk; F	*Macromol.*, **22**, 517	1989	Hamburg U.

TABLE 10
Polyarletherketones containing seven or more rings in the backbone of the repeat unit

Polymer structure	Tg (°C)	Tm (°C)	Synthesis	Reference	Year	Organization
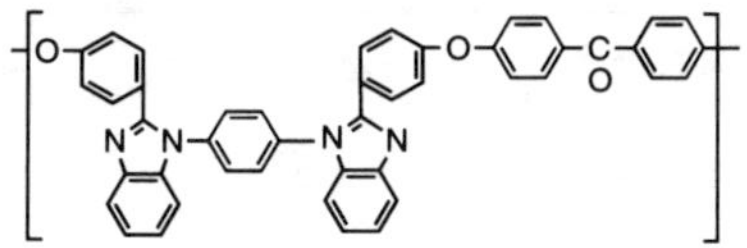	242		N-9; K_2CO_3, Sulfolane; F	*Polym. Prep.*, **33**(1), 1098	1992	NASA/ Langley
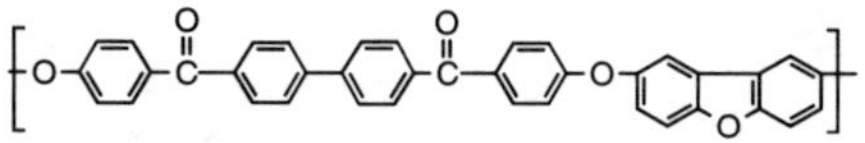	225		A-12; Bulk, CsF	*J. Chem. Soc., Chem. Commun.*, 336	1990	ICI
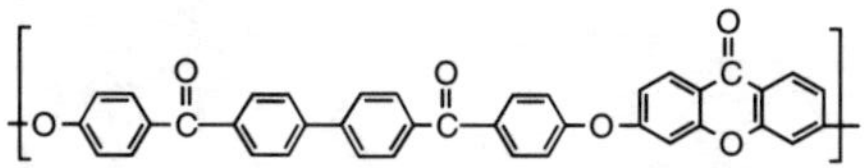			A-12; Bulk, CsF	*J. Chem. Soc., Chem. Commun.*, 336	1990	ICI
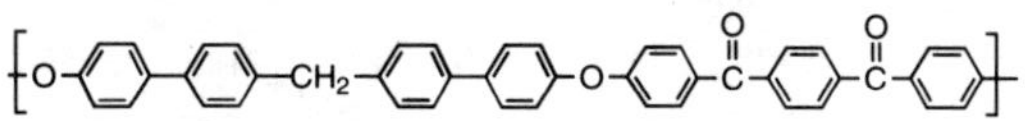	195		N-9; K_2CO_3, DMAC; Cl	*Polym.*, **29**, 358	1988	NASA/ Langley
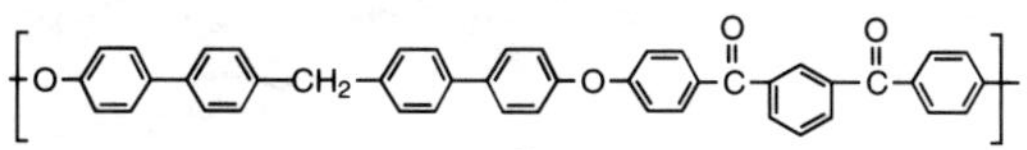	184		N-9; K_2CO_3, DMAC; Cl	*Polym.*, **29**, 358	1988	NASA/ Langley
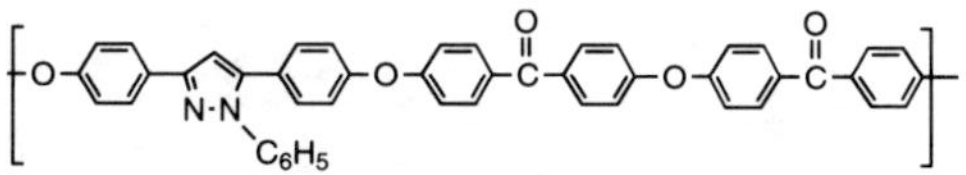	190		N-9; K_2CO_3, DMAC, NMP; F	*Polym. Prep.*, **32**(1), 619	1991	Va. Comm. U.
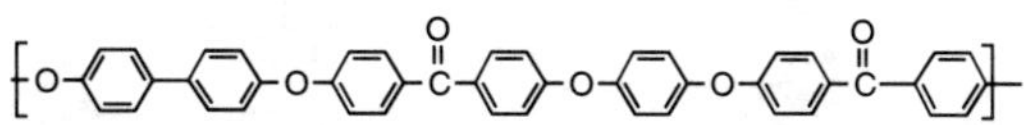	159 173	383 370	E-5; TFMSA	*Polym.*,**29**, 1902; GB Pat. 2116990	1988 1983	ICI ICI
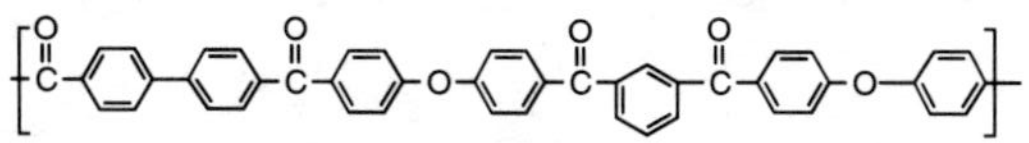	163	387	E-3; $AlCl_3$, $(CH_3)_2SO_2$, CH_2Cl_2	US Pat. 4826947	1989	Raychem

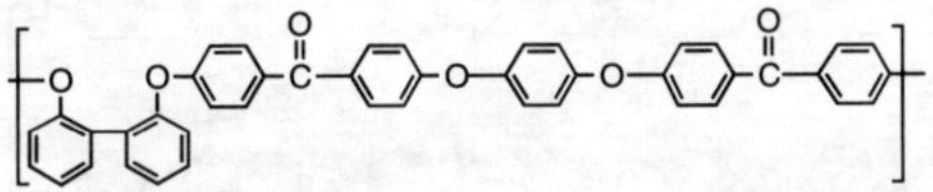	141		N-11; CsF, Bulk; F	*New Polym. Mater.*, **1**, 127	1988	Hamburg U.
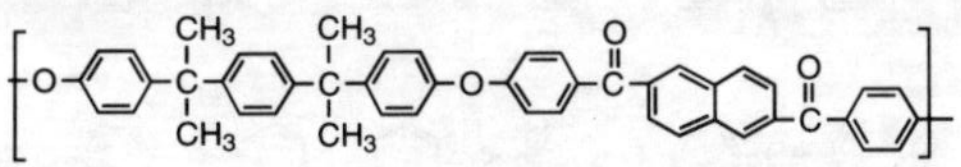	185		N-9; K_2CO_3, DMAC; F	*Polym.*, **29**, 358	1988	NASA Langley
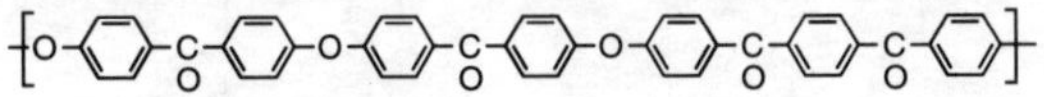		376	N-9; K_2CO_3, Diphenyl sulfone; Cl	Jap. Pat. 1065130	1989	Asahi Chemical
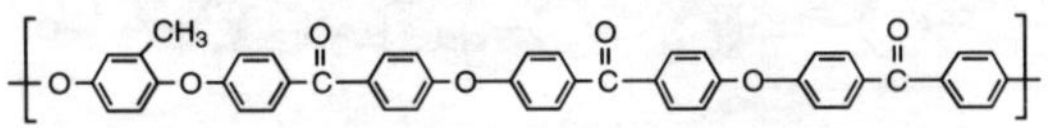	139	280	N-11; CsF, Bulk; F	*New Polym. Mater.*, **1**, 127	1988	Hamburg U.
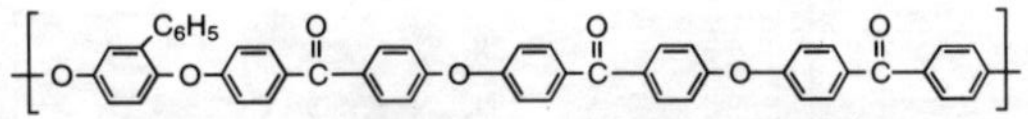	149		N-11; CsF, Bulk; F	*New Polym. Mater.*, **1**, 127	1988	Hamburg U.
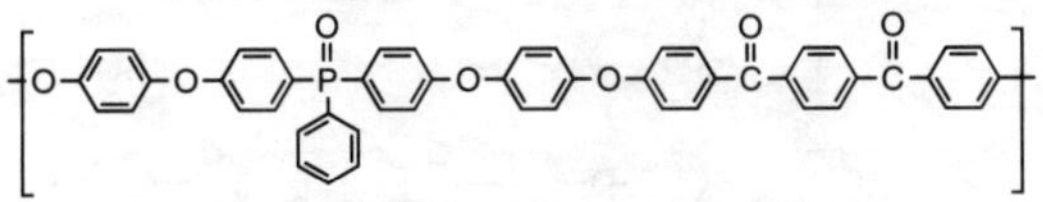	187		N-9; K_2CO_3, DMAC; F	*Polym. Prep.*, **33**(2), 241	1992	NASA/ Langley
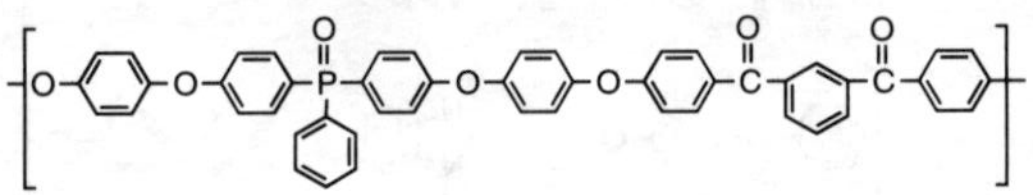	171		N-9; K_2CO_3, DMAC; F	*Polym. Prep.*, **33**(2), 241	1992	NASA/ Langley
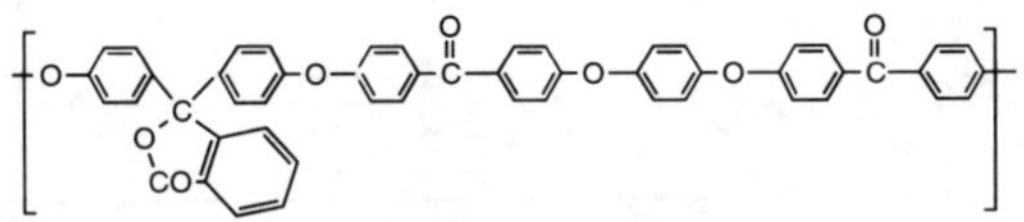	184		N-11; CsF, Bulk; F	*New Polym. Mater.*, **1**, 127	1988	Hamburg U.
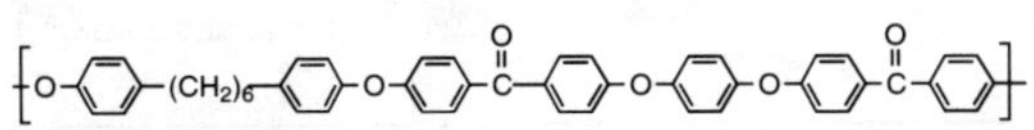	88	274	N-11; CsF, Bulk; F	*Macromol.*, **22**, 517	1989	Hamburg U.

TABLE 10—*continued*

Polymer structure	Tg (°C)	Tm (°C)	Synthesis	Reference	Year	Organization
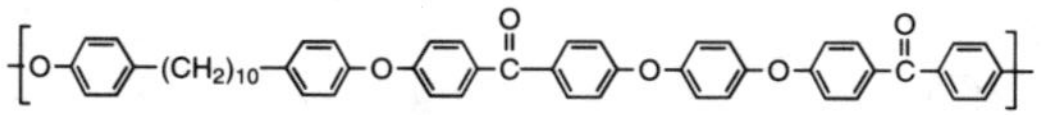	70	247	N-11; CsF, Bulk; F	*Macromol.*, **22**, 517	1989	Hamburg U.
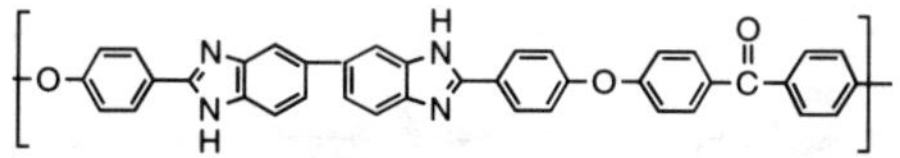	307		N-9; K_2CO_3, DMAC; F	*Polym. Prep.*, **32**(3),195	1991	NASA/ Langley
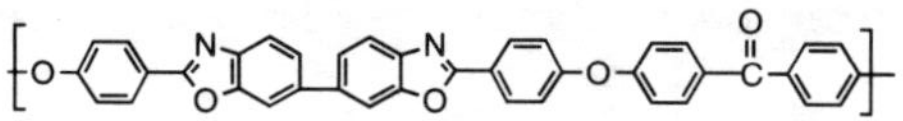	249	425	N-9; K_2CO_3, DMAC, Diphenyl sulfone; F	*Polym. Prep.*, **32**(1), 646	1991	NASA/ Langley
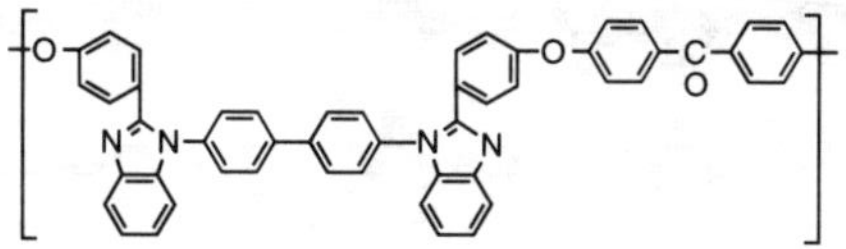	268		N-9; K_2CO_3, Sulfolane; F	*Polym. Prep.*, **33**(1), 1098	1992	NASA/ Langley
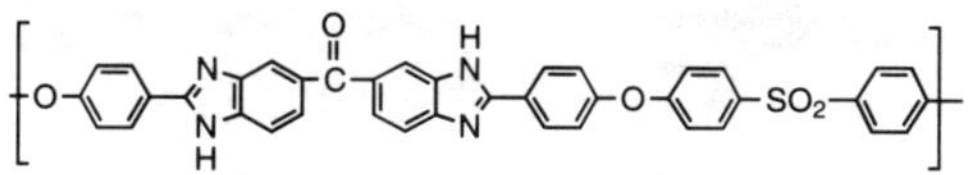			N-9; K_2CO_3, DMAC; F	*Polym. Prep.*, **32**(3), 195	1991	NASA/ Langley
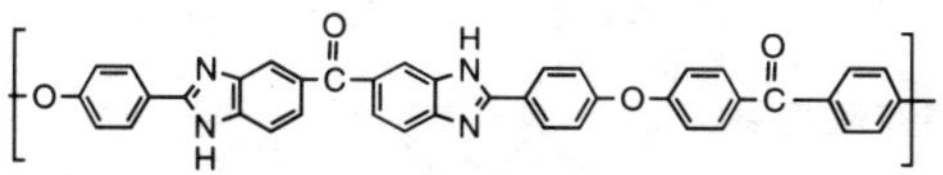			N-9; K_2CO_3, DMAC; F	*Polym. Prep.*, **32**(3), 195	1991	NASA/ Langley
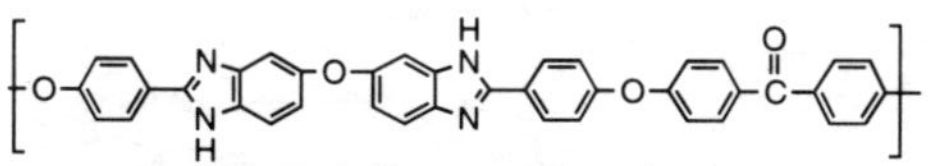	294		N-9; K_2CO_3, DMAC; F	*Polym. Prep.*, **32**(3), 195	1991	NASA/ Langley
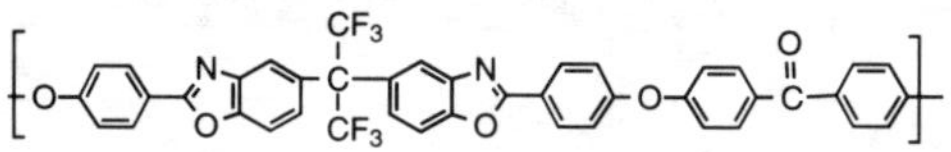	254		N-9; K_2CO_3, DMAC, Diphenyl sulfone; F	*Polym. Prep.*, **32**(1), 646	1991	NASA/ Langley

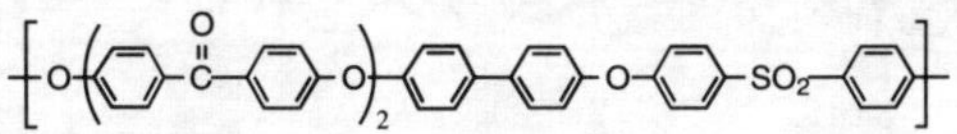	201			SAMPE, **32**, 382	1987	ICI
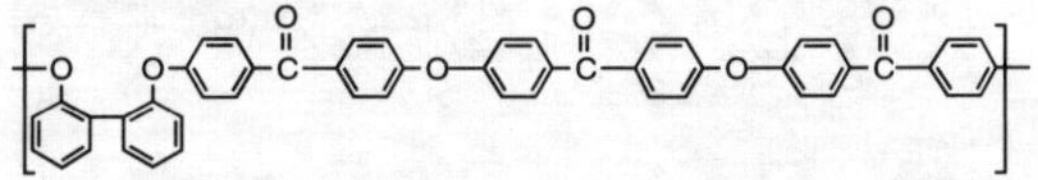	147		N-11; CsF, Bulk; F	*New Polym. Mater*, . **1**, 127	1988	Hamburg U.
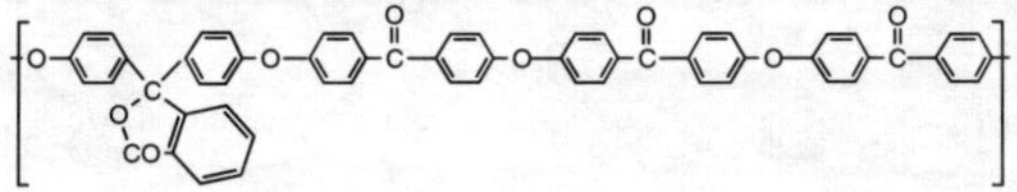	192		N-11; CsF, Bulk; F	*New Polym. Mater*., **1**, 127	1988	Hamburg U.
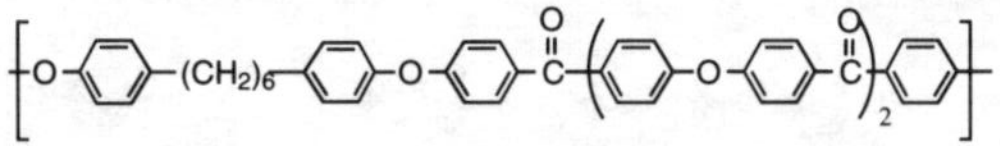	109	300	N-11; CsF, Bulk; F	*Macromol*., **22**, 517	1989	Hamburg U.
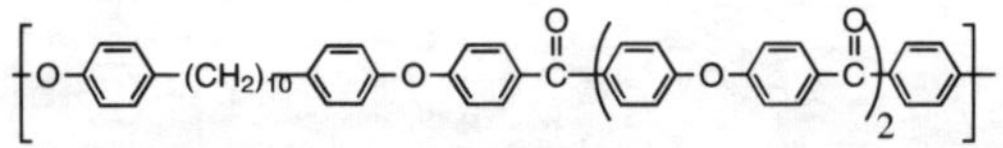	89	253	N-11; CsF, Bulk; F	*Macromol*., **22**, 517	1989	Hamburg U.
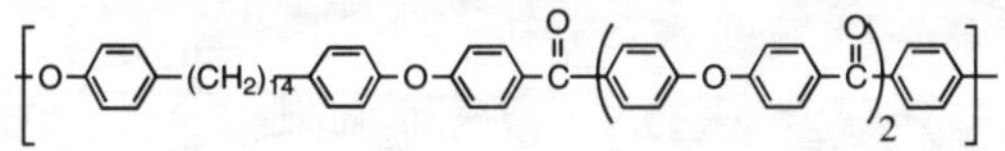	70	250	N-11; CsF, Bulk; F	*Macromol*., **22**, 517	1989	Hamburg U.
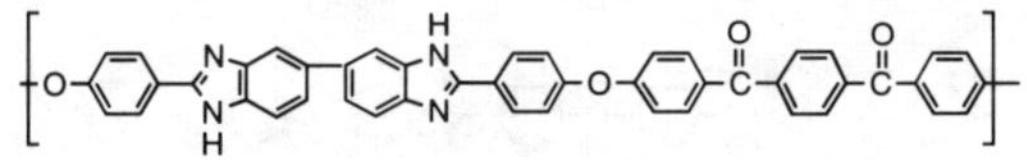	295		N-9; K_2CO_3, DMAC; F	*Polym. Prep*., **32**(3), 195	1991	NASA/ Langley
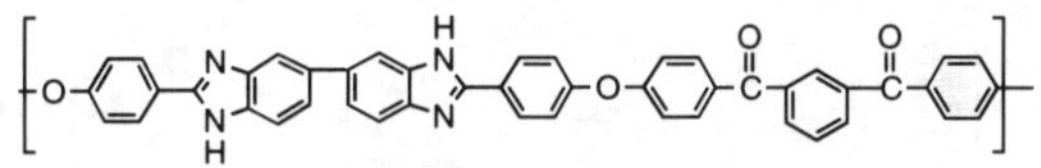	276		N-9; K_2CO_3, DMAC; F	*Polym. Prep*., **32**(3), 195; *Polym. Prep*., **33**(1), 412; *Polym. Prep*., 34(1), 525	1991 1992 1993	NASA/ Langley
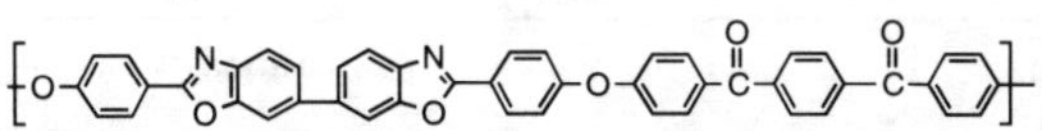	237	405, 446	N-9; K_2CO_3, DMAC, Diphenyl sulfone; F	*Polym. Prep*., **32**(1), 646	1991	NASA/ Langley

TABLE 10—*continued*

Polymer structure	Tg (°C)	Tm (°C)	Synthesis	Reference	Year	Organization
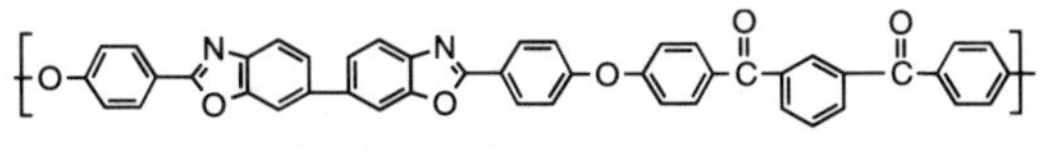	215	373	N-9; K_2CO_3, DMAC, Diphenyl sulfone; F	*Polym.* Prep., **32**(1), 646	1991	NASA/ Langley
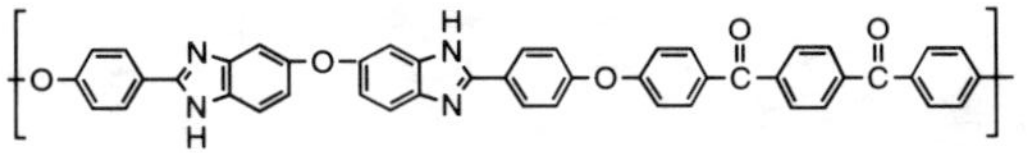	282		N-9; K_2CO_3, DMAC; F	*Polym.* Prep., **32**(3), 195	1991	NASA/ Langley
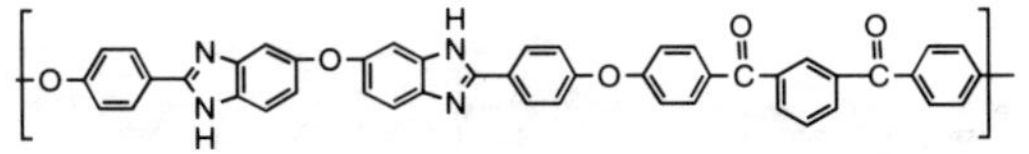	269		N-9; K_2CO_3, DMAC; F	*Polym. Prep.*, **32**(3), 195	1991	NASA/ Langley
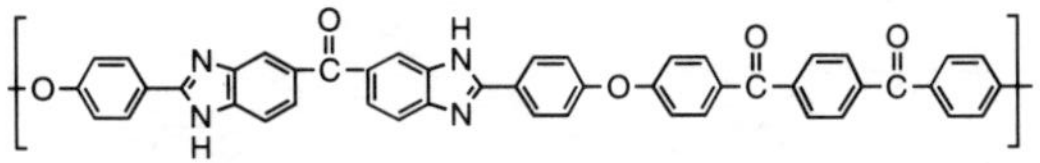	276		N-9; K_2CO_3, DMAC; F	*Polym. Prep.*, **32**(3), 195	1991	NASA/ Langley
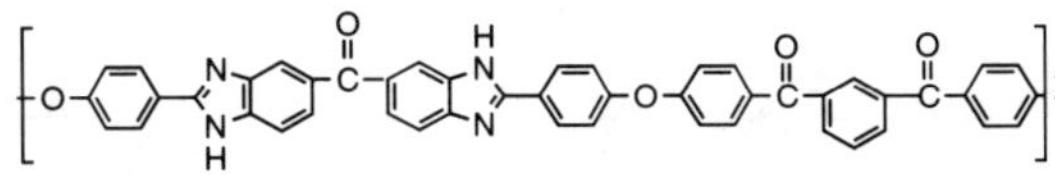	264		N-9; K_2CO_3, DMAC; F	*Polym. Prep.*, **32**(3), 195	1991	NASA/ Langley
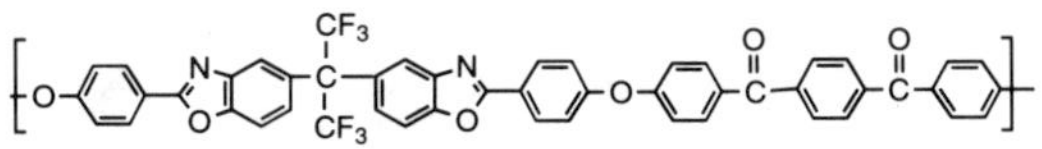	249		N-9; K_2CO_3, DMAC, Diphenyl sulfone; F	*Polym. Prep.*, **32**(1), 646	1991	NASA/ Langley
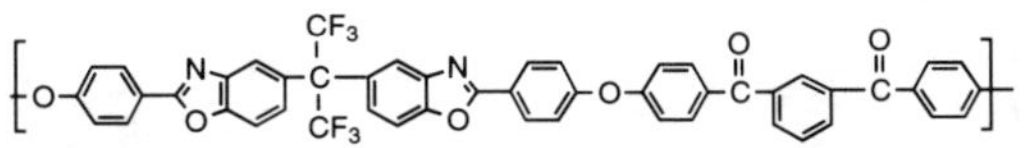	227		N-9; K_2CO_3, DMAC, Diphenyl sulfone; F	*Polym. Prep.*, **32**(1), 646	1991	NASA/ Langley
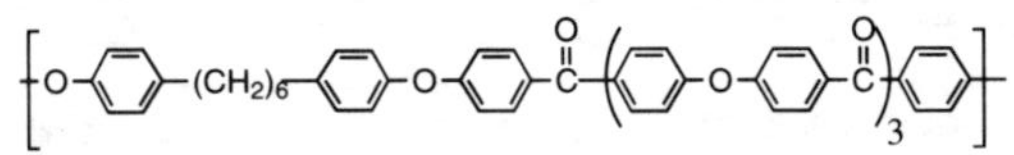	102	279	N-11; CsF, Bulk; F	*Macromol.*, **22**, 517	1989	Hamburg U.
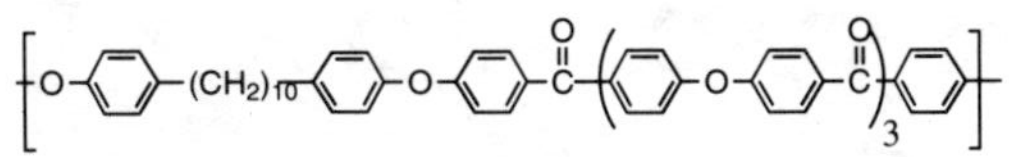	87	266	N-11; CsF, Bulk; F	*Macromol.*, **22**, 517	1989	Hamburg U.
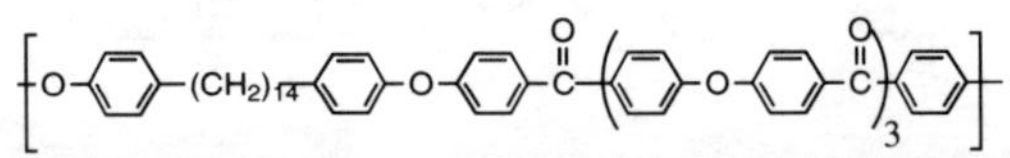	91	284	N-11; CsF, Bulk; F	*Macromol.*, **22**, 517	1989	Hamburg U.

TABLE 11
Polyaryletherimides containing five or fewer rings in the backbone of the repeat unit

Polymer structure	Tg (°C)	Tm (°C)	Synthesis	Reference	Year	Organization
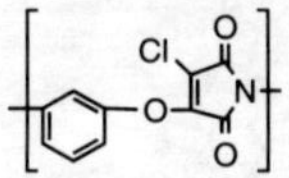	154		N-1; DMF; Cl	*J. Polym. Sci., Polym. Chem.*, **11**, 561	1973	GE
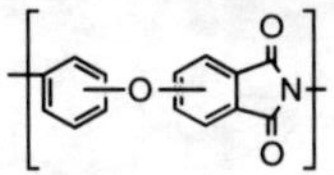			N-14; DMSO, THF; NO_2	*J. Polym. Mater.*, **2**, 115	1985	
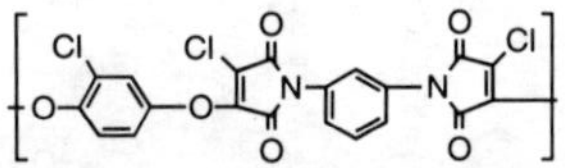	194		N-1; DMF; Cl	*J. Polym. Sci., Polym. Chem*., **11**, 561	1973	GE
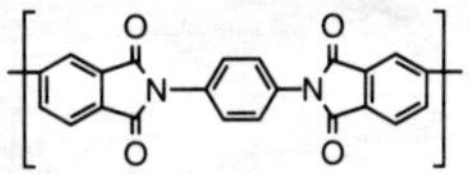 Upilex-S™	550			PC		Ube
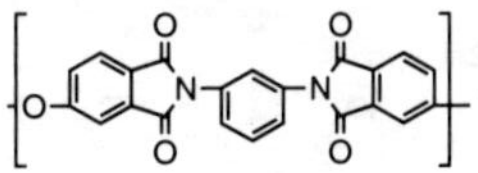		490		*Ency. Polym. Sci. Eng.*,**12**, 364	1988	
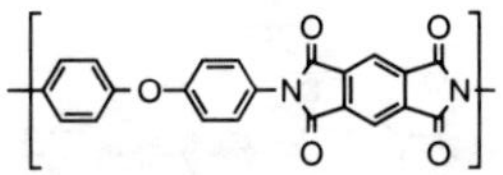			N-15	*Polym. Prep.*, **34**(1), 525	1993	DuPont
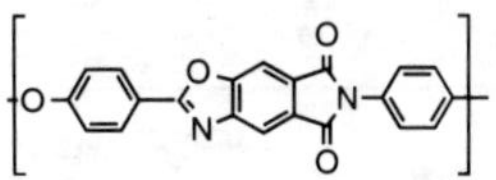		470	N-14; NaH, NMP, CHP; F	*Polym. Prep.*, **33**(2), 142	1992	U. So. Miss.

TABLE 11—*continued*

Polymer structure	Tg (°C)	Tm (°C)	Synthesis	Reference	Year	Organization
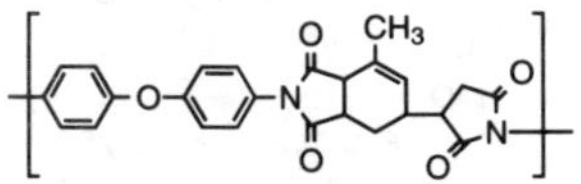			N-15	PC		DIC
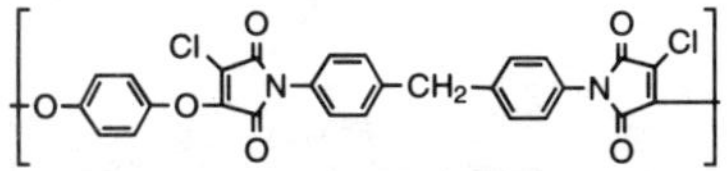	187		N-1; DMSO; Cl	*J. Polym. Sci., Polym. Chem.*, **11**, 561	1973	GE
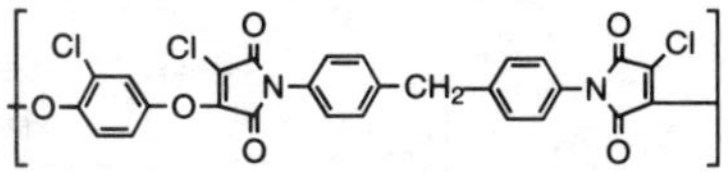	184		N-1; DMF; Cl	*J. Polym. Sci., Polym. Chem.*, **11**, 561	1973	GE
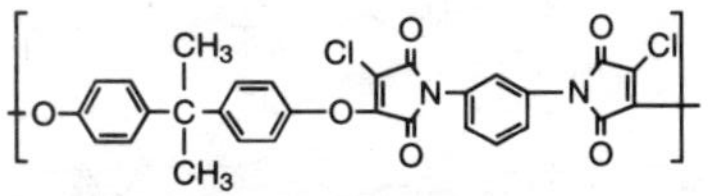	167		N-1; DMF; Cl	*J. Polym. Sci., Polym. Chem.*, **11**, 561	1973	GE
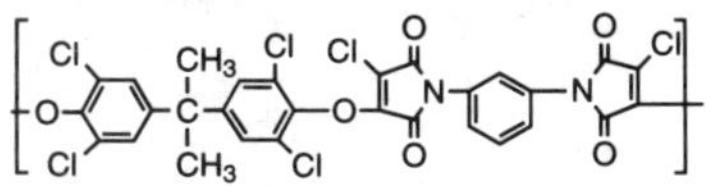	222		N-1; DMF; Cl	*J. Polym. Sci., Polym. Chem.*, **11**, 561	1973	GE
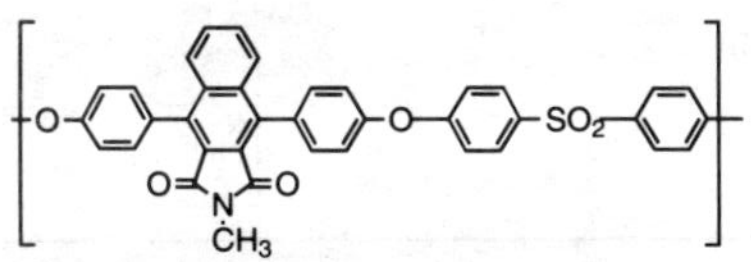	314		N-2; NMP, Sulfolane; F	*Polym. Prep.*, **32**(2), 164	1991	McGill U.
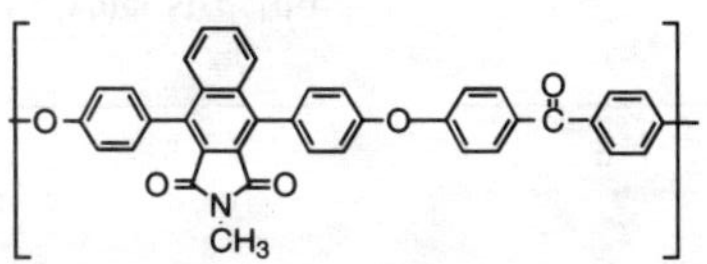	283		N-9; NMP, Sulfolane; F	*Polym. Prep.*, **32**(2), 164	1991	McGill U.

Structure					
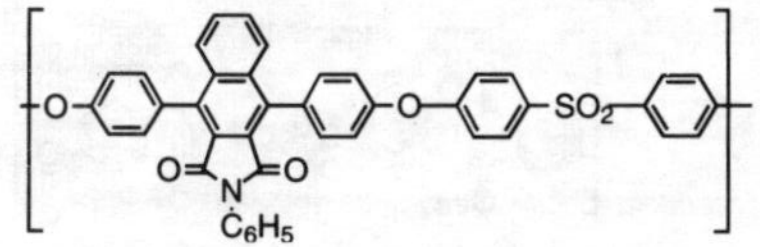	310	N-2; NMP, Sulfolane; F	*Polym. Prep.*, **32**(2), 164	1991	McGill U.
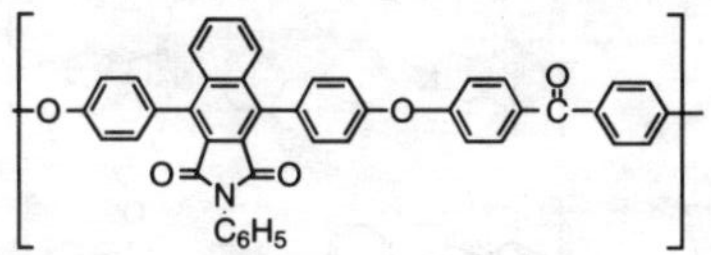	275	N-9; NMP, Sulfolane; F	*Polym. Prep.*, **32**(2), 164	1991	McGill U.
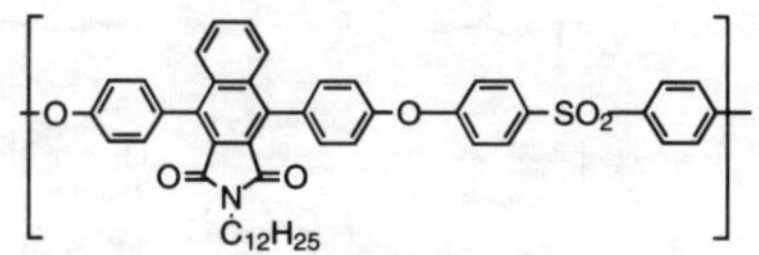	167	N-2; NMP, Sulfolane; F	*Polym. Prep.*, **32**(2), 164	1991	McGill U.
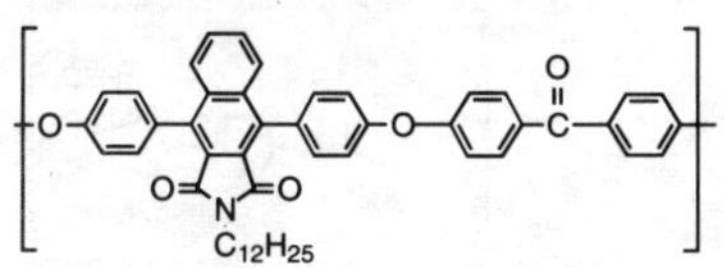	147	N-2; NMP, Sulfolane; F	*Polym. Prep.*, **32**(2), 164	1991	McGill U.
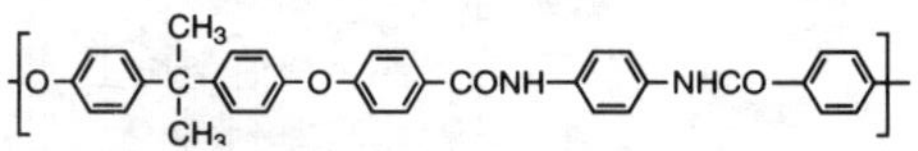	230	N-14; K_2CO_3, NMP, CHP; F	*Polym. Prep.*, **33**(1), 1014	1992	IBM
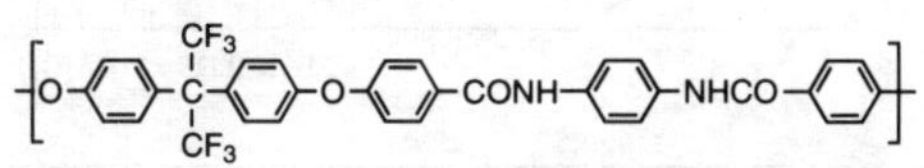	218	N-14; K_2CO_3, NMP, CHP; F	*Polym. Prep.*, **33**(1), 1014	1992	IBM

TABLE 12
Polyarylethermides containing six rings in the backbone of the repeat unit

Polymer structure	Tg (°C)	Tm (°C)	Synthesis	Reference	Year	Organization
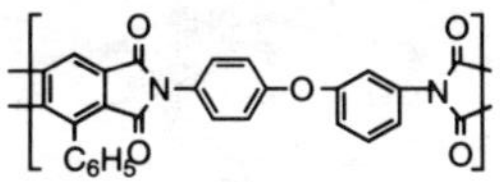	318		N-15; m-Cresol	*J. Polym. Sci., Polym. Chem.*, **31**, 141	1993	U. Cal. St. Barbara
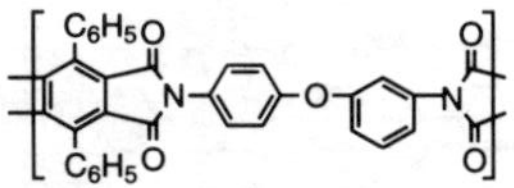	338		N-15; m-Cresol	*J. Polym. Sci., Polym. Chem.*, **31**, 141	1993	U. Cal. St. Barbara
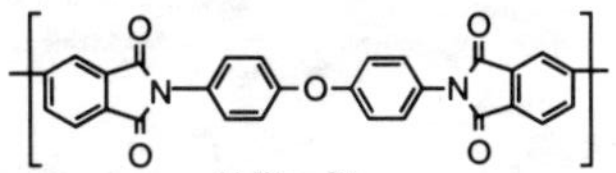 U-Pirex™			N-15	PC		Ube
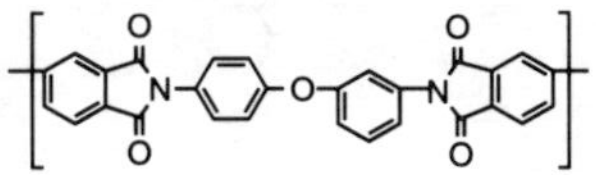	273	415	N-15; m-Cresol	*J. Polym. Sci.,Polym. Chem.*, **31**, 141	1993	U. Cal. St. Barbara
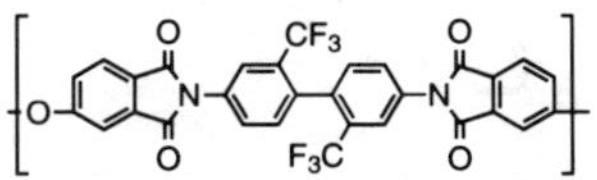	275		N-15; m-Cresol, Isoquinoline	*Polym. Prep.*, **31**(1), 342	1990	U. Akron
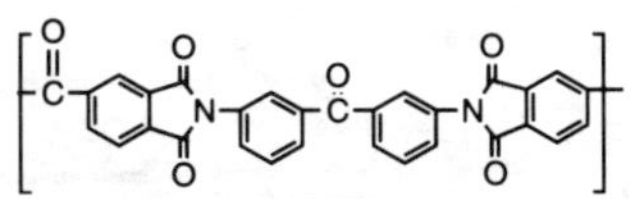 LARC-TPI	264		N-15	*J. Polym. Sci.,Polym. Chem.*, **14**, 2275	1976	NASA/ Langley
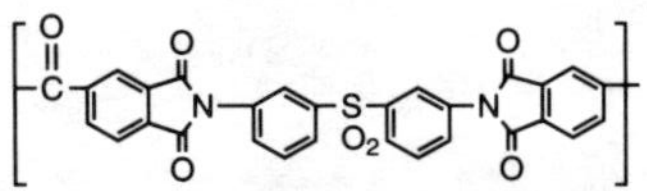 $PISO_2$	273		N-15	US Pats. 4398021, 4489027	1983 1984	NASA/ Langley

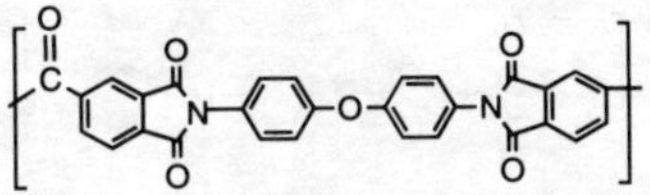 Skybond™	283	N-15 N-15	PC *J. Appl. Polym. Sci.*, **47**, 753	1993	Monsanto U. Mass.
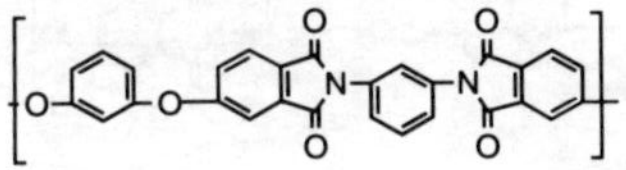	224	N-15; m-Cresol, Toluene	*Polym. Prep.*, **24**(2), 312; *J. Polym. Sci., Polym. Symp.*, **74**, 93	1983 1986	GE
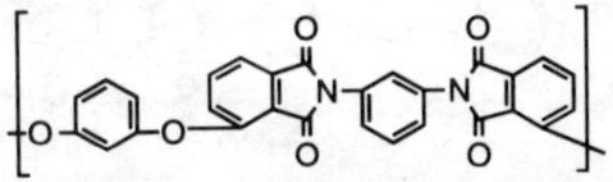	241	N-15; m-Cresol, Toluene	*Polym. Prep.*, **24**(2), 312; *J. Polym. Sci., Polym. Symp.*, **74**, 93	1983 1986	GE
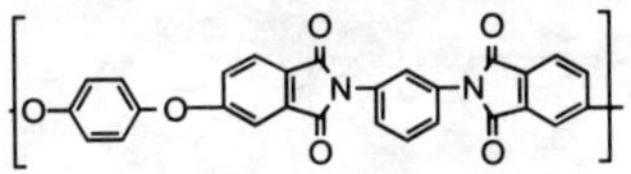	255	N-15; m-Cresol, Toluene	*Polym. Prep.*, **24**(2), 312, *J. Polym. Sci.; Polym. Symp.*, **74,** 93	1983 1986	GE
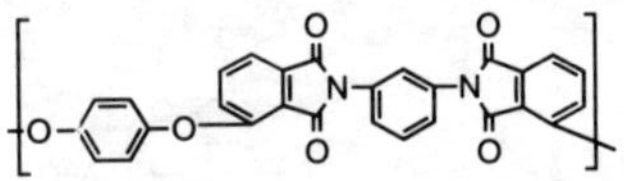	259	N-15; m-Cresol, Toluene	*Polym. Prep.*, **24**(2), 312, J. *Polym. Sci.; Polym. Symp.*, **74,** 93	1983 1986	GE
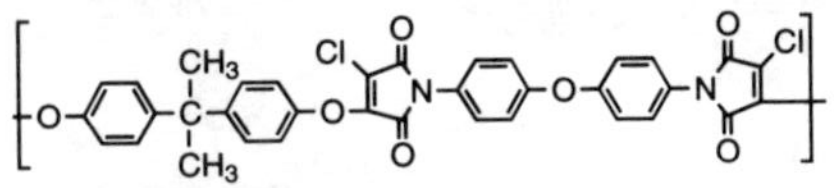	174	N-1; DMF; Cl	*J. Polym. Sci., Polym. Chem.*, **11**, 561	1973	GE
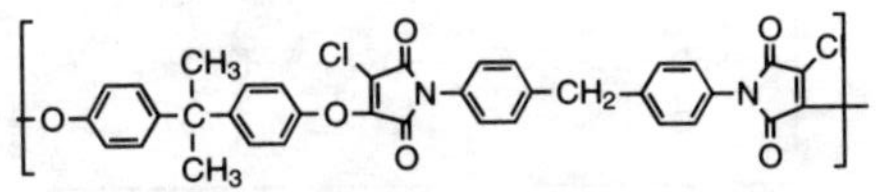	187	N-1; DMSO; Cl	*J. Polym. Sci., Polym. Chem.*, **11**, 561	1973	GE
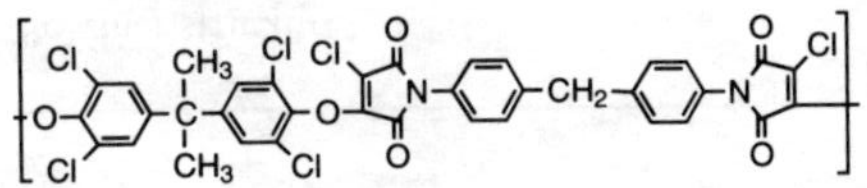	215	N-1; DMF; Cl	*J. Polym. Sci., Polym. Chem.*, **11**, 561	1973	GE

TABLE 12—*continued*

Polymer structure	Tg (°C)	Tm (°C)	Synthesis	Reference	Year	Organization
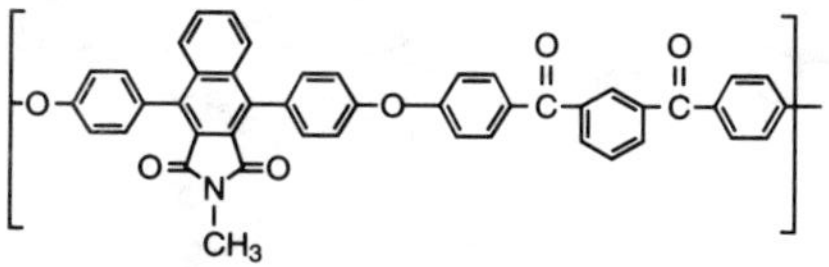	251		N-9; NMP, Sulfolane; F	*Polym. Prep.*, **32**(2), 164	1991	McGill U.
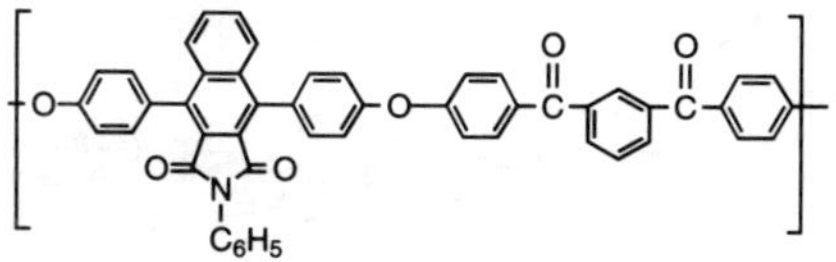	245		N-9; NMP, Sulfolane; F	*Polym. Prep.*, **32**(2), 164	1991	McGill U.
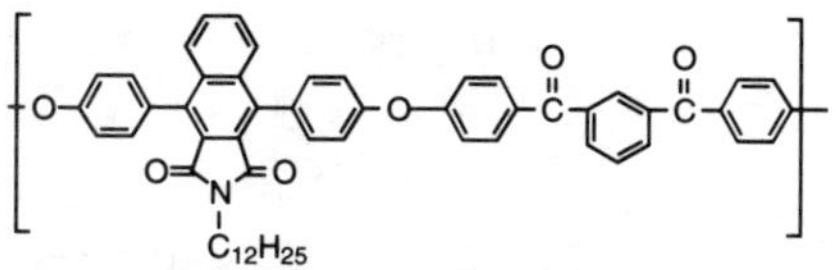	140		N-9; NMP, Sulfolane; F	*Polym. Prep.*, **32**(2), 164	1991	McGill U.
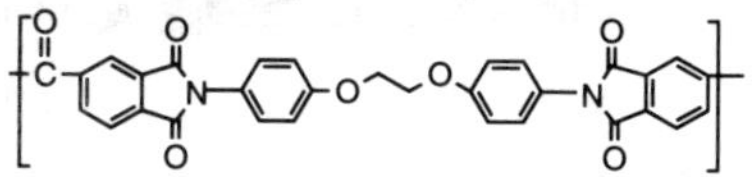	246	460	N-15, NMP	*Polym. Prep.*, **26**(2), 287; **25** (1), 160	1985 1984	NASA/ Langley; U. Akron
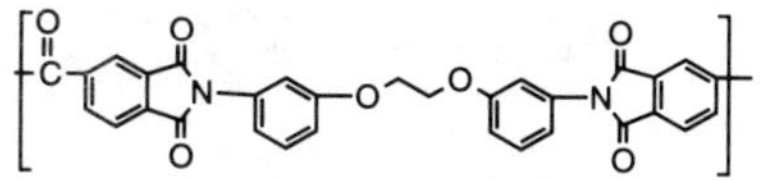	191	290	N-15, NMP	*Polym. Prep.*, **26**(2), 287; **25** (1), 160	1985	NASA/ Langley; U. Akron
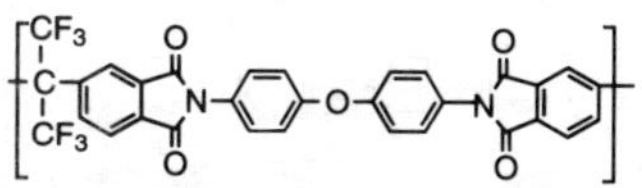	297		N-15; NMP, Ac_2O	*J. Appl. Polym. Sci.*, **47**, 1961	1993	Nitto Denko Corp.
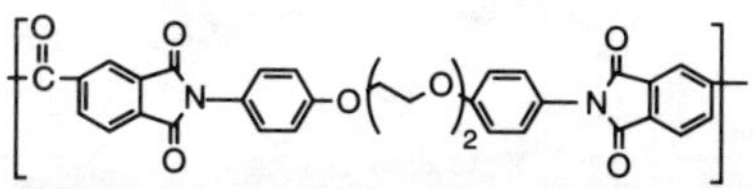	235	410	N-15; NMP	*Polym. Prep.*, **26**(2), 287; **25** (1), 160	1985 1984	NASA/| Langley; U. Akron

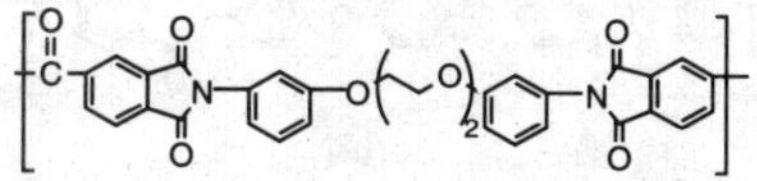	155	236	N-15; NMP	*Polym. Prep.*, **26**(2), 287; **25**(1), 160	1985 1984	NASA/ Langley; U. Akron
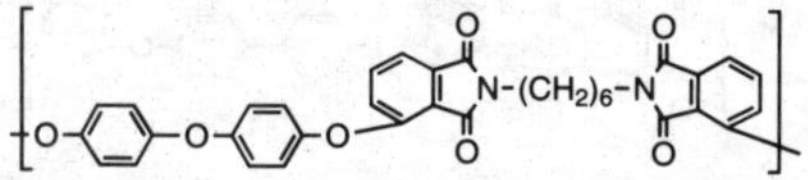	128		N-14; DMSO; NO2	*J. Polym. Sci., Polym. Chem.*, **19**, 1635; **18** 3069	1981; 1980	GE
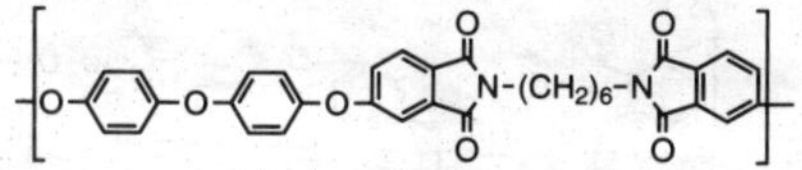	128			*Ency. Polym. Sci. Eng.*, **12**, 364	1988	
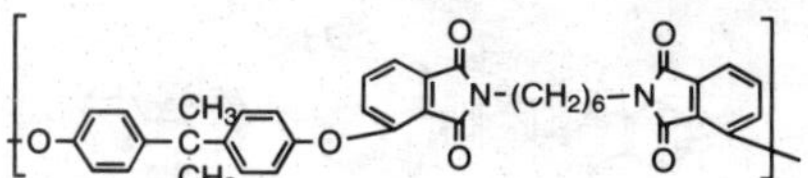	135		N-14; DMSO; NO2	*J. Polym. Sci., Polym. Chem. Ed.*, **18**, 3069	1980	GE
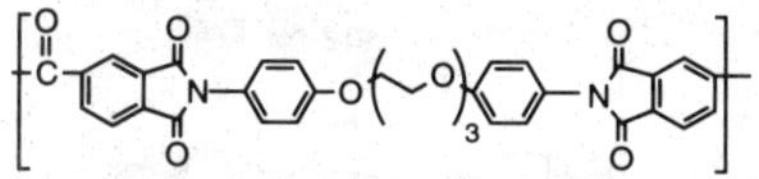	200	395	N-15, NMP	*Polym. Prep.*, **26**(2), 287; **25**(1), 160	1985 1984	NASA/ Langley; U. Akron

TABLE 13
Polyaryletherimides containing seven or more rings in the backbone of the repeat unit.

Polymer structure	Tg (°C)	Tm (°C)	Synthesis	Reference	Year	Organization
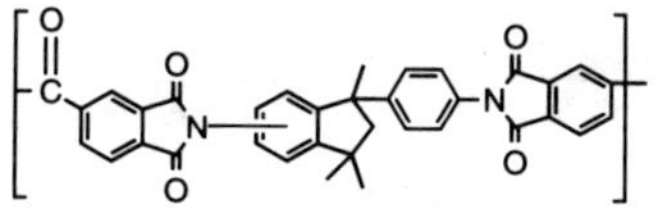 XU-218			N-15	*High Perf. Polym.: Their Origin & Devel.* Elsevier, p. 317	1986	Ciba-Geigy
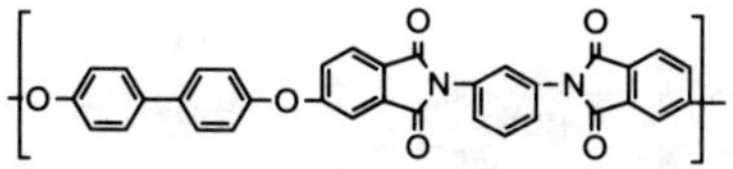 PEI-1	247		N-15; m-Cresol, Toluene	*Ency. Polym. Sci. Eng.*, **12**, 364; *Polym. Prep.*, **24**(2), 312	1988 1983	GE
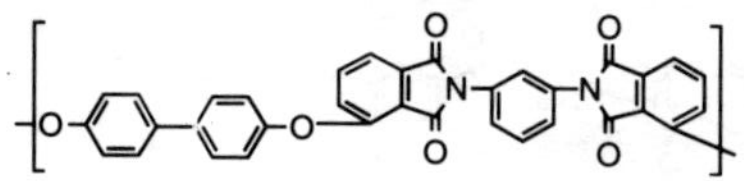	275		N-15; m-Cresol, Toluene	*Polym. Prep.*, **24**(2), 312; *J. Polym. Sci., Polym. Symp.*, **74**, 93	1983 1986	GE
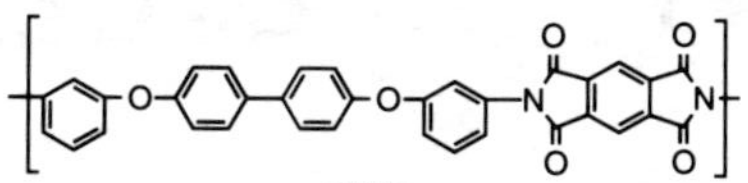 N-TPI	250	380	N-15	*Polym.*, **34**, 3315	1993	Mitsui Toatsu
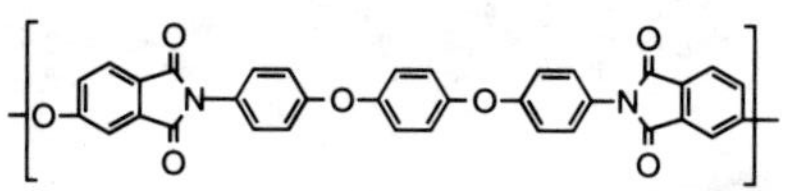		450		*Ency. Polym. Sci. Eng.*, **12**, 364	1988	
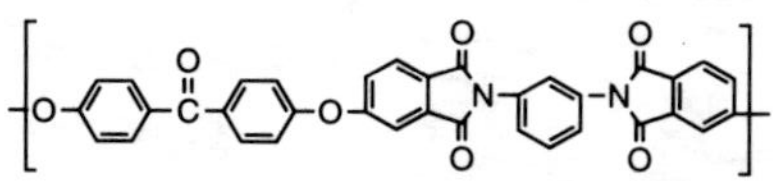	239		N-15; m-Cresol, Toluene	*Ency. Polym. Sci. Eng.*, **12**, 364; *Polym. Prep.*, **24**(2), 312	1988 1983	

Structure	T_g	Method; Solvent	Reference	Year	Company
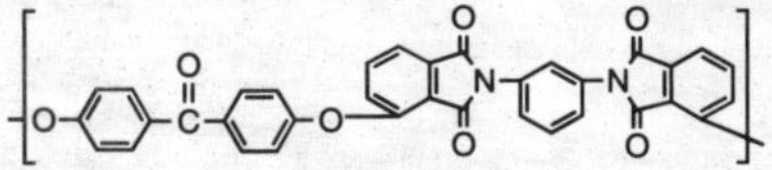	248	N-15; m-Cresol, Toluene	*Polym. Prep.*, **24**(2), 312;	1983	GE
			J. Polym. Sci., Polym. Symp., **74**, 93	1986	
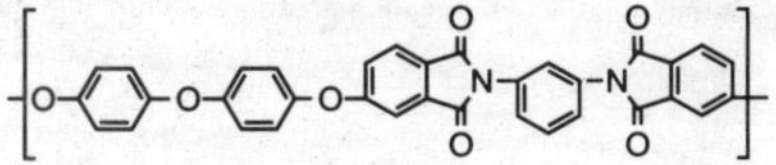	227	N-15; m-Cresol, Toluene	*Ency. Polym. Sci. Eng.*, **12**, 364;	1988	
			Polym. Prep., **24**(2), 312	1983	GE
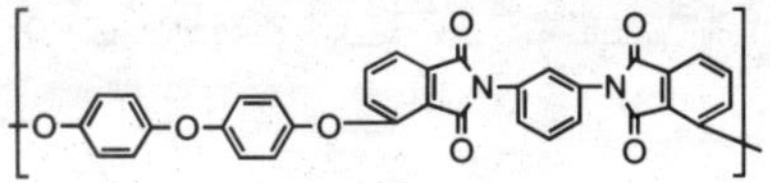	238	N-15; m-Cresol, Toluene	*Polym. Prep.*, **24**(2), 312;	1983	GE
	232		*J. Polym. Sci., Polym. Symp.*, **74**, 93;	1986	
		N-14; DMF	*J. Polym. Sci., Polym. Chem.*, **18**, 3069	1980	
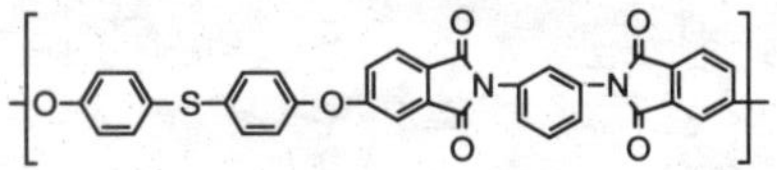	209	N-15; m-Cresol, Toluene	*Polym. Prep.*, **24**(2), 312;	1983	GE
	209		*J. Polym. Sci., Polym. Symp.*, **74**, 93;	1986	
		N-14; DMAC	*J. Polym. Sci., Polym. Chem.*, **18**, 3069	1980	
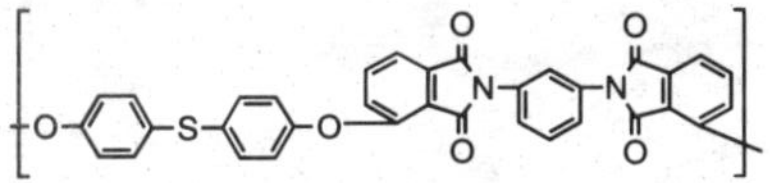	231	N-15; m-Cresol, Toluene	*Polym. Prep.*, **24**(2), 312;	1983	GE
	209		*J. Polym. Sci., Polym. Symp.*, **74**, 93;	1986	
		N-14; DMAC	*J. Polym. Sci., Polym. Chem.*, **18**, 3069	1980	
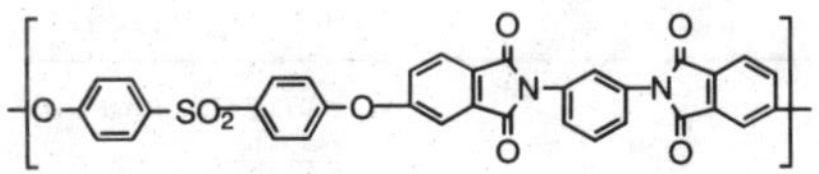	265	N-15; m-Cresol, Toluene	*Polym. Prep.*, **24**(2), 312;	1983	GE
			J. Polym. Sci., Polym. Symp., **74**, 93	1986	

TABLE 13—*continued*

Polymer structure	Tg (°C)	Tm (°C)	Synthesis	Reference	Year	Organization
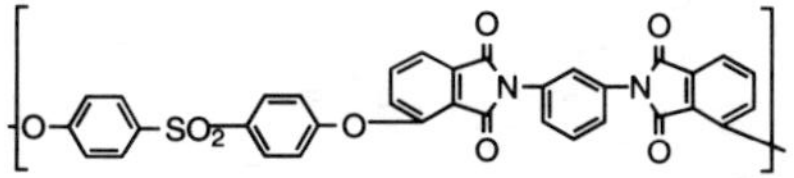	266		N-15; m-Cresol, Toluene	*Polym. Prep.*, **24**(2), 312; *J. Polym. Sci., Polym. Symp.*, **74**, 93	1983 1986	GE
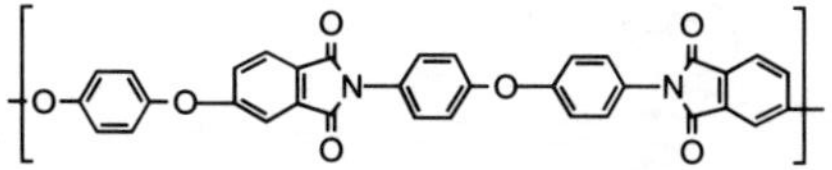	237		N-15; m-Cresol, Toluene	*Polym. Prep.*, **24**(2), 312; *J. Polym. Sci., Polym. Symp.*, **74**, 93	1983 1986	GE
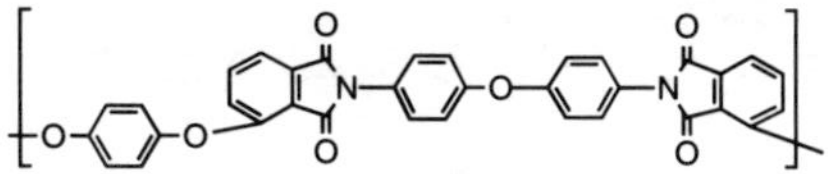	263		N-15; m-Cresol, Toluene	*Polym. Prep.*, **24**(2), 312; *J. Polym. Sci., Polym. Symp.*, **74**, 93;	1982 1986	GE
	237		N-14; DMF, DMSO; NO_2	*J. Polym. Sci., Polym. Chem.*, **18**, 3069; **19**,1635.	1980 1981	GE
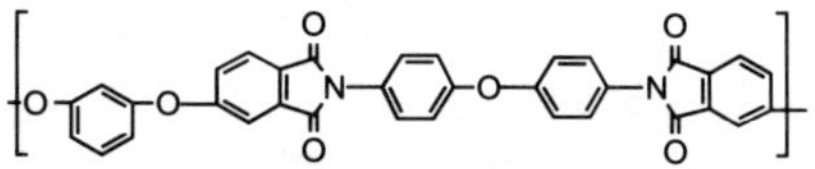	209		N-15; m-Cresol, Toluene	*Polym. Prep.*, **24**(2), 312; *J. Polym. Sci., Polym. Symp.*, **74**, 93;	1983	GE
			N-14; DMF	*J. Polym. Sci., Polym. Chem.*, **18**, 3069	1980	
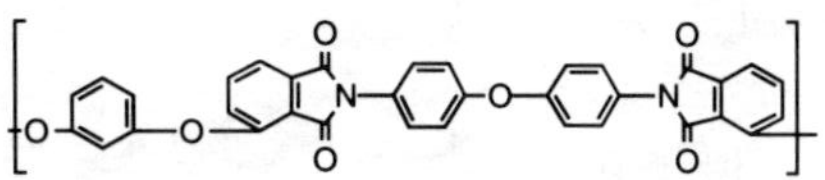	226		N-15; m-Cresol, Toluene	*Polym. Prep.*, **24**(2), 312; *J. Polym. Sci., Polym. Symp.*, **74**, 93;	1983 1986	GE
	226		N-14; DMF; NO_2	*J. Polym. Sci., Polym. Chem.*, **18**, 3069; **19**, 1635	1980 1981	

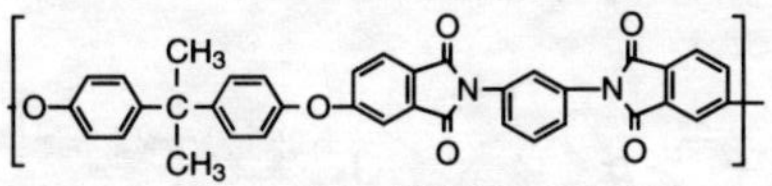 Ultem™ 1000	215	N-15; m-Cresol, Toluene	*Polym. Prep.*, **24**(2), 312; *J. Polym. Sci., Polym. Symp.*, **74**, 93	1983 1986	GE
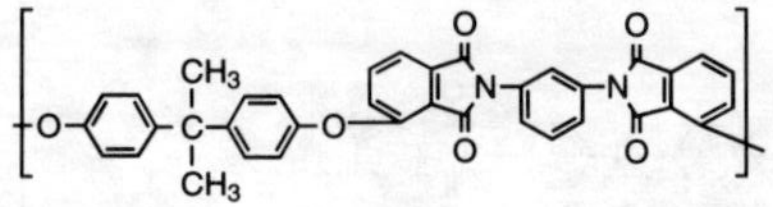	236	N-15; m-Cresol, Toluene	*Polym. Prep.*, **24**(2), 312; *J. Polym. Sci., Polym. Symp.*, **74**, 93	1983 1986	GE
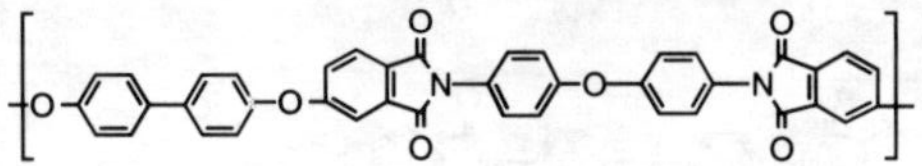	229	N-15; m-Cresol, Toluene	*Polym. Prep.*, **24**(2), 312; *J. Polym. Sci., Polym. Symp.*, **74**, 93	1983 1986	GE
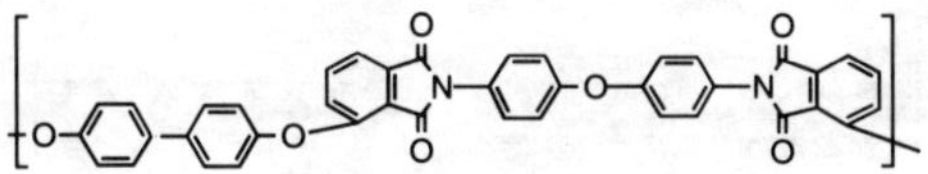	277	N-15; m-Cresol, Toluene	*Polym. Prep.*, **24**(2), 312; *J. Polym. Sci., Polym. Symp.*, **74**, 93;	1983 1986	GE
	277	N-14; DMSO; NO_2	*J. Polym. Sci., Polym. Chem.*, **18**, 3069; **19**, 1635	1980 1981	
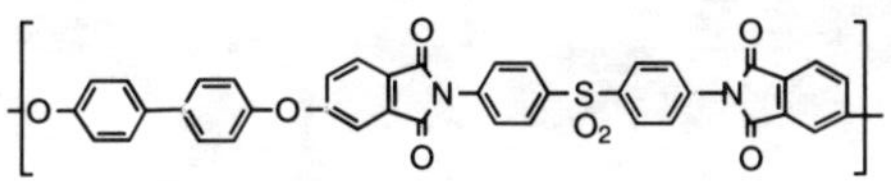	270	N-15	PC		GE
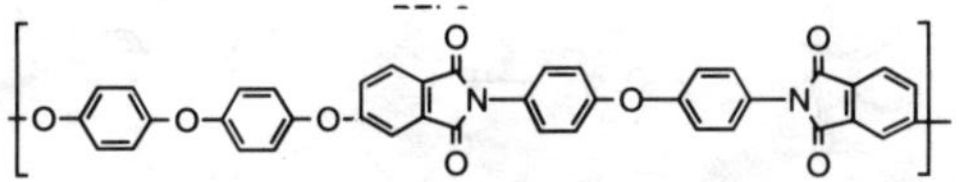	215	N-14; DMF; NO_2	*J. Polym. Sci., Polym. Chem.*, **19**, 1635; **18**, 3069	1981 1980	GE
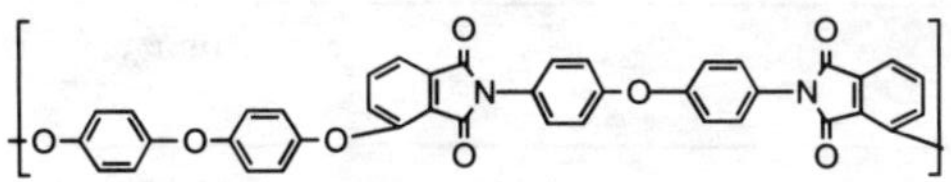	239	N-14; DMF; NO_2	*J. Polym. Sci., Polym. Chem.*, **19**, 1635; **18**, 3069	1981 1980	GE

TABLE 13—*continued*

Polymer structure	Tg (°C)	Tm (°C)	Synthesis	Reference	Year	Organization
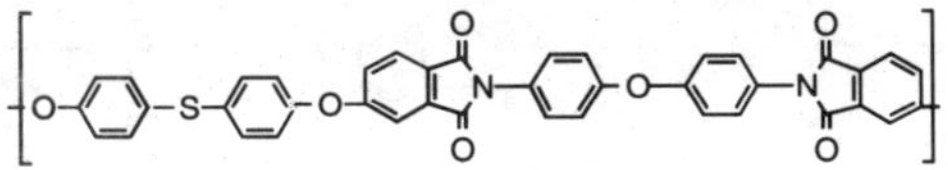	212		N-15; m-Cresol, Toluene	*Polym. Prep.*, **24**(2), 312;	1983	GE
				J. Polym. Sci., Polym. Symp., **74**, 93;	1986	
			N-14; DMAC	*J. Polym. Sci., Polym. Chem.*, **18**, 3069	1980	
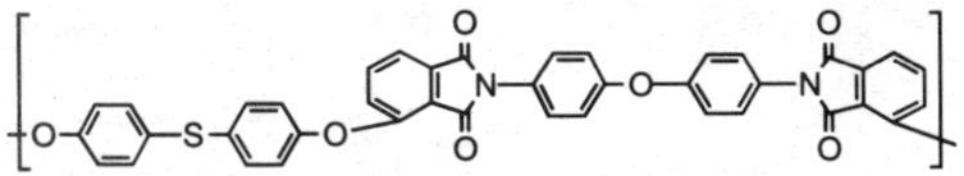	234		N-14; DMSO; NO_2	*J. Polym. Sci., Polym. Chem.*, **19**, 1635; **18**, 3069	1981 1980	GE
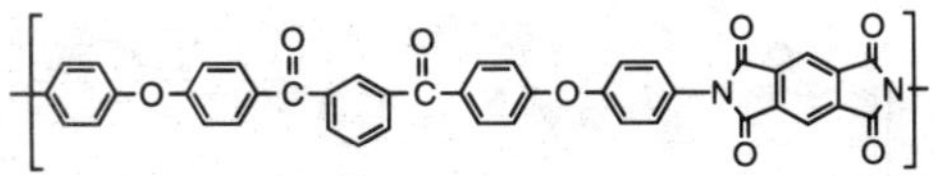	247	442	N-15; DMAC	*J. Polym. Sci., Polym. Chem.*, **25**,1093	1987	NASA/ Langley
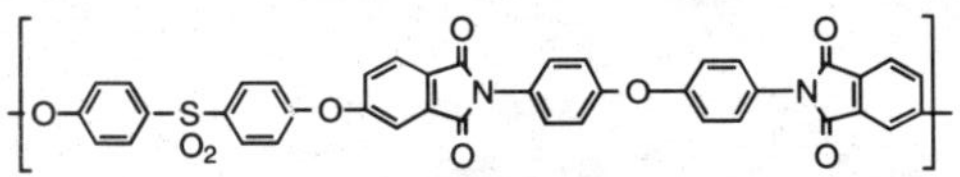	260		N-15; m-Cresol, Toluene	*Polym. Prep.*, **24**(2), 312;	1983	GE
				J. Polym. Sci., Polym. Symp., **74**, 93	1986	
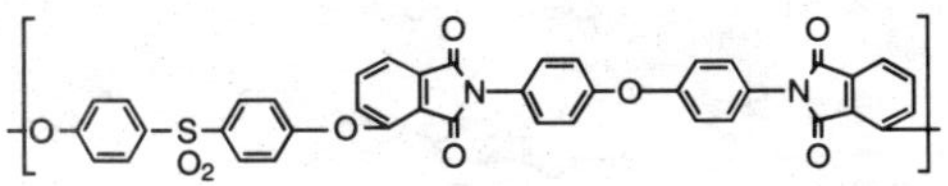	267		N-15; m-Cresol, Toluene	*Polym. Prep.*, **24**(2), 312;	1983	GE
				J. Polym. Sci., Polym. Symp., **74**, 93	1986	
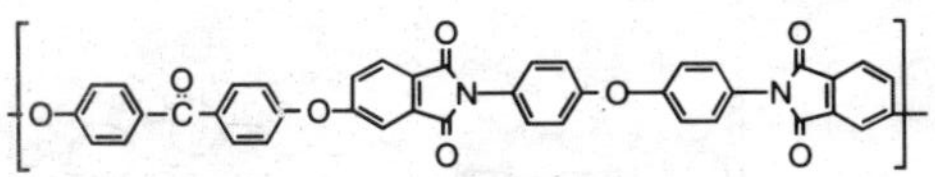	210		N-15; m-Cresol, Toluene	*Polym. Prep.*, **24**(2), 312;	1983	GE
				J. Polym. Sci., Polym. Symp., **74**, 93	1986	
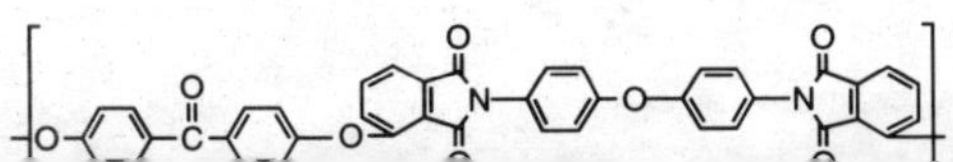	252		N-15; m-Cresol, Toluene	*Polym. Prep.*, **24**(2), 312;	1983	GE
				J. Polym. Sci., Polym.	1986	

Structure				Reference	Year	
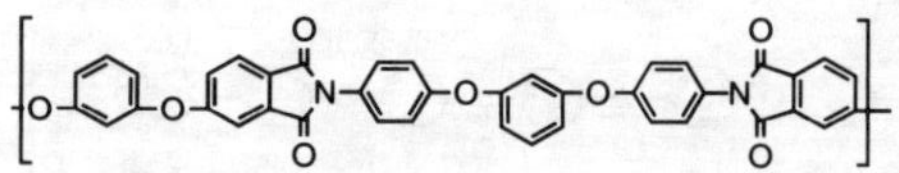	188		N-15; m-Cresol, Toluene	*Polym. Prep.*, **24**(2), 312; *J. Polym. Sci., Polym. Symp.*, **74**, 93	1983 1986	GE
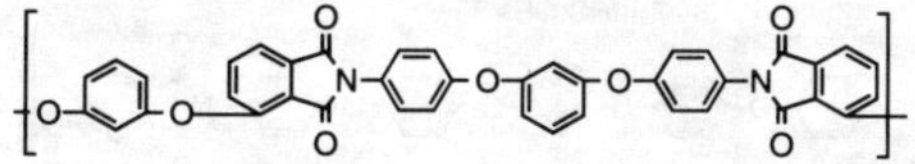	193		N-15; m-Cresol, Toluene	*Polym. Prep.*, **24**(2), 312; *J. Polym. Sci., Polym. Symp.*, **74**, 93	1983 1986	GE
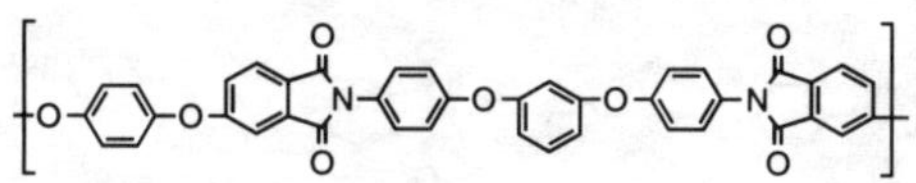	199	330	N-15; m-Cresol, Toluene	*Polym. Prep.*, **24**(2), 312; *J. Polym. Sci., Polym. Symp.*, **74**, 93	1983 1986	GE
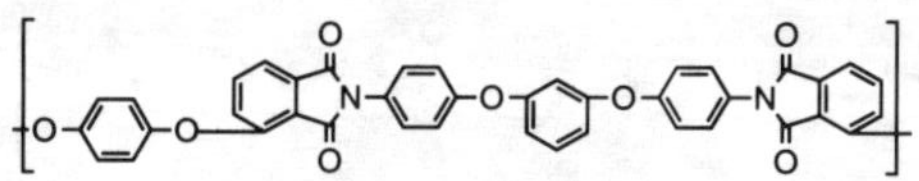	214		N-15; m-Cresol, Toluene	*Polym. Prep.*, **24**(2), 312; *J. Polym. Sci., Polym. Symp.*, **74**, 93	1983 1986	GE
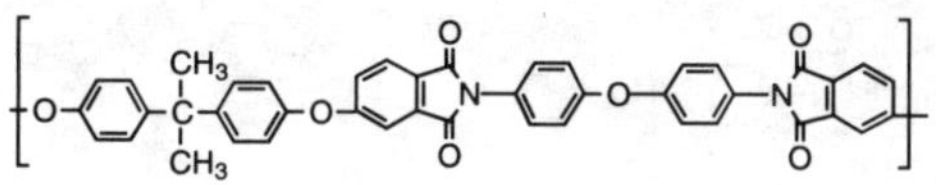	223		N-15; m-Cresol, Toluene	*Polym. Prep.*, 24(2), 312; *J. Polym. Sci., Polym. Symp.*, **74**, 93;	1983 1986	GE
	196		N-14; DMSO; NO_2	*J. Polym. Sci., Polym. Chem.*, **18**, 3069; **19**, 1635	1980 1981	
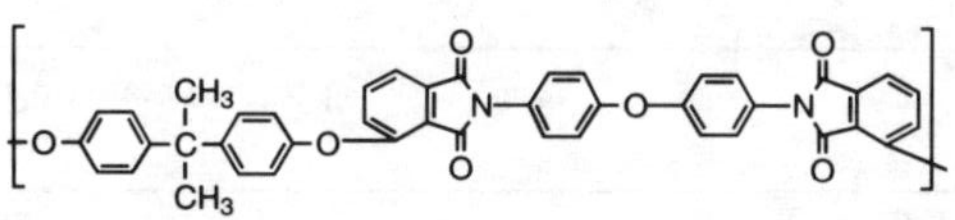	235		N-15; m-Cresol, Toluene	*Polym. Prep.*, **24**(2), 312; *J. Polym. Sci., Polym. Symp.*, **74**, 93;	1983 1986	GE
	226		N-14; DMSO; NO_2	*J. Polym. Sci., Polym. Chem.*, **18**, 3069; **19**, 1635	1980 1981	

TABLE 13—*continued*

Polymer structure	Tg (°C)	Tm (°C)	Synthesis	Reference	Year	Organization
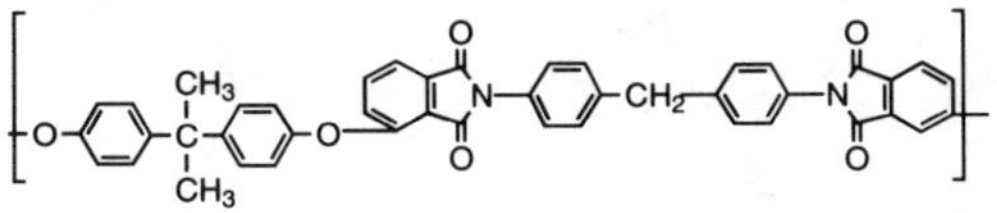	224		N-14; DMF; NO_2	*J. Polym. Sci., Polym. Chem.*, **19**,1635	1981	GE
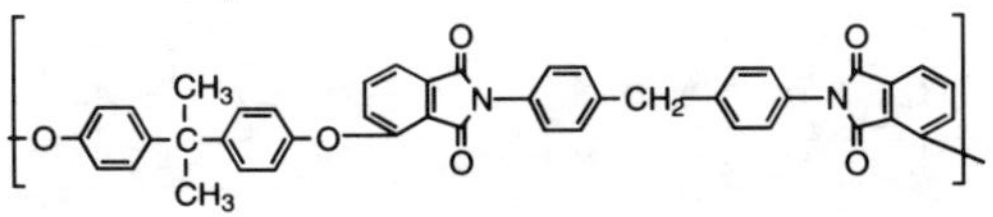	230		N-14; DMSO; NO_2	*J. Polym. Sci., Polym. Chem.*, **18**, 3069; **19**, 1635	1980 1981	
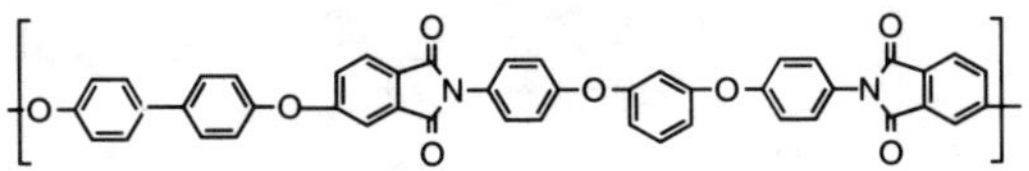	205	343	N-15; m-Cresol, Toluene	*Polym. Prep.*, **24**(2), 312; *J. Polym. Sci., Polym. Symp.*, **74**, 93	1983 1986	GE
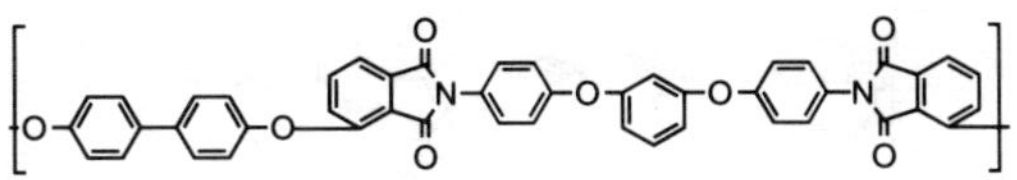	224		N-15; m-Cresol, Toluene	*Polym. Prep.*, **24**(2), 312; *J. Polym. Sci., Polym. Symp.*, **74**, 93	1983 1986	GE
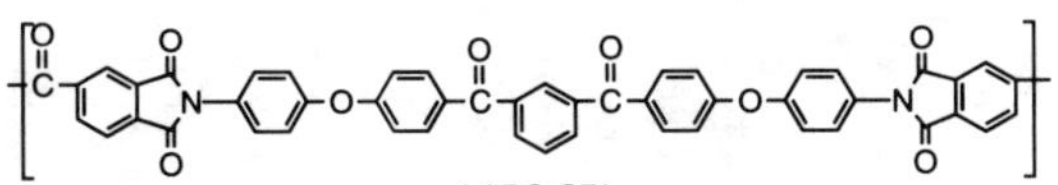 LARC-CPI	222	350	N-15; DMAC	*J. Polym. Sci., Polym. Chem.*, **25**,1093; *Polym. Prep.*, **33** (1),409	1987 1992	NASA/ Langley
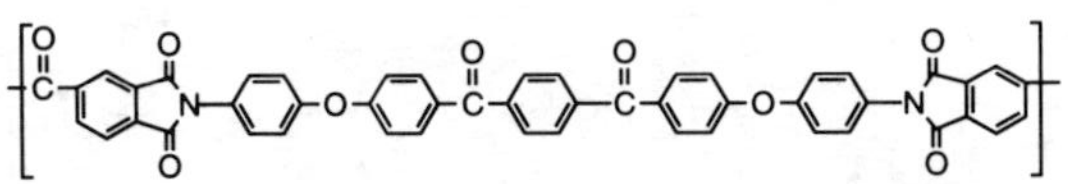	233	427	N-15; DMAC	*J. Polym. Sci., Polym. Chem.*, **25**,1093	1987	NASA/ Langley
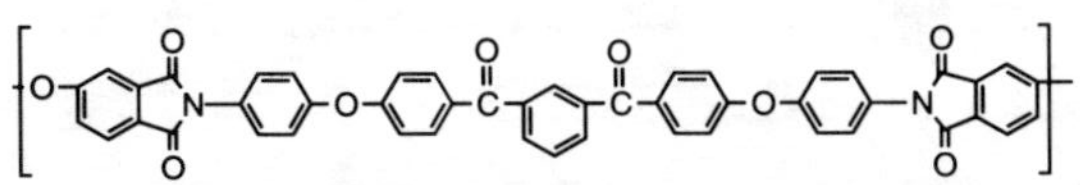	208		N-15; DMAC	*J. Polym. Sci., Polym. Chem.*, **25**,1093	1987	NASA/ Langley

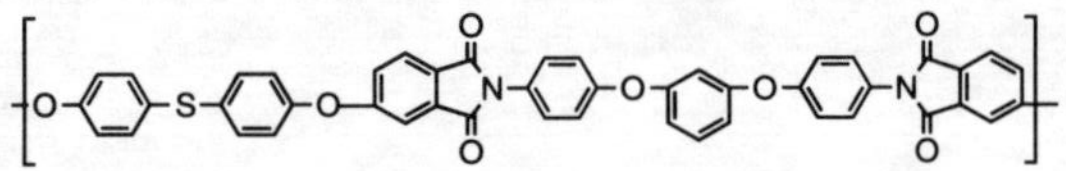	178	N-15; m-Cresol, Toluene	*Polym. Prep.*, **24**(2), 312; *J. Polym. Sci., Polym. Symp.*, **74**, 93	1983 1986	GE
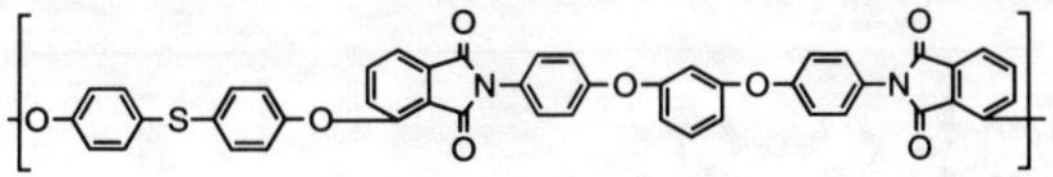	202	N-15; m-Cresol, Toluene	*Polym. Prep.*, **24**(2), 312; *J. Polym. Sci., Polym. Symp.*, **74**, 93	1983 1986	GE
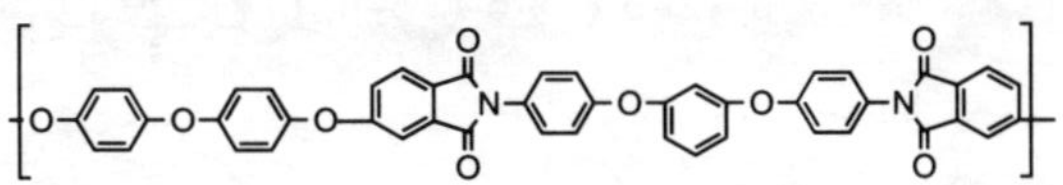	184	N-15; m-Cresol, Toluene	*Polym. Prep.*, **24**(2), 312; *J. Polym. Sci., Polym. Symp.*, **74**, 93	1983 1986	GE
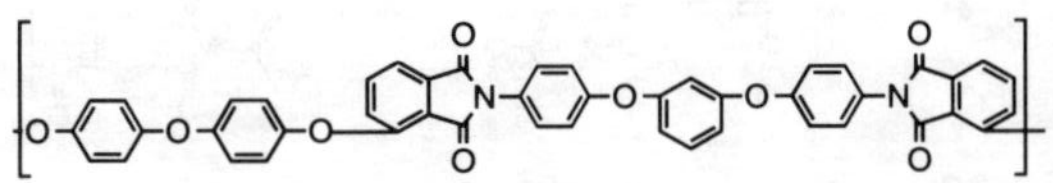	198	N-15; m-Cresol, Toluene	*Polym. Prep.*, **24**(2), 312; *J. Polym. Sci., Polym. Symp.*, **74**, 93	1983 1986	GE
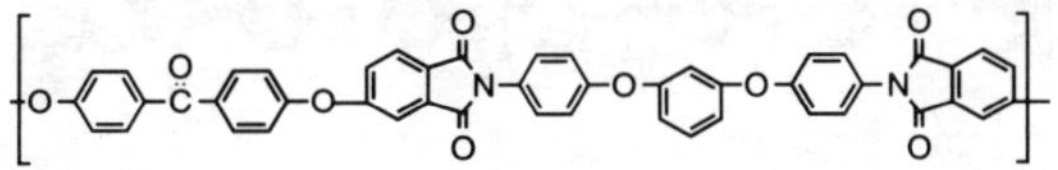	194	N-15; m-Cresol, Toluene	*Polym. Prep.*, **24**(2), 312; *J. Polym. Sci., Polym. Symp.*, **74**, 93	1983 1986	GE
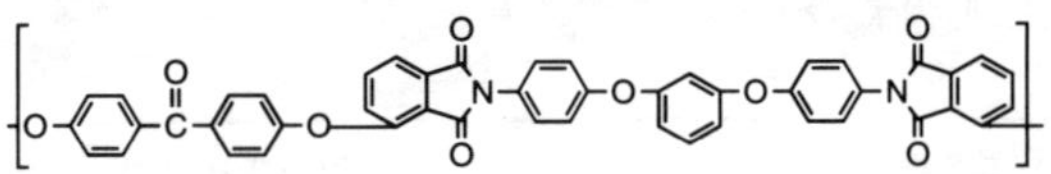	216	N-15; m-Cresol, Toluene	*Polym. Prep.*, **24**(2), 312; *J. Polym. Sci., Polym. Symp.*, **74**, 93	1983 1986	GE
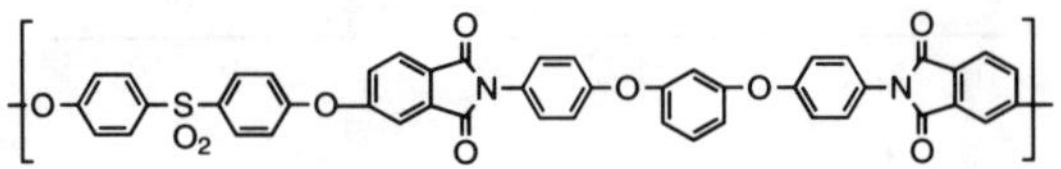	219	N-15; m-Cresol, Toluene	*Polym. Prep.* , **24**(2), 312; *J. Polym. Sci., Polym. Symp.*, **74**, 93	1983 1986	GE

TABLE 13—*continued*

Polymer structure	Tg (°C)	Tm (°C)	Synthesis	Reference	Year	Organization
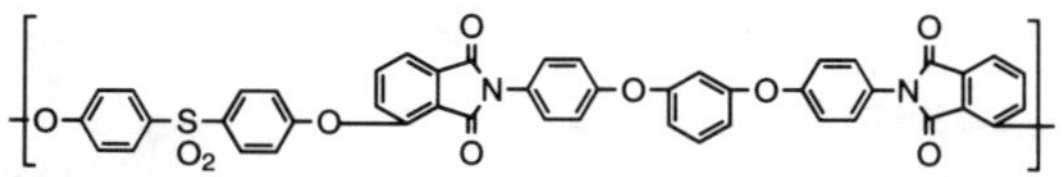	230		N-15; m-Cresol, Toluene	*Polym. Prep.*, **24**(2), 312; *J. Polym. Sci., Polym. Symp.*, **74**, 93	1983 1986	GE
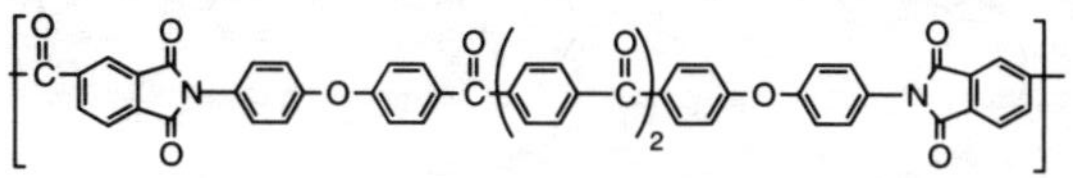	233	422	N-15; DMAC	*J. Polym. Sci., Polym. Chem.*, **25**,1093	1987	NASA/ Langley
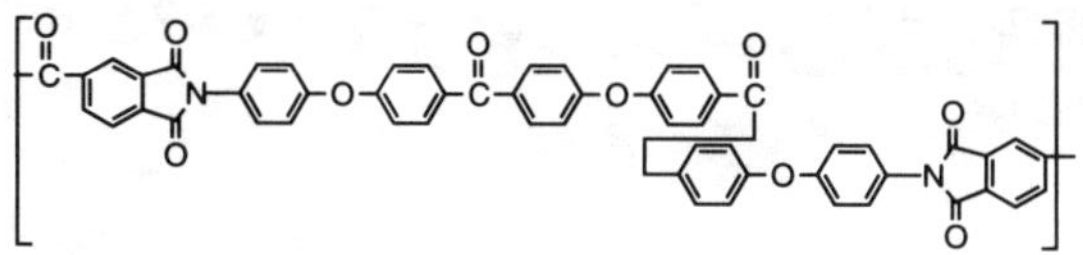	215	418	N-15; DMAC	*J. Polym. Sci., Polym. Chem.*, **25**,1093	1987	NASA/ Langley
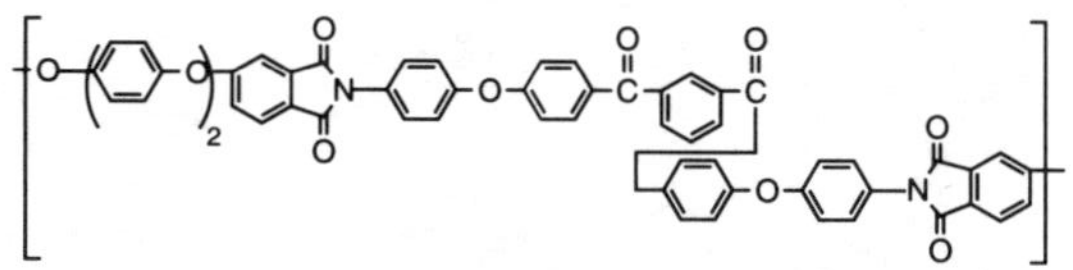	192		N-15; DMAC	*J. Polym. Sci., Polym. Chem.*, **25**,1093	1987	NASA/ Langley

TABLE 14
Polyarylethers containing three or fewer rings in the backbone of the repeat unit

Polymer structure	Tg (°C)	Tm (°C)	Synthesis	Reference	Year	Organization
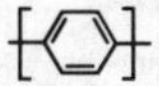		530				ICI
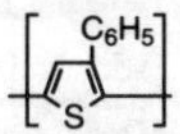	177		A-19; DMAC; Cl	*Macromol.*, **24**, 2694	1991	Yamagata U.
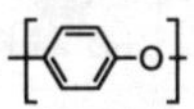	85	285	A-16; Dimethoxy-benzene; Br	*Prog. Polym. Sci.*, **12**, 111;	1986	Ind. Inst. Tech.; ICI
				Polym., **21**, 577	1980	
				Eur. Polym. J., **4**, 275	1968	
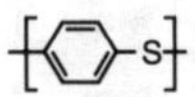	92	295		*Prog. Polym. Sci.*, **12**, 111	1986	Ind. Inst. Tech.; ICI
				Polym., **21**, 577	1980	
Ryton™	83	283		*Ency. Chem. Tech.*, **18**, 594	1982	Philips
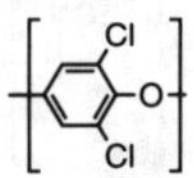	269		A-16; Tetramethyl urea; Br	*J. Polym. Sci.*, **58**, 469	1972	
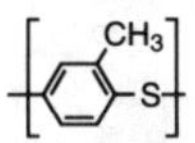	95		A-16; Quinoline, Pyridine; Br	*J. Appl. Polym. Sci.*, **24**, 1145	1979	
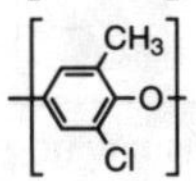			A-14; Pyridine, $C_6H_5NO_2$	*J. Polym. Sci.*, A, **58**, 581	1962	GE
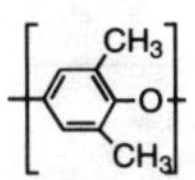	207			*Polym. Prep.*, **32**(3), 376	1991	No. Carolina St. U.

TABLE 14—*continued*

Polymer structure	Tg (°C)	Tm (°C)	Synthesis	Reference	Year	Organization
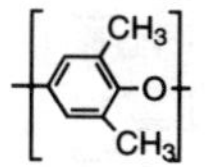 PPO™	211		A-16; Benzene, water; Br	*J. Appl. Polym. Sci.*, **16**, 461	1972	
	208	260	A-14; Pyridine, Nitrobenzene	*J. Polym. Sci.*, A, **58**, 581	1962	GE
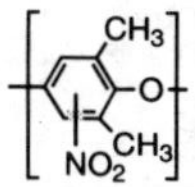	233		Nitration (15%) of PPO	*J. Memb. Sci.*, **72**, 91	1992	No. Carolina St. U.
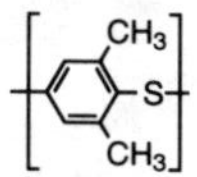	180		A-16; Quinoline, Pyridine; Br	*J. Appl. Polym. Sci.*, **24**, 1145	1979	
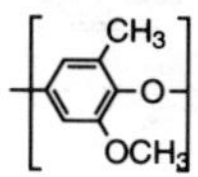			A-14; Pyridine, $C_6H_5NO_2$	*J. Polym. Sci.*, A, **58**, 581	1962	GE
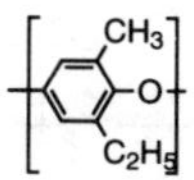			A-14; Pyridine, $C_6H_5NO_2$	*J. Polym. Sci.*, A, **58**, 581	1962	GE
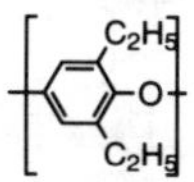			A-14; Pyridine, $C_6H_5NO_2$	*J. Polym. Sci.*, A, **58**, 581	1962	GE
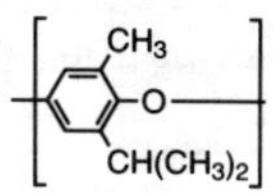			A-14; Pyridine, $C_6H_5NO_2$	*J. Polym. Sci.*, A, **58**, 581	1962	GE
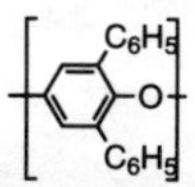	235	480	A-14; Pyridine, $C_6H_5NO_2$	*J. Polym. Sci.*, A, **58**, 581; *J. Polym. Sci., Polym. Chem.*, **31**, 1907; **31**,	1962 1993	GE

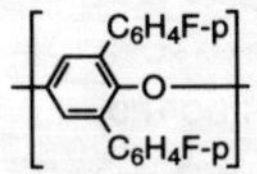	213	351	A-14; Cu-TMEDA, *o*-DCB	*J. Polym. Sci., Polym. Chem.*, **31**, 1261; **31**, 1907; **31**, 2015	1993	McGill U.
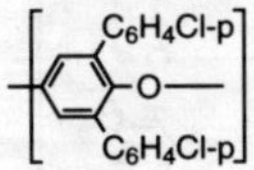	233	420	A-14; Cu-TMEDA, *o*-DCB	*J. Polym. Sci., Polym. Chem.*, **31**, 1261	1993	McGill U.
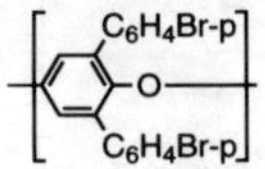	243	404	A-14; Cu-TMEDA, *o*-DCB	*J. Polym. Sci., Polym. Chem.*, **31**, 1261	1993	McGill U.
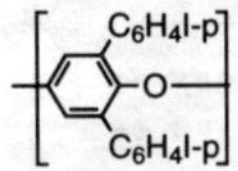	262		A-14; Cu-TMEDA, o-DCB	*J. Polym. Sci., Polym. Chem.*, **31**, 1261	1993	McGill U.
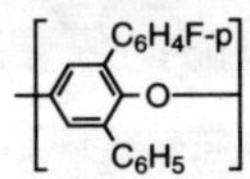	196	427	A-14; Cu-TMEDA, *o*-DCB	*J. Polym Sci., Polym Chem.*, **31**, 1907; **31**, 2015	1993	McGill
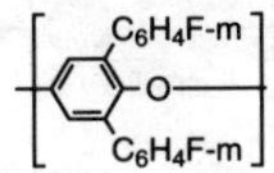	192	352	A-14; Cu-TMEDA, *o*-DCB	*J. Polym Sci., Polym Chem.*, **31**, 1907; **31**, 2015	1993	McGill
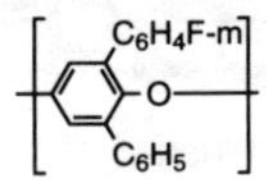	208	474	A-14; Cu-TMEDA, *o*-DCB	*J. Polym Sci., Polym Chem.*, **31**, 1907; **31**, 2015	1993	McGill
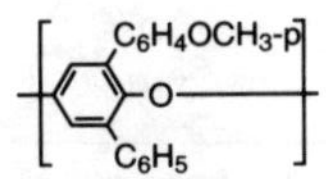	217		A-14; Cu-TMEDA, *o*-DCB	*J. Polym. Sci., Polym. Chem.*, **31**, 1261	1993	McGill U.
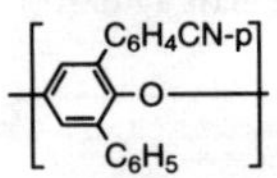	241		A-14; Cu-TMEDA, *o*-DCB	*J. Polym. Sci., Polym. Chem.*, **31**, 1261	1993	McGill U.

TABLE 14—*continued*

Polymer structure	Tg (°C)	Tm (°C)	Synthesis	Reference	Year	Organization
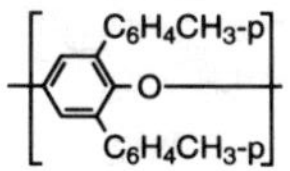	221	321	A-14; Cu-TMEDA, *o*-DCB	*J. Polym. Sci., Polym. Chem.*, **31**, 1261	1993	McGill U.
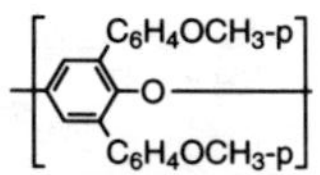	220	356	A-14; Cu-TMEDA, *o*-DCB	*J. Polym. Sci., Polym. Chem.*, **31**, 1261	1993	McGill U
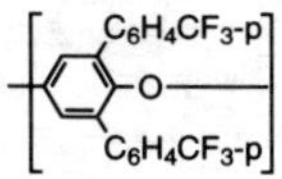		384	A-14; Cu-TMEDA, *o*-DCB	*J. Polym Sci., Polym Chem.*, **31**, 1907; **31**, 2015	1993	McGill U.
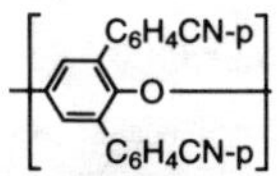	298		A-14; Cu-TMEDA, *o*-DCB	*J. Polym. Sci., Polym. Chem.*, **31**, 1261	1993	McGill U.
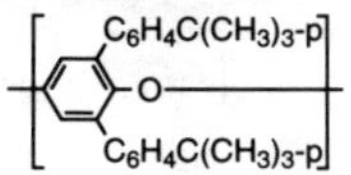	248	483	A-14; Cu-TMEDA, *o*-DCB	*J. Polym. Sci., Polym. Chem.*, **31**, 1261	1993	McGill U.
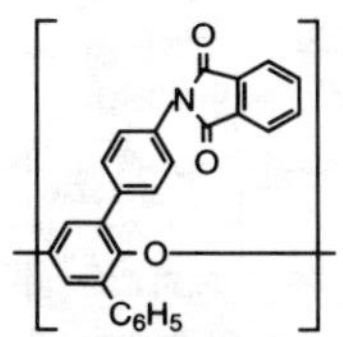	275		A-14; Cu-TMEDA, *o*-DCB	*J. Polym. Sci., Polym. Chem.*, **31**, 1261	1993	McGill U.
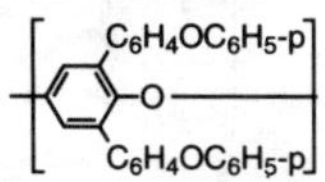	141		A-14; Cu-TMEDA, *o*-DCB	*J. Polym. Sci., Polym. Chem.*, **31**, 1261	1993	McGill U.

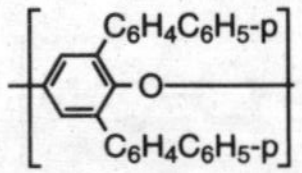	265	>500	A-14; Cu-TMEDA, o-DCB	*J. Polym. Sci., Polym. Chem.*, **31**, 1261	1993	McGill U.
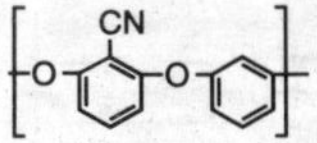 PEN™	145	340	N-17, Na_2CO_3, NMP; Cl	S. Matsuo, PC		Idemitsu Kosan
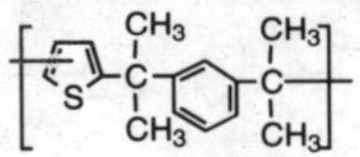	120		E-9; Trityl Cl, $AlCl_3$, Nitrobenzene, CH_2Cl_2	*J. Polym. Sci.*, A-1, **10**, 2365	1972	DuPont
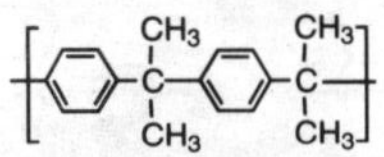	120		E-9; Trityl Cl, $AlCl_3$, Nitrobenzene, CH_2Cl_2	*J. Polym. Sci.*, A-1, **10**, 2365	1972	DuPont
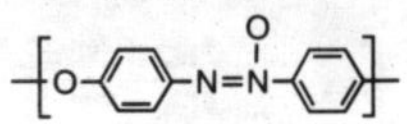			N-1; NaOH, DMSO; Cl	*J. Polym. Sci., Polym. Chem. Ed.*, **24,** 2447	1986	
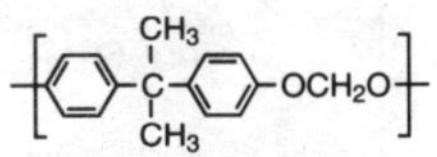	89		N-1; KOH, CH_2Cl_2; Cl	*Polym. Prep.*, **34**(1), 417	1993	Kanagawa U.
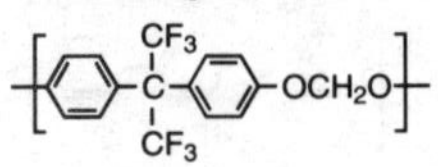	127		N-1; KOH, CH_2Cl_2; Cl	*Polym. Prep.*, **34**(1), 417	1993	Kanagawa U.
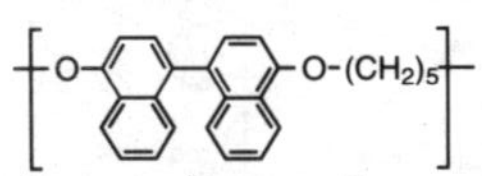			A-17; $FeCl_3$, Nitrobenzene	*Polym. Prep.*, **32**(1), 391	1991	Case West. Reserve U.
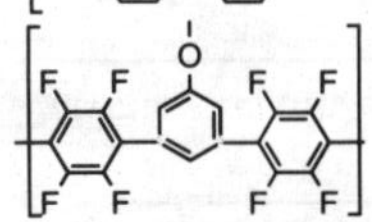	152		N-17; NaH, DMAC; F	*Polym. Prep.*, **34**(1), 58	1993	AT&T

TABLE 14—*continued*

Polymer structure	Tg (°C)	Tm (°C)	Synthesis	Reference	Year	Organization
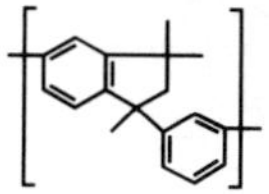		280–320	E-9; Trityl Cl, $AlCl_3$, Nitrobenzene, CH_2Cl_2	*J. Polym. Sci.*, A-1, **10**, 2365	1972	DuPont
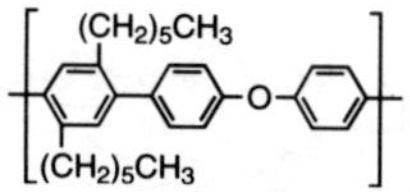			A-18; Na_2CO_3, Pd(0), Toluene, H_20; Br	*Makromol. Chem., Rapid Comm.*, **11**, 535	1990	U. Karlsruhe
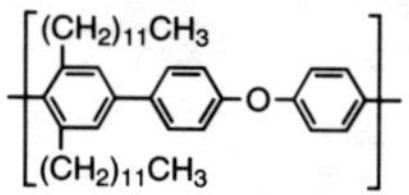			A-18; Na_2CO_3, Pd(0), Toluene, H_20; Br	*Makromol. Chem., Rapid Comm.*, **11**, 535	1990	U. Karlsruhe
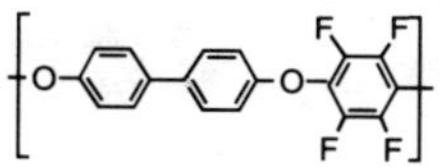		>280	N-18; DMAC; F	*Polym. Prep.*, **22**(2), 383	1981	U. Texas, San Antonio
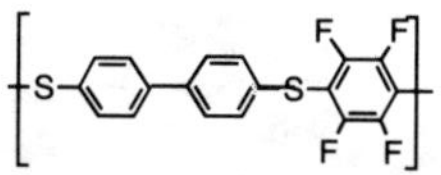	190	300	N-18; DMAC; F	*Polym. Prep.*, **22**(2), 383	1981	U. Texas, San Antonio
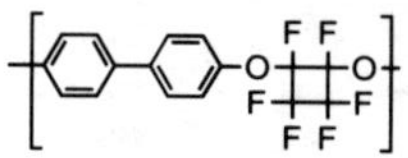	165		Vinyl ether Dimerization	*Polym. Prep.*, **34**(1), 413	1993	Dow Chemical Co.
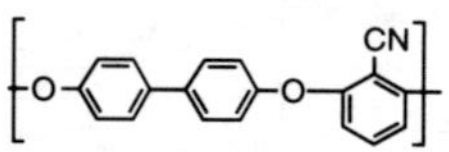	221		N-17; K_2CO_3, NMP; F, Cl	Jap. Pat. 60-14739	1985	Idemitsu Ind.
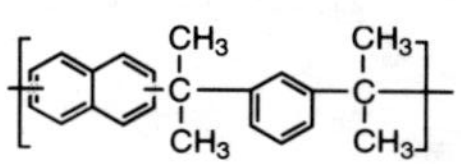	190		E-9; Trityl Cl, $AlCl_3$, Nitrobenzene, CH_2Cl_2	*J. Polym. Sci.*, A-1, **10**, 2365	1972	DuPont

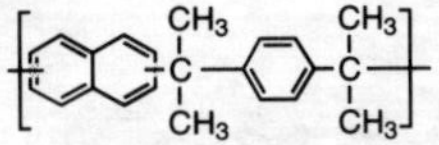	220		E-9; Trityl Cl, $AlCl_3$, Nitrobenzene, CH_2Cl_2	*J. Polym. Sci.*, A-1, **10**, 2365	1972	DuPont
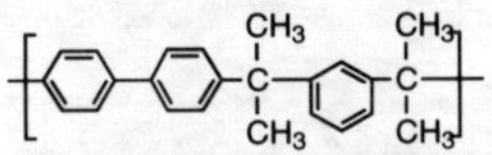	140		E-9; Trityl Cl, $AlCl_3$, Nitrobenzene, CH_2Cl_2	*J. Polym. Sci.*, A-1, **10**, 2365	1972	DuPont
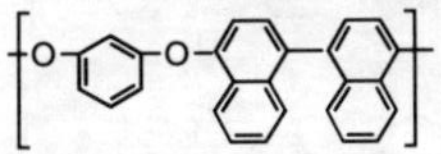	218		A-17; $FeCl_3$, Nitrobenzene	*Macromol.*, **25**, 64	1992	Case West. Res.U.
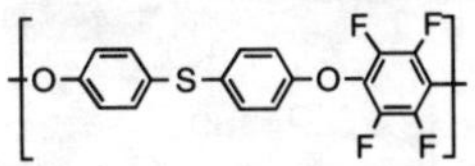			N-18; DMAC; F	*Polym. Prep.*, **21**(2), 164	1980	U. Texas, San Antonio
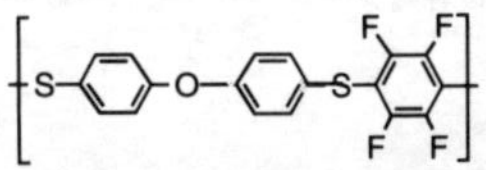	135	300	N-18; DMAC; F	*Polym. Prep.*, **22**(2), 383	1981	U. Texas, San Antonio
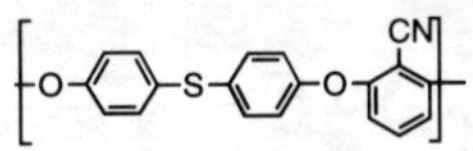	147		N-1; NaOH, DMSO; NO_2	*J. Polym. Sci., Polym. Chem. Ed.*, **18**, 3069	1980	GE
	152		N-17; K_2CO_3, NMP; Cl	*Polym. Prep.*, **23**(1), 284	1982	Va. P. I. & State U.
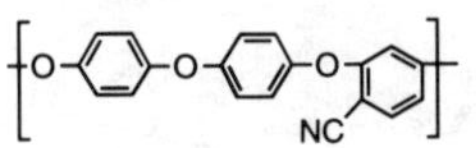	136		N-1; NaOH, DMSO; NO_2	*J. Polym. Sci., Polym. Chem. Ed.*, **18**, 3069	1980	GE
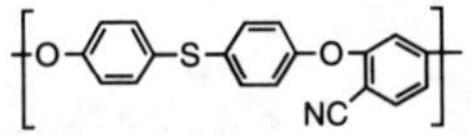	134		N-1; NaOH, DMSO; NO_2	*J. Polym. Sci., Polym. Chem. Ed.*, **18**, 3069	1980	GE
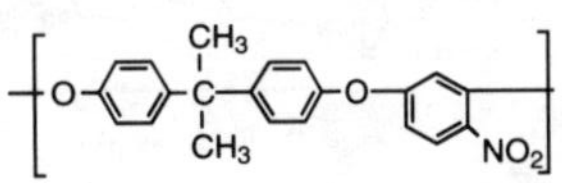	150		N-1; NaOH, DMSO, Sulfolane; Cl	*J.Polym.Sci.*, A-1, **5**, 2375	1967	Union Carbide
				Ency. Polym. Sci. & Eng., **13**,196	1988	

TABLE 14—*continued*

Polymer structure	Tg (°C)	Tm (°C)	Synthesis	Reference	Year	Organization
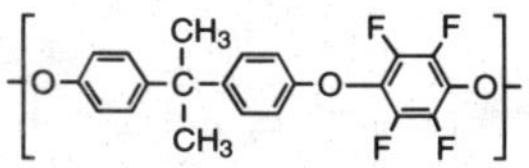	153		N-18; DMAC; F	*Polym. Prep.*, **21**(2),164	1980	U. Texas, San Antonio
				Polym. Prep., **22**(2), 385		
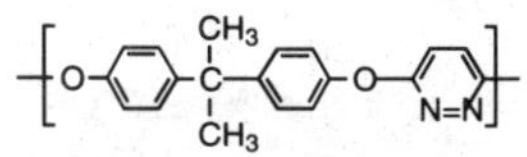	180	250	N-1; NaOH, DMSO, Sulfolane; Cl	*J.Polym. Sci.*, A-1, **5**, 2375	1967	Union Carbide
			N-19; K_2CO_3, NMP; Cl	*Makromol. Chem.*, **191**, 2027	1990	Hamburg U
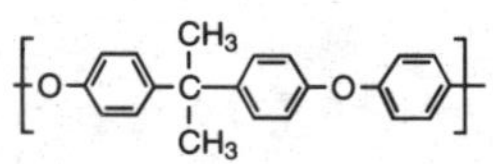	135		A-16; NaOH, CuCl, Pyridine, Benzo-phenone; Br	US Pat. 3332909	1967	Union Carbide
			N-1; NaOH, Cu_2O, DMSO; Br	*J. Polym. Sci.*, C, No. **16**, 715	1967	ICI
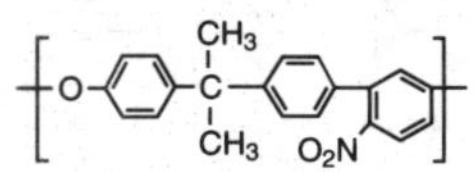	150		N-1; KOH, DMSO; Cl	GB Pat. 1078234	1967	Union Carbide
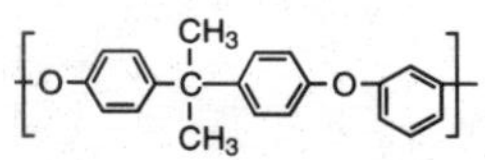			A-16; NaOH, CuCl, Pyridine, Benzo-phenone; Br	US Pat. 3332909	1967	Union Carbide
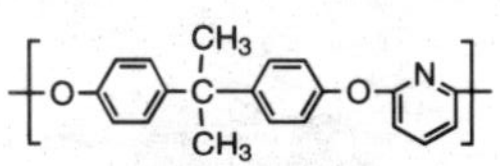			N-19; K_2CO_3, NMP; Cl	*Makromol. Chem.* **191**, 2027	1990	Hamburg U.
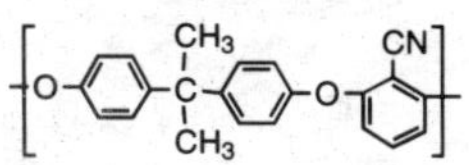	173		N-17; K_2CO_3, NMP; Cl	*Polym. Prep.*, **23**(1), 284	1982	Va. P. I. & State U.
	173		N-20; DMSO; NO_2	*J. Polym. Sci., Polym. Chem.*, **18**, 3069	1980	
			N-19; K_2CO_3, NMP; Cl	*Makromol. Chem.* **191**, 2027	1990	Hamburg U.

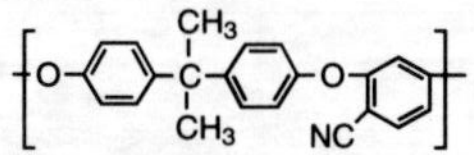	141	N-1; NaOH , DMSO; NO_2	*J. Polym. Sci., Polym. Chem.*, **18**, 3069	1980	GE
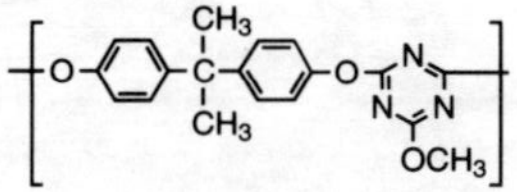		N-1; NaOH, $CHCl_3$; Cl	*Eur. Polym. J.*, **20**, 519	1984	
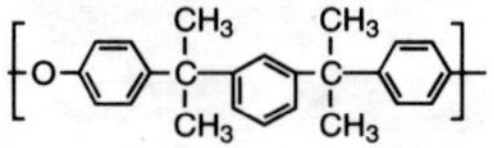	130	E-9; Trityl Cl, $AlCl_3$, Nitrobenzene, CH_2Cl_2	*J. Polym. Sci.*, A-1, **10**, 2365	1972	DuPont
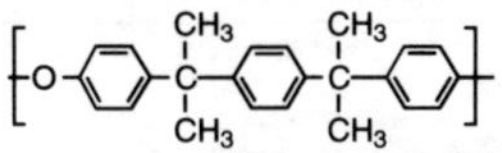	150	E-9; Trityl Cl, $AlCl_3$, Nitrobenzene, CH_2Cl_2	*J. Polym. Sci.*, A-1, **10**, 2365	1972	DuPont
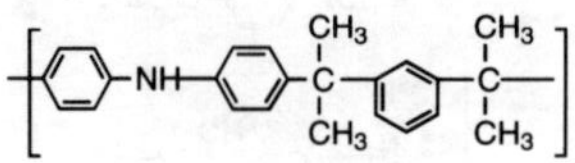	170	E-9; Trityl Cl, $AlCl_3$, Nitrobenzene, CH_2Cl_2	*J. Polym. Sci.*, A-1, **10**, 2365	1972	DuPont
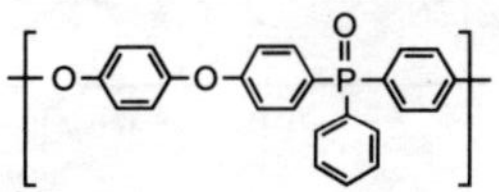	215	N-17; K_2CO_3, NMP; F	*Polym. Prep.*, **32**(1), 93	1991	Va. P. I. & State U.
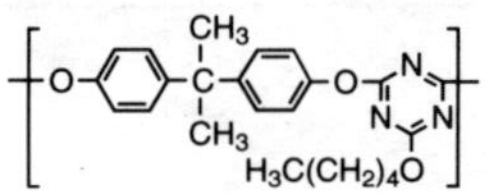		N-1; NaOH, $CHCl_3$; Cl	*J. Polym. Sci.*, A-1, **7**, 2749	1969	
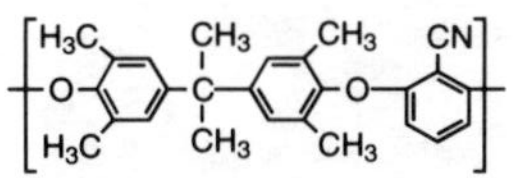		N-17; K_2CO_3, NMP; F	*Polym. Prep.*, **25**(2), 19	1984	Va. P. I. & State U.

TABLE 14—*continued*

Polymer structure	Tg (°C)	Tm (°C)	Synthesis	Reference	Year	Organization
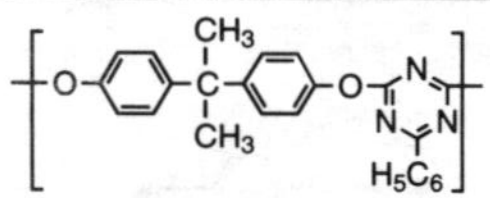			N-1; NaOH, Nitrobenzene; Cl	*J. Polym. Sci.*, A-1, **7,** 3089	1969	
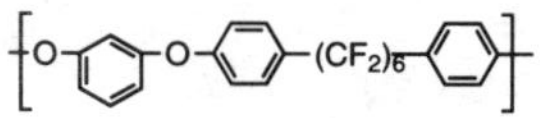	49		N-17; K_2CO_3, NMP; F	*Macromol.* , **23**, 5371	1990	IBM

TABLE 15
Polyarylethers containing four rings in the backbone of the repeat unit

Polymer structure	Tg (°C)	Tm (°C)	Synthesis	Reference	Year	Organization
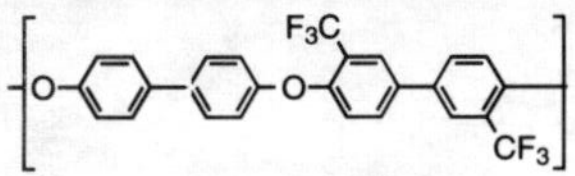	200		N-20; K_2CO_3, NMP; NO_2	*Polym. Prep.*, **34**(1), 415	1993	IBM
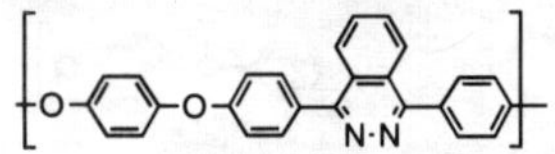	260		Post Reaction on PAEK	*Macromol.*, **25**, 1025	1992	McGill U.
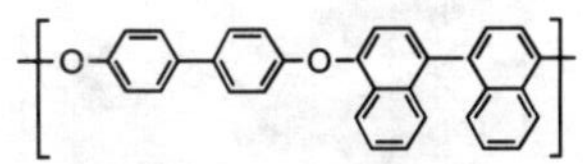	235		A-17; $FeCl_3$, Nitrobenzene	*Macromol.*, **25**, 64	1992	Case West. Res.U.
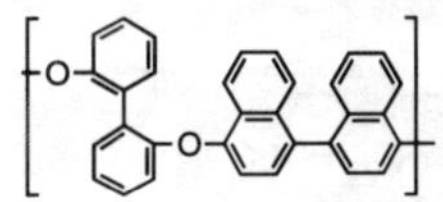	236		A-17; $FeCl_3$, Nitrobenzene	*Macromol.*, **25**, 64	1992	Case West. Res.U.
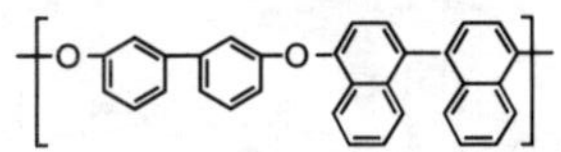	215		A-17; $FeCl_3$, Nitrobenzene	*Macromol.*, **25**, 64	1992	Case West. Res.U.
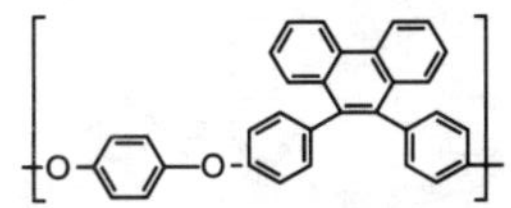	242		Post reaction on PAEK	*Polym. Prep.*, **33**(2), 352	1992	Carleton U.
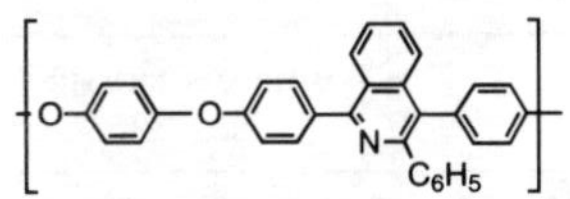	220		Post Reaction on PAEK	*Macromol.*, **25**, 1033	1992	McGill U.

TABLE 15—*continued*

Polymer structure	Tg (°C)	Tm (°C)	Synthesis	Reference	Year	Organization
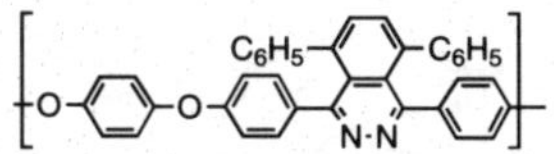	270		Post Reaction on PAEK	*Macromol.*, **25**, 1025	1992	McGill U.
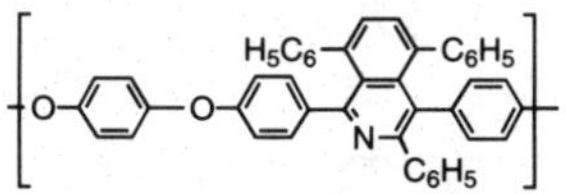	260		Post Reaction on PAEK	*Macromol.*, **25**, 1033	1992	McGill U.
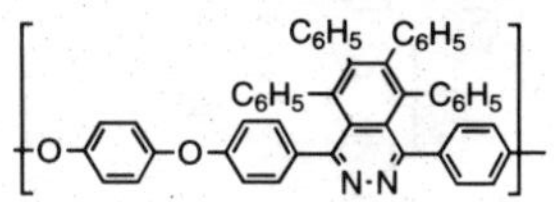	296		Post Reacton on PAEK	*Macromol.*, **24**(9), 2640	1991	McGill U.
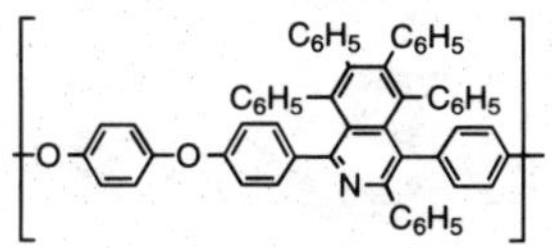	297		Post reaction on PAEK	*Macromol.*, **24**(9), 2643	1991	McGill U.
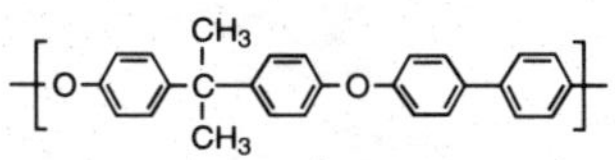	175		A-16; NaOH, CuCl, Pyridine, Benzophenone; Br	US Pat. 3332909	1967	Union Carbide
	154		A-16; Br	*Appl. Polym. Symp.* No. **26**, 373	1975	Union Carbide
	133		A-19; Ni(0), DMAC; Cl	*J. Polym. Sci., Polym. Chem.*, **31**, 923	1993	Case W. R. U.
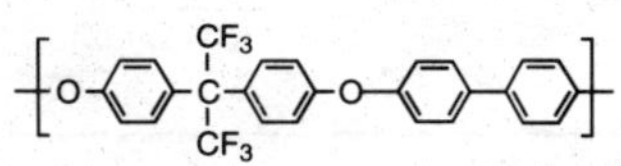	190		A-16; NaOH, CuCl, Pyridine, Benzophenone; Br	US Pat. 3332909	1967	Union Carbide
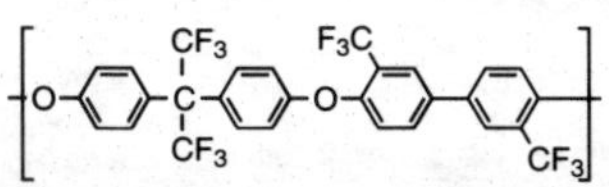	167		N-20; K_2CO_3, NMP; NO_2	*Polym. Prep.*, **34**(1), 415	1993	IBM

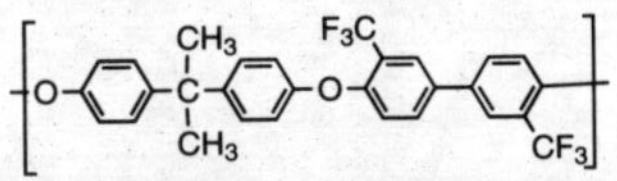	167	N-20; K_2CO_3, NMP; NO_2	*Polym. Prep.*, **34**(1), 415	1993	IBM
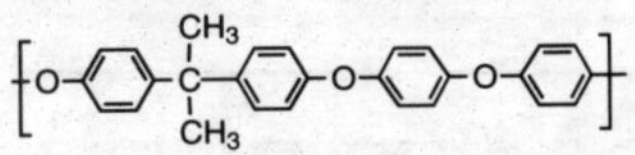	125	A-16; NaOH, CuCl, Pyridine, Benzo-phenone; Br	US Pat. 3332909	1967	Union Carbide
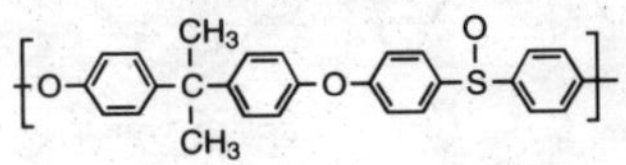	165	N-1; NaOH, DMSO; Cl	*J. Polym. Sci.*, A-1, **5**, 2375	1967	Union Carbide
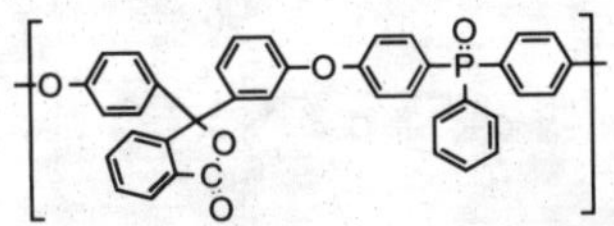	279	N-17; K_2CO_3, NMP, DMAC; F	*Polym. Prep.*, **33**(2), 231; **34**(1), 310	1992 1993	Va. P. I & State U.
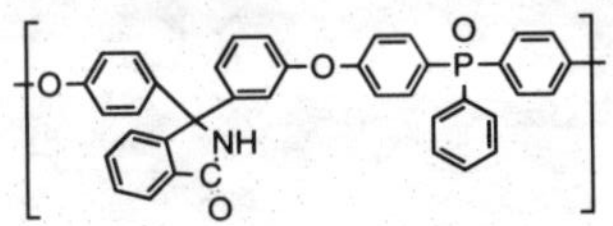	294	N-17; K_2CO_3, NMP, DMAC; F	*Polym. Prep.*, **33**(2), 231; **34**(1), 310	1992 1993	Va. P. I & State U.
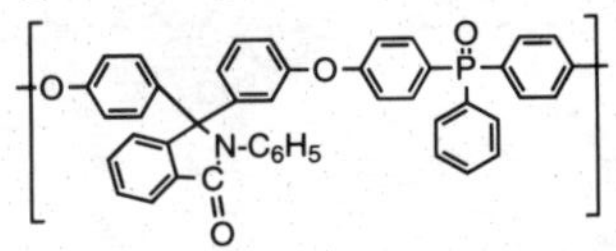	273	N-17; K_2CO_3, NMP, DMAC; F	*Polym. Prep.*, **33**(2), 231; **34**(1), 310	1992 1993	Va. P. I & State U.
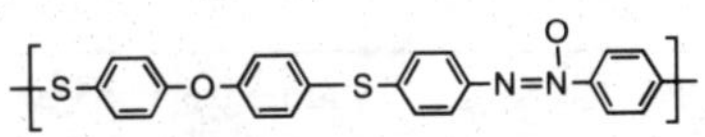		N-1; NaOH, DMSO; Cl	*J. Polym. Sci., Polym. Chem.*, **24,** 2447	1986	
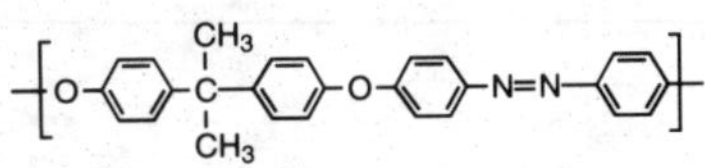	175	N-1; NaOH, DMSO, Sulfolane; Cl,F	*J.Polym.Sci.*, A-1, **5**, 2375; *Ency. Polym. Sci. & Eng.*, **13**,196	1967 1988	Union Carbide

TABLE 15—*continued*

Polymer structure	Tg (°C)	Tm (°C)	Synthesis	Reference	Year	Organization
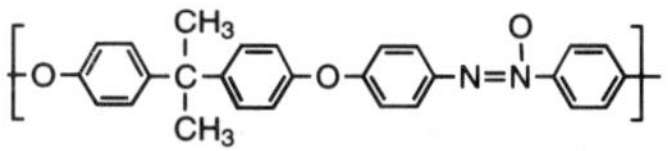			N-1; NaOH, DMSO; Cl	*J. Pol. Sci., Polym. Chem.*, **24**, 2447	1986	
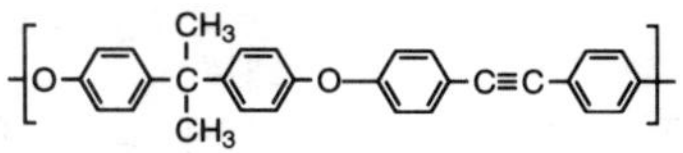	163		N-17; K_2CO_3, NMP, Toluene; F	*Macromol.*, **26**, 1777	1993	McGill U.
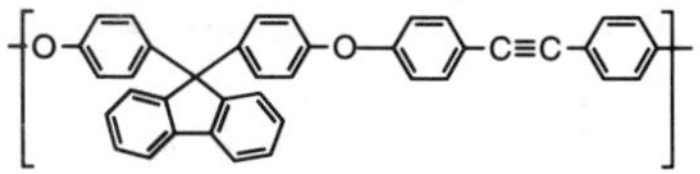	264		N-17; K_2CO_3, NMP, Toluene; F	*Macromol.*, **26**, 1777	1993	McGill U.
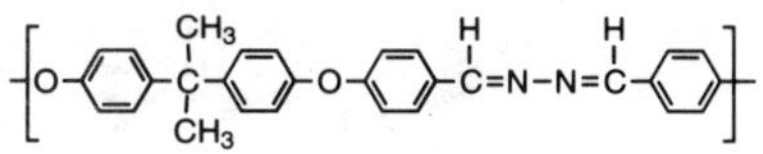	163		N-17; K_2CO_3, DMPU; F	*Polym. Prep.*, **34**(1), 521	1993	IBM
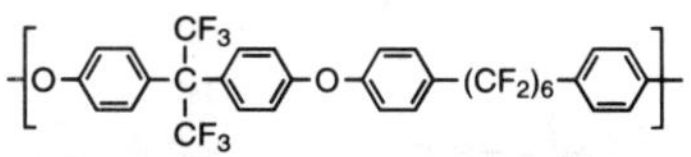	95		N-17; K_2CO_3, NMP; F	*Macromol.*, **23**, 5371	1990	IBM
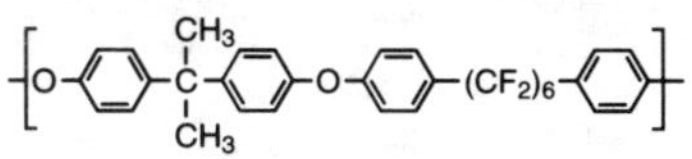	93		N-17; K_2CO_3, NMP; F	*Macromol.*, **23**, 5371	1990	IBM

TABLE 16
Polyarylethers containing five rings in the backbone of the repeat unit

Polymer structure	Tg (°C)	Tm (°C)	Synthesis	Reference	Year	Organization
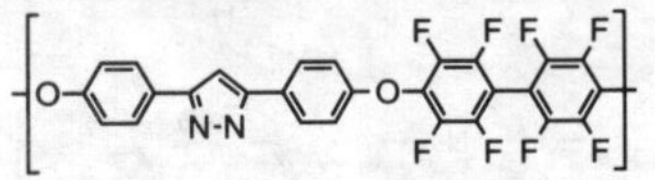	196		N-9; K_2CO_3, DMAC; F	*Polym. Prep.*, **34**(1), 441	1993	Va. Comm. U.
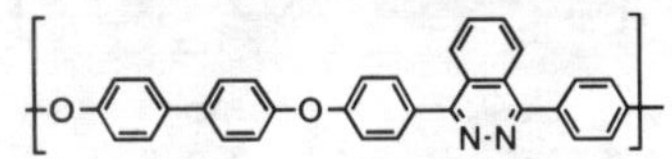	255		Post Reaction on PAEK	*Macromol.*, **25**, 1025	1992	McGill U.
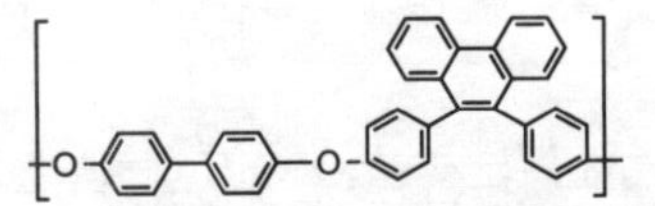	251		Post Reaction on PAEK	*Polym. Prep.*, **33**(2), 351	1992	Carleton U.
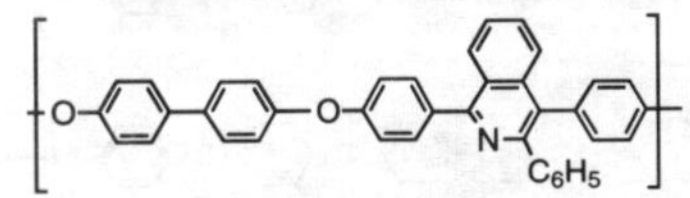	230		Post Reaction on PAEK	*Macromol.*, **25**, 1033	1992	McGill U.
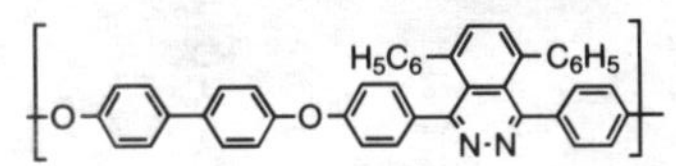	300		Post Reaction on PAEK	*Macromol.*, **25**, 1025	1992	McGill U.
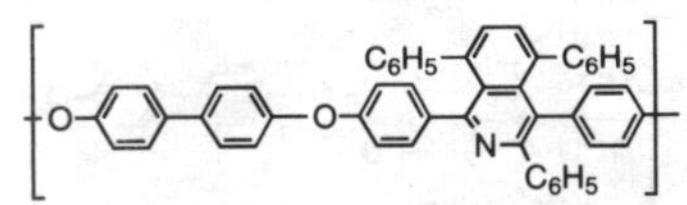	275		Post Reaction on PAEK	*Macromol.*, **25**, 1033	1992	McGill U.
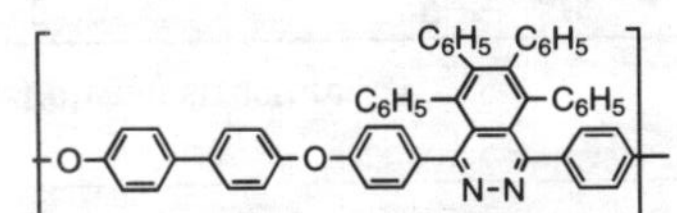	340		Post Reacton on PAEK	*Macromol.*, **24**(9), 2640	1991	McGill U.

TABLE 16—*continued*

Polymer structure	Tg (°C)	Tm (°C)	Synthesis	Reference	Year	Organization
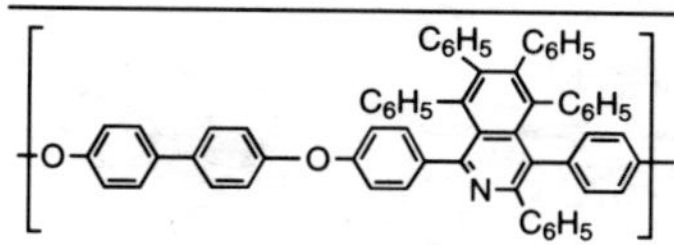	320		Post reaction on PAEK	*Macromol.* **24**(9), 2643	1991	McGill U.
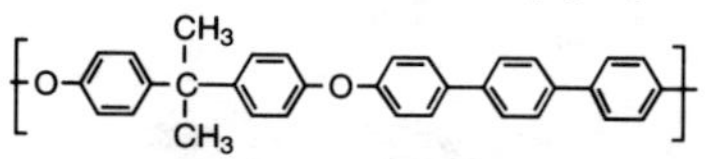	182		A-16; Br	*Appl. Polym. Symp.* No. **26**, 373	1975	Union Carbide
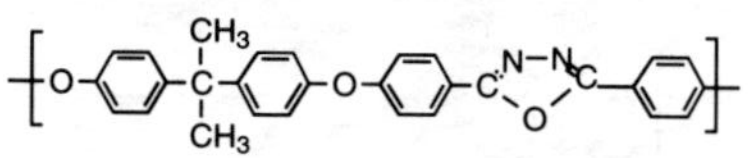	180		N-1; NaOH, DMSO, Sulfolane; Cl,F	*J.Polym.Sci.*, A-1, **5**, 2375	1967	Union Carbide
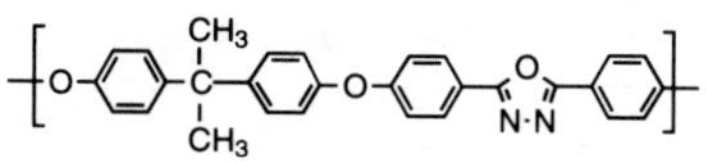	201		N-17; K_2CO_3, NMP, CHP; F	*Polym. Bull.*, **25**, 543	1991	IBM, San Jose
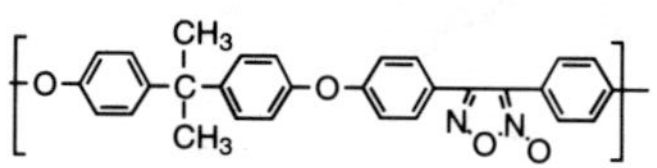			N-1; NaOH, DMSO; Cl	*Eur. Polym. J.*, **22**, 447	1986	
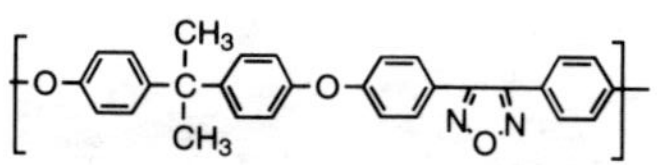			N-1; NaOH, DMSO; Cl	*Eur. Polym. J.*, **22**, 447	1986	
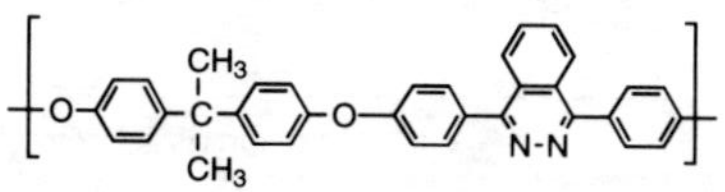	236		Post Reaction on PAEK	*Macromol.*, **24**(9), 2640	1991	McGill U.
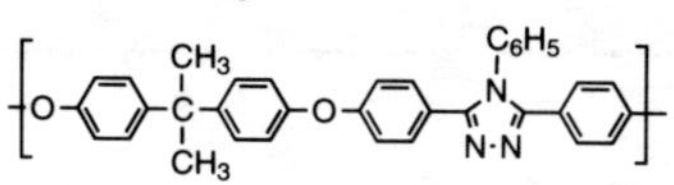	207		N-17; K_2CO_3, DMPU; F	*Polym. Prep.*, **34**(1), 481	1993	IBM

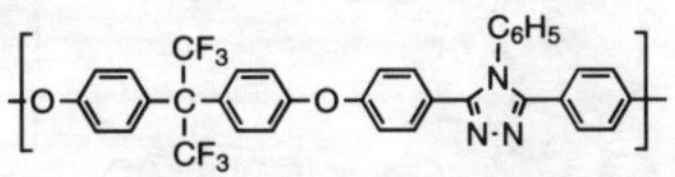	203	N-17; K_2CO_3, DMPU; F	*Polym. Prep.*, **34**(1), 481	1993	IBM
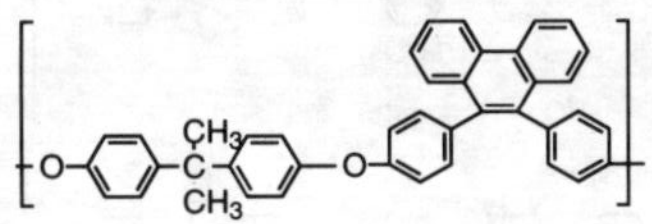	228	Post Reaction on PAEK	*Polym. Prep.*, **33**(2), 351	1992	Carleton U
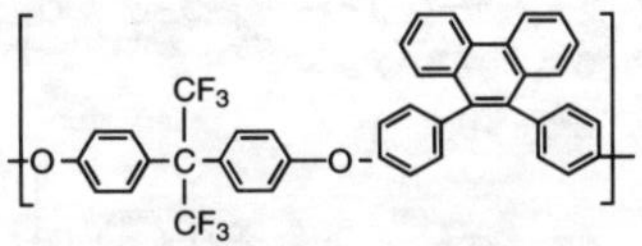	230	Post Reaction on PAEK	*Polym. Prep.*, **33**(2), 351	1992	Carleton U.
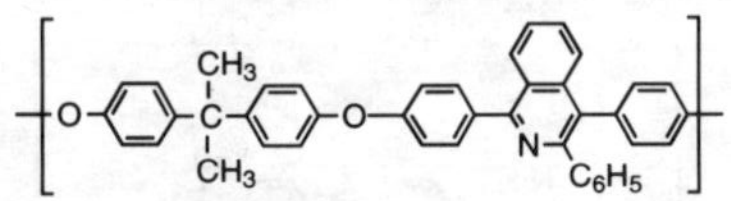	226	Post reaction on PAEK	*Macromol.*, **24**(9), 2643	1991	McGill U.
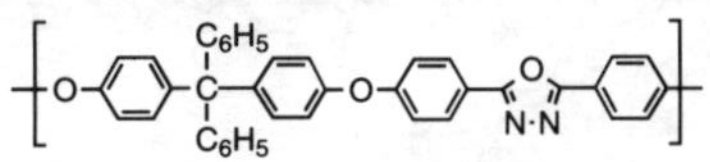	190	N-17; K_2CO_3, NMP, CHP; F	*Polym. Bull.*, **25**, 543	1991	IBM, San Jose
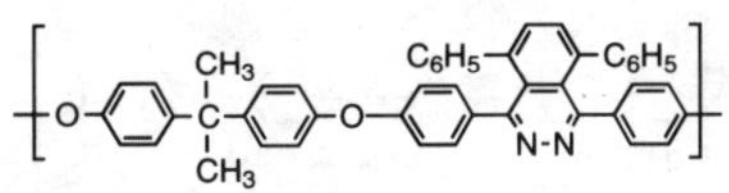	250	Post Reaction on PAEK	*Macromol.*, **24**(9), 2640	1991	McGill U.
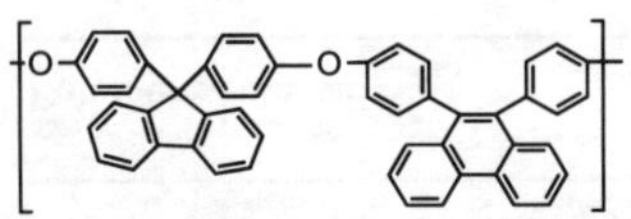		Post Reaction on PAEK	*Polym. Prep.*, **33**(2), 351	1992	Carleton U.

TABLE 16—*continued*

Polymer structure	Tg (°C)	Tm (°C)	Synthesis	Reference	Year	Organization
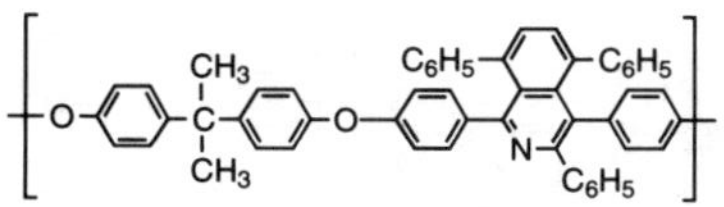	234		Post reaction on PAEK	*Macromol.*, **24**(9), 2643	1991	McGill U.
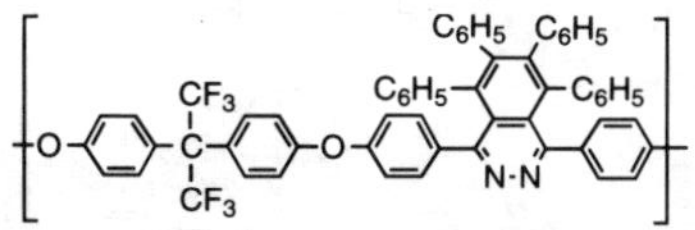	293		Post Reacton on PAEK	*Macromol.*, **24**(9),2640	1991	McGill U.
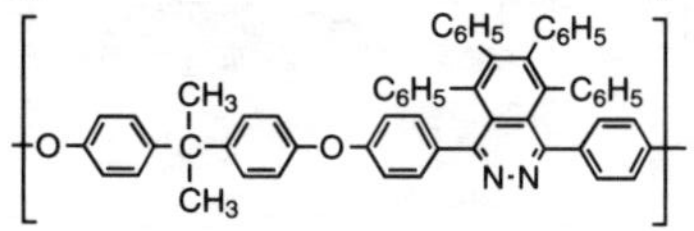	283		Post Reaction on PAEK	*Macromol.*, **24**(9), 2640	1991	McGill U.
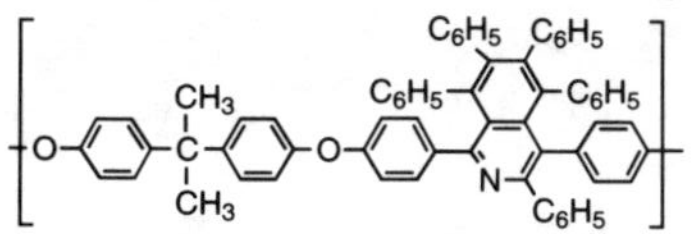	280		Post reaction on PAEK	*Macromol.*, **24**(9), 2643	1991	McGill U.
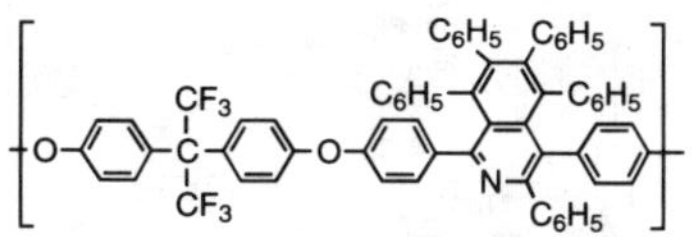	294		Post reaction on PAEK	*Macromol.*, **24**(9), 2643	1991	McGill U.
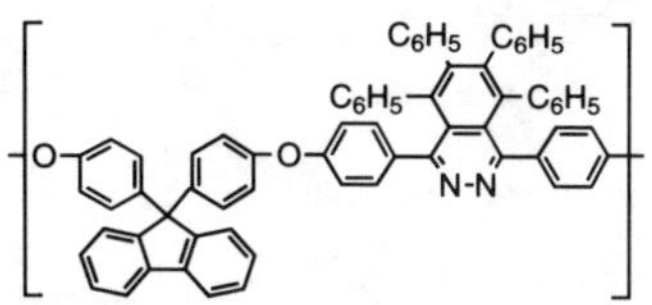	340		Post Reacton on PAEK	*Macromol.*, **24**(9), 2640	1991	McGill U.

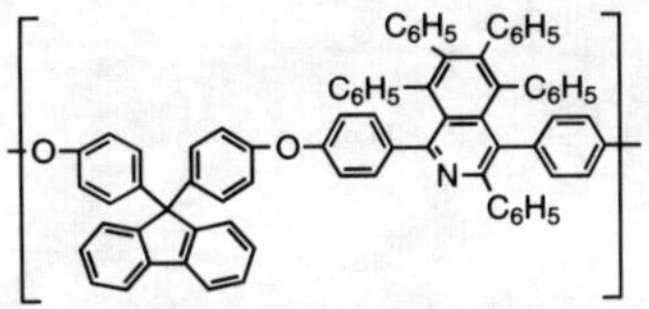	323	Post reaction on PAEK	*Macromol.*, **24**(9), 2643	1991	McGill U.
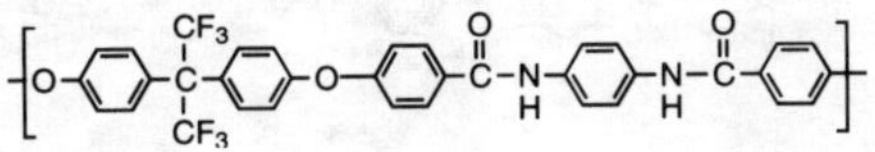	218	N-17; K_2CO_3, NMP, CHP; F	*Makromol. Chem., Macromol. Symp.*, **54/55**, 313	1992	IBM
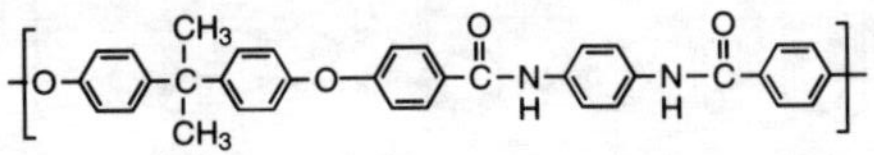	230	N-17; K_2CO_3, NMP, CHP; F	*Makromol. Chem., Macromol. Symp.*, **54/55**, 313	1992	IBM
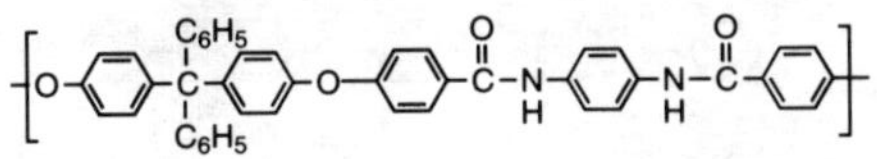	254	N-17; K_2CO_3, NMP, CHP; F	*Makromol. Chem., Macromol. Symp.*, **54/55**, 313	1992	IBM

TABLE 17
Polyarylethers containing six or more rings in the backbone of the repeat unit

Polymer structure	Tg (°C)	Tm (°C)	Synthesis	Reference	Year	Organization
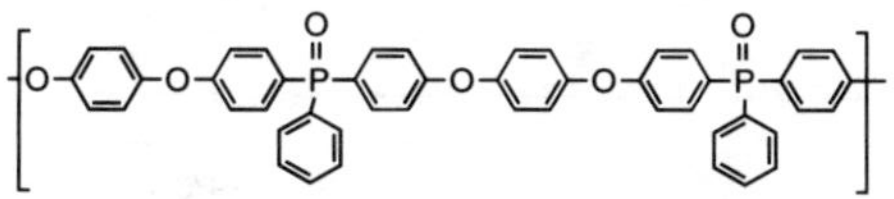	213		N-17; K_2CO_3, DMAC; F	*Polym. Prep.*, **33**(2), 241	1992	NASA/ Langley
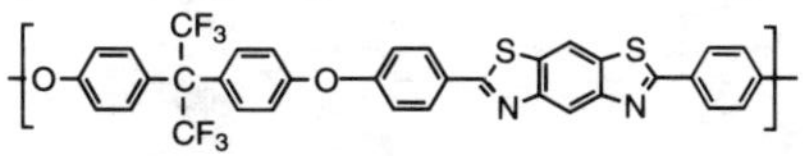			N-17; K_2CO_3,NMP, CHP; F	*Makromol. Chem., Macromol. Symp.*, **54/55**, 313	1992	IBM
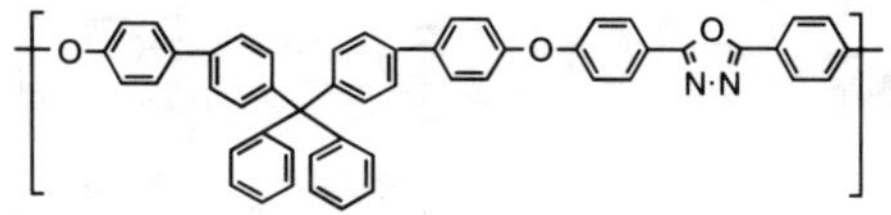	220		N-17; K_2CO_3, NMP, CHP; F	*Polym. Bull.*, **25**, 543	1991	IBM, San Jose
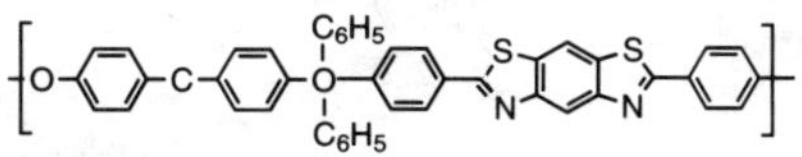			N-17; K_2CO_3, NMP, CHP; F	*Makromol. Chem., Macromol. Symp.*, **54/55**, 313	1992	IBM
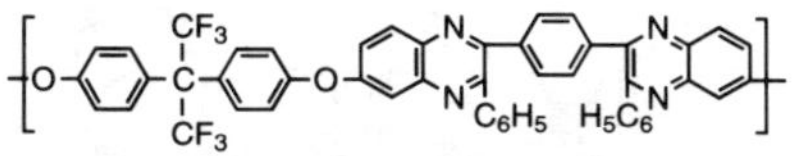	255		N-17; K_2CO_3, NMP; F	*Makromol. Chem., Macromol. Symp.*, **54/55**, 313	1992	IBM
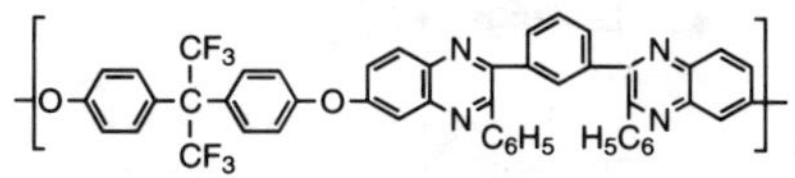	240		N-17; K_2CO_3, NMP; F	*Makromol. Chem., Macromol. Symp.*, **54/55**, 313	1992	IBM
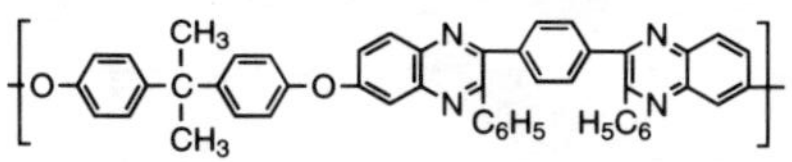	255		N-17; K_2CO_3, NMP; F	*Makromol. Chem., Macromol. Symp.*, **54/55**, 313	1992	IBM

Structure					
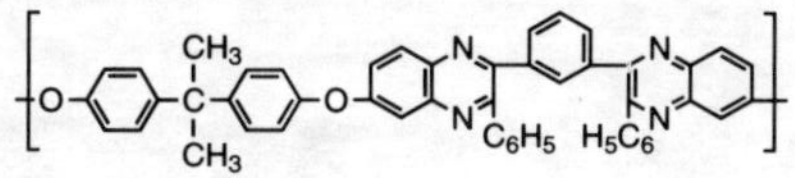	230	N-17; K_2CO_3, NMP; F	*Makromol. Chem., Macromol. Symp.*, **54/55**, 313	1992	IBM
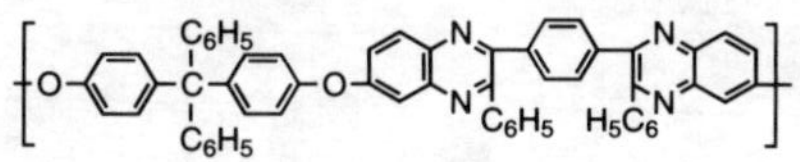	270	N-17; K_2CO_3, NMP; F	*Makromol. Chem., Macromol. Symp.*, **54/55**, 313	1992	IBM
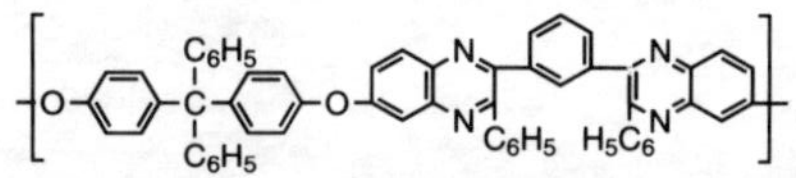	250	N-17; K_2CO_3, NMP; F	*Makromol. Chem., Macromol. Symp.*, **54/55**, 313	1992	IBM
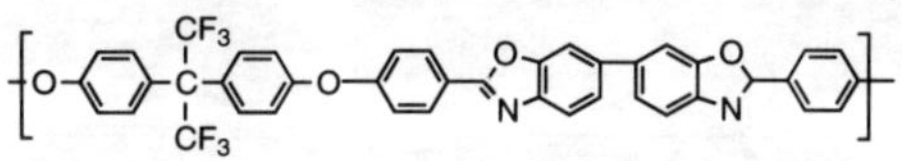	230	N-17; K_2CO_3, NMP; F	*Makromol. Chem., Macromol. Symp.*, **54/55**, 313	1992	IBM
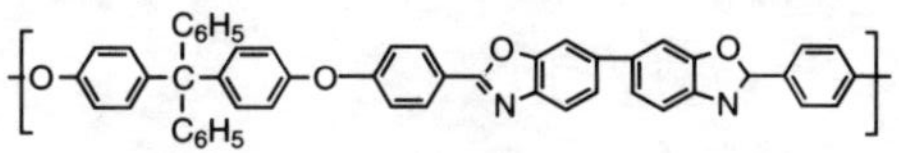	245	N-17; K_2CO_3, NMP; F	*Makromol. Chem., Macromol. Symp.*, **54/55**, 313	1992	IBM
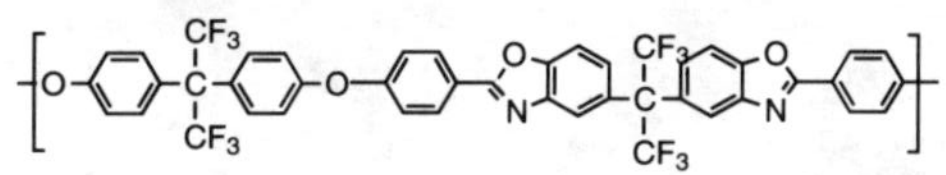	241	N-17; K_2CO_3, NMP; F	*Makromol. Chem., Macromol. Symp.*, **54/55**, 313	1992	IBM
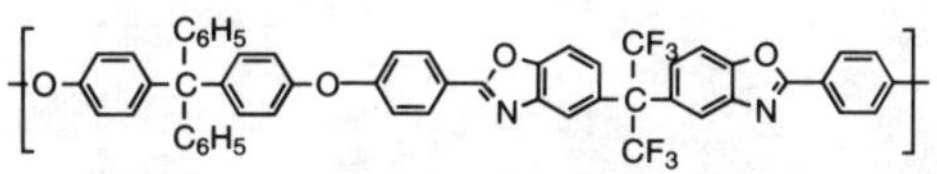	259	N-17; K_2CO_3, NMP; F	*Makromol. Chem., Macromol. Symp.*, **54/55**, 313	1992	IBM

CHAPTER 4

Polyarylethers Derived From Bisphenol A

Although the advent of the era of high performance, engineering plastics can be attributed to many factors, the commercial availability of Bisphenol A ranks as one of the most important. Bisphenol A was originally developed for use as the diphenol component in the diglycidyl ether type of epoxy resins during the 1950's. Its common name derives from the fact that it is synthesized by the condensation of phenol and acetone, the latter reactant providing the "A" in its name. More rigorously, it is known as 2,2-*bis*(4-hydroxyphenyl)propane.

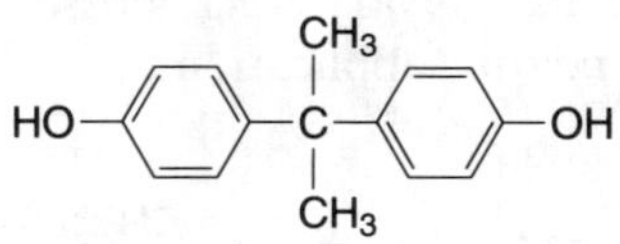

Bisphenol A

The synthesis of Bisphenol A is shown in the accompanying equation. The process of choice employs a cationic, sulfonic acid-based, ion exchange resin as the catalyst in conjunction with a mercaptan promoter (*1*). This process is preferred over conventional mineral acid catalysis for efficiency and environmental reasons. Very high yields of very pure 4,4′-isomer are obtained on a commercial scale. The availability of high purity Bisphenol A at reasonable costs led to exhaustive investigations of its utility in all kinds of polymers.

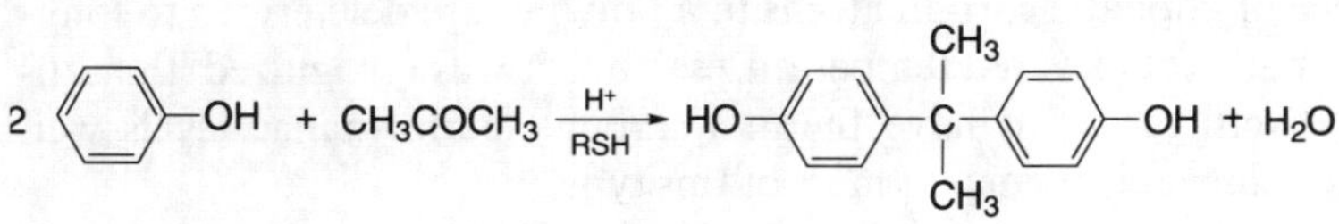

(IV-1)

As noted above, the original impetus for the commercialization of Bisphenol A was for use in epoxy resins, which are thermoset polymers (*2*). The purity of the Bisphenol A used in epoxy resin manufacture is not as critical as that used for the synthesis of thermoplastic polymers. The largest volume epoxy resin is comprised of the diglycidyl ether of Bisphenol A and oligomers thereof, viz.,

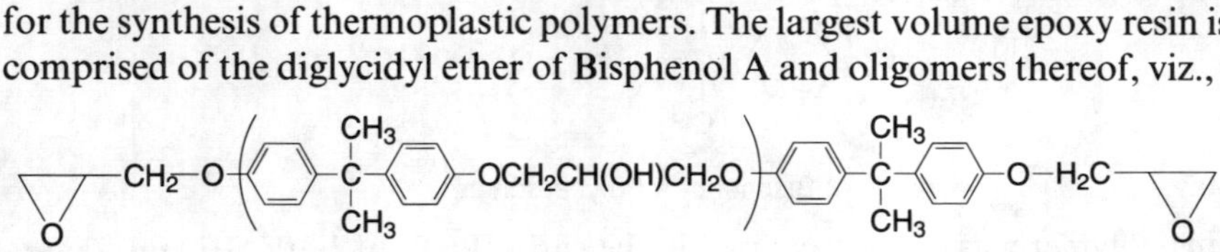

(IV-2)

It is polymerized via the epoxy groups to yield thermoset compositions that are actually aryl-alkyl polyethers as shown in the accompanying representative, polymerized structure.

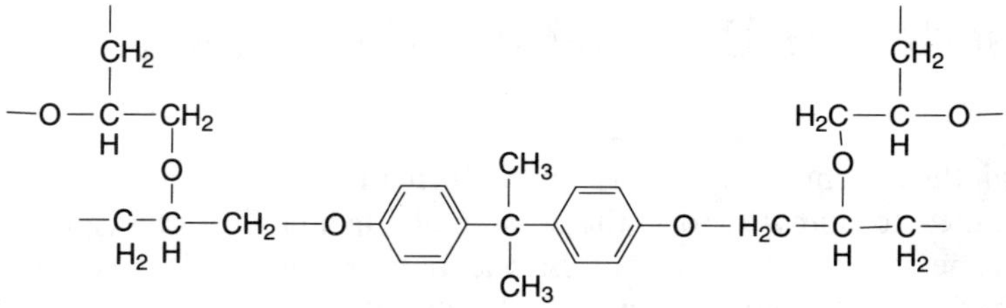

(IV-3)

A high molecular weight "linear epoxy resin" of the polyhydroxyether family known commercially as Phenoxy™ was introduced by the Union Carbide Corporation in the 1960's. Its structure and synthesis is shown in the accompanying equation (*3*). This "polyarylether" possesses a Tg of about 100° C and has utility primarily in coating applications.

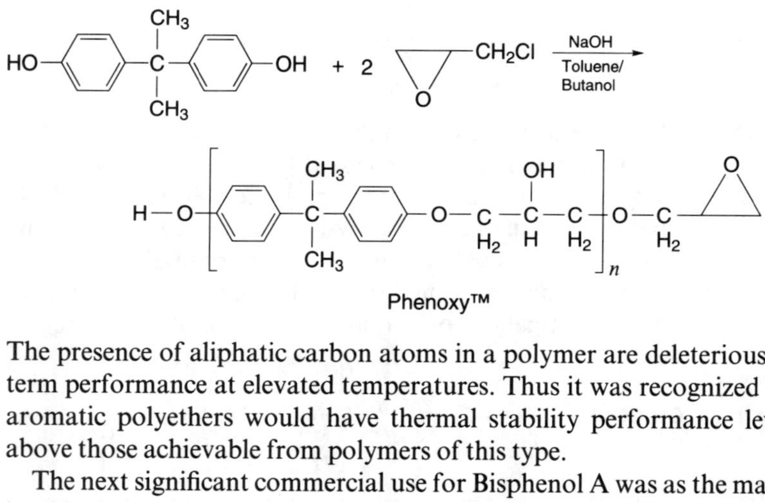

(IV-4)

Phenoxy™

The presence of aliphatic carbon atoms in a polymer are deleterious to long-term performance at elevated temperatures. Thus it was recognized that all-aromatic polyethers would have thermal stability performance levels well above those achievable from polymers of this type.

The next significant commercial use for Bisphenol A was as the main building block for the polycarbonate family of polymers (*4, 5*). Phosgenation of Bisphenol A produces Bisphenol A Polycarbonate.

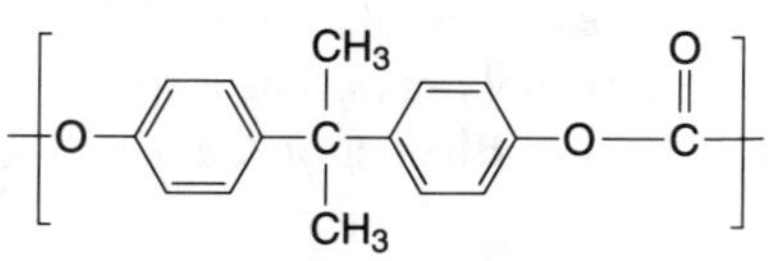

Bisphenol A Polycarbonate

This polymer was commercialized in the early 1960's by both GE and Bayer. This thermoplastic polymer possessed a glass transition temperature of 150°

C, which provided a significant jump in use temperatures over the injection moldable thermoplastics of that time. Many of the improvements in the color and quality of Bisphenol A Polycarbonate that have occurred since its commercial introduction have been due to improvements in the purity and quality of the Bisphenol A monomer. Polycarbonate-grade 4,4′-Bisphenol A is now one of the purest organic chemicals available commercially. An important deficiency of the polycarbonate polymers is the hydrolyzable nature of the carbonate linkage. This property is improved upon in many members of the polyarylether family because of the absence of hydrolyzable groups.

The development of nucleophilic displacement chemistry on haloaromatic substrates led to the discovery and commercialization of Polysulfone (see Chapter I). This polymer is made from Bisphenol A and 4,4′-dichlorodiphenyl sulfone. Its structure is shown below.

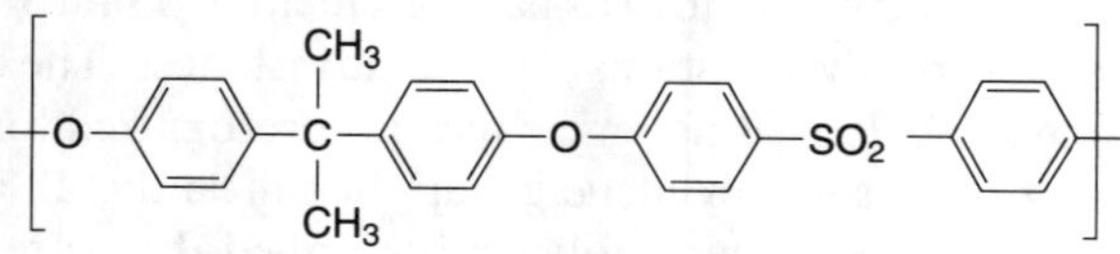

Polysulfone

Polysulfone possesses a glass transition temperature of 185° C, which represents a step jump over that of Bisphenol A Polycarbonate. It also exhibits a secondary (β) relaxation at –100° C. This transition has been shown to be due to 180° phenyl ring flips (*6*). A gamma-transition at 166° K has also been identified (*7*). Polysulfone's high glass transition temperature coupled with the absence of hydrolyzable groups in the backbone, opened up many new applications to injection moldable thermoplastics. High glass transition temperatures in themselves do not necessarily provide useful service lives at high temperatures. This is also a function of the stability of the structural features of the repeat unit. In this regard, Bisphenol A possesses only aromatic and primary aliphatic hydrogen atoms, two of the most thermo-oxidatively stable states of hydrogen.

Although many, many other bisphenols have been prepared and evaluated in polymers, none offer the combination of desirable structural features and economics of Bisphenol A. Thus, the growth of the higher end of the engineering plastics businesses has been intertwined with the growth and availability of Bisphenol A.

Finally, Bisphenol A is one of the key reactants in the polyetherimide known as Ultem™ 1000. Its structure is shown below. This polymer possesses a glass transition temperature of 220°C, which again is a significant increase over Polysulfone's 185°C. As pointed out earlier, this progression of new materials with increasing Tg's led to an increasing number of applications for the new entrant, as well as the existing engineering plastics. Other non Bisphenol A based polymers, such as Polyethersulfone, with Tg's in the 220° C range are also available and also contributed to the growth of the engineering plastics field.

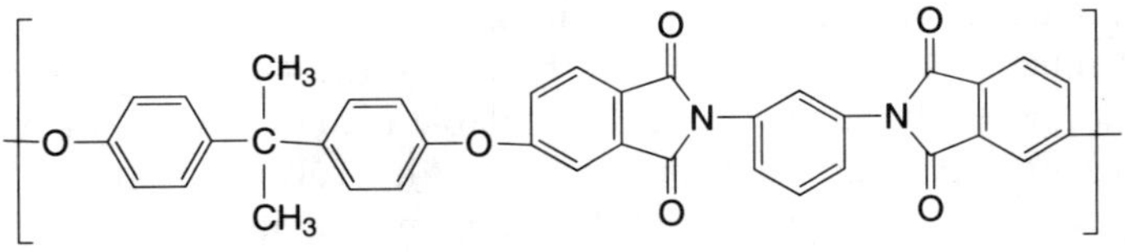

Ultem™ 1000

Table 18 contains all of the Bisphenol A based polymers that are reported on in this book. The glass transition temperatures of the entries range from 89°C to 283°C. Interestingly, only six of the approximately seventy entries were crystalline, or, at least were described as having a melting point. Whether they were truly crystalline may be questionable in some instances. The lack of crystallinity in Bisphenol A derived polymers has been recognized for some time and is attributed to the isopropylidene group. Significantly, all of the commercialized Bisphenol A-containing polymers described above are amorphous materials. To reduce or eliminate crystallinity in a given polymer, addition of Bisphenol A (if the chemistry is compatible) may be worth a try.

References

1. Farnham, A. G., Apel, F. N., and Bender, H. L., US Patent 3242219 (1966). Union Carbide Corporation.
2. Lee, H., and Neville, K., "*Handbook of Epoxy Resins*". McGraw-Hill Book Co., New York, NY. 1967, 1982.
3. Reinking, N. H., Barnabeo, A E., and Hale, W. F., *J. Appl. Polym. Sci.*, **7**, 2135 (1963). Also see p. 467 in "*Preparative Methods of Polymer Chemistry*", Sorenson, W.R., and Campbell, T. W., 2nd Edition Interscience Publishers, Inc. New York.
4. Sikdar, S. K., Chemtech, February, 1987, pp. 112–118.
5. Wood, A. S., *Modern Plastics*, November, 1989, p. 34.
6. Dumais, J. J., Cholli, A. L., Jelinski, L. W., Hedrick, J. L., and McGrath, J. E., *Macromolecules*, **19**, 1884 (1986).
7. Fried, J. R., Letton, A., and Welsh, W. J., *Polymer*, **31**, 1032 (1990).

TABLE 18
Polyarylethers derived from Bisphenol A (2,2-bis(4-hydroxyphenyl) propane)

Polymer structure	Tg (°C)	Tm (°C)	Synthesis	Reference	Year	Organization
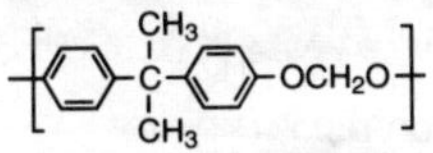	89		N-1; KOH, CH_2Cl_2; Cl	*Polym. Prep.*, **34**(1), 417	1993	Kanagawa U.
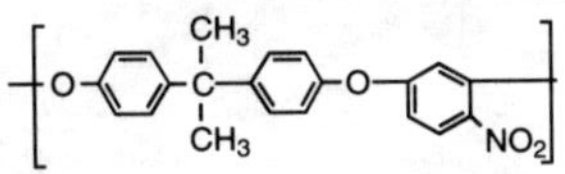	150		N-1; NaOH, DMSO, Sulfolane; Cl	*J.Polym.Sci.*, A-1, **5**, 2375;	1967	Union Carbide
				Ency. Polym. Sci. & Eng., **13**,196;	1988	
				GB 1078234	1967	
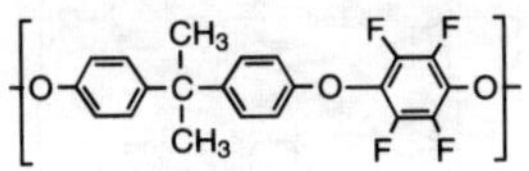	153		N-18; DMAC; F	*Polym. Prep.*, **21**(2), 164	1980	U. Texas, San Antonio
				Polym. Prep., **22**(2), 385		
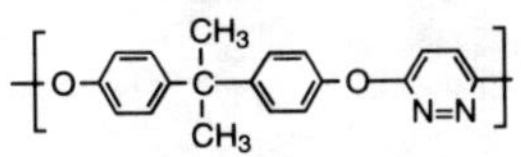	180	250	N-1; NaOH, DMSO, Sulfolane; Cl	*J.Polym.Sci.*, A-1, **5**, 2375	1967	Union Carbide
			N-19; K_2CO_3, NMP; Cl	*Makromol. Chem.*, **191**, 2027	1990	Hamburg U.
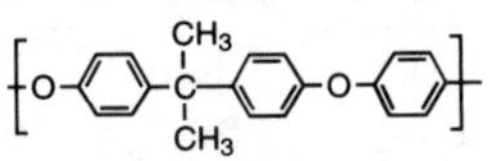			N-1; NaOH, Cu_2O, DMSO; Br	*J. Polym. Sci.*, C, No. **16**, 715	1967	ICI
	135		A-16; NaOH, CuCl, Pyridine, Benzophenone; Br	US Pat. 3332909	1967	Union Carbide
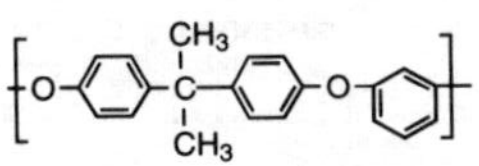			A-16; NaOH, CuCl, Pyridine, Benzophenone; Br	US Pat. 3332909	1967	Union Carbide

TABLE 18—*continued*

Polymer structure	Tg (°C)	Tm (°C)	Synthesis	Reference	Year	Organization
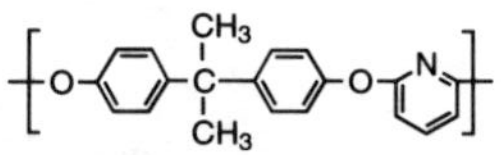			N-19; K_2CO_3, NMP; Cl	*Makromol. Chem.*, **191**, 2027	1990	Hamburg U.
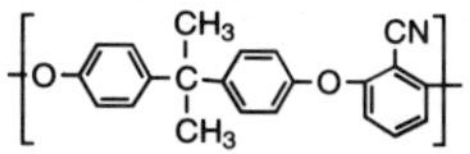	173		N-17; K_2CO_3, NMP; Cl	*Polym. Prep.*, **23**(1), 284	1982	Va. P. I. & State U
	173		N-20; DMSO; NO_2	*J. Polym. Sci., Polym. Chem.*, **18**, 3069	1980	
			N-19; K_2CO_3, NMP; Cl	*Makromol. Chem.*, **191**, 2027	1990	Hamburg U.
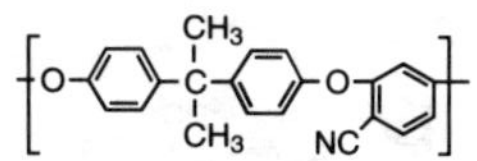	141		N-1; NaOH, DMSO; NO_2	*J. Polym. Sci., Polym. Chem.*, **18**, 3096	1980	GE
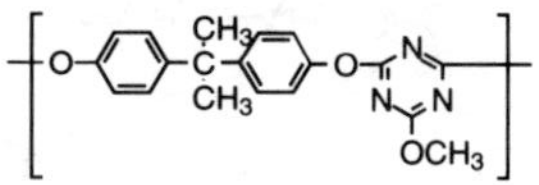			N-1; NaOH, $CHCl_3$; Cl	*Eur. Polym. J.*, **20**, 519	1984	
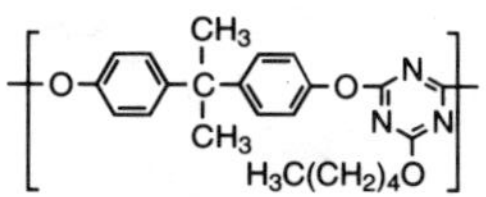			N-1; NaOH, $CHCl_3$; Cl	*J. Polym. Sci.*, A-1, **7**, 2749	1969	
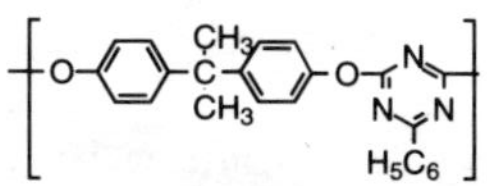			N-1; NaOH, Nitrobenzene; Cl	*J. Polym. Sci.*, A-1, **7**, 3089	1969	
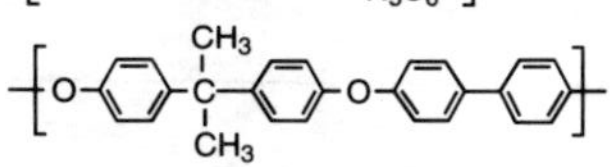	175		A-16; NaOH, CuCl, Pyridine, Benzophenone; Br	US Pat. 3332909	1967	Union Carbide
	154		A-16; Br	*Appl. Polym. Symp.* No. **26**, 373	1975	Union Carbide

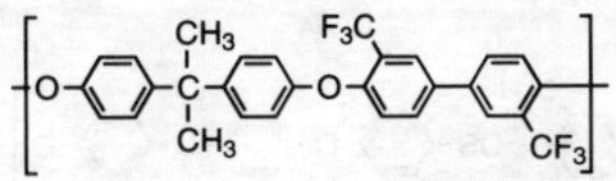	167	N-20; K_2CO_3, NMP; NO_2	*Polym. Prep.*, **34**(1), 415	1993	IBM
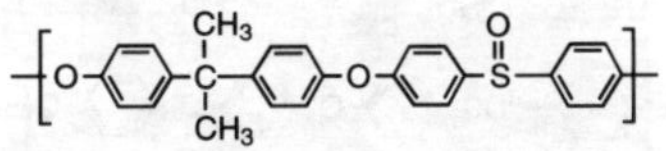	165	N-1; NaOH, DMSO, Sulfolane; Cl, F	*J.Polym.Sci.*, A-1, **5**, 2375;	1967	Union Carbide
			Ency. Polym. Sci. & Eng., **13**, 196	1988	
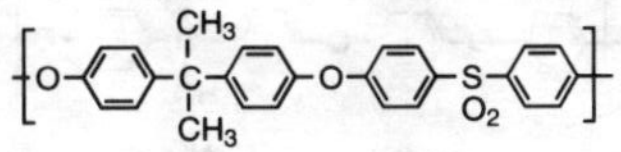 Udel™ Polysulfone	195	N-1; DMSO, Sulfolane; Cl, F Thermal Stability	*J. Polym. Sci.*, A-1, **5**, 2375, 2399	1967	Union Carbide
	184	N-1; NaOH, DMSO; Cl	*Appl. Polym. Symp.*, No. **26**, 373	1975	Union Carbide
		N-2; K_2CO_3, DMSO; NO_2	Jap. Pat. 1087628	1989	Toyo Soda
		N-2; K_2CO_3, NMP; Cl	*Polym. Prep.*, **23**(1), 284	1982	Va. P. I. & State U.
		N-6; K_2CO_3, NMP; Cl	*Makromol. Chem.*, **191**, 2027	1990	Hamburg U.
	186	Steam Sterilization Permeability Data	*Polym. News*, **11**, 359;	1986	Union Carbide
			J. App. Polym. Sci., **28**, 2925;	1983	U. Texas
			Polym., **32**, 840	1991	
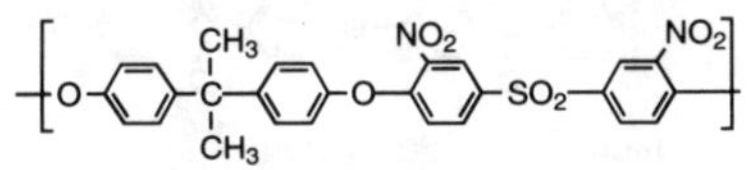	138	N-2; K_2CO_3, Sulfolane; Cl	*J. Polym. Sci., Polym. Lett.*, **17**, 85	1979	
			Makromol. Chem., **179**, 2989	1978	
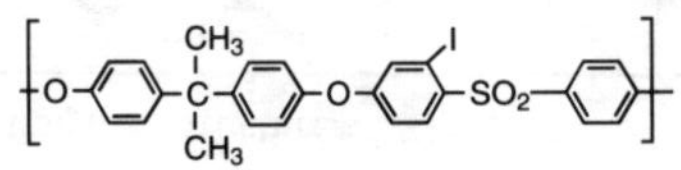		Post Reaction on PSF	US Pat. 4797457; US Pat. 4833219	1989	Nat. Res. Council, Canada
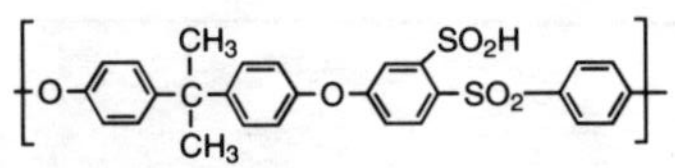		Post Reaction on PSF	US Pat. 4797457; US Pat. 4833219	1989	Nat. Res. Council, Canada

TABLE 18—*continued*

Polymer structure	Tg (°C)	Tm (°C)	Synthesis	Reference	Year	Organization
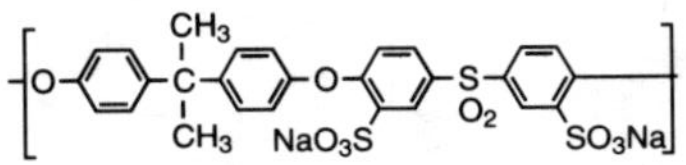	192		N-2; K_2CO_3, DMAC; Cl	*J. Polym. Sci., Polym. Chem.*, **31**, 853	1993	Nippon Steel
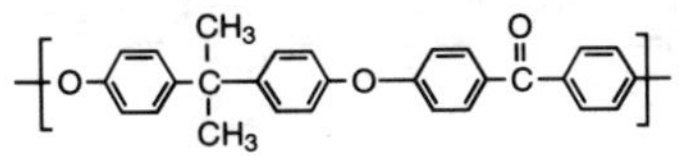	155	(248)	N-8; NaOH, DMSO, Sulfolane; F	*J. Polym. Sci.*, A-1, **5**, 2375;	1967	Union Carbide
				Ency. Polym. Sci. & Eng., **13**, 196;	1988	
		190–220		*Makromol. Chem.*, **130**, 45	1969	Bayer A-G
	155		N-9; K_2CO_3, DMAC; Cl	*Polym.*, **29**, 358	1988	NASA/ Langley
	154		N-11; Bulk; F	*Polym.*, **25**(8), 1151;	1984	Hamburg U.
	152			US Pat. 4331798	1982	ICI
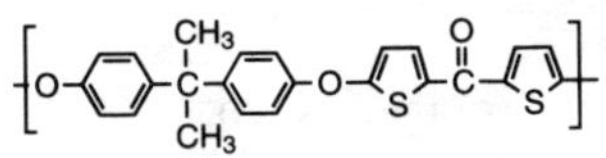	124		N-9; K_2CO_3, NMP; Cl	*Polym. Prep.*, **33**(1), 418; **33**(2), 170	1992	U. No Carolina
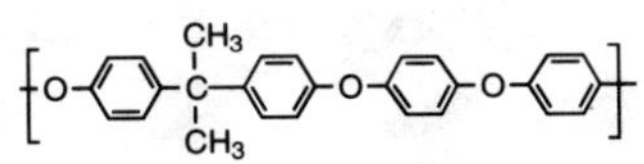	125		A-16; NaOH, CuCl, Pyridine, Benzo-phenone; Br	US Pat. 3332909	1967	Union Carbide
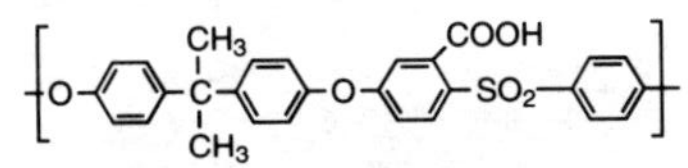	210		Post reaction on PSF	*Polym. Bull.*, **29**, 521	1992	U. of Singapore
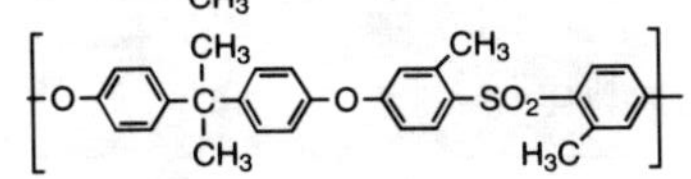			Post Reaction on PSF	US Pat. 4797457; US Pat. 4833219	1989	Nat. Res. Council, Canada

					Canada
	220		*Polym. Bull.* **29**, 521	1992	U. of Singapore
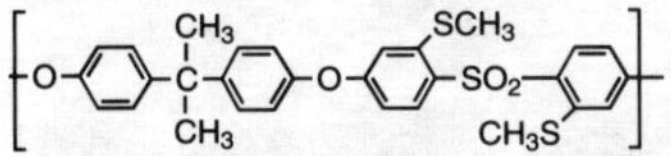		Post Reaction on PSF	US Pat. 4797457; US Pat. 4833219	1989	Nat. Res. Council, Canada
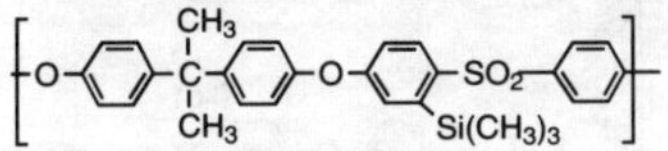	171	Post Reaction on PSF	*Polym. Prep.*, **32**(2), 258	1991	Amoco
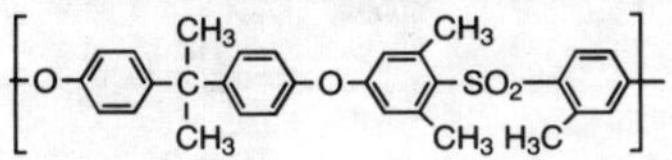		Post Reaction on PSF	US Pat. 4797457; US Pat. 4833219	1989	Nat. Res Council, Canada
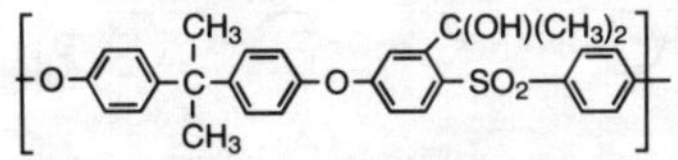		Post Reaction on PSF	US Pat. 4797457; US Pat. 4833219	1989	Nat. Res. Council, Canada
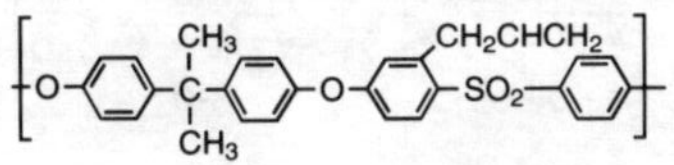		Post Reaction on PSF	US Pat. 4797457; US Pat. 4833219	1989	Nat. Res. Council, Canada
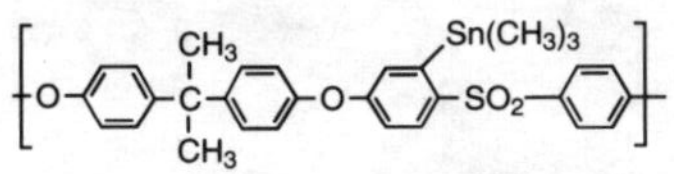		Post Reaction on PSF	US Pat. 4797457; US Pat. 4833219	1989	Nat. Res. Council, Canada
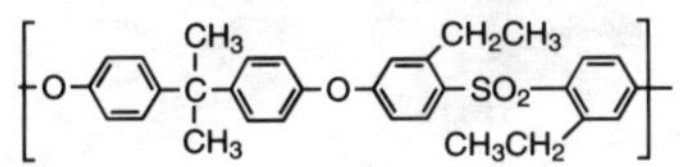		Post Reaction on PSF	US Pat. 4797457; US Pat. 4833219	1989	Nat. Res. Council, Canada
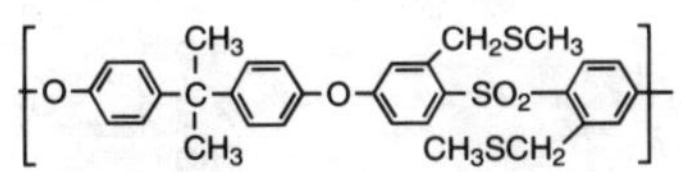		Post Reaction on PSF	US Pat. 4797457; US Pat. 4833219	1989	Nat. Res. Council, Canada
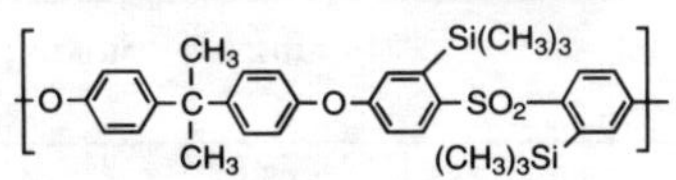		Post Reaction on PSF	US Pat. 4797457; US Pat. 4833219	1989	Nat. Res Council, Canada

TABLE 18—*continued*

Polymer structure	Tg (°C)	Tm (°C)	Synthesis	Reference	Year	Organization
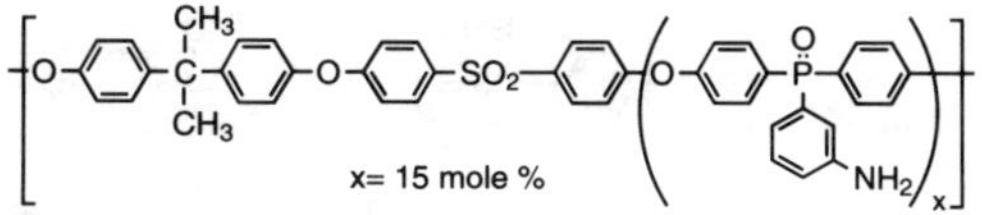	198		N-2; K_2CO_3, NMP; F, Cl	*Polym. Bull.*, **29**, 477	1992	Va. P. I. & State. U.
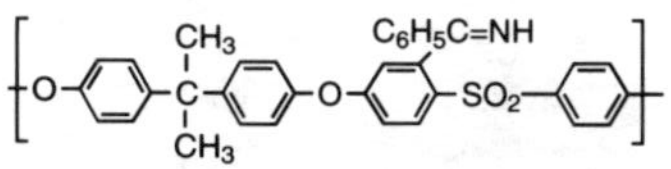			Post Reaction on PSF	US Pat. 4797457; US Pat. 4833219	1989	Nat. Res. Council, Canada
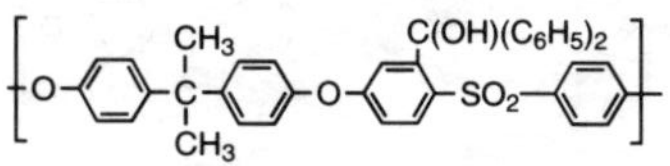			Post Reaction on PSF	US Pat. 4797457; US Pat. 4833219	1989	Nat. Res Council, Canada
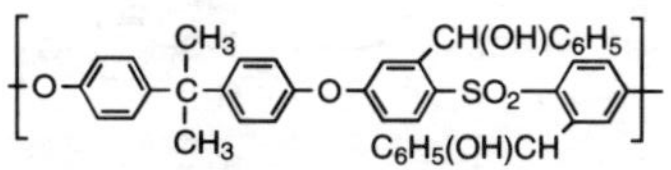			Post Reaction on PSF	US Pat. 4797457; US Pat. 4833219	1989	Nat. Res. Council, Canada
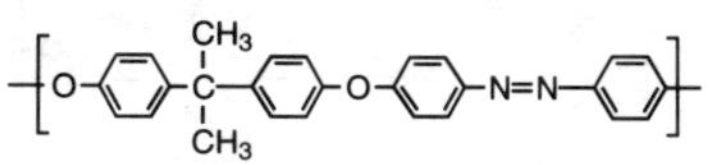	175		N-1; NaOH, DMSO, Sulfolane; Cl,F	*J.Polym.Sci.* A-1, **5**, 2375; *Ency. Polym. Sci. & Eng.*, **13**,196	1967 1988	UnionCarbide
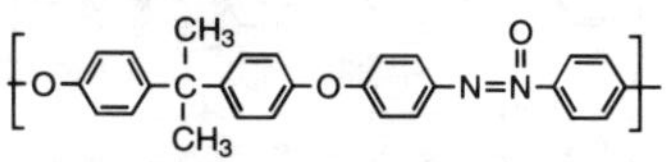			N-1; NaOH, DMSO; Cl	*J. Pol. Sci., Polym. Chem.*, **24,** 2447	1986	
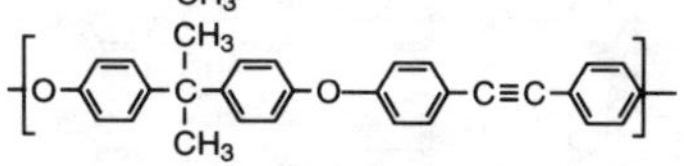	163		N-17; K_2CO_3, NMP, Toluene; F	*Macromol.*, **26**, 1777	1993	McGill U.
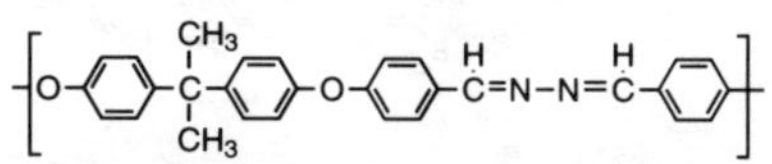	163		N-17; K_2CO_3, DMPU; F	*Polym. Prep.*, **34**(1), 521	1993	IBM

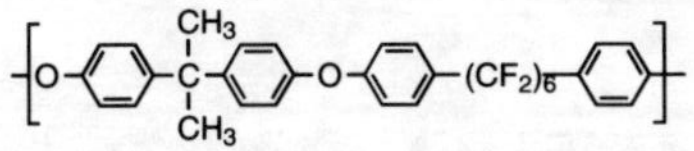	93	N-17; K_2CO_3, NMP; F	*Macromol.*, **23**, 5371	1990	IBM
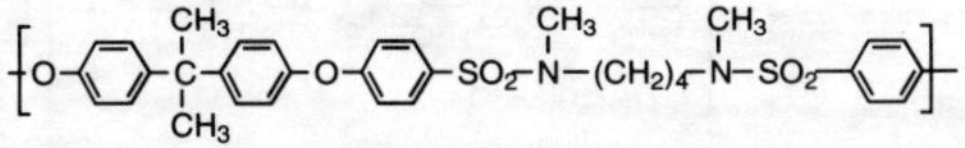	120	N-1; NaOH, DMSO, Sulfolane; Cl, F	*J. Polym. Sci.*, A-1, **5**, 2375	1967	Union Carbide
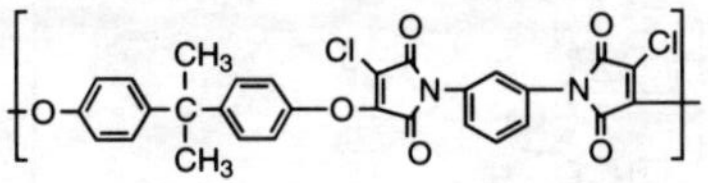	167	N-1; DMF; Cl	*J. Polym. Sci., Polym. Chem.*, **11**, 561	1973	GE
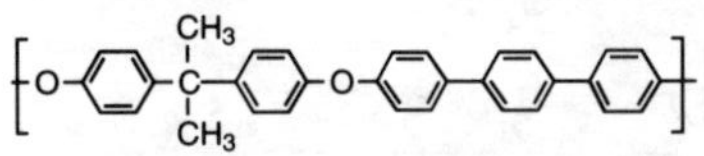	182	A-16; Br	*Appl. Polym. Symp.* No. **26**, 373	1975	Union Carbide
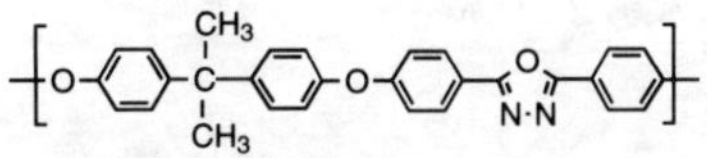	180	N-1; NaOH, DMSO, Sulfolane; Cl,F	*J.Polym.Sci.*, A-1, **5**, 2375	1967	Union Carbide
	210	N-17; K_2CO_3, NMP, CHP; F	*Polym. Bull.*, **25**, 543	1991	IBM, San Jose
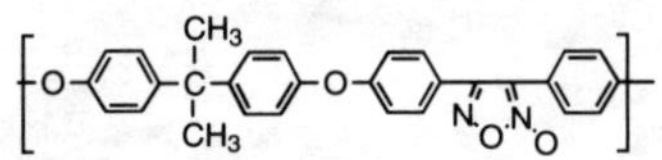		N-1; NaOH, DMSO; Cl	*Eur. Polym. J.*, **22**, 447	1986	
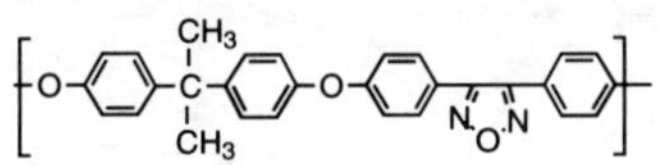		N-1; NaOH, DMSO; Cl	*Eur. Polym. J.*, **22**, 447	1986	
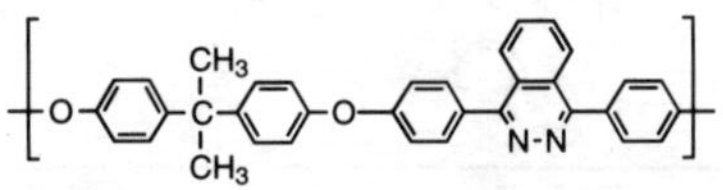	236	Post Reaction on PAEK	*Macromol.*, **24**(9), 2640	1991	McGill U.
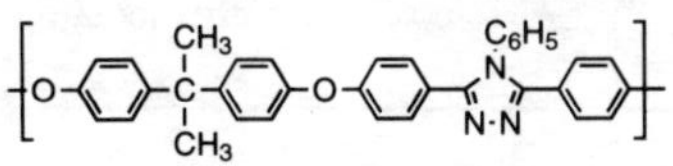	207	N-17; K_2CO_3, DMPU; F	*Polym. Prep.*, **34**(1), 481	1993	IBM

TABLE 18—*continued*

Polymer structure	Tg (°C)	Tm (°C)	Synthesis	Reference	Year	Organization
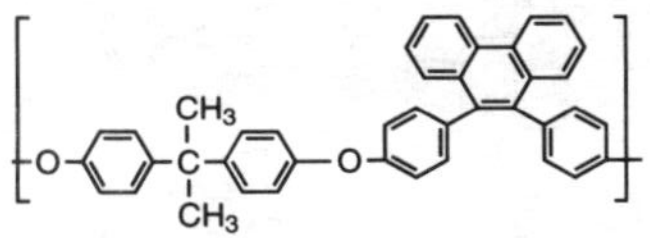	228		Post Reaction on PAEK	*Polym. Prep.*, **33**(2), 351	1992	Carleton U.
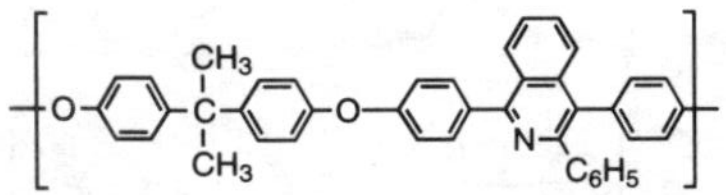	226		Post reaction on PAEK	*Macromol.*, **24**(9), 2643	1991	McGill U.
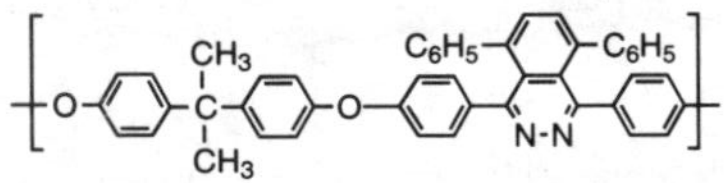	250		Post Reaction on PAEK	*Macromol.*, **24**(9), 2640	1991	McGill U.
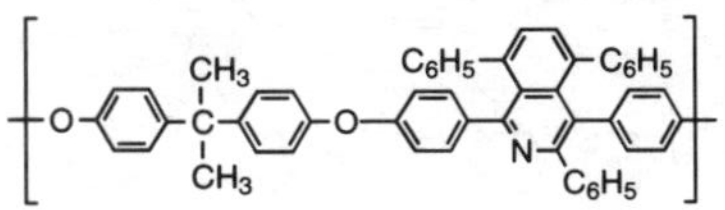	234		Post reaction on PAEK	*Macromol.*, **24**(9), 2643	1991	McGill U.
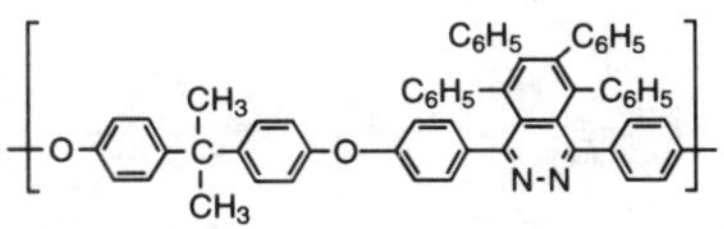	283		Post Reaction on PAEK	*Macromol.*, **24**(9), 2640	1991	McGill U.
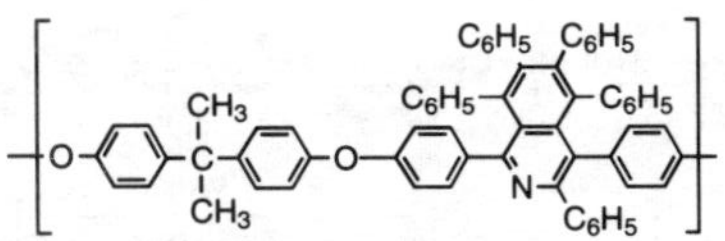	280		Post reaction on PAEK	*Macromol.*, **24**(9), 2643	1991	McGill U.

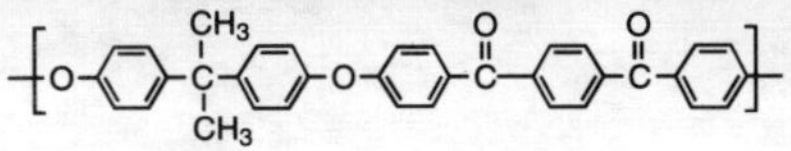	165		N-8; NaOH, DMSO, Sulfolane; Cl, F	*J. Polym. Sci.*, A-1, **5**, 2375	1967	Union Carbide
	166		N-9; K_2CO_3, NMP; F	*Polym. Prep.*, **32**(2), 172	1991	U. No. Carolina
	170		N-9; K_2CO_3, NMP, DMAC; Cl	*Polym.*, **26**,174; **29**, 358	1985 1988	NASA/ Langley
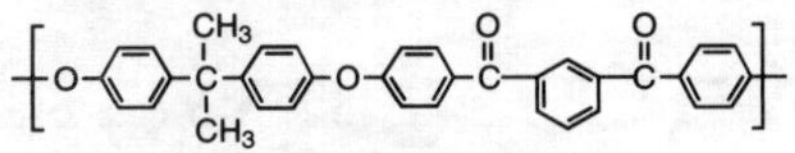	152		N-9; K_2CO_3, NMP; F	*Polym. Prep.*, **32**(2) 172	1991	U. No. Carolina
	152		N-9; K_2CO_3, DMAC; Cl, F	*Polym.*, **29**, 358	1988	NASA/ Langley
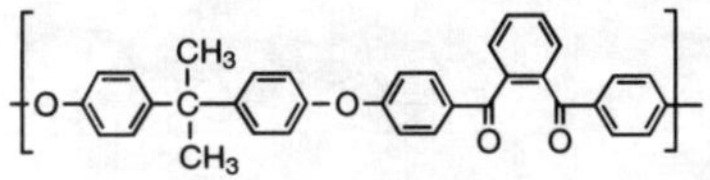	160 182		N-9; K_2CO_3, DMAC; F	*Macromol.*, **4**(9), 2637	1991	McGill U.
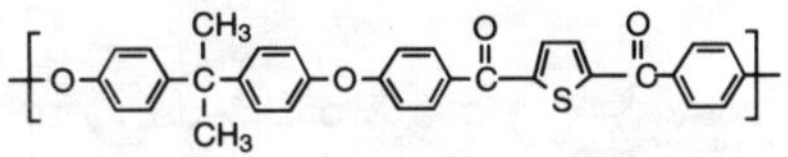	158 147		N-9; K_2CO_3, NMP; F	*Polym. Prep.*, **32**(2) 172; **33**(1), 485	1991 1992	U. No Carolina; U Florida
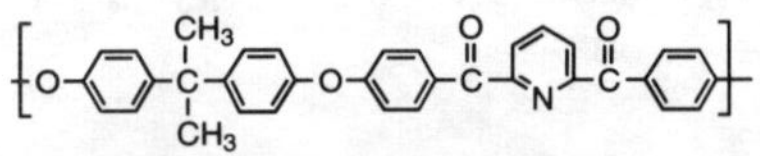	132		N-9; K_2CO_3, DMAC; F	*Polym.*, **29**, 358	1988	NASA/ Langley
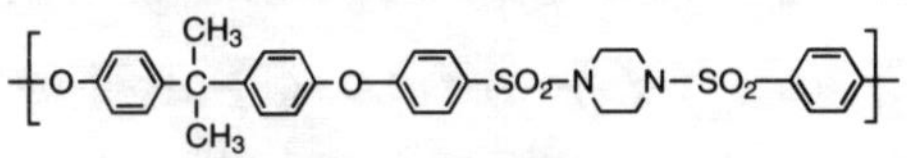		255–260	N-1; NaOH, DMSO; Cl	US Pat. 4108837	1978	Union Carbide
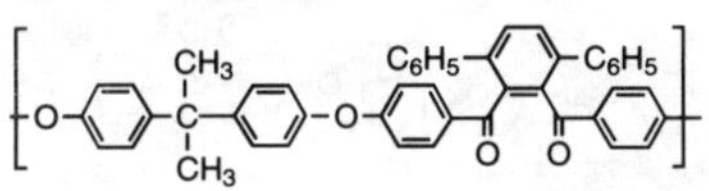	221		N-9; K_2CO_3, DMAC; F	*Macromol.*, **4**(9), 2637	1991	McGill U.
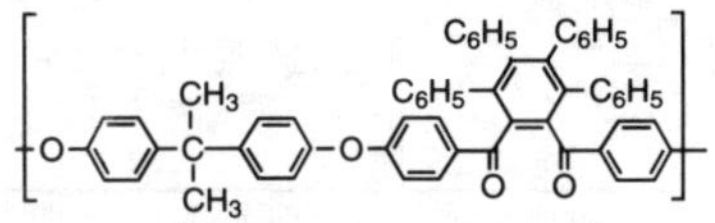	268		N-9; K_2CO_3, DMAC; F	*Macromol.*, **4**(9), 2637	1991	McGill U.
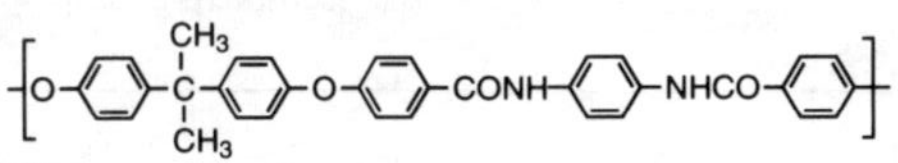	230		N-14; K_2CO_3, NMP, CHP; F	*Makromol. Chem.*, *Macromol. Symp.*, **54/55**, 313	1992	IBM

TABLE 18—*continued*

Polymer structure	Tg (°C)	Tm (°C)	Synthesis	Reference	Year	Organization
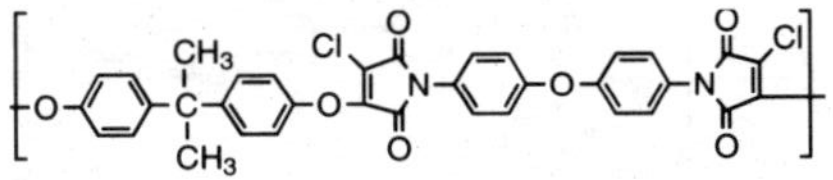	174		N-1; DMF; Cl	*J. Polym. Sci., Polym. Chem. Ed.*, **11**, 561	1973	GE
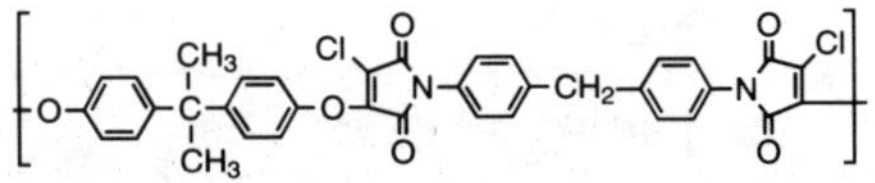	187		N-1; DMSO; Cl	*J. Polym. Sci., Polym. Chem. Ed.*, **11**, 561	1973	GE
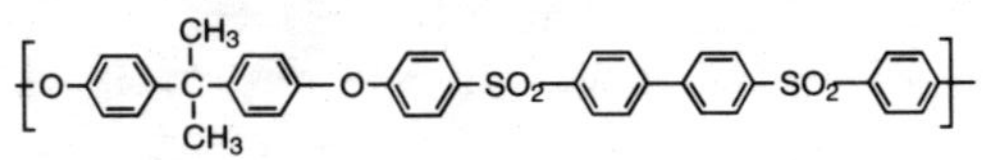	233–242		N-2; K_2CO_3, Sulfolane; F, Cl	*Polym.*,**18**(4), 370	1977	ICI
	240		N-2; K_2CO_3, Sulfolane; Cl;	*Polym. Prep.*, **32**(2), 166;	1991	Amoco
			Permeability	*J. Polym. Sci.*, **B**, *Polym. Phys.*, **31**, 983	1993	U. Texas
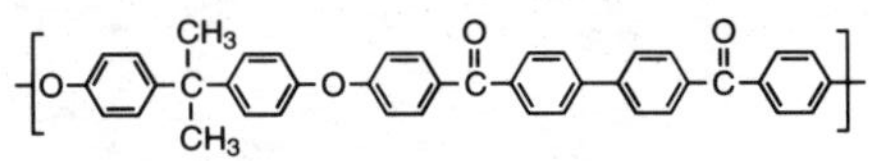	219		N-9 Permeability	*J. Polym. Sci.*, **B**, *Polym. Phys.*, **31**, 983	1993	U. Texas
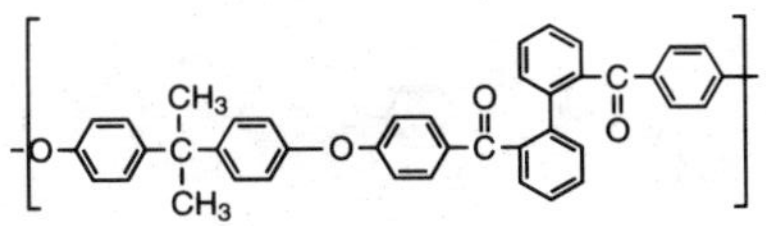	172		N-9; K_2CO_3, NMP; F	*Polym. Prep.*, **33**(2), 351	1992	Carleton U.
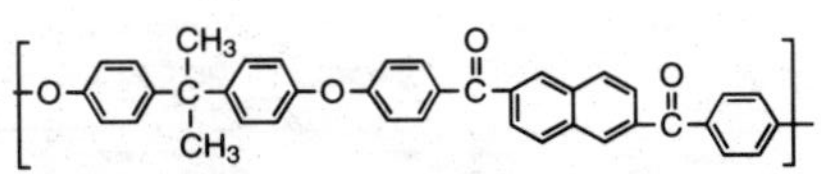	185	220	N-9; K_2CO_3, DMAC; F	*Polym.*, **29**, 358	1988	NASA/ Langley
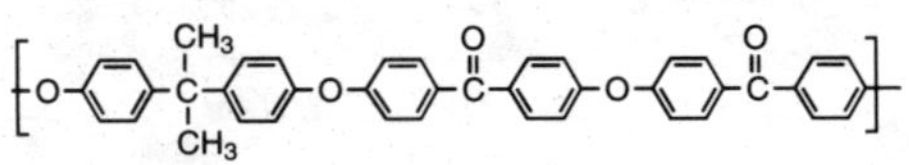	162		N-9; K_2CO_3, DMAC; F	*Polym.*, **29**, 358	1988	NASA/ Langley

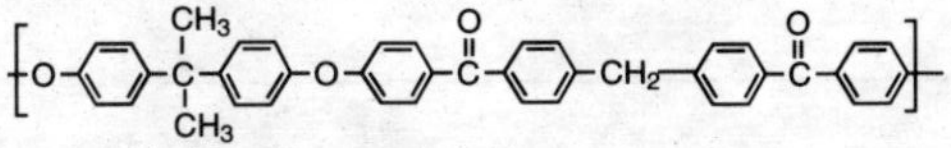	155		N-9; K_2CO_3, DMAC; F	*Polym.*, **29**, 358	1988	NASA/ Langley
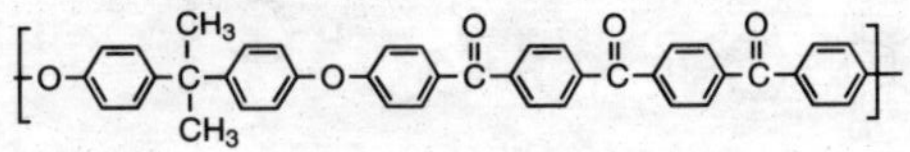	155	283	N-9; K_2CO_3, DMAC; F	*Polym.*, **29**, 358	1988	NASA/ Langley
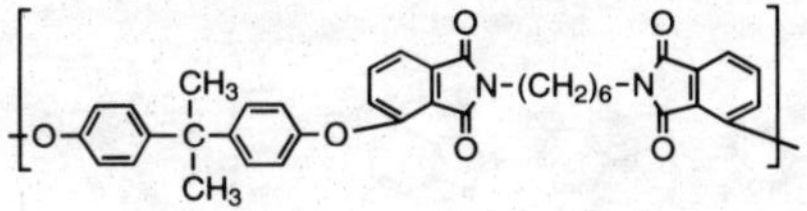	135		N-14; DMSO; NO_2	*J. Polym. Sci., Polym. Chem. Ed.*, **18**, 3069	1980	GE
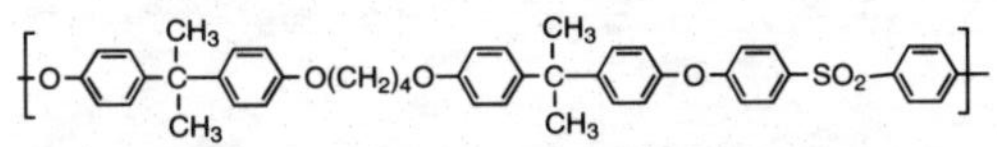	140		N-1; NaOH, DMSO, Sulfolane; Cl	*J. Polym. Sci.*, A-1, **5**, 2375	1967	Union Carbide
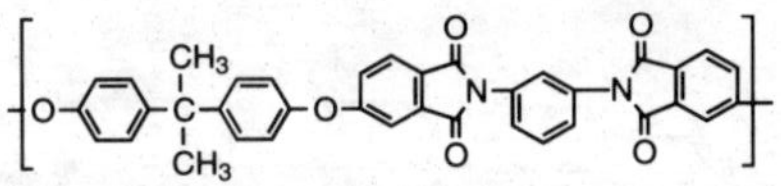 Ultem™ 1000	215		N-15; m-Cresol, Toluene	*Polym. Prep.*, **24**(2), 312; *J. Polym. Sci., Polym. Symp.*, **74**, 93	1983 1986	GE
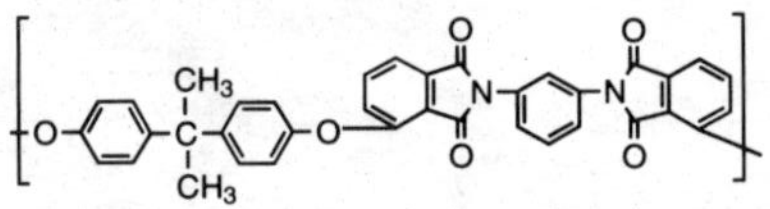	236		N-15; m-Cresol, Toluene	*Polym. Prep.*, **24**(2), 312; *J. Polym. Sci., Polym. Symp.*, **74**, 93	1983 1986	GE
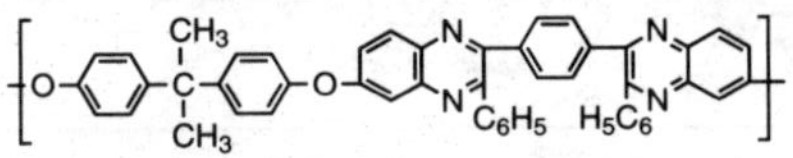	255		N-17; K_2CO_3, NMP; F	*Makromol. Chem., Macromol. Symp.*, **54/55**, 313	1992	IBM
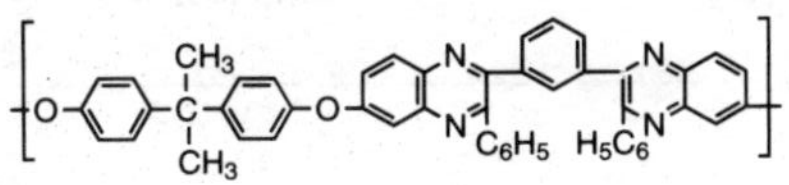	230		N-17; K_2CO_3, NMP; F	*Makromol. Chem., Macromol. Symp.*, **54/55**, 313	1992	IBM

TABLE 18—*continued*

Polymer structure	Tg (°C)	Tm (°C)	Synthesis	Reference	Year	Organization
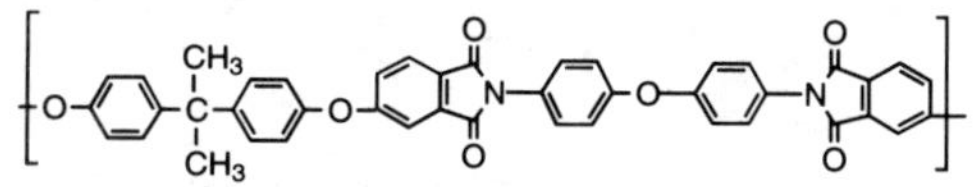	223		N-15; m-Cresol, Toluene	*Polym. Prep.*, **24**(2), 312; *J. Polym. Sci., Polym. Symp.*, **74**, 93;	1983 1986	GE
	196		N-14; DMSO; NO_2	*J. Polym. Sci., Polym. Chem.*, **18**, 3069; **19**, 1635	1980 1981	
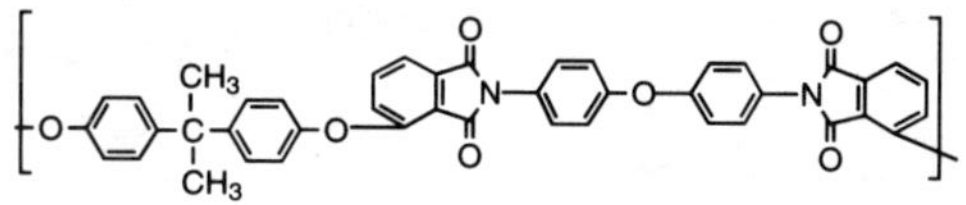	235		N-15; m-Cresol, Toluene	*Polym. Prep.*, **24**(2), 312; *J. Polym. Sci., Polym. Symp.*, **74**, 93;	1983 1986	GE
	226		N-14; DMSO; NO_2	*J. Polym. Sci., Polym. Chem.*, **18**, 3069; **19**, 1635	1980 1981	
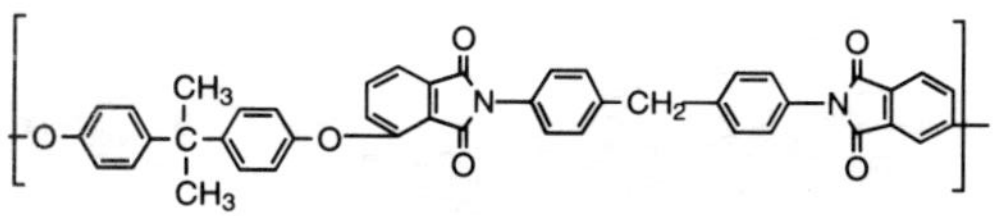	224		N-14; DMF; NO_2	*J. Polym. Sci., Polym. Chem.*, **19**,1635	1981	GE
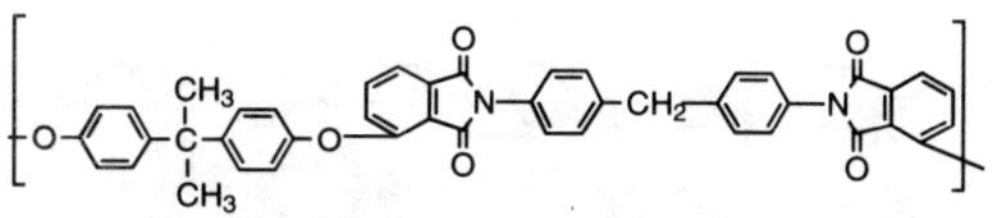	230		N-14; DMSO; NO_2	*J. Polym. Sci., Polym. Chem.*, **18**, 3069; **19**, 1635	1980 1981	

CHAPTER 5

Polyarylethers Derived From 4,4′-Disubstituted Diphenyl Sulfones

This Chapter covers the polyarylethersulfones that have been made from 4,4′-disubstituted diphenylsulfones. Table 19 lists the polyarylethersulfones that fit this category. The polymers listed in this table contain either of the following two structures in their repeat units, viz.,

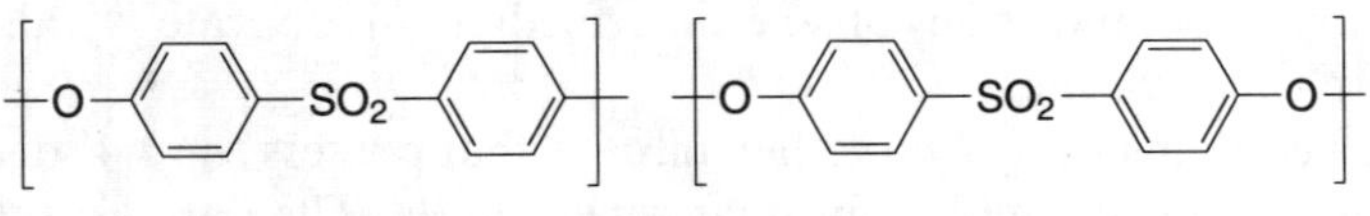

(V-1)

The vast majority of the polyarylethersulfones that have been synthesized have used 4,4′-dichlorodiphenyl sulfone (DCDPS) as the sulfone-supplying monomer. It is the "workhorse" monomer of the polyarylethersulfone field. Scanning the 'Synthesis' column in Table 19 for the leaving group, confirms the wide applicability of 4,4′-dichlorodiphenyl sulfone. This dichloro monomer is produced on a large scale and is used to make all of the commercialized polyarylethersulfones. The basic chemistry that is employed to make DCDPS involves the coupling of chlorobenzene by sulfur trioxide to yield the dichloro compound. The low cost and ready availability of these starting materials

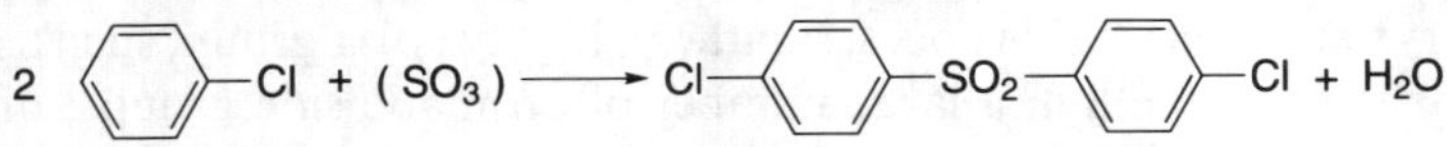

(V-2)

underscore the desirability of using DCDPS for polyarylethersulfone manufacture. The coupling chemistry can be carried out using a wide variety of co-reagents and reaction conditions. However, the purity and non-4,4′-isomer content of the product is critical and must be carefully controlled and monitored to produce a very high purity product. Amoco Chemical Company's patent (*1, 2*) for a vapor phase process was described as a major process innovation that has the environmental benefit of low emissions, is energy efficient and has replaced an older thionyl chloride process. This process first became commercially operational in 1991 at a new world-scale sulfone monomer plant in Augusta, Georgia. In addition to environmental benefits, the new process cuts energy and production costs. Based on an estimated 1991 production of polysulfone polymers of almost 28 million pounds (*3*), the 1991 production of 4,4′-dichlorodiphenyl sulfone was probably about twenty million pounds.

4,4′-Difluorodiphenyl sulfone and the analogous dinitro monomer can also be employed to make polyarylethersulfones on a laboratory scale. The difluoro monomer is available from chemical supply houses and is frequently used to make small samples of polyarylethersulfones. The higher reactivity of the difluoro monomer facilitates the synthesis of high polymer from unreactive or hindered bisphenols when using nucleophilic methods. Thus, it is frequently used for polyarylethersulfone synthesis. The difluorodiphenyl sulfone monomer has proven too costly to be used on a large commercial scale. It can be made by several different methods including reaction of fluorobenzene with sulfur trioxide. Displacement of the chloro groups in 4,4′-dichlorodiphenyl sulfone by potassium fluoride can be effected at high temperatures in bulk. An additional route is from 4,4′-diaminodiphenyl sulfone using the Schiemann reaction. Diazotization, reaction with fluoboric acid, followed by decomposition of the intermediate fluoborate yields 4,4′-difluorodiphenyl sulfone.

Table 19 contains the polyaryethersulfones that possess the 4,4′-disubstituted diphenyl sulfone moiety in their repeat units. They are arranged in order of an increasing number of rings in the backbone of the repeat unit. This is the standard convention that has been used in this book. Most of the polymers have been made from a nucleophilic polymerization of a bisphenol and the 4,4′-dihalodiphenyl sulfone. The carbonate process (Method N-2) has been used most frequently, especially in recent years. For the most part, these polymers are amorphous materials. The glass transition temperatures of these polymers can range up to about 325°C depending on the others groups that are present. Only very few polymers with the 4,4′-disubstituted diphenyl sulfone moiety in the backbone were reported to be crystalline. The polymers that are crystalline possess either a highly polar group, such as carbonyl or trifluoromethyl; a large number of aromatic side groups; or terphenyl or similar rigid groups in the backbone. The lack of crystallinity in these sulfone-containing polymers parallels that discussed in Chapter IV for Bisphenol A-containing polymers. Thus, it is not surprising that Polysulfone has never been crystallized. Incorporating the 4,4’-diphenyl sulfone moiety in a polymer has a high probability of producing an amorphous product.

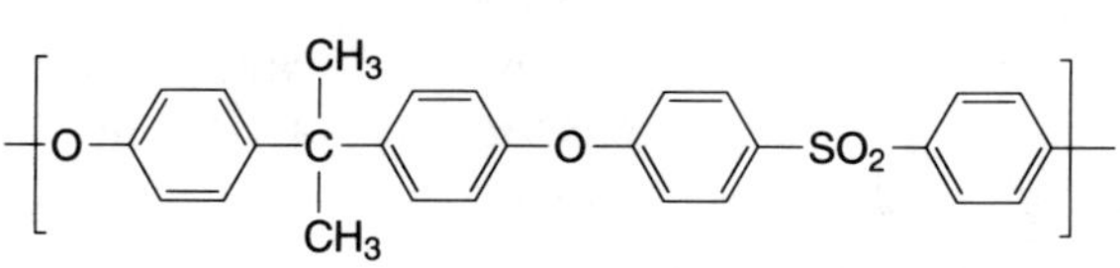

(V-3)

Polysulfone

References

1. Steiner, U. A., and Reichle, W. T., U. S. Patent 4937387 (1990). Amoco Corp.
2. R & D Council of New Jersey, News Release, 5/3/92. Newsletter No. 105, October, 1992.
3. Reisch, M. S., *Chemical and Engineering News*, August 30, 1993.

TABLE 19
Polyarylethers derived from 4,4′-disubstituted (halo, nitro) diphenyl sulfones

Polymer structue	Tg (°C)	Tm (°C)	Synthesis	Reference	Year	Organization
< = Three Rings						
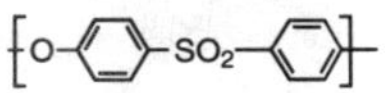	230		N-1; DMSO; F, Cl, Br;	*Br. Polym. J.*, **4**, 391;	1972	ICI
Victrex™ PES			Kinetics	*Polym.*, **13**, 465;	1972	
				ACS Symp.Ser., **4**, 63;	1974	
			Steam Sterilization	*Polym., News*, **11**, 359	1986	Union Carbide
		250–265	E-1; $AlCl_3$, $C_6H_5NO_2$	*J.Polym.Sci.*, A-1, **4**,	1966	Monsanto
			E-2; PPA	722;		
			E-1; Bulk, $FeCl_3$	*Polym.*, **6**, 589	1965	ICI
	250		E-1; Dimethyl Sulfone, $FeCl_3$	*Chem. Ind.*, (15), 461	1968	ICI
			E-1; BF_3, HF	US Pat. 3441538	1969	DuPont
			E-2; NMP	Jap. Pat. 02011634	1990	New Japan Chem.
			E-2; CH_3SO_3H, P_2O_5	Jap. Pat. 60-228541	1985	Idemitsu Inds.
			A-6; K Phenate, Bulk	Eur. Pat. 218064	1987	Bayer, A. G.
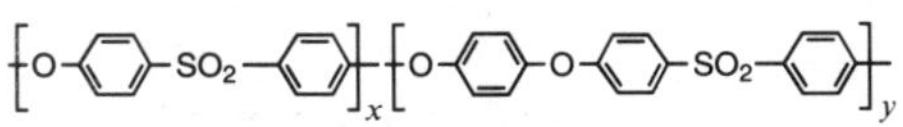 Radel ™ A-100-x/y=4.7	220		N-2; NMP, Sulfolane; Cl	*High Perf. Polym.*, Elsevier, p.157; J.	1986	Union Carbide
			Gas Permeability	*Polym. Sci., B; Polym. Physics*, **31**, 891;	1993	
				J. Memb. Sci., **78**, 123	1993	
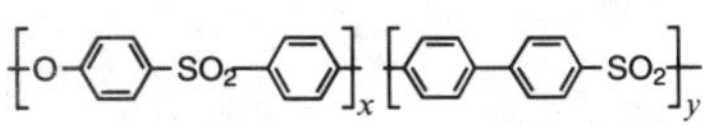 Victrex™ 720P- x>>y	250		E-1;	*Chimia*, **28**(9), 561;	1974	ICI
				High Perf. Polym., Elsevier, p. 171	1986	
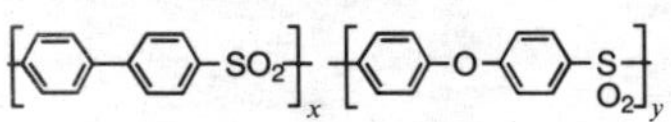 Astrel™ 360- x:y = 60-70: 30-40	285		E-1	*Chimia*, **28**(9), 561	1974	3M

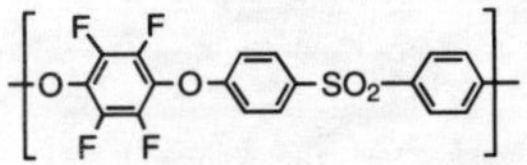			N-5; DMAC; F	*Polym. Prep.*, **21**(2), 164	1980	
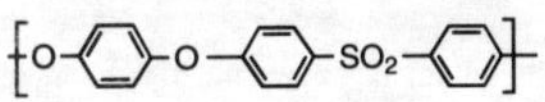	210	310	N-1; NaOH, DMSO, Sulfolane; Cl	*J. Polym. Sci.*, A-1, **5**, 2375;	1967;	UnionCarbide
				Ency. Polym. Sci. & Eng., **13**,196	1988	
	217	310	N-2; K_2CO_3, NMP; Cl; Kinetics	*J. Polym. Sci.,Polym. Chem.*, **23**, 287;	1986	Va. P. I. & State U.
				Polym., **25**, 1827	1984	
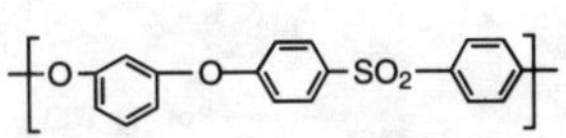			N-1; NaOH, DMSO	Eur. Pat. 0161576	1986	Sumitomo Chemical Co.
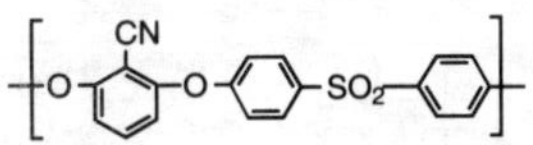	222		N-2; K_2CO_3, NMP; Cl	*Polym. Prep.*, **23**(1), 284	1982	Va. Polytech. & S. U.
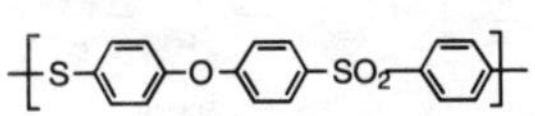	188		N-2; K_2CO_3, Sulfolane; Cl	Can. Pat. 946091	1974	Uniroyal
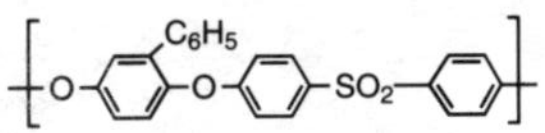			N-6; K_2CO_3, NMP; Cl	*Makromol. Chem.*, **191**, 2027	1990	Hamburg U.
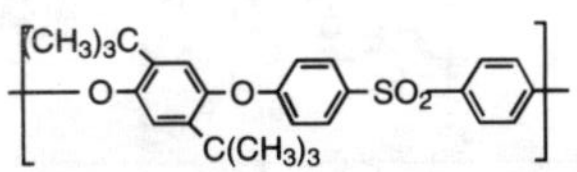	225–230	387	N-2; K_2CO_3, Diphenyl sulfone; Cl	US Pat. 4849503	1989	Amoco
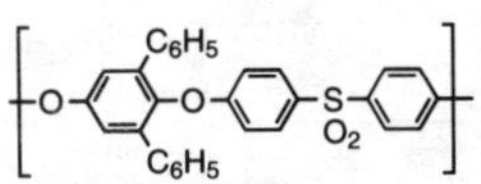	227		N-2; K_2CO_3, NMP; F	*J. M. S.-Pure Appl. Chem.*, **A29**(12), 1141	1992	McGill U

TABLE 19—*continued*

Polymer structue	Tg (°C)	Tm (°C)	Synthesis	Reference	Year	Organization
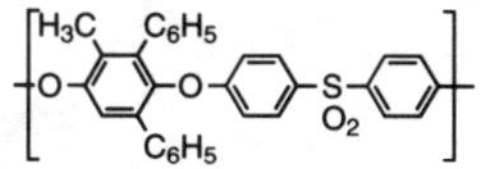	235		N-2; K_2CO_3, NMP; F	*J. M. S.-Pure Appl. Chem.*, **A29**(12), 1141	1992	McGill U.
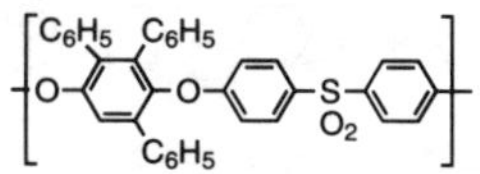	240		N-2; K_2CO_3, NMP; F	*J. M. S.-Pure Appl. Chem.*, **A29**(12), 1141	1992	McGill U.
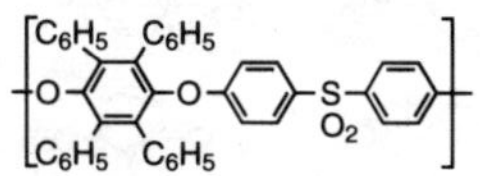	265	411	N-2; K_2CO_3, NMP; F	*J. M. S.-Pure Appl. Chem.*, **A29**(12), 1141	1992	McGill U.
Four Rings						
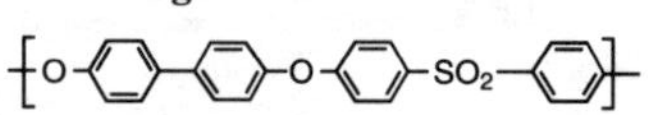 Radel™ R	220		N-1; NaOH, DMSO, Sulfolane; Cl	*Ency. Polym. Sci & Eng.*, **13**,196	1988	Union Carbide
	220		N-2; K_2CO_3, NMP, Sulfolane; Cl	*High Perf. Polym.*, Elsevier, p. 157;	1986	Amoco/Union Carbide
				Polym. Prep., **32**(2),166;	1991	
	232			*J. Polym. Sci., Polym. Chem.*, **23**, 287	1986	Va. P. I. & State U.
			N-6; CsPhenate, Bulk, Phenol	US Pat. 4794155	1988	Dow
	215		A-3; DMAC; Cl	*J. Polym. Sci., Polym. Chem.*, **28**, 367;	1990	Amoco/Union Carbide
	190			*Polym. J.*, **23**, 297	1991	U. Yamagata
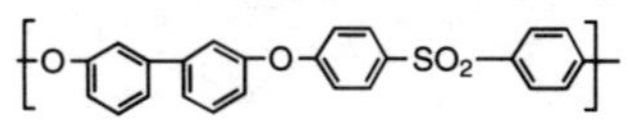	142		A-3; DMAC; Cl	*Polym. J.*, **23**, 297	1991	U. Yamagata

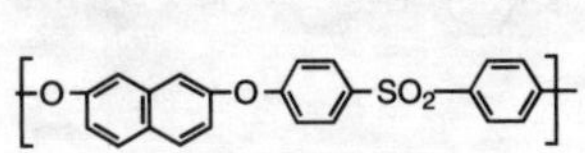	215 225	N-2; K_2CO_3, Sulfolane; Cl	Jap. Pat. 61-275326 *J. Polym. Sci., B, Polym Phys.*, **31**, 1061	1986 1993	Idemitsu Kosan U. Texas
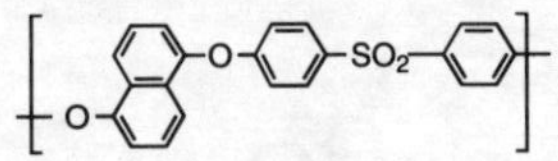	238 240	N-2; K_2CO_3, Sulfolane; Cl	Jap. Pat. 61-275326 *J. Polym. Sci., B, Polym Phys.*, **31**, 1061	1986 1993	Idemitsu Kosan U. Texas
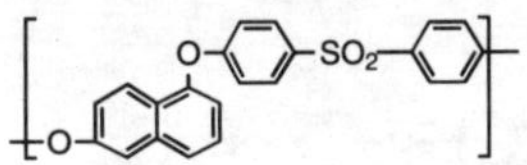	221	N-2; K_2CO_3, Sulfolane; Cl	Jap. Pat. 61-275326	1986	Idemitsu Kosan
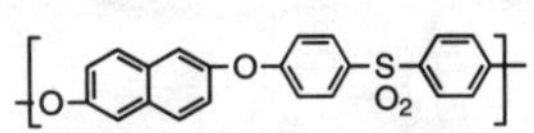	227	N-2; Permeability	*J. Polym. Sci., B, Polym. Phys.*, **31**, 1061	1993	U. Texas
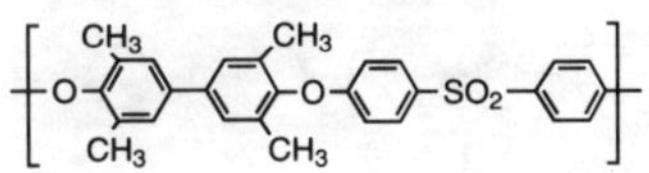	262	N-1; NaOH, DMSO; Cl	*Appl. Polym. Symp.*, No. **26**, 373	1975	Union Carbide
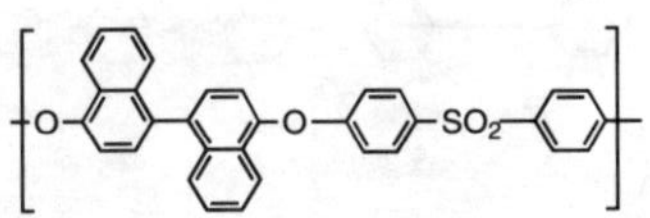	281	A-2; $FeCl_3$, Nitrobenzene	*J. Polym. Sci., A*, **29**(7), 965; Eur. Pat. 282096	1991 1988	Case West. Res. U.
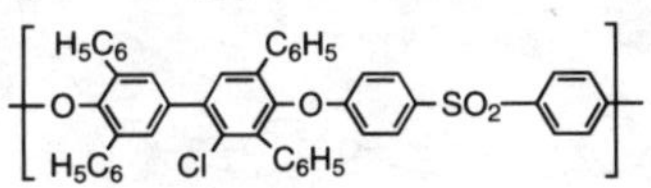	280	N-2; K_2CO_3, NMP; F	*Polym. Prep.*, **32**(1), 389	1991	McGill U.
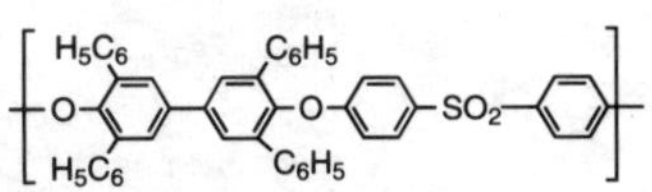	270	N-2; K_2CO_3, NMP; F	*Polym. Prep.*, **32**(1), 389	1991	McGill U.
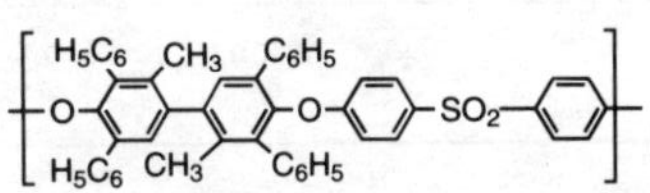	285	N-2; K_2CO_3, NMP; F F	*Polym. Prep.*, **32**(1), 389	1991	McGill U.

TABLE 19—*continued*

Polymer structue	Tg (°C)	Tm (°C)	Synthesis	Reference	Year	Organization
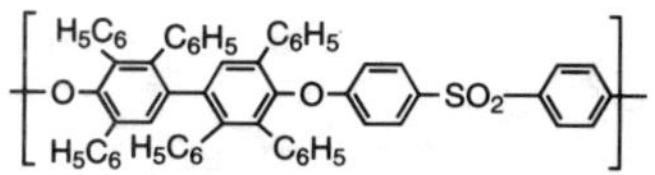	265		N-2; K_2CO_3, NMP; F	*Polym. Prep.*, **32**(1), 389	1991	McGill U.
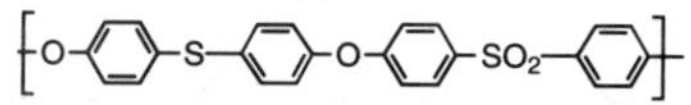	175		N-1; NaOH, DMSO, Sulfolane; Cl, F	*J.Polym.Sci.*, A-1, **5**, 2375;	1967	Union Carbide
				Ency.Polym.Sci & Eng., **13**,196;	1988	
			N-2; K_2CO_3, NMP; Cl	*Polym. Prep.*, **23**(1),284	1982	Va. P. I. & State U.
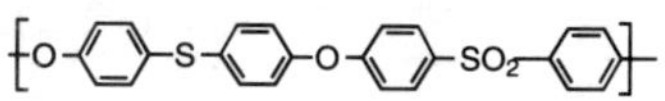	180		N-2; K_2CO_3, DMAC; Cl Kinetics	*Polym.*, **25**, 1827	1984	Va. P. I. & State. U.
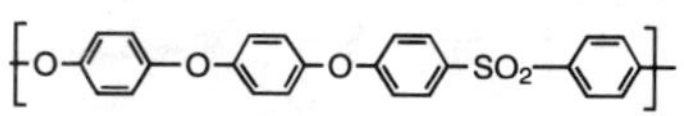	180		N-1; NaOH, DMSO, Sulfolane; Cl, F	*J.Polym.Sci.*, A-1, **5**, 2375;	1967	Union Carbide
				Ency. Polym. Sci. & Eng., **13**, 196	1988	
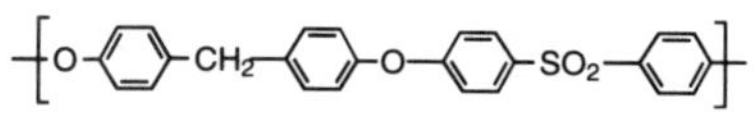	180		N-1; NaOH, DMSO, Sulfolane; Cl, F	*J. Polym. Sci.*, A-1, **5**, 2375	1967	Union Carbide
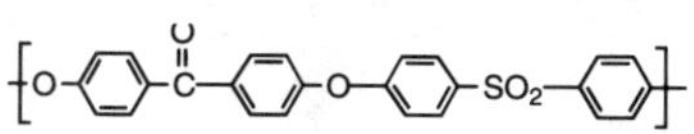	205		N-1; NaOH, DMSO, Sulfolane; F	*J. Polym. Sci.*, A-1, **5**, 2375;	1967	Union Carbide
				Ency. Polym. Sci. & Eng.,**13**,196	1988	
			N-1; DMSO; NO_2	*Makromol. Chem.*, **130**, 45	1969	Bayer, A-G
	185		N-6; Bulk; F	*Polym.*, **25**(8), 1151	1984	Hamburg U.
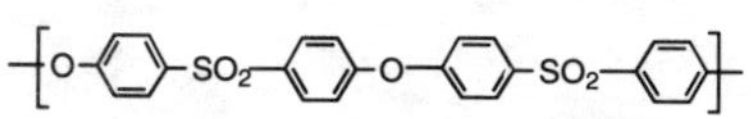	245		N-1; NaOH, DMSO, Sulfolane; Cl, F	*J. Polym. Sci.*, A-1, **5**, 2375;	1967	Union Carbide
				Ency. Polym. Sci. & Eng., **13**, 196;	1988	

		N-2; K_2CO_3, NMP; Cl	*Polym. Prep.*, **23**(1), 284	1982	Va. P. I. & State U.
	220	E-1; $FeCl_3$, Nitrobenzene, Dimethyl sulfone	*J. Polym. Sci.*, A-1, **8**, 2035	1970	3M
	250	E-1; $FeCl_3$, Bulk	*J. Polym. Sci.*, C, No.**16**, 715	1967	ICI
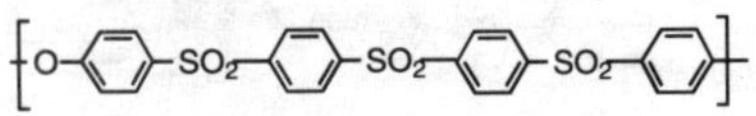		E-1; $FeCl_3$, Bulk	*J. Polym. Sci.*, C, No.**16**, 715	1967	ICI
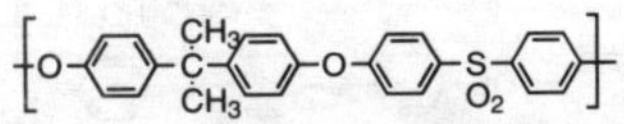 Udel™ Polysulfone	184, 195	N-1; DMSO, Sulfolane; Cl, F	*J. Polym. Sci.*, A-1, **5**, 2375, 2399;	1967	Union Carbide
		Thermal Stability	*Appl. Polym. Symp.*, No. **26**, 373	1975	
		N-2; K_2CO_3, DMSO; NO_2;	Jap. Pat. 1087628	1989	Toyo Soda
		N-2; K_2CO_3, NMP; Cl	*Polym. Prep.*, **23**(1), 284	1982	Va. P. I. & State U.
		N-6; K_2CO_3, NMP; Cl	*Makromol. Chem.*, **191**, 2027	1990	Hamburg U.
	186	Steam Sterilization	*Polym. News*, **11**, 359;	1986	Union Carbide
		Permeability Data	*J. App. Polym. Sci.*, **28**, 2925;	1983	U. Texas
			Polym., **32**, 840	1991	
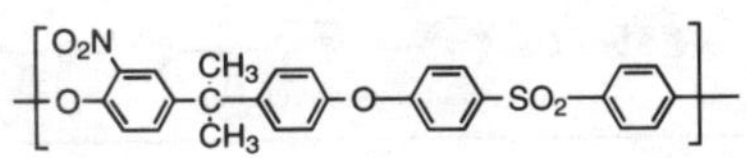	190	Post Reaction on PSF Permeation Data	*J. Memb. Sci.*, **72**, 91	1992	N. C. State U.
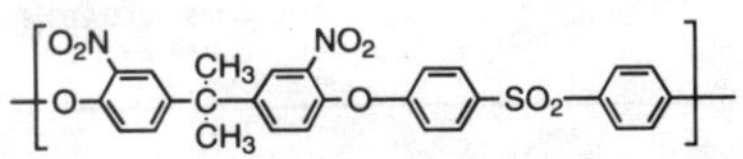	195	Post Reaction on PSF Permeation Data	*J. Memb. Sci.*, **74**, 193	1992	U. of Twente

TABLE 19—*continued*

Polymer structue	Tg (°C)	Tm (°C)	Synthesis	Reference	Year	Organization
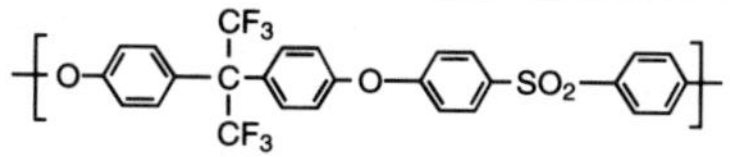	205	255	N-1; NaOH, DMSO, Sulfolane; Cl,F	*J.Polym. Sci.*, A-1, **5**, 2375;	1967	Union Carbide
				Ency. Polym. Sci. & Eng., **13**,196	1988	
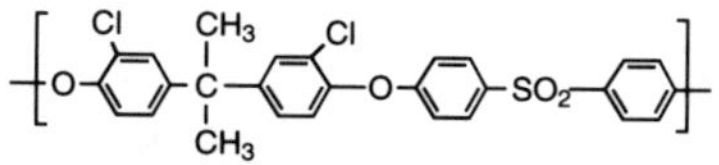	205		N-1; NaOH, DMSO, Sulfolane; Cl	*J. Polym. Sci.*, A-1, **5**, 2375;	1967	Union Carbide
				Polym. Prep., **16**(1), 476	1975	
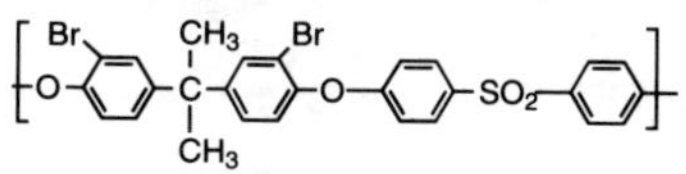			Bromination of Udel 3500	*Polym.*, **30**, 1137	1989	Nat. Res. Council, Canada
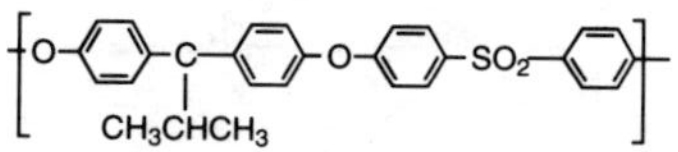	200		N-1; NaOH, DMSO, Sulfolane; Cl, F	*J. Polym. Sci.*, A-1, **5**, 2375	1967	Union Carbide
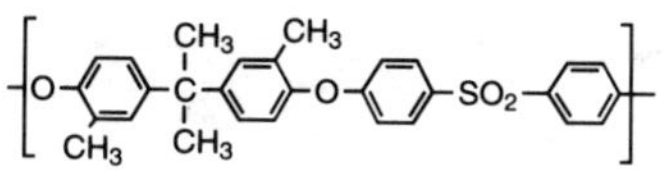	178		N-1; NaOH, DMSO; Cl	*Appl. Polym. Symp.*, No. **26**, 373	1975	Union Carbide
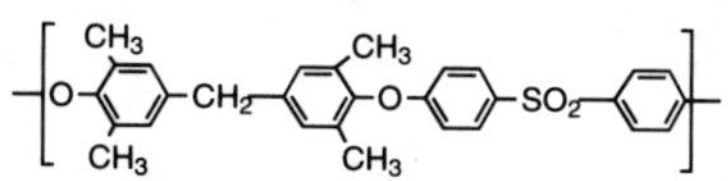	228		N-1; NaOH, DMSO; Cl	US Pat. 4156068	1979	
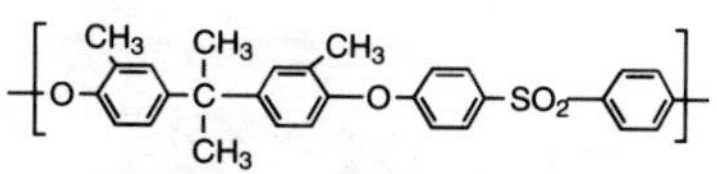	178		N-1; NaOH, DMSO; Cl, Permeability Data	*Polym. Prep.*, **16**(1), 476	1975	Union Carbide
	180			*Polymer*, **32**, 840	1991	U. Texas
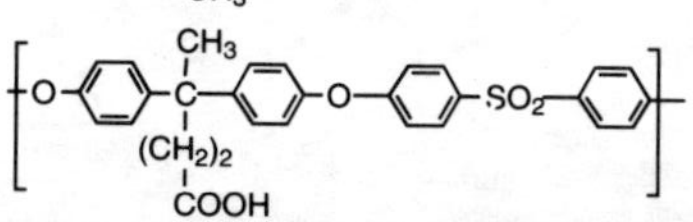			N-1; NaOH, DMSO; Cl	*Adv. Chem. Ser.*, **176**, 559	1979	Union Carbide

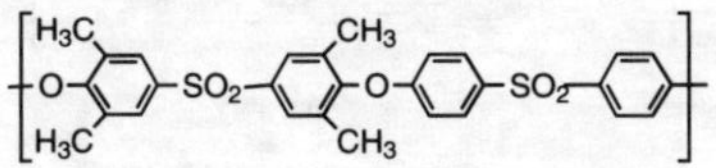	265	N-2; K_2CO_3, Sulfolane; Cl	US Pat. 4473684	1984	Amoco
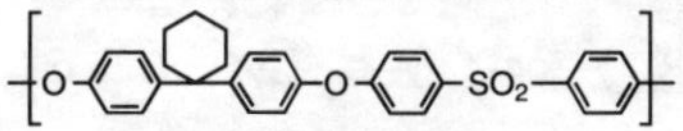	205	N-1; NaOH, DMSO, Sulfolane; Cl, F	*J. Polym. Sci.*, A-1, **5**, 2375; *Ency. Polym. Sci. & Eng.*, **13**,196	1967 1988	Union Carbide
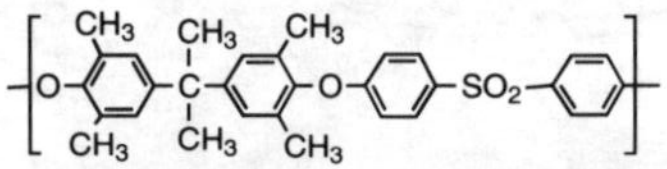	235	N-1; NaOH, DMSO, Sulfolane; Cl	*J. Polym. Sci.*, A-1, **5**, 2375; *Appl. Polym. Symp.*, No. **26**, 373;	1967 1975	Union Carbide
	242	Permeability Data	*Polym.*, **32**, 840	1991	U. Texas
	235	N-2; K_2CO_3, NMP: F	*Polym. Prep.*, **25**(2), 19	1984	Va. P. I. & State. U.
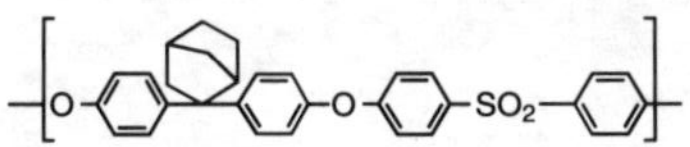	250	N-1; NaOH, DMSO, Sulfolane; Cl, F	*J. Polym. Sci.*, A-1, **5**, 2375	1967	Union Carbide
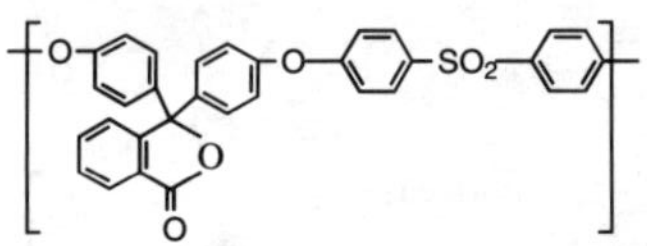		N-6; K_2CO_3, NMP; Cl	*Makromol. Chem.*, **191**, 2027	1990	Hamburg U.
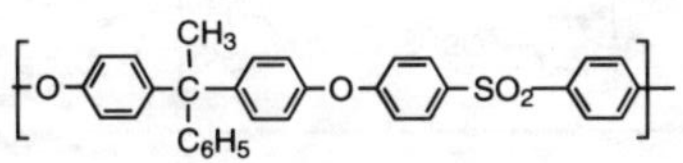	200	N-1; NaOH, DMSO, Sulfolane; Cl,F	*J. Polym. Sci.*, A-1, **5**, 2375	1967	Union Carbide
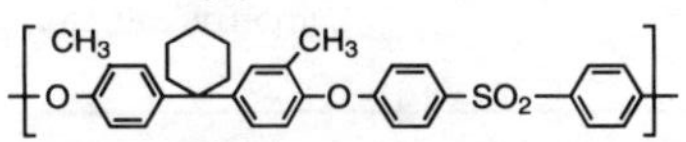	197	N-1, N-2; DMSO, NMP; Cl Permeabilty Data	*Polym.*, **32**, 840	1991	U. Texas

TABLE 19—*continued*

Polymer structue	Tg (°C)	Tm (°C)	Synthesis	Reference	Year	Organization
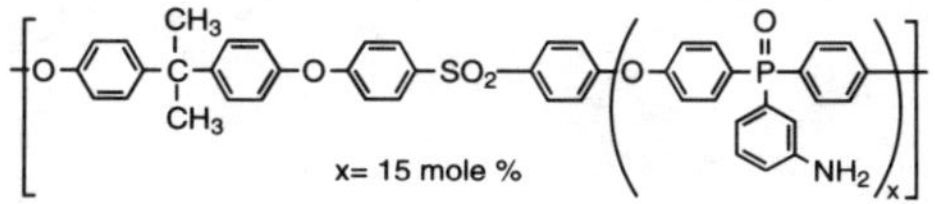	198		N-2; K_2CO_3, NMP; F, Cl	*Polym. Bull.*, **29**, 477	1992	Va. P. I. & State. U.
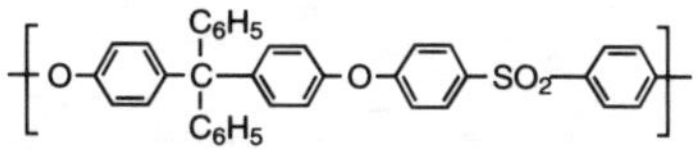	230		N-1; NaOH, DMSO, Sulfolane; Cl, F	*J. Polym. Sci.* A-1, **5,** 2375; *Ency. Polym. Sci. & Eng.*,**13,**196	1967 1988	Union Carbide
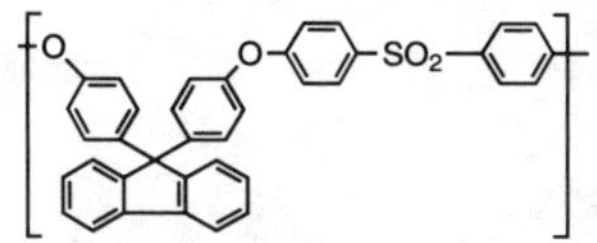	280 280		N-2; K_2CO_3, DMAC; Cl	*Polym.*, **29**,358; US Pat. 4806618	1988 1989	NASA, Langley; Central Glass (Jap.)
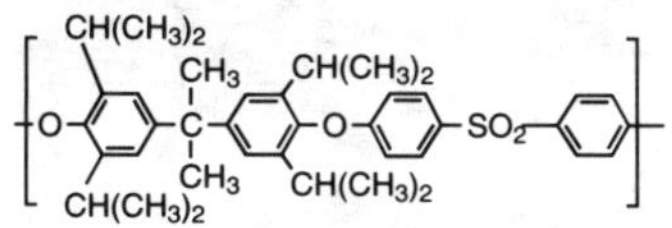	175		N-1; NaOH, DMSO; Cl	*Appl. Polym. Symp.*, No. **26**, 373	1975	Union Carbide
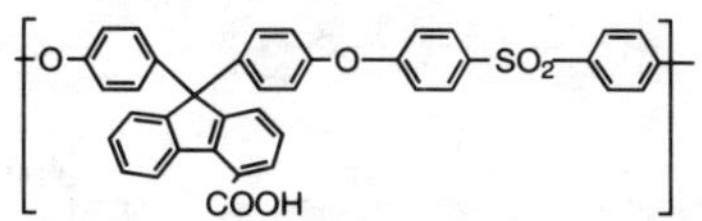			N-1; KOH, DMSO; F, Cl	*Proc. Int. Membrane* Conf. p. 187	1986	Nat. Res. Council, Canada
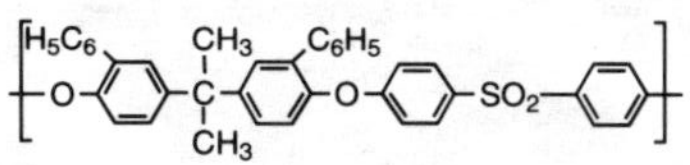	191		N-1; NaOH, NMP; Cl	Jap. Pat. 1098629	1989	Idemitsu Kosan

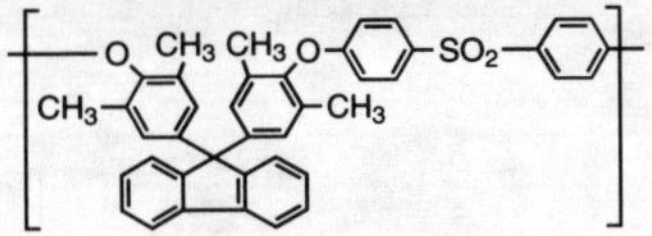	310	N-2; K_2CO_3, DMAC; Cl, F	*Polym.*, **29**, 358	1988	NASA/ Langley
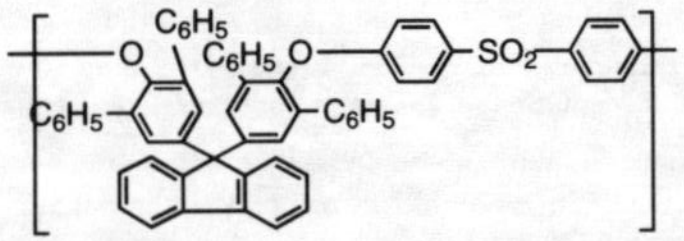	262	N-2; K_2CO_3, DMAC; F	*J. Polym. Sci.*, A, **29**(7), 1045	1991	McGill U.
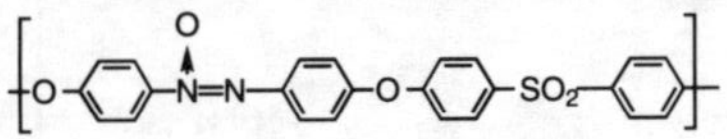		N-1; NaOH, DMSO; Cl	*Prog. Polym. Sci.*, **12**, 111	1986	Ind. Inst. . Tech
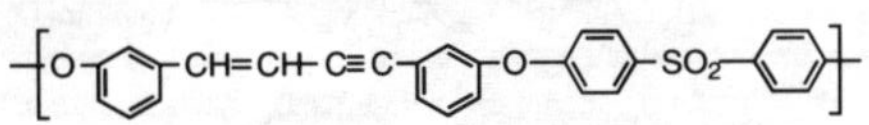	179	N-2; K_2CO_3, DMSO, Sulfolane; F	*Polym. Prep.*, **20**(1), 211	1979	
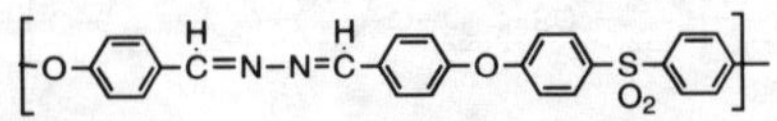	151	N-2; K_2CO_3, DMPU; F	*Polym. Prep.*, **34**(1), 521	1993	IBM
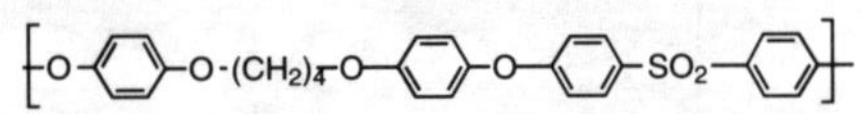	116	N; NaOH, DMSO; Cl	Eur. Pat. 327501	1989	Ciba-Geigy
Five Rings					
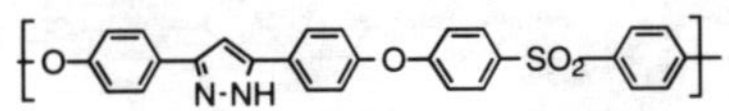	260	N-2; K_2CO_3, DMAC, NMP; Cl	*Polym. Prep.*, **32**(1), 619	1991	Va. Comm. U.
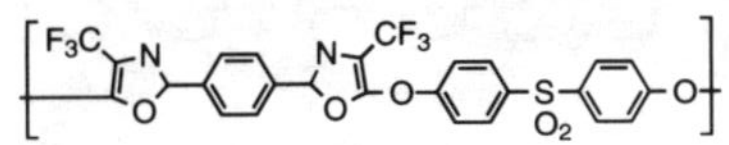	191	N-2; $(Et)_3N$, NMP; F	*Polym. Prep.*, **34**(1), 429	1993	U. Bayreuth
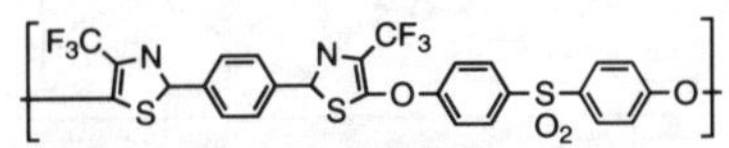	203	N-2; $(Et)_3N$, NMP; F	*Polym. Prep.*, **34**(1), 429	1993	U. Bayreuth
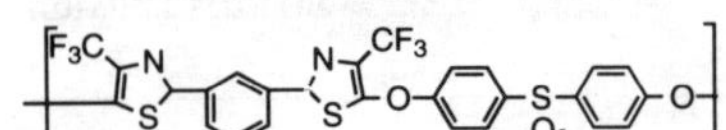		N-2; $(Et)_3N$, NMP; F	*Polym. Prep.*, **34**(1), 429	1993	U. Bayreuth

TABLE 19—*continued*

Polymer structue	Tg (°C)	Tm (°C)	Synthesis	Reference	Year	Organization
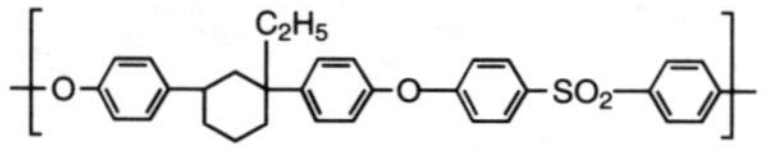	230		N-1; NaOH, DMSO; Cl	GB Pat. 1078234	1967	Union Carbide
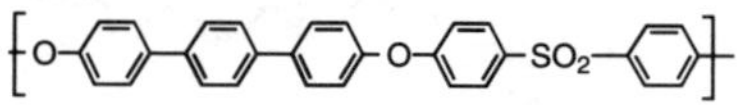	251	389	N-2; Diphenyl sulfone; Cl, F	*Bull. Soc. Chim. Belg.*, **98**, 667	1989	ICI
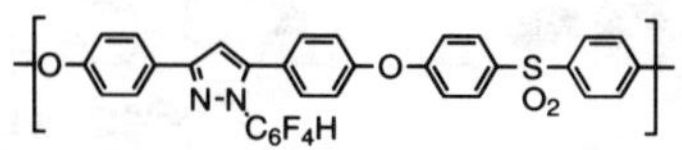	230		N-9; K_2CO_3, DMAC; F	*Polym. Prep.*, **34**(1), 441	1993	Va. Comm. U.
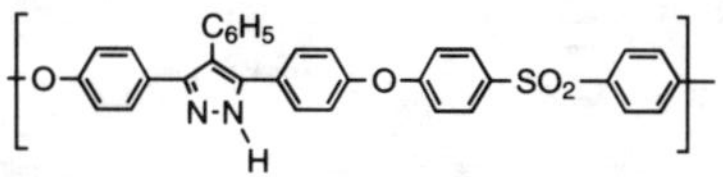	237		N-2; K_2CO_3, DMAC; Cl	*Polym. Prep.*, **32**(2), 160	1991	Va. Comm. U.; NASA/ Langley
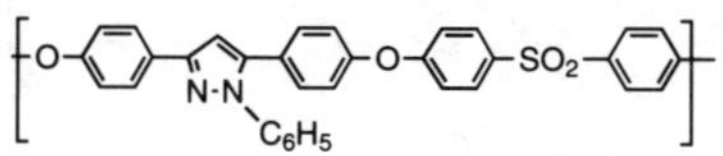	235		N-2; K_2CO_3, DMAC, NMP; F	*Polym. Prep.*, **32**(1), 619	1991	Va. Comm. U.
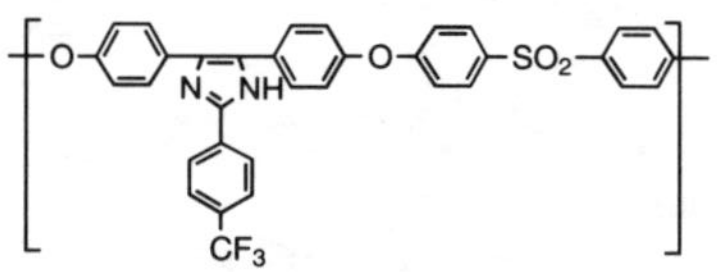	284		N-2; K_2CO_3, DMAC; F	*Polym. Prep.*, **32**(2), 162	1991	NASA/ Langley
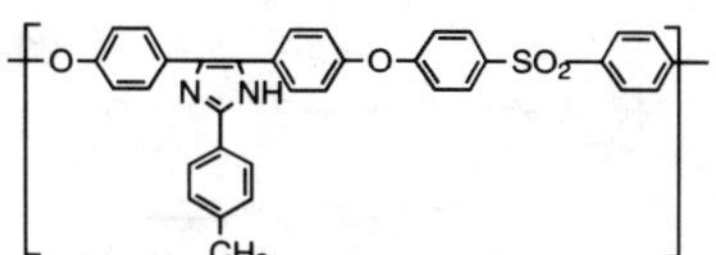	275		N-2; K_2CO_3, DMAC; F	*Polym. Prep.*, **32**(2), 162	1991	NASA/ Langley

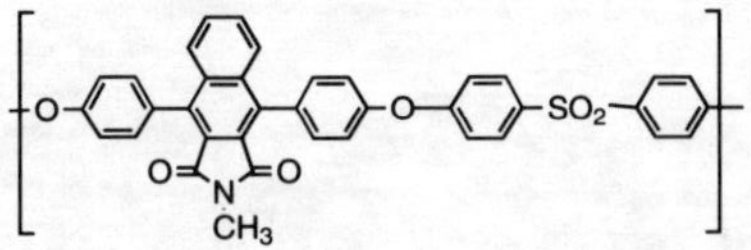	314	N-2; NMP, Sulfolane; F	*Polym. Prep.*, **32**(2), 164	1991	McGill U.
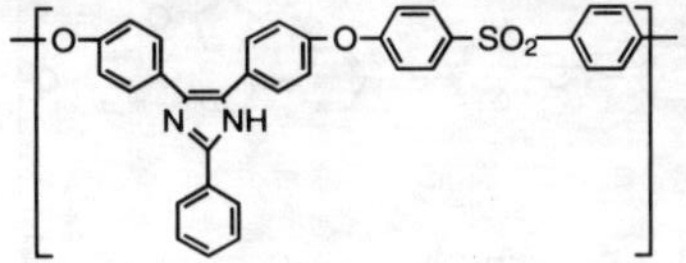	277	N-2; K_2CO_3, DMAC; F	NASA Tech. Briefs, Aug., p. 48	1990	NASA/ Langley
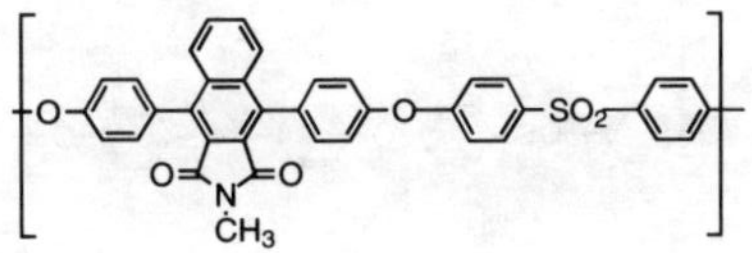	314	N-2; NMP, Sulfolane; F	*Polym. Prep.*, **32**(2), 164	1991	McGill U.
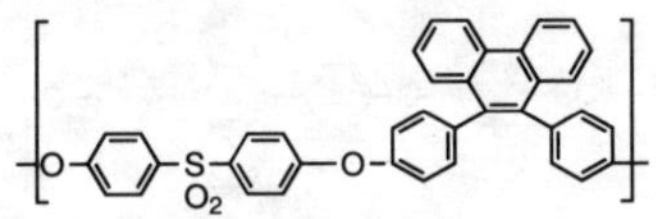	238	Post Reaction on PAEK	*Polym. Prep.*, **33**(2), 351	1992	Carleton U.
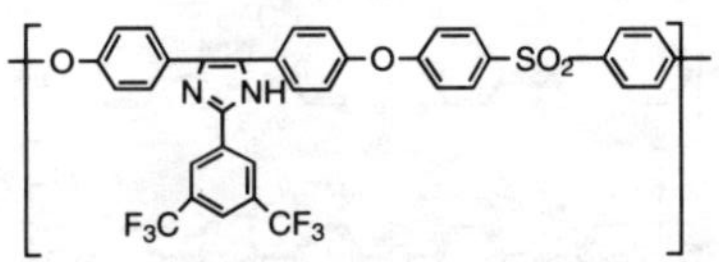	295	N-2; K_2CO_3, DMAC; F	*Polym. Prep.*, **32**(2), 162	1991	NASA/ Langley
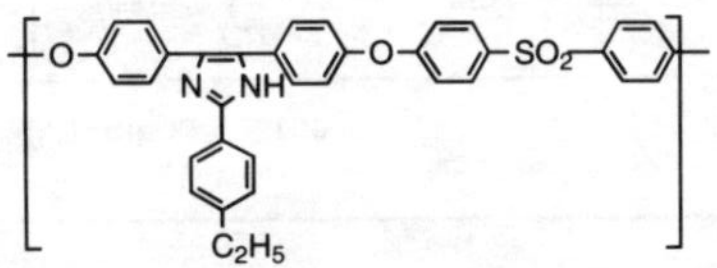	271	N-2; K_2CO_3, DMAC; F	*Polym. Prep.*, **32**(2), 162	1991	NASA/ Langley

TABLE 19—*continued*

Polymer structue	Tg (°C)	Tm (°C)	Synthesis	Reference	Year	Organization
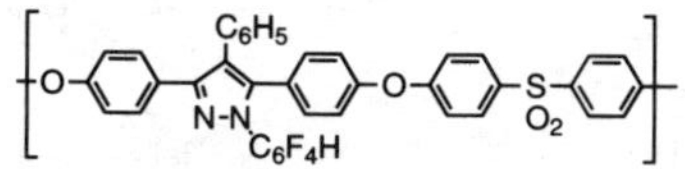	221		N-9; K_2CO_3, DMAC; F	*Polym. Prep.*, **34**(1), 441	1993	Va. Comm. U.
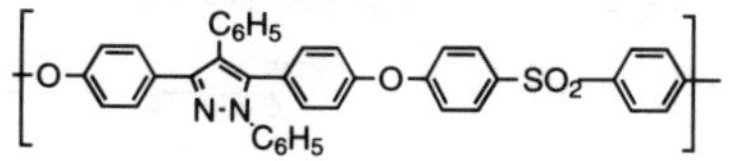	243		N-2; K_2CO_3, DMAC; F	*Polym. Prep.*, **32**(2), 160	1991	Va. Comm. U.; NASA/ Langley
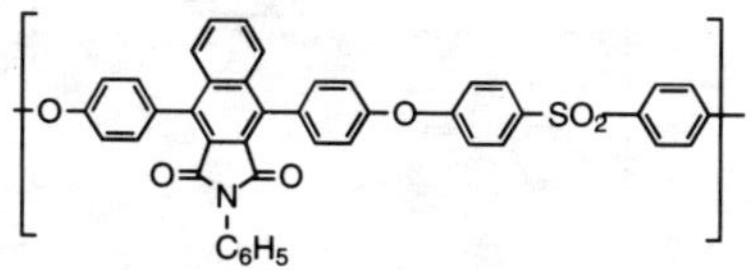	310		N-2; NMP, Sulfolane; F	*Polym. Prep.*, **32**(2), 164	1991	McGill U.
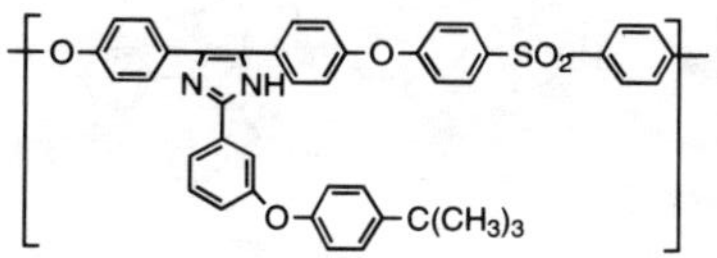	235		N-2; K_2CO_3, DMAC; F	*Polym. Prep.*, **32**(2), 162	1991	NASA/ Langley
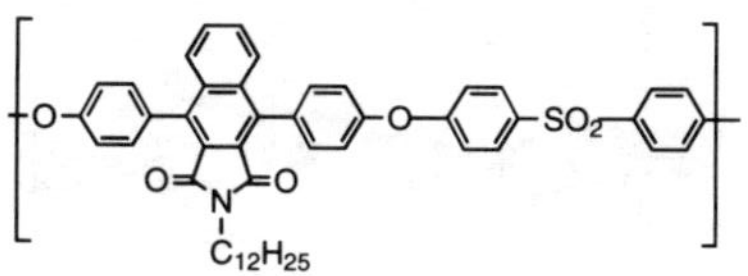	167		N-2; NMP, Sulfolane; F	*Polym. Prep.*, **32**(2), 164	1991	McGill U.

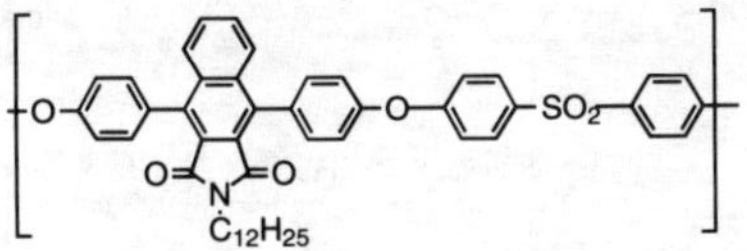	147		N-2; NMP, Sulfolane; F	*Polym. Prep.*, **32**(2), 164	1991	McGill U.
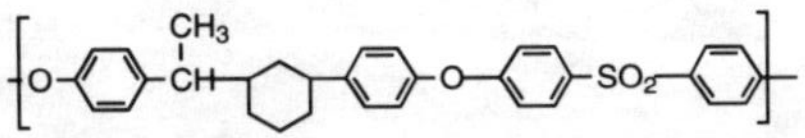	230		N-1; NaOH, DMSO, Sulfolane; Cl, F	*J. Polym.. Sci.*, A-1, **5**, 2375	1967	Union Carbide
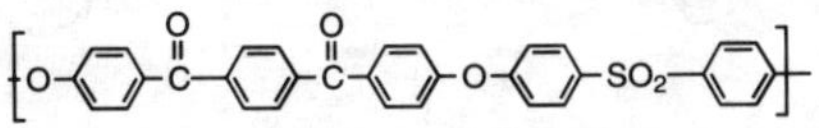	175		N-2; K_2CO_3, NMP; Cl	*CA*, **96**, 7266; Ger. Pat. 3014230	1982	BASF
	196	(214)	E-5; TFMSA, P_2O_5; H	US Pat. 4861856	1989	Eastman Kodak
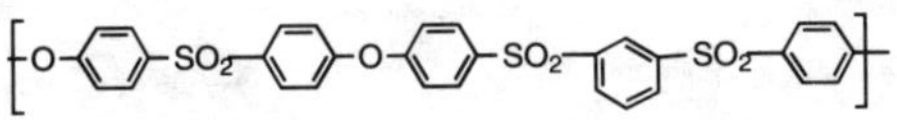			E-1; $FeCl_3$, Bulk	*J. Polym. Sci.*, C, No.**16**, 715	1967	ICI
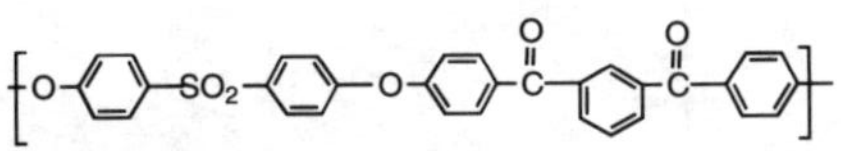	172		N-2; K_2CO_3, Diphenyl sulfone; F	*Bull. Soc. Chim. Belg.*, **98**, 667	1989	ICI
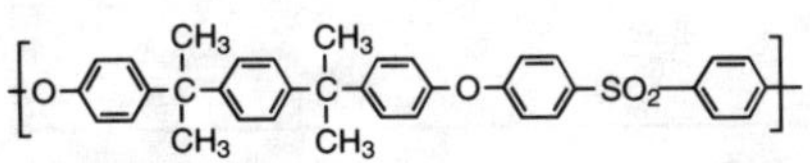	158 (HDT)	4.1% Cryst.	N-1; NaOH, Sulfolane; Cl	*Polym. Rxns. & New Polymers*, ACS No.**129**, 131	1973	Uniroyal
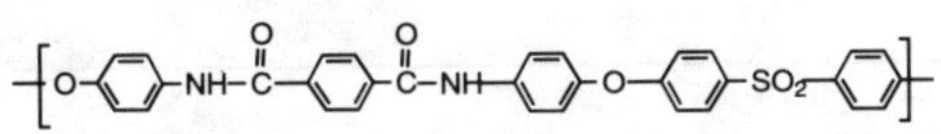			N-2; K_2CO_3, NMP; Cl	*CA*, **95**, 8035; Ger. Pat. 2940190	1981	BASF

TABLE 19—*continued*

Polymer structue	Tg (°C)	Tm (°C)	Synthesis	Reference	Year	Organization
Six Rings						
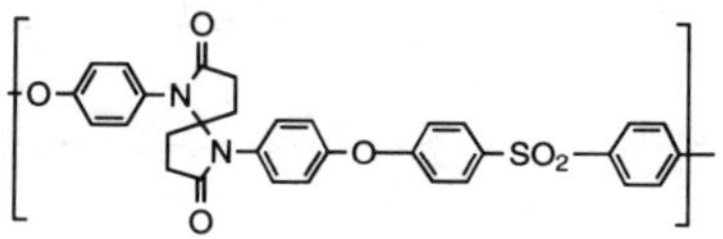	272		N-2; Cl	*Polym. Prep.*, **32**(2), 180	1991	Shell
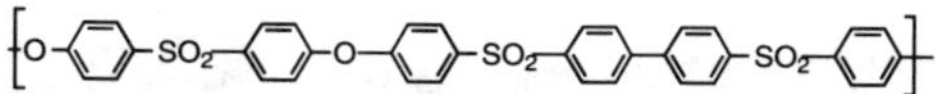 Victrex™ HTA	265		N-2; K_2CO_3, Sulfolane; F, Cl	*Polym.*, **18**(4), 359; **18**(4), 370	1977	ICI
				Polym. Prep., **32**(2), 166	1991	Amoco
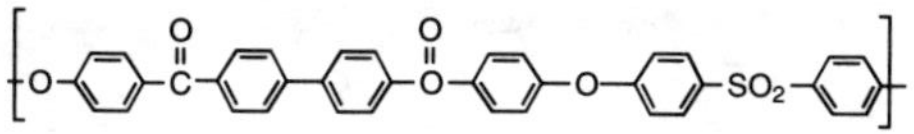			A-5; CsF, Bulk	*J. Chem. Soc., Chem. Commun.*, 336;	1990	ICI
	219		Permeability	*J. Polym. Sci.,B, Polym. Phys.* **31**, 983	1993	U. Texas
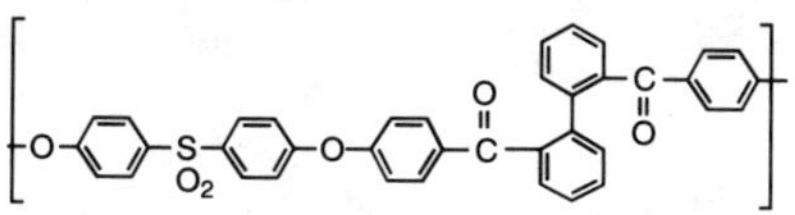	201		N-2; K_2CO_3, NMP; F	*Polym. Prep.*, **33**(2), 351	1992	Carleton U.
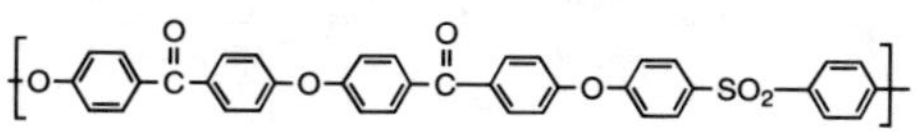			N-2; K_2CO_3, Sulfolane; Cl	*CA*, **87**, 136615; Ger. Pat. 2549529	1977	BASF
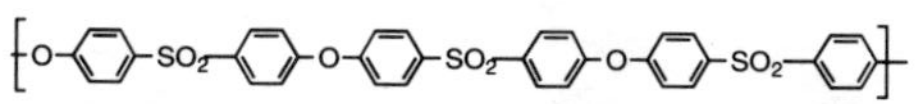			E-1; $FeCl_3$, Bulk	*J. Polym. Sci.*, C, No.**16**, 715	1967	ICI
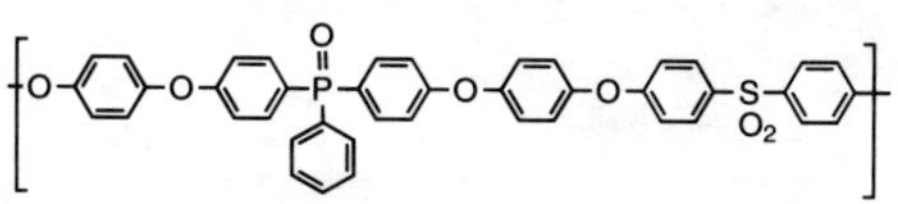	207		N-2; K_2CO_3, DMAC; F	*Polym. Prep.*, **33**(2), 241	1992	NASA/ Langley

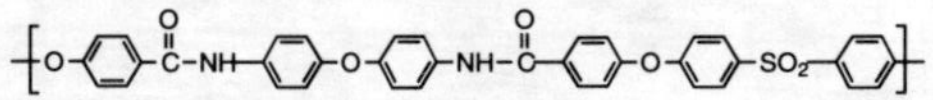		N-2; K_2CO_3, Sulfolane; Cl	*CA*, **87**, 102832; Ger. Pat. 2557652	1977	BASF
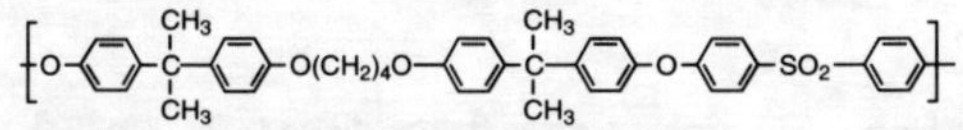	140	N-1; NaOH, DMSO, Sulfolane; Cl	*J. Polym. Sci.*, A-1, **5**, 2375	1967	Union Carbide
>= Six Rings					
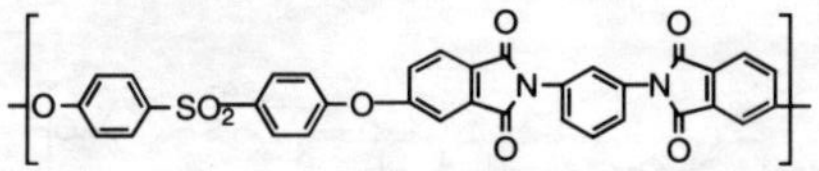	265	N-15; m-Cresol, Toluene	*Polym. Prep.*, **24**(2), 312; *J. Polym. Sci., Polym. Symp.*, **74**, 93	1983 1986	GE
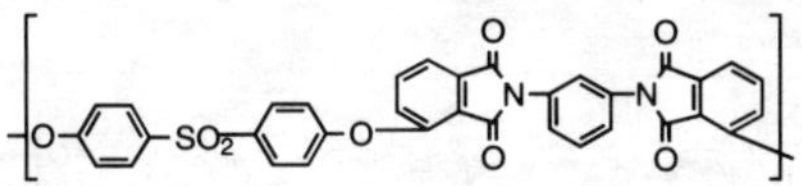	266	N-15; m-Cresol, Toluene	*Polym. Prep.*, **24**(2), 312; *J. Polym. Sci., Polym. Symp.* **74**, 93	1983 1986	GE
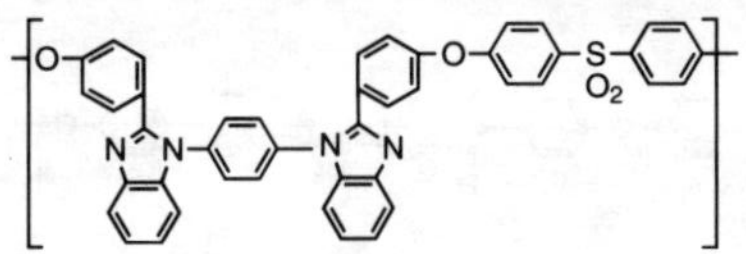	270	N-2; K_2CO_3, Sulfolane; F	*Polym. Prep.*, **33**(1), 1098	1992	NASA/ Langley
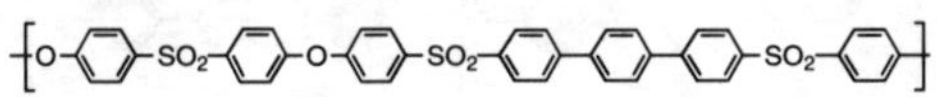	290	N-2; K_2CO_3, Sulfolane, Diphenyl sulfone; F, Cl. E-1	*Polym.*, **18**(4), 370; *Bull. Soc. Chim. Belg.*, **98**, 667; *Polym.*, 15, 456	1977 1989 1974	ICI
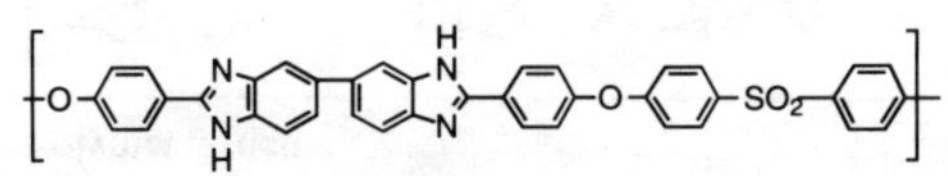	352	N-2; K_2CO_3, DMAC; F	*Polym. Prep.*, **32**(3), 195	1991	NASA/ Langley
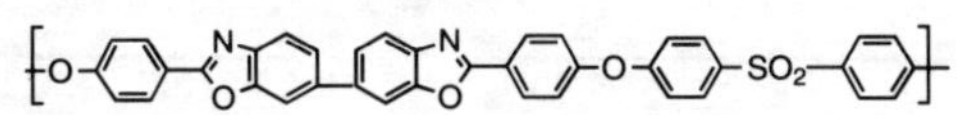	247	N-2; K_2CO_3, DMAC, Diphenyl sulfone; F	*Polym. Prep.*, **32**(1), 646	1991	NASA/ Langley

TABLE 19—*continued*

Polymer structue	Tg (°C)	Tm (°C)	Synthesis	Reference	Year	Organization
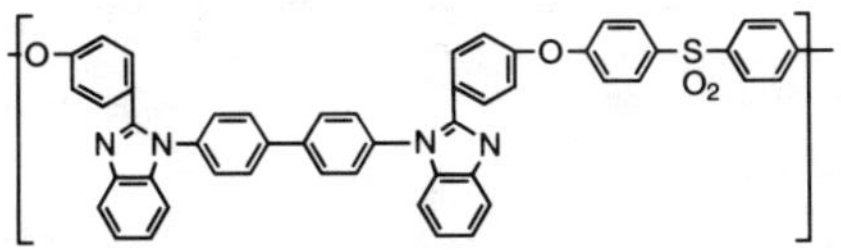	289		N-2; K_2CO_3, Sulfolane; F	*Polym. Prep.*, **33**(1), 1098	1992	NASA/ Langley
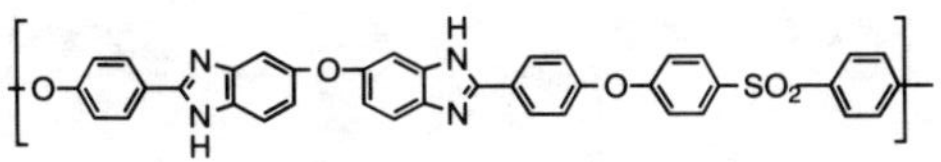	322		N-2; K_2CO_3, DMAC; F	*Polym. Prep.*, **32**(3), 195	1991	NASA/ Langley
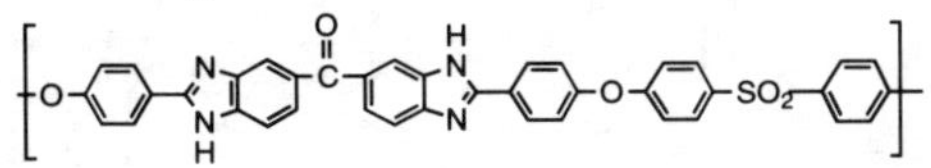			N-2; N-9; K_2CO_3, DMAC; F	*Polym. Prep.*, **32**(3), 195	1991	NASA/ Langley
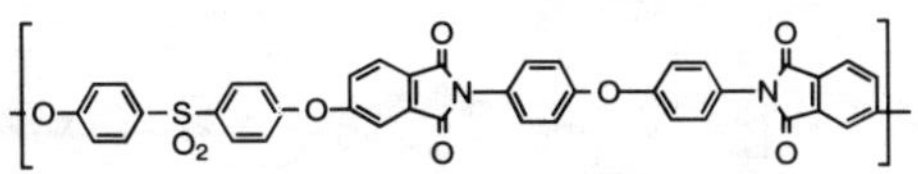	260		N-15; m-Cresol, Toluene	*Polym. Prep.*, **24**(2), 312; *J. Polym. Sci., Polym. Symp.*, **74**, 93	1983 1986	GE
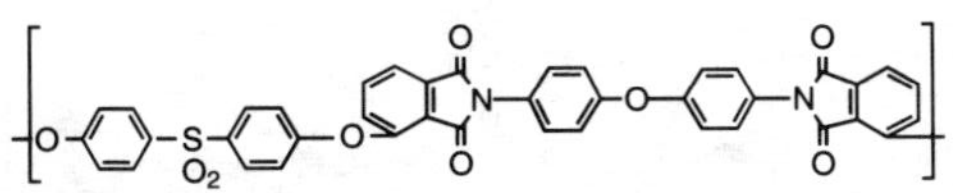	267		N-15; m-Cresol, Toluene	*Polym. Prep.*, **24**(2), 312; *J. Polym. Sci., Polym. Symp.*, **74**, 93	1983 1986	GE
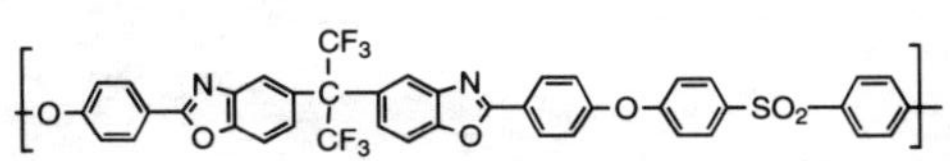	285		N-2; K_2CO_3, DMAC, Diphenyl sulfone; F	*Polym. Prep.*, **32**(1), 646	1991	NASA/ Langley

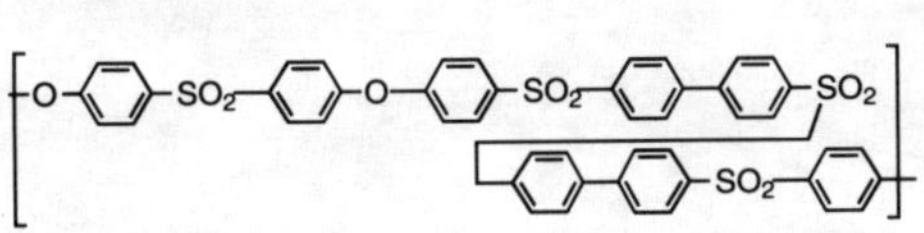	284–289	N-2; Sulfolane; F, Cl	*Polym.*, **18**(4), 359; **18**(4), 370	1977	ICI
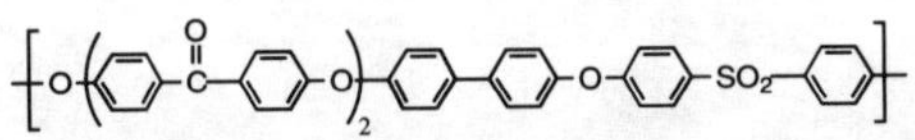	201		*SAMPE*, **32**, 382	1987	ICI
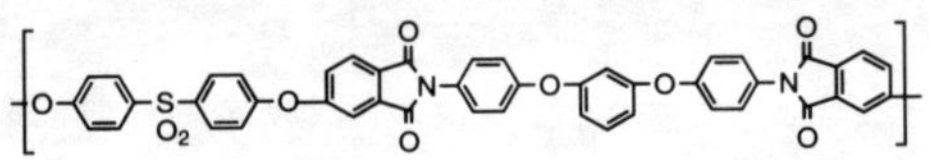	219	N-15; m-Cresol, Toluene	*Polym. Prep.*, **24**(2), 312; *J. Polym. Sci., Polym. Symp.*, **74**, 93	1983 1986	GE
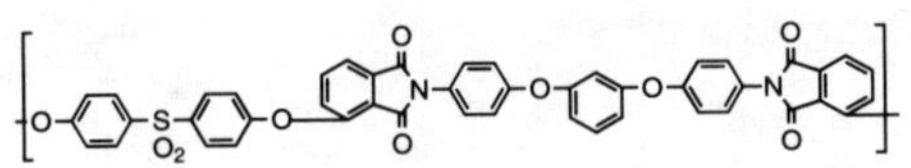	230	N-15; m-Cresol, Toluene	*Polym. Prep.*, **24**(2), 312; *J. Polym. Sci., Polym. Symp.*, **74**, 93	1983 1986	GE

CHAPTER 6

Crystalline Polyarylethers

Tables 20 and 21 contain those polymers for which a melting point has been reported. Although a melting point implies crystallinity, there may be some examples in these tables that are not actually crystalline. For example, a melting point has been reported for Polyethersulfone prepared by electrophilic routes. This polymer which has been commercially available by nucleophilic routes is well known to be amorphous. The entries in table 20 are for all those crystalline polymers that do not contain a ketone group. Since the vast majority of the polyaryletherketones are crystalline, they are reported on separately in Table 21. They are arranged in table 21 in order of increasing melting points. There is very little information available on the degree of crystallization, (except for PEEK and PEK), for any of these polymers.

The transition temperatures for polyarylethers were chosen as the only polymer properties to be included in the tables in this book. This decision was made in part for data handling reasons, but more importantly because they, more than any other property, serve to define or classify a polymer. Knowledge of a polymer's glass transition temperature (Tg) and its melting point (Tm) allows one to immediately rank its potential applicability among all other polymers.The basic classification is, of course, whether a polymer is amorphous or crystalline. Next, the actual value for the glass transition temperature of an amorphous polymer orders or positions it among other amorphous polymers. The glass transition temperature of an amorphous polymer represents an absolute upper use temperature. Practically though, its use temperature will be up to ten to fifty degrees below Tg depending on the load the polymer is subject to in a given application. Reinforcing fibers will improve an amorphous polymer's heat distortion temperature somewhat, but not above its Tg in any case. Obviously, crystalline polymers have an innate potential for having higher use temperatures than amorphous polymers. It has been recognized for many years that the ratio of the glass transition temperature to melting point (in degrees K) has about the same value for many polymers: Tg/ Tm = 2/3 (1). Although this ratio can be as low as 0.56 and as high as 0.76, most polymers exhibit a ratio of 2/3. The majority of the crystalline polyarylethers in Tables 20 and 21 fall in this range. PEEK with a Tg of 144°C and a melting point of 335°C exhibits a Tg/Tm ratio of 0.68. A polymer with a Tg of 200°C (473°K) would be estimated to have a melting point of 432°C (706°K) using the 2/3 ratio. Crystalline polymers have the potential of being used at temperatures up to near their melting points, but this is dependent on their degree of crystallinity and whether they are under stress. Since many polymers are composites of amorphous and crystalline phases, the amounts of each phase will determine properties. Other details such as the nature of the crystal structure and the size and number of spherulites also play a role. Orientation of crystalline polymers can increase the degree of crystallinity in a polymer and improve its use temperature and its mechanical properties. Additionally, fiber reinforcement (glass and carbon) of crystalline polymers is widely used to enhance the high temperature performance of crystalline polymers. In many instances, satisfactory performance at temperatures near the polymer's melting point can be achieved by composites of this type.

Some additional generalizations regarding the effects of crystallinity on polymer properties are worth noting. The term crystalline polymer is used in this context simply to denote polymers with any level of crystallinity. The density of a crystalline polymer is higher than an amorphous polymer which could present an economic disadvantage in an end use application. A fabricated part made from a higher density material will use more pounds of polymer and, hence, cost more than the same part made from a lower density material. Crystalline polymers are less soluble than amorphous polymers. Plasticization, which can have beneficial effects such as improving melt flow and toughness, is more difficult for crystalline polymers. Reduced solubility also manifests itself in an improved level of environmental stress crack resistance for crystalline polymers. This improves as the level of crystallinity increases because failure is believed to proceed by attack or dissolution of the amorphous phase by the environmental agent being tested. Permeation (the product of solubility and diffusivity) of liquids and gases through crystalline polymers is decreased since solubility is usually confined to the amorphous regions of the polymer. However, greater selectivity in separating components of a mixture usually accompanies lower permeation rates. Several recent studies (*2–8*) have been published on the gas permeation properties of polyarylethers. Finally, crystalline plastics tend to be more opaque than amorphous plastics because of the difference in refractive indices of the amorphous and crystalline phases. However, this can be (partially) ameliorated by reducing the spherulite size in the polymer.

A number of the mechanical properties of a polymer can be influenced by its degree of crystallinity as well as molecular weight. These include tensile properties, such as strength and modulus or stiffness, break elongation, impact strength, flex life and hardness. Generalizations about the specific direction of the effects to be expected can be misleading. On the one hand, an increasing degree of crystallinity can lead to a stiffer, more brittle and harder material. On the other hand, if the details of the crystallinity (spherulite size, etc.) are optimized and controlled, tough, more useful products are obtained from crystalline polymers. For example, it is well known that when orientation processes are applied to crystalline polymers very high strength levels are achievable that are the basis for the commercial successes. High performance fibers are a case in point. While crystallinity in a polymer can confer desirable property attributes, it should be kept in mind that they don't appear automatically. In other words, the details of the crystallinity need to be optimized to the same degree as other basic polymer properties, such as molecular weight.

The crystallization behavior of several polyaryletherketones has been the subject of many studies. PEEK, which is commercially available from ICI, Ltd., has received the most attention. Its crystallization behavior from the melt (*8–10*) and solution (*11*) has been reported. PEEK has been investigated extensively as a semi-crystalline matrix resin in carbon fiber composites. The impetus for these studies was the hope that thermoplastic matrix resins would offer fabrication and performance advantages in advanced composites over the state of the art thermoset matrices. The crystalline morphology of PEEK when incorporated into this type of composite has been studied and reported (*12–15*). The structures of crystalline PEK (*16–18*) and PEKK (*19*) have also been studied.

References

1. Van Krevelen, D. W., and Hoftyzer, P. J., "Properties of Polymers", Elsevier Publishing Co., New York. 1976, p. 122.
2. Aitken, C. L., Paul, D. R., and Mohanty, D. K., J. *Polym. Sci.: Part B: Polym. Physics*, **31**, 983 (1993); *ibid.*, **31**, 1061 (1993).
3. McHattie, J. S., Koros, W.J., and Paul, D. R., *Polymer*, **32**, 840 (1991).
4. Ghosal, K., Chern, R. T., and Freeman, B. D., *J. Polym. Sci.: Part B: Polym. Physics*, **31**, 891 (1993).
5. Ghosal, K., and Chern, R. T., *J. Membrane Sci.*, **72**, 91 (1992).
6. Kamps, K. M. P., Teunis, H. A., Wessling, M., and Smolders, C. A., *J. Membrane Sci.*, **74**, 193 (1992).
7. Matsumoto, K., and Xu, P., *J. Appl. Polym. Sci.*, **47**, 1961 (1993).
8. Bassett, D. C., Olley, R. H., and Al Raheil, I. A. M., *Polymer*, **29**, 1745 (1988).
9. Cebe, P., and Hong, S.-D., *Polymer*, **27**, 1183 (1986).
10. Nguyen, H. X., and Ishida, H., Polymer, **27**,1400 (1986).
11. Lovinger, A. J., and Davis, D. D., *Macromolecules*, **19**, 1861 (1986).
12. Cebe, P., Lowry, L., Chung, S. Y., Yavrouian, A., and Gupta, A., *J. Appl. Polym. Sci.*, **34**, 2273 (1987).
13. Wakelyn, N. T., *J. Polym. Sci.: Part A: Polym. Chem.*, **24**, 2101 (1986).
14. Blundell, D. J., Chalmers, J. M., Mackenzie, M. W., and Gaskin, W. F., *SAMPE Quart.*, **16**, 22 (1985).
15. Blundell, D. J., and Osborn, B. N., *SAMPE Quart.*, 17, 1 (1985).
16. Hay, J. N., Kemmish, D. J., Langford, J. I., and Rae, A. I. M., *Polym. Comm.*, **26**, 283 (1985).
17. Hay, J. N., and Kemmish, D. J., *Polym. Comm.*, **30**, 77 (1989).
18. Jog, J. P., and Nadkarni, V. M., *J. Appl. Polym. Sci.*, **32**, 3317 (1986).
19. Avakian, P., Gardner, K. H., and Matheson, R. R., *J. Polym. Sci.: Part C: Polym. Lett.*, **28**, 243 (1990).

TABLE 20
Crystalline polyarylethers (non-ketone) arranged in order of increasing melting points within each sub-group

Polymer structure	Tg (°C)	Tm (°C)	Synthesis	Reference	Year	Organization
Polyarylethersulfones						
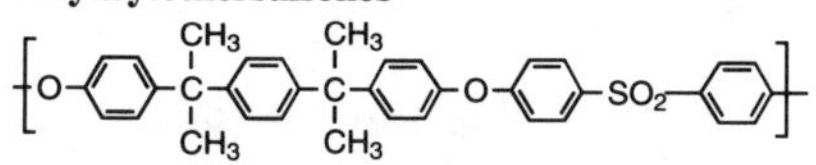	158 (HDT)	4.1% Cryst.	N-1; NaOH, Sulfolane; Cl	*Polym. Rxns. & New Polymers*, ACS, No.1**29**,131	1973	Uniroyal
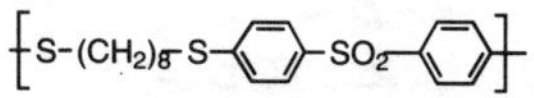		120–125		*J. Org. Chem.*, **29**, 207	1964	
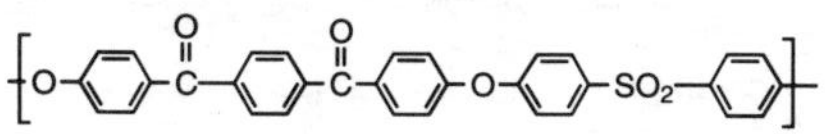	196	214	E-5; TFMSA, P_2O_5	US Pat. 4861856	1989	Eastman Kodak
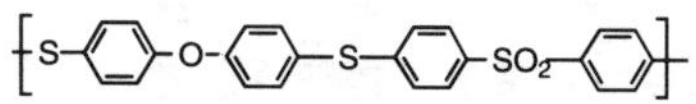		225–245	N-2; K_2CO_3, DMF; Cl	*Makromol. Chem.*, **140**, 83	1970	Celanese
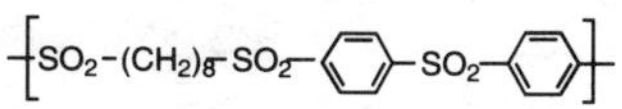		250–255		*J. Org. Chem.*, **29**, 207	1964	
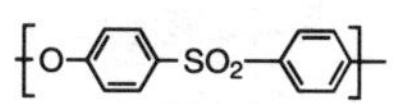 Victrex™ PES		250–265	E-1; $AlCl_3$, $C_6H_5NO_2$ E-2; PPA E-1; Bulk, $FeCl_3$	*J.Polym.Sci.*, A-1, **4**, 722; *Polym.*, **6**, 589	1966 1965	Monsanto ICI
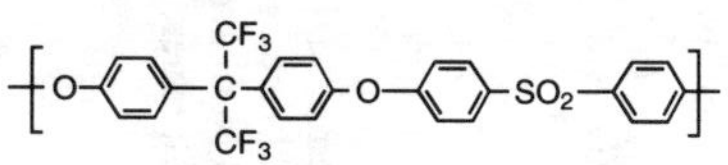	205	255	N-1; NaOH, DMSO, Sulfolane; Cl,F	*J.Polym. Sci.*, A-1, **5**, 2375; *Ency. Polym. Sci. & Eng.*, **13**,196	1967 1988	Union Carbide
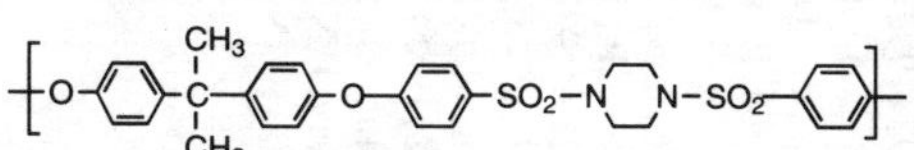		255–260	N-1; NaOH, DMSO; Cl	US Pat. 4108837	1978	Union Carbide
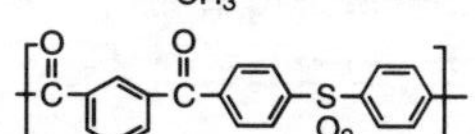	195	257	N-13; NaH, DMF; F	*Polym. Prep.*, **34**(1), 570	1993	Va. P. I. & State U.
		286	N-2; K_2CO_3, DMF; Cl	*Makromol. Chem.*,	1970	Celanese

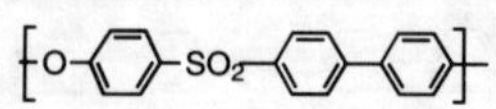	216	310	N-1; NaOH, DMSO; Cl	*Polym.*, **22**, 1096	1981	ICI
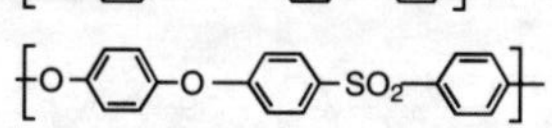	210	310	N-1; NaOH, DMSO, Sulfolane; Cl	*J. Polym. Sci.*, A-1, **5**, 2375;	1967	UnionCarbide
				Ency. Polym. Sci. & Eng., **13**, 196	1988	
	210	310	N-2; K_2CO_3, DMAC; Cl Kinetics	*Polym.*, **25,** 1827	1984	Va. P. I. & State U.
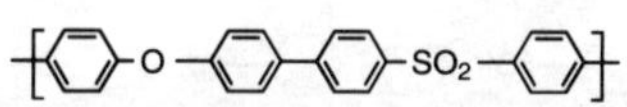	168	320	A-4;Sn/Pd, DMAC; Br	*J. Polym.Sci., Polym. Chem.*, **30**, 2511	1992	U. of East Anglia
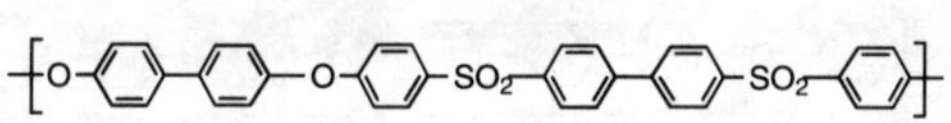	270	336	N-2; K_2CO_3, Sulfolane; Cl, F	*Polym. Prep.*, **32**(2), 166;	1991	Amoco
	280			*Polym.*, **18**(4), 370	1977	ICI
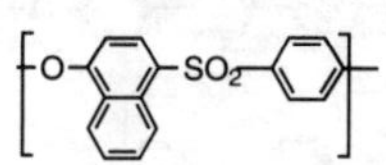	238	367	N-1; Melt, Sulfolane; F, Cl	*Polym.*,**15**, 456; **18**, 359; **18**, 370	1974; 1977	ICI
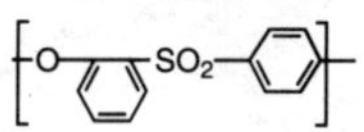	221	368	N-1; Melt,Sulfolane; F, Cl	*Polym.*,**15**, 456; **18**, 359;**18**, 370	1974 1977	ICI
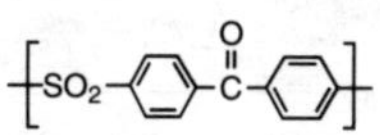		375	Oxidation of sulfide polymer	*Brit. Polym. J.*, **4**, 391	1972	ICI
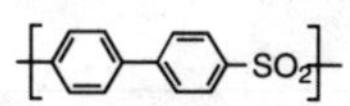		375, 410	E-1; Chlorobiphenyl, Bulk, $FeCl_3$	*Chem. Ind.* (15), 461; *Polym.*, **6**, 589	1968 1965	ICI

TABLE 20—*continued*

Polymer structure	Tg (°C)	Tm (°C)	Synthesis	Reference	Year	Organization
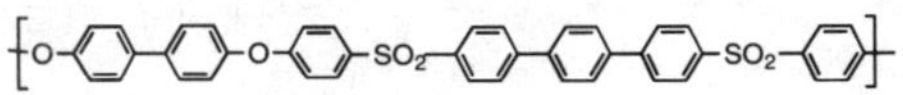	264	383	N-2; K_2CO_3, Diphenyl sulfone; Cl	*Bull. Soc. Chim. Belg.*, **98**, 667	1989	ICI
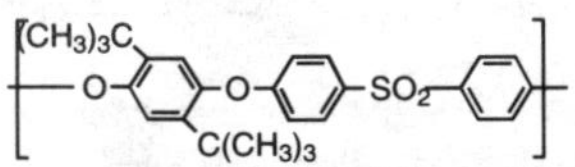	225–230	387	N-2; K_2CO_3, Diphenyl sulfone; Cl	US Pat. 4849503	1989	Amoco
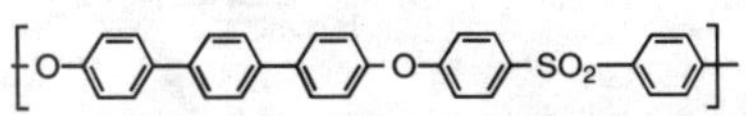	251	389	N-2; Diphenyl sulfone; Cl, F	*Bull. Soc. Chim. Belg.*, **98**, 667	1989	ICI
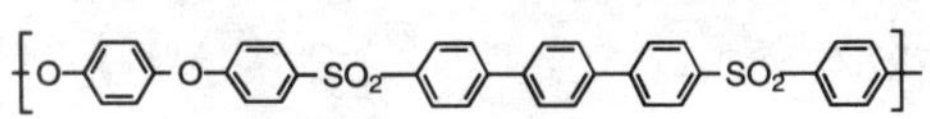	259	393	N-2; K_2CO_3, Diphenyl sulfone; Cl	*Bull. Soc. Chim. Belg.*, **98**, 667; Eur. Pat. 331305	1989	ICI
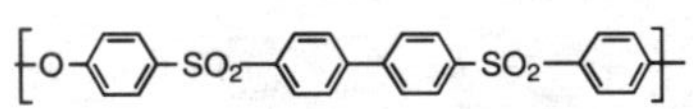	276–289	395	N-1; Melt,Sulfolane; F, Cl	*Polym.*, **18(4),** 359; **18**(4), 370	1977	ICI
	275	Sl. Cryst.	E-1; FeCl3 Nitrobenzene, Dimethyl sulfone	*J.Polym.Sci.*, A-1, **8**, 2035	1970	3M
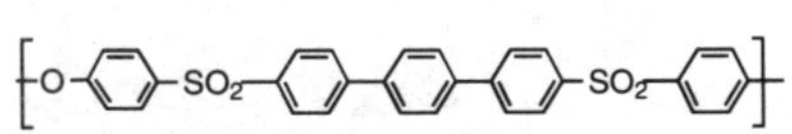	311	398	N-2; Diphenyl sulfone; Cl	*Bull. Soc. Chim. Belg.*, **98**, 667	1989	ICI
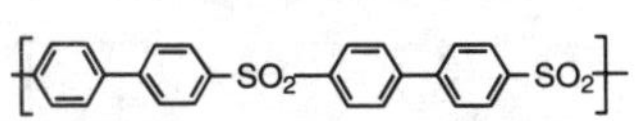		400 (>)	E-1; FeCl3, Nitrobenzene, Dimethyl sulfone	*J.Polym.Sci.*, A-1, **8**, 2035	1970	3M
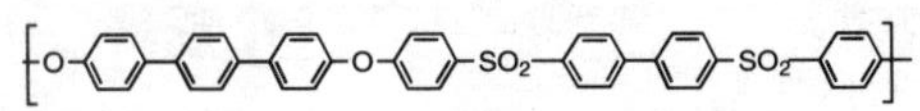	263	408	N-2; K_2CO_3, Diphenyl sulfone; Cl	*Bull. Soc. Chim. Belg.*, **98**, 667	1989	ICI
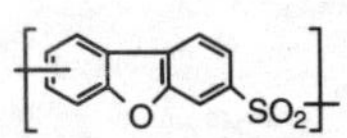		410–440	E-1	*Adv. Macromol. Chem.*, **1**, 335	1968	3M

Structure						
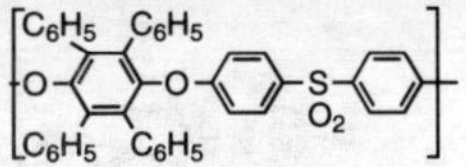	265	411	N-2; K_2CO_3, NMP; F	*J. M. S.-Pure Appl. Chem.*, **A29**(12), 1141	1992	McGill U
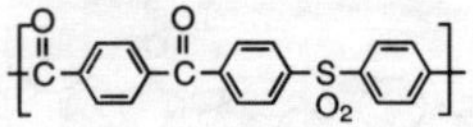	228	414	N-13; NaH, DMF; F	*Polym. Prep.*, **34**(1), 570	1993	Va. P. I. & State U.
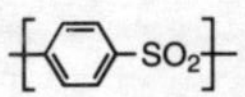	350	520		*Chimia*, **28**(9), 561	1974	ICI
Polyaryletherimides						
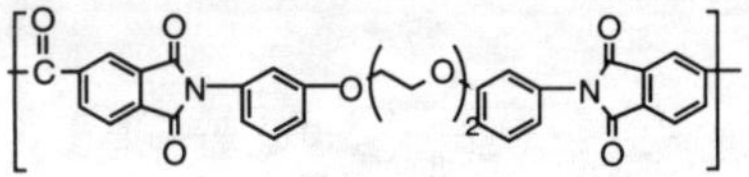	155	236	N-15; NMP	*Polym. Prep.*, **26**(2), 287; **25** (1), 160	1985 1984	NASA/ Langley; U. Akron
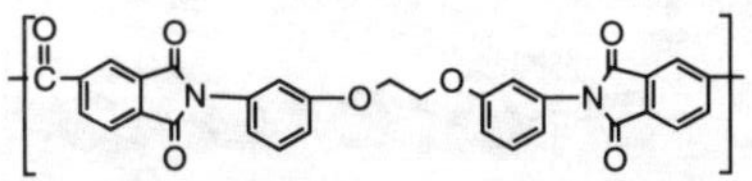	191	290	N-15; NMP	*Polym. Prep.*, **26**(2), 287; **25** (1), 160	1985 1984	NASA/ Langley; U. Akron
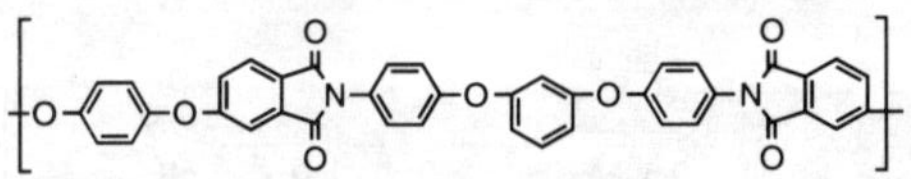	199	330	N-15; m-Cresol, Toluene	*Polym. Prep.*, **24**(2), 312; *J. Polym. Sci., Polym. Symp.*, **74**, 93	1983 1986	GE
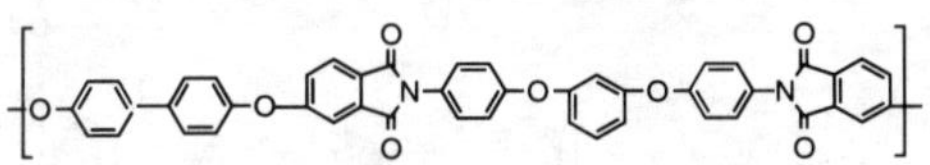	205	343	N-15; m-Cresol, Toluene	*Polym. Prep.*, **24**(2), 312; *J. Polym. Sci., Polym. Symp.*, **74**, 93	1983 1986	GE
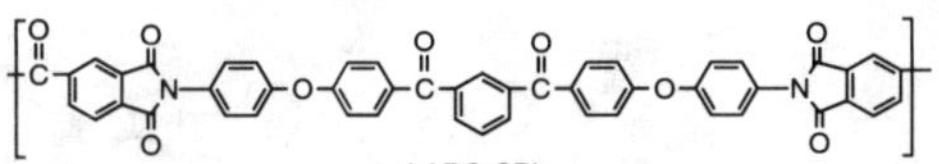 LARC-CPI	222	350	N-15	*J. Polym. Sci., Polym. Chem .*, **25**,1093; *Polym. Prep.*, **33** (1), 409	1987 1992	NASA/ Langley
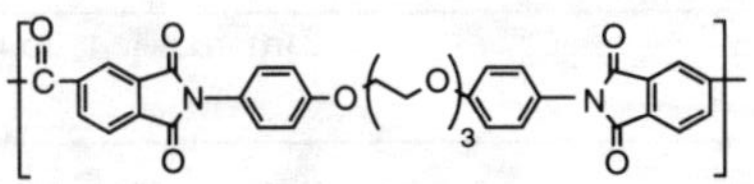	200	395	N-15; NMP	*Polym. Prep.*, **26**(2), 287; **25**(1), 160	1985 1984	NASA/ Langley; U. Akron

TABLE 20—*continued*

Polymer structure	Tg (°C)	Tm (°C)	Synthesis	Reference	Year	Organization
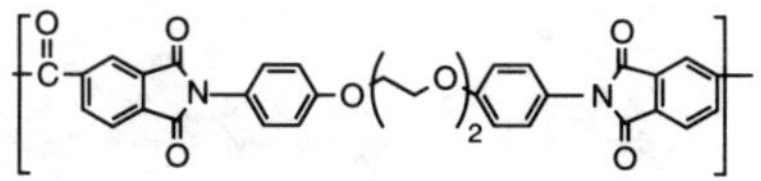	235	410	N-15; NMP	*Polym. Prep.*, **26**(2), 287; **25**(1), 160	1985 1984	NASA/ Langley; U. Akron
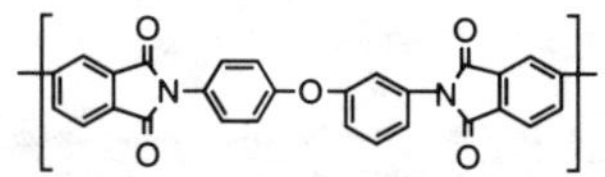	273	415	N-15; m-Cresol	*J. Polym. Sci., Polym. Chem.*, **31**, 141	1993	U. Cal. St Barbara
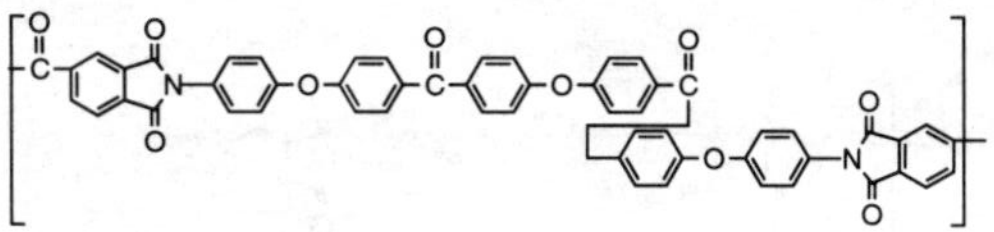	215	418	N-15	*J. Polym. Sci., Polym. Chem.*, **25**,1093	1987	NASA/ Langley
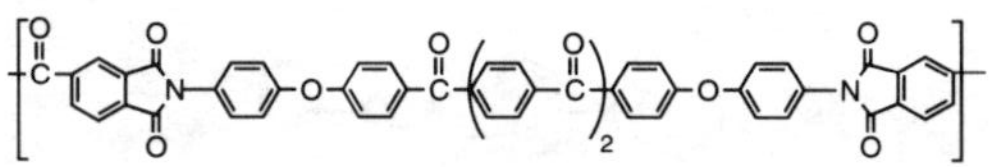	233	422	N-15	*J. Polym. Sci., Polym. Chem.*, **25**,1093	1987	NASA/ Langley
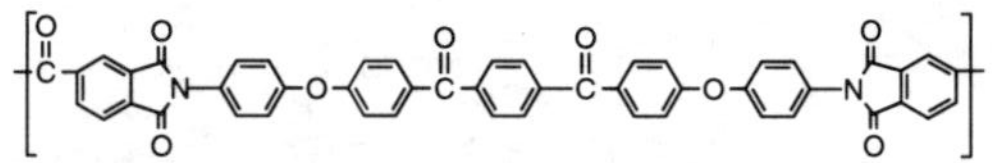	233	427	N-15	*J. Polym. Sci., Polym. Chem.*, **25**, 1093	1987	NASA/ Langley
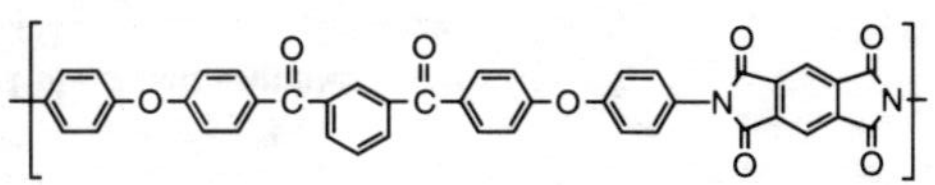	247	442	N-15	*J. Polym. Sci., Polym. Chem.*, **25**,1093	1987	NASA/ Langley
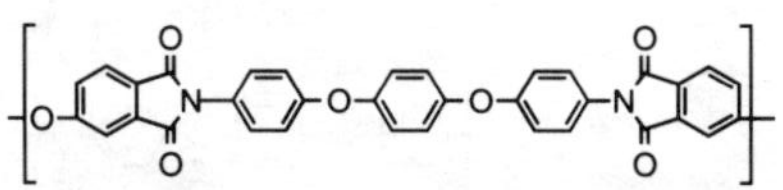		450		*Ency. Polym. Sci. & Eng.*, **12**, 364	1988	
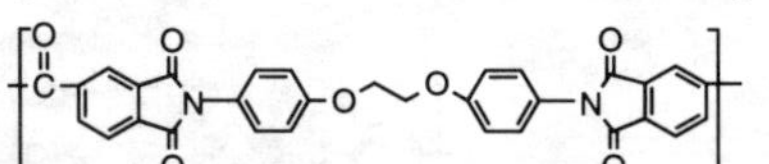	246	460	N-15; NMP	*Polym. Prep.*, **26**(2), 287; **25**(1), 160	1985 1984	NASA/ Langley; U. Akron

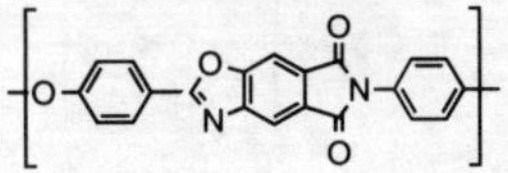		470	N-14; NaH, NMP, CHP; F	*Polym. Prep.*, **33**(2), 142	1992	U. So. Miss
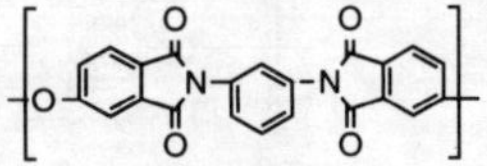		490		*Ency. Polym. Sci. & Eng.*,**12**, 364	1988	
Polyarylethers						
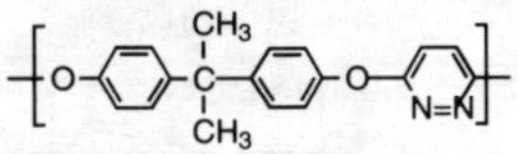	180	250	N-1; NaOH, DMSO, Sulfolane; Cl	*J.Polym.Sci.*, A-1, **5**, 2375	1967	Union Carbide
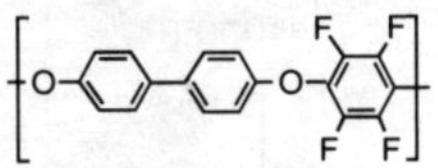		280 (>)	N-18; DMAC; F	*Polym. Prep.*, **22**(2), 383	1981	U. Texas, San Antonio
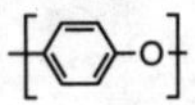	85	285	A-16; Dimethoxy benzene; Br	*Prog.Polym. Sci.*, **12**,111; *Polym.*, **21**, 577 *Eur. Polym. J.*,**4**, 275	1986 1980 1968	Ind. Inst. Tech.; ICI
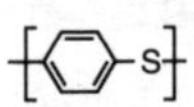 Ryton™	92	295		*Prog.Polym. Sci.*, **12**,111 *Polym.*, **21**, 577 *Ency. Chem. Tech.*, **18**, 893	1986 1980 1982	Ind. Inst. Tech.; ICI
	83	283				Philips
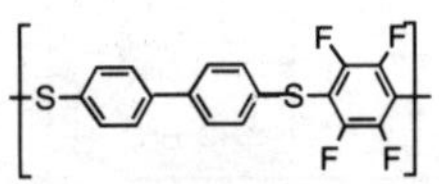	190	300	N-18; DMAC; F	*Polym. Prep.*, **22**(2), 383	1981	U. Texas, San Antonio
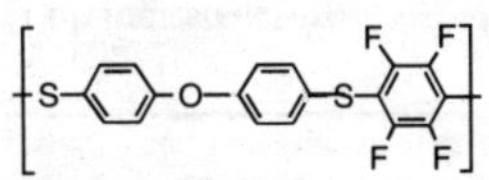	135	300	N-18; DMAC; F	*Polym. Prep.*, **22**(2), 383	1981	U. Texas, San Antonio

TABLE 20—*continued*

Polymer structure	Tg (°C)	Tm (°C)	Synthesis	Reference	Year	Organization
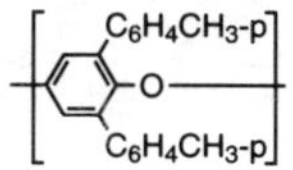	221	321	A-14; Cu-TMEDA, o-DCB	*J. Polym. Sci., Polym. Chem.*, **31**, 1261	1993	McGill U
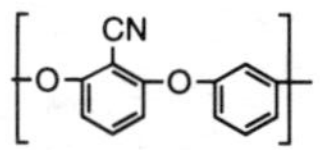	145	340	N-17	PC		Idemitsu Kosan
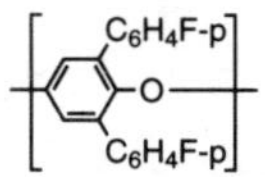	213	351	A-14; Cu-TMEDA, *o*-DCB	*J. Polym. Sci., Polym. Chem.*, **31**, 1261; **31**, 1907; **31**, 2015	1993	McGill U.
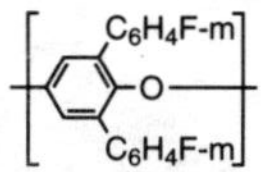	192	352	A-14; Cu-TMEDA, *o*-DCB	*J. Polym. Sci., Polym. Chem.*, **31**, 1907; **31**, 2015	1993	McGill
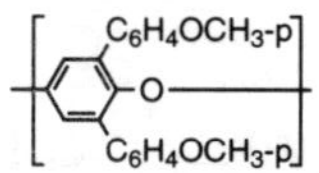	220	356	A-14; Cu-TMEDA, *o*-DCB	*J. Polym. Sci., Polym. Chem.*, **31**, 1261	1993	McGill U.
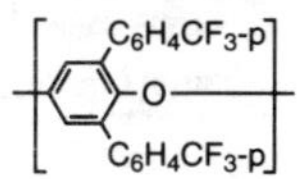		384	A-14; Cu-TMEDA, *o*-DCB	*J. Polym. Sci., Polym. Chem.*, **31**, 1907; **31**, 2015	1993	McGill U
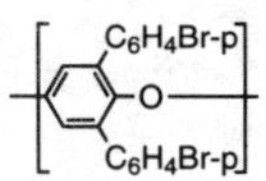	243	404	A-14; Cu-TMEDA, *o*-DCB	*J. Polym. Sci.,Polym. Chem.*, **31**, 1261	1993	McGill U.
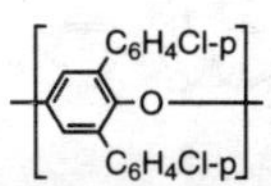	233	420	A-14; Cu-TMEDA, *o*-DCB	*J. Polym. Sci., Polym. Chem.*, **31**, 1261	1993	McGill U.

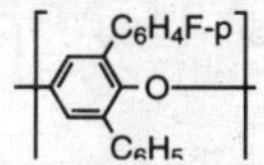	196	427	A-14; Cu-TMEDA, *o*-DCB	*J. Polym. Sci., Polym. Chem.*, **31**, 1907; **31**, 2015	1993	McGill
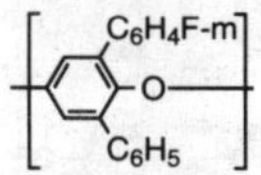	208	474	A-14; Cu-TMEDA, *o*-DCB	*J. Polym. Sci., Polym. Chem.*, **31**, 1907; **31**, 2015	1993	McGill
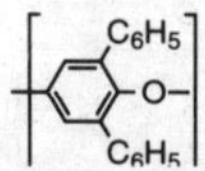	235	480	A-14; Pyridine, $C_6H_5NO_2$	*J. Polym. Sci.*, A, **58**, 581; *J. Polym. Sci., Polym. Chem.*, **31**,1907	1962 1993	GE
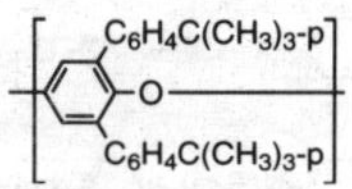	248	483	A-14; Cu-TMEDA, *o*-DCB	*J. Polym. Sci., Polym. Chem.*, **31**, 1261	1993	McGill U.
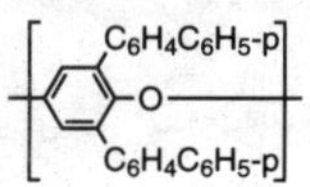	265	500 (>)	A-14; Cu-TMEDA, *o*-DCB	*J. Polym. Sci., Polym. Chem.*, **31**, 1261	1993	McGill U.
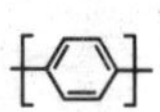		530				ICI

TABLE 21
Crystalline polyaryletherketones arranged in order of increasing melting points

Polymer structure	Tg (°C)	Tm (°C)	Synthesis	Reference	Year	Organization
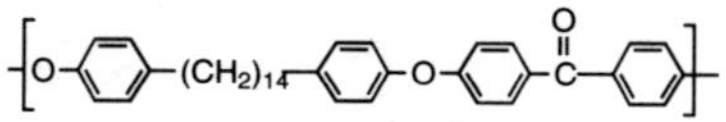	22	130	N-11; CsF, Bulk; F	*Macromol.*, **22**, 517	1989	Hamburg U.
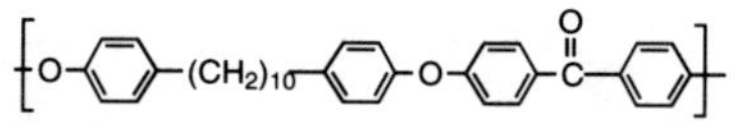	41	176	N-11; CsF, Bulk; F	*Macromol.*, **22**, 517	1989	Hamburg U.
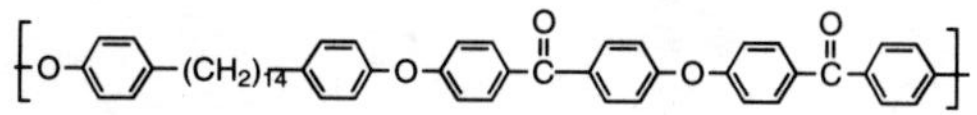	26	192	N-11; CsF, Bulk; F	*Macromol.*, **22**, 517	1989	Hamburg U.
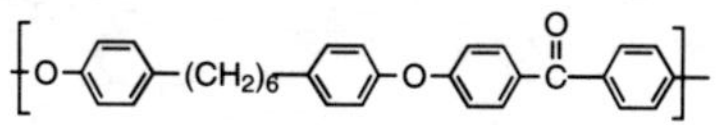	66	207	N-11; CsF, Bulk; F	*Macromol.*, **22**, 517	1989	Hamburg U.
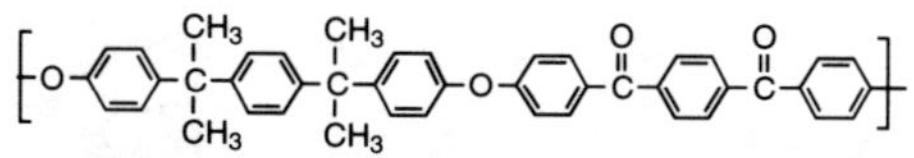	168	207	N-9; K_2CO_3, DMAC; Cl	*Polym.*, **29**, 358	1988	NASA/ Langley
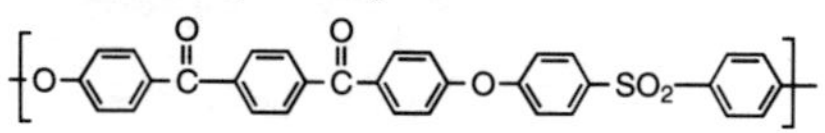	196	214	E-5; TFMSA, P_2O_5	US Pat. 4861856	1989	Eastman Kodak
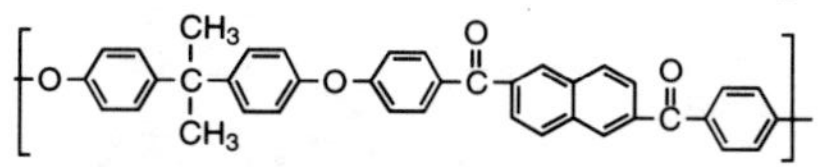	185	220	N-9; K_2CO_3, DMAC; F	*Polym.*, **29**, 358	1988	NASA/ Langley
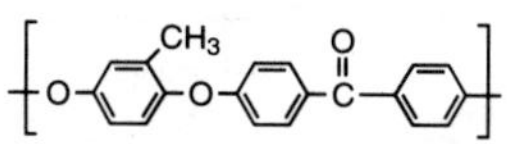	141	223	N-11; CsF, Bulk; F	*New Polym. Mater.*, **1**, 127	1988	Hamburg U.
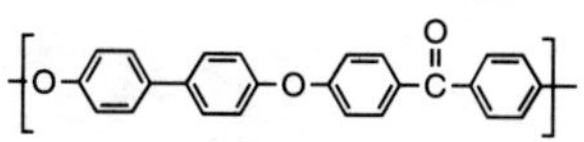		236–255	N-8; NaOH, DMSO; NO_2	*Makromol. Chem.*, **130**, 45	1969	Bayer A-G

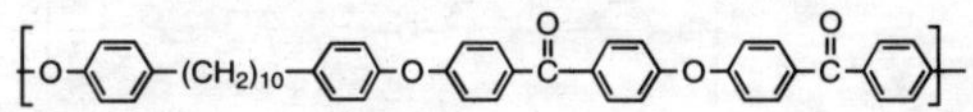	72	241	N-11; CsF, Bulk; F	*Macromol.*, **22**, 517	1989	Hamburg U.
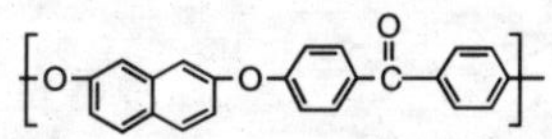	154 153	243 242	N-11; Bulk; F	*Polym.*, **25**(8),1151; US Pat. 4331798	1984 1982	Hamburg U. ICI
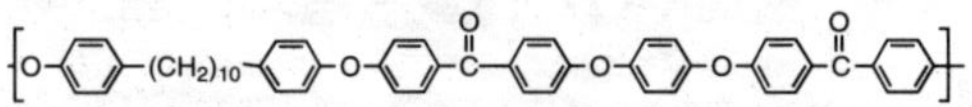	70	247	N-11; CsF, Bulk; F	*Macromol.*, **22**, 517	1989	Hamburg U.
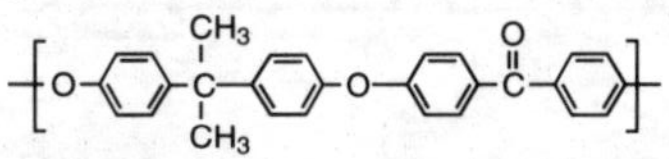	155	248	N-8; NaOH, DMSO, Sulfolane; F	*J. Polym. Sci.*, A-1, **5**, 2375;	1967	Union Carbide
				Ency. Polym. Sci. & Eng., **13**,196;	1988	
		190–220		*Makromol. Chem.*, **130**, 45	1969	Bayer A-G
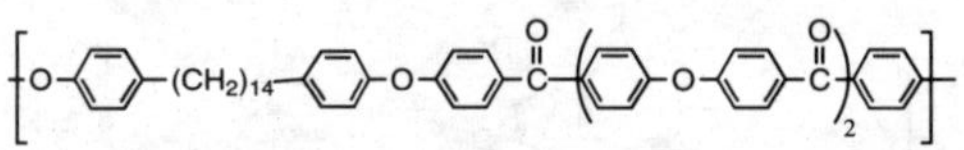	70	250	N-11; CsF, Bulk; F	*Macromol.*, **22**, 517	1989	Hamburg U.
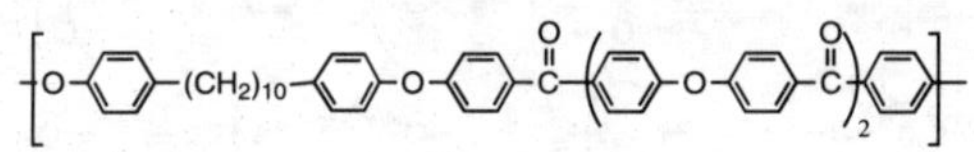	89	253	N-11; CsF, Bulk; F	*Macromol.*, **22**, 517	1989	Hamburg U.
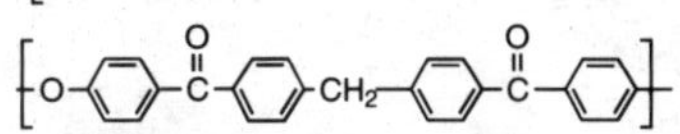		254–256	E-3; AlCl3	GB Pat. 971227	1964	
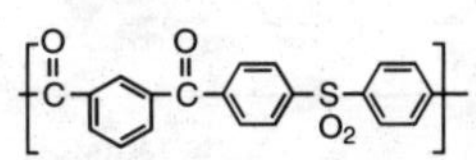	195	257	N-13; NaH, DMF; F	*Polym. Prep.*, **34**(1), 570	1993	Va. P. I. & State U.
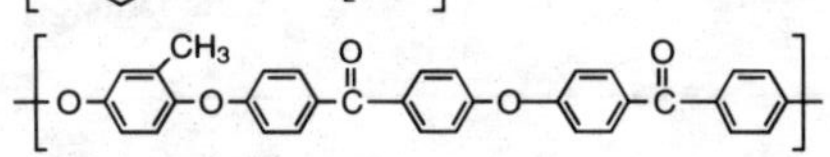	150	259	N-11; CsF, Bulk; F	*New Polym. Mater.*, **1**, 127	1988	Hamburg U.

TABLE 21—*continued*

Polymer structure	Tg (°C)	Tm (°C)	Synthesis	Reference	Year	Organization
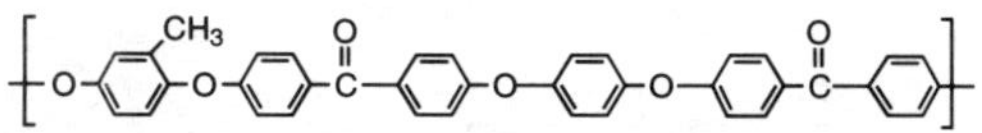	121	266	N-11; CsF, Bulk; F	*New Polym. Mater.*, **1**, 127	1988	Hamburg U.
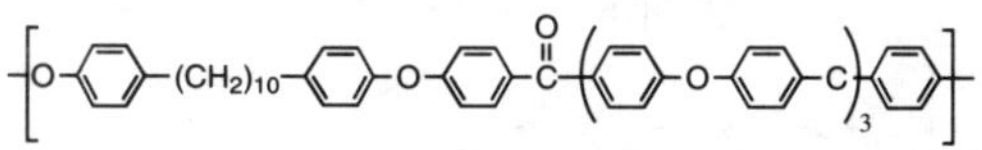	87	266	N-11; CsF, Bulk; F	*Macromol.*, **22**, 517	1989	Hamburg U.
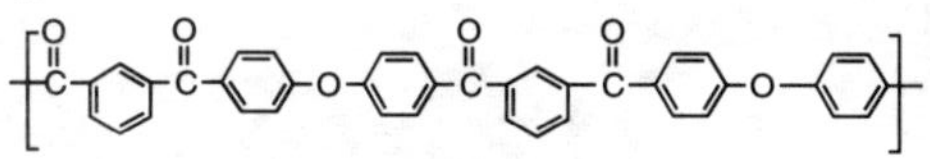	158	269	E-5; TFMSA, P_2O_5	US Pat. 4839459	1989	Eastman Kodak
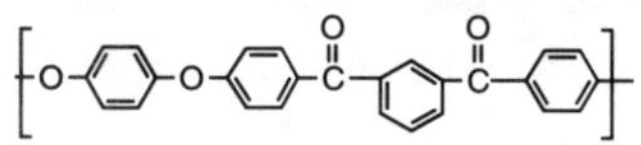	153	271	E-4; HF, BF_3	US Pat. 3956240	1976	Raychem
	156	224	E-5; TFMSA, P_2O_5	US Pat. 4820794	1989	Eastman Kodak
	136	277	N-9; K_2CO_3, Diphenyl sulfone; F	*Bull. Soc. Chim. Belg.*, **98**, 667	1989	ICI
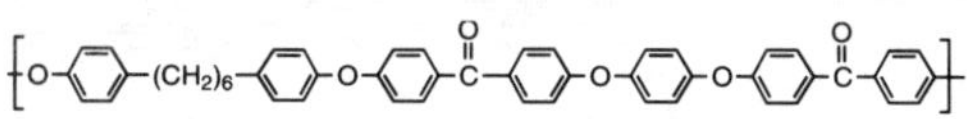	88	274	N-11; CsF, Bulk; F	*Macromol.*, **22**, 517	1989	Hamburg U.
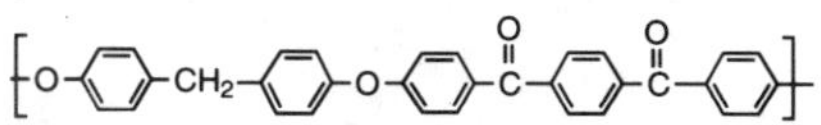	120	275	N-9; K_2CO_3, DMAC; Cl	*Polym.*, **29**, 358	1988	NASA/ Langley
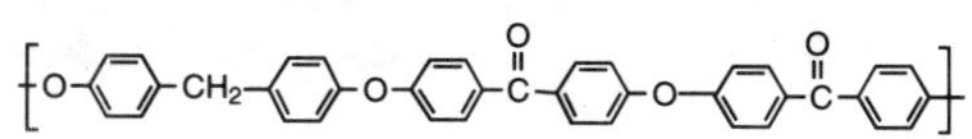	128	275	N-9; K_2CO_3, DMAC; F	*Polym.*, **29**, 358	1988	NASA/ Langley
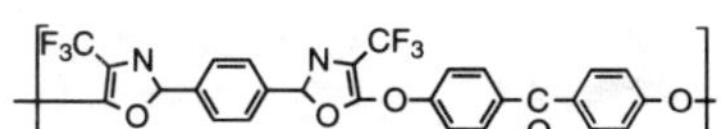	167	278	N-9; $(Et)_3N$, NMP; F	*Polym. Prep.*, **34**(1), 429	1993	U. Bayreuth
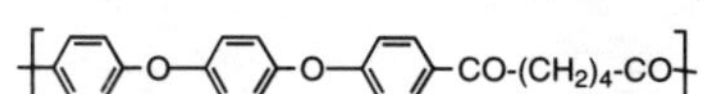	92	279	E-5; TFMSA, P_2O_5	US Pat. 4820794	1989	Eastman Kodak
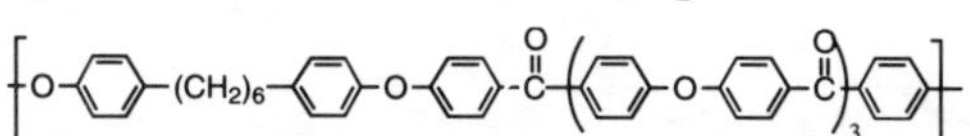	102	279	N-11; CsF, Bulk; F	*Macromol.*, **22**, 517	1989	Hamburg U.

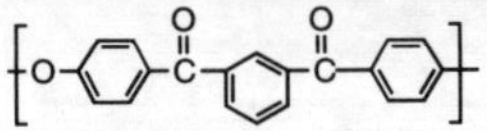	175 139	280 285	E-5; TFMSA N-9; Diphenyl sulfone; F. E; $AlCl_3$, Dichloroethane	US Pat. 4361693 *Bull. Soc. Chim. Belg.*, **98**, 667; *Gaofenzi Tongxun*, **6**, 442	1982 1989 1984	 ICI
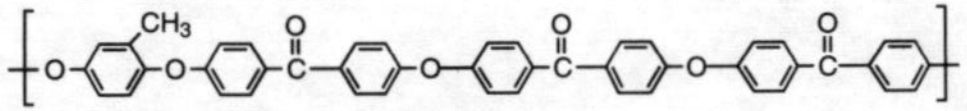	139	280	N-11; CsF, Bulk; F	*New Polym. Mater.*, **1**, 127	1988	Hamburg U.
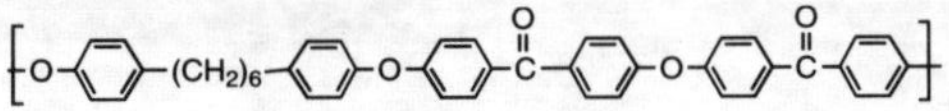	88	282	N-11; CsF, Bulk; F	*Macromol.*, **22**, 517	1989	Hamburg U.
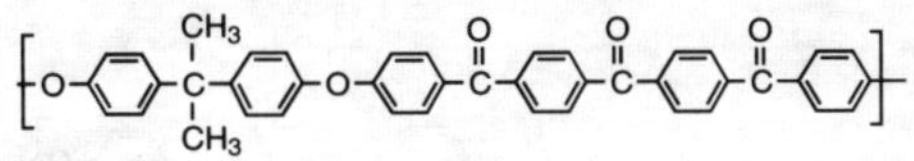	155	283	N-9; K_2CO_3, DMAC;	*Polym.*, **29**, 358	1988	NASA/ Langley
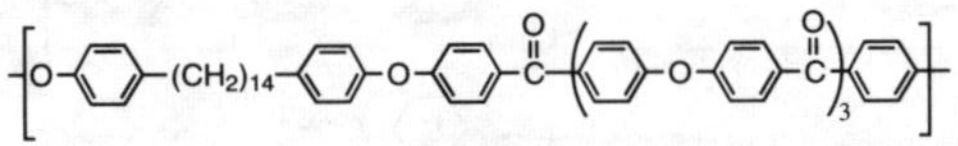	91	284	N-11; CsF, Bulk; F	*Macromol.*, **22**, 517	1989	Hamburg U.
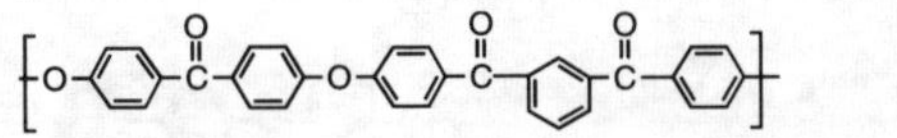	152	285	N-9; K_2CO_3, Diphenyl sulfone; F	*Bull. Soc. Chim. Belg.*, **98**, 667	1989	ICI
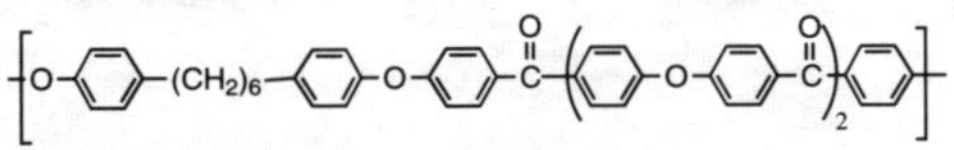	109	300	N-11; CsF, Bulk; F	*Macromol.*, **22**, 517	1989	Hamburg U.
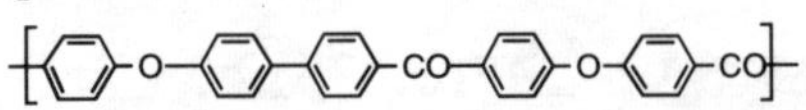		300–350	E-4; BF_3, HF	US Pat. 3956240	1976	Raychem
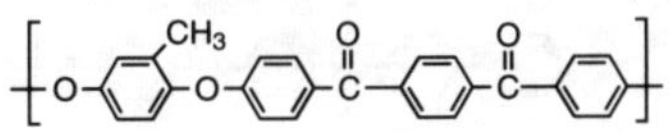	159	305	N-11; CsF, Bulk; F	*New Polym. Mater.*, **1**, 127	1988	Hamburg U.
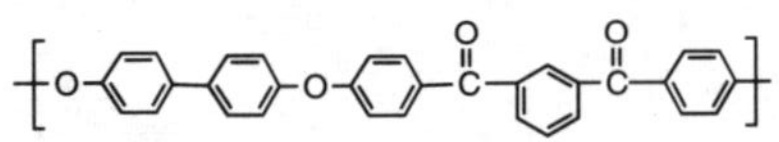	160	305	N-9; K_2CO_3, Diphenyl sulfone; F	*Bull. Soc. Chim. Belg.*, **98**, 667	1989	ICI
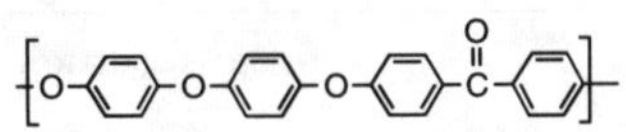	150 129	315 324	N-8; DMSO, Sulfolane; F. Blends	*J. Polym. Sci.*, A-1, **5**, 2375; *J. Polym. Sci.*, B, *Polym. Phys.*, **25**, 311	1967 1987	Union Carbide

TABLE 21—*continued*

Polymer structure	Tg (°C)	Tm (°C)	Synthesis	Reference	Year	Organization
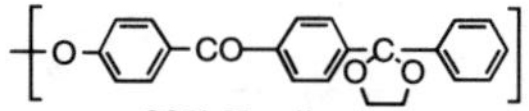 20% Ketalization		315	Post Ketalization on PEKK	US Pat. 3734888	1973	DuPont
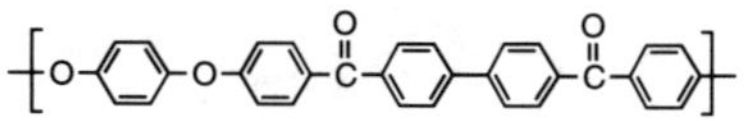		323	A-9; DMAC; Cl	*Macromol.*, **23**, 926	1990	Yamagata U.
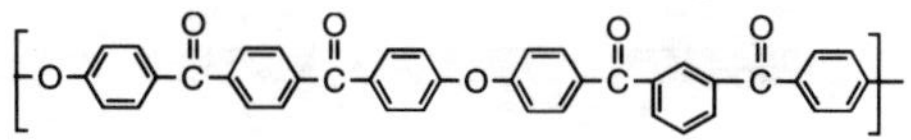		325	E-3; $AlCl_3$, o-Dichlorobenzene	*Polym. Prep.*, **32**(3), 255, 258; Eur. Pat. 192260; US Pat. 4816556	1991 1986	DuPont
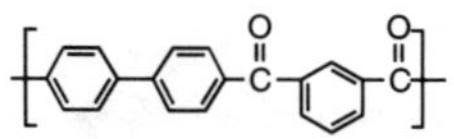	201	332	E-5; TFMSA, P_2O_5	US Pat. 4820794	1989	Eastman Kodak
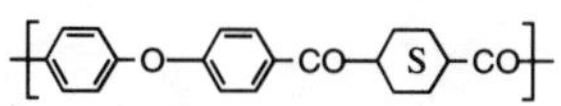	170	332	E-5; TFMSA, P_2O_5	US Pat. 4820794	1989	Eastman Kodak
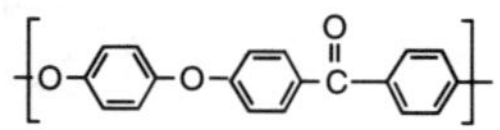 Victrex™ PEEK	144 152	335	N-9; K_2CO_3, Diphenyl sulfone; F	*Polym. Prep.*, **20**(1),191 *Polym.*, **21**, 577; **22**,1096; Mod. Plast., p. 86, 11/81;	1979	ICI
	152	334	E-5 ; TFMSA Morphology	*Polym. Prep.*, **25**(2),17; *Polym.*, **24**, 953; **29**,1902	1984 1983	ICI
	160	350	N-8; NaOH, DMSO, Sulfolane; F	*J.Polym.Sci.*, A-1, **5**, 2375	1967	Union Carbide
		345	N-11; CsF, Bulk	*Polym.*, **25**, 1151;	1984	Hamburg U.
	344			US Pat. 4331798	1982	ICI
	144	335	A-11, Ru/C, DMF, DMSO; Cl	*J. Chem. Soc., Chem. Commun.*, (19), 1338	1985	ICI

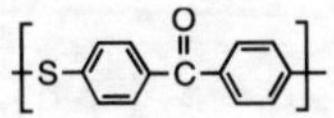	152	335	N; Na_2S, CHP; F	*J. Polym. Sci., Polym. Chem.*, **27**, 661	1989	Dow
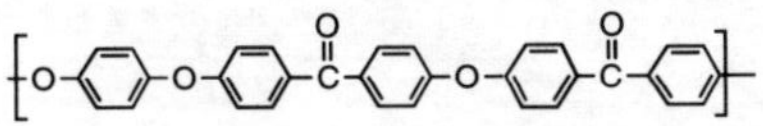	143	336	N-9; K_2CO_3, CHP; F	*J. Polym. Sci*, A ,**27**, 661	1989	Dow
		340	E-4; HF, BF_3	US Pat. 3956240; GB Pat. 3956240	1976 1976	Raychem
	151	345	E-5 ; TFMSA	GB Pat. 2116990	1983	ICI
			Blends	*J. Polym. Sci.*, B, *Polym. Phys.*, **25**,311	1987	Union Carbide
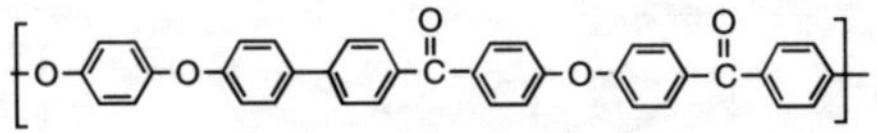	165	348	E-5; TFMSA	GB Pat. 2116990	1983	ICI
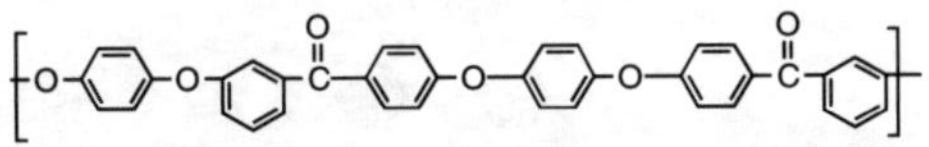	147	350	E-5; P_2O_5, CH_3SO_3H	*Macromol.*, **20**, 2675	1987	Yamagata U.
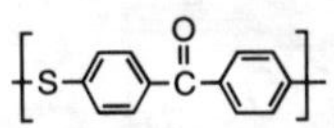		352	N	*Brit. Polym. J.*, **4**, 391	1972	ICI
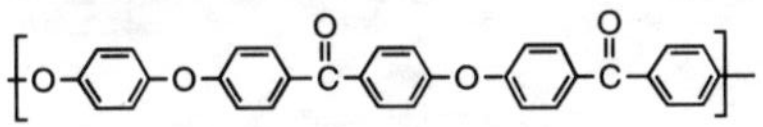	165	352	E-5; P_2O_5, CH_3SO_3H	*Macromol.*, **20**, 2675	1987	Yamagata U.
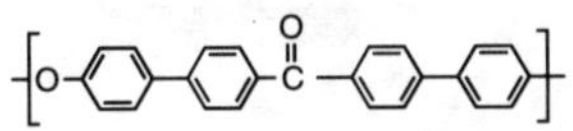	152	354	A-10; DMAC; Br	*J. Polym. Sci., Polym. Chem.*, **30**, 2511	1992	U. East Anglia
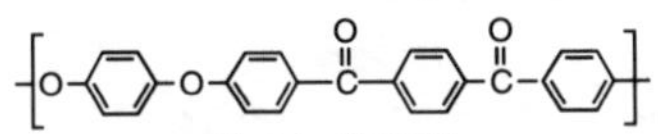 Hostatec™ X915	154	358	N-9; K_2CO_3, Diphenyl sulfone; F	*Polym.*, **22**, 1096;	1981	ICI
			Blends	*J. Polym. Sci.*, B, *Polym. Phys.*, **25**, 311	1987	Union Carbide
		360	E-4; HF, BF_3	US Pat. 3956240	1976	Raychem
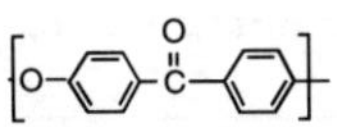 Stilan™	154	367	N-9; K_2CO_3, Diphenyl sulfone; F	*Polym.*,**18**, 365; **21**, 577; **22**,1096	1977 1980 1981	ICI

TABLE 21—*continued*

Polymer structure	Tg((°C)	Tm (°C)	Synthesis	Reference	Year	Organization
	163	361	E-4; HF, BF_3	US Pat. 3441538	1969	DuPont
	170	365	E-4; HF, BF_3	US Pats. 3953400; 4229564; 4247682	1976 1980 1981	Raychem
	162	363	E-5; TFMSA, P_2O_5	US Pat. 4820794	1989	Eastman Kodak
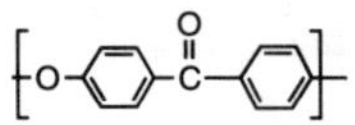 from 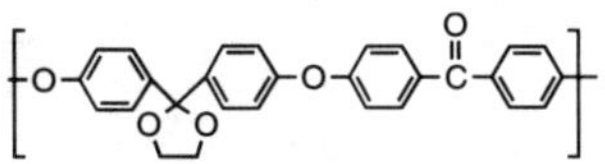	156	366	N-13; Heating the polyketal with acid.	*Macromol.*, **20**,1204 US Pat. 4882397	1987 1989	Amoco
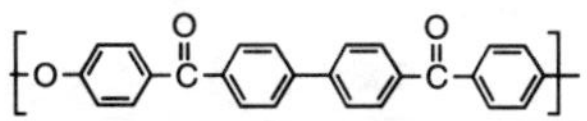		361	A-9; DMAC; Cl	*Macromol.*, **23**, 926	1990	Yamagata U.
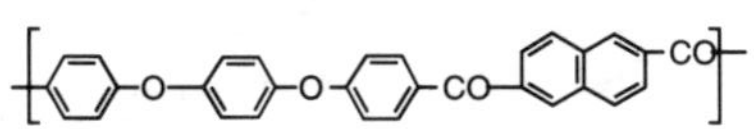	184	361	E-5; TFMSA, P_2O_5	US Pat. 4820794	1989	Eastman Kodak
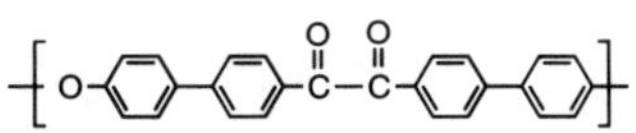	142	362–378	A-10; DMAC; Br	*J. Polym. Sci., Polym. Chem.*, **30**, 2511	1992	U. East Anglia
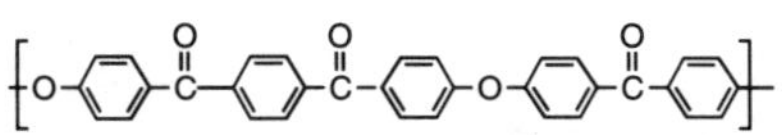		370	E-4; HF, BF_3	US Pat. 3956240	1976	Raychem
			E-3; $AlCl_3$, $(CH_3)_2SO_2$, CH_2Cl_2	US Pats. 4843131; 4912181	1989 1990	
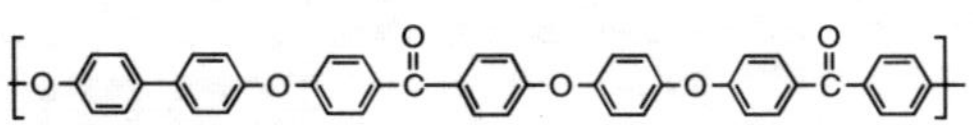	173	370	E-5; TFMSA	GB Pat. 2116990	1983	ICI
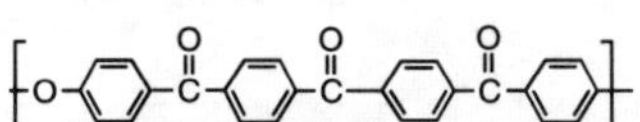		370–380	E-3; AlCl3	GB Pat. 971227	1964	

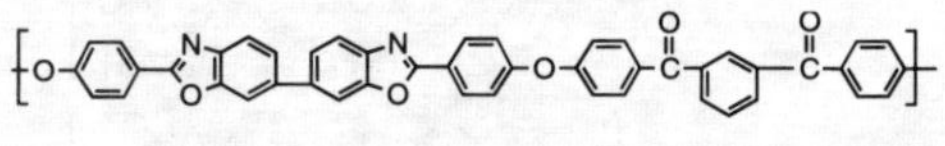	215	373	N-9; K_2CO_3, DMAC, Diphenyl sulfone; F	*Polym. Prep.*, **32**(1), 646	1991	NASA/ Langley
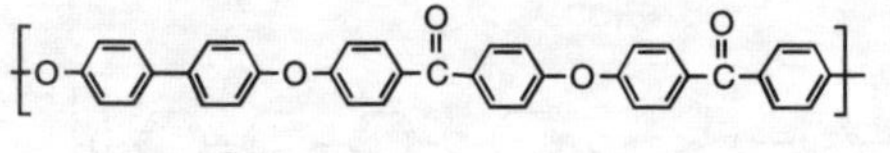	172	374	E-5; TFMSA	GB Pat. 2116990	1983	ICI
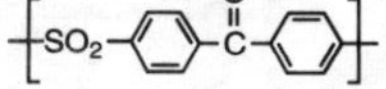		375	Oxidation of sulfide polymer	*Brit. Polym. J.*, **4**, 391	1972	ICI
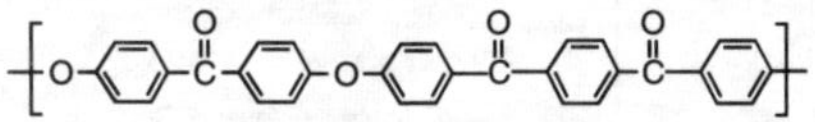 Ultrapek™ KR 4177	175	375	E-3; $AlCl_3$, CH_2Cl_2, Dimethyl sulfone	Ger. Pat. 3829520; *Kunst. Ger. Plastics*, **79**, 75	1990 1989	BASF
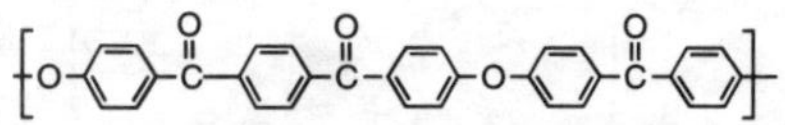		383	N-9; K_2CO_3, Diphenyl sulfone; F	*Polym.*, **22**, 1096; *J. Polym. Sci.*, B,	1981	ICI
	160	384	Blends	*Polym. Phys.*, **25**, 311	1987	Union Carbide
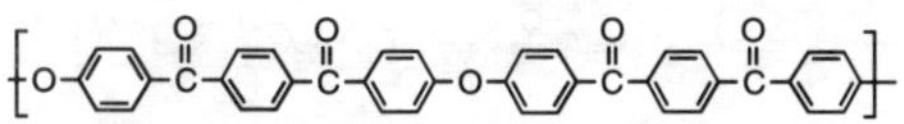		375	E-3; AlCl3, o-Di-chlorobenzene	*Polym. Prep.*, **32**(3), 255, 258	1991	DuPont
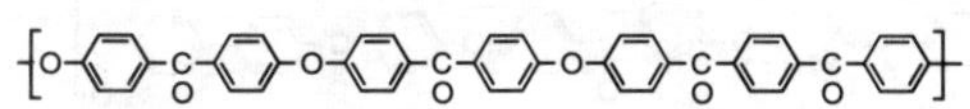		376	N-9; K_2CO_3, Diphenyl sulfone; Cl	Jap. Pat. 1065130	1989	Asahi Chemical
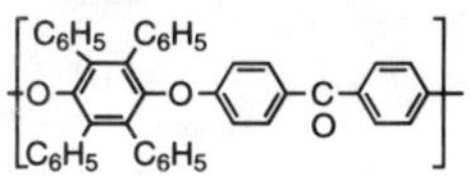	222	383	N-9; K_2CO_3, NMP; F	*J. M. S.-Pure Appl. Chem.*, **A29**(12), 1141	1992	McGill U
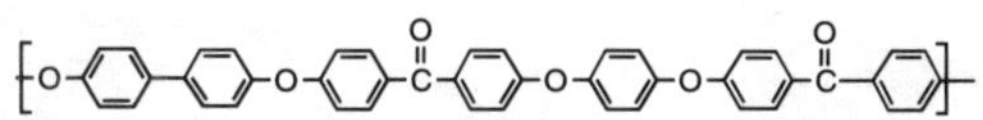	159	383	E-5; TFMSA	*Polym.*, **29**, 1902	1988	ICI
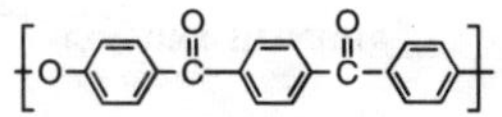		384	N-9; K_2CO_3, Diphenyl sulfone; F	*Polym.*, **22**, 1096	1981	ICI
	185	385	E-3; AlCl3, o-Di-chlorobenzene	Preprints, Japan-US Pol. Symp.	1985	DuPont

TABLE 21—*continued*

Polymer structure	Tg (°C)	Tm (°C)	Synthesis	Reference	Year	Organization
				Polym. Prep., **32**(2), 262, 264	1991	
		389	E-3; $AlCl_3$, $(CH_3)_2SO_2$, CH_2Cl_2	US Pat. 4826947	1989	Raychem
	165	391	E-3 Blends	*J. Polym. Sci.*, B, *Polym. Phys.*, **25**, 311	1987	Union Carbide
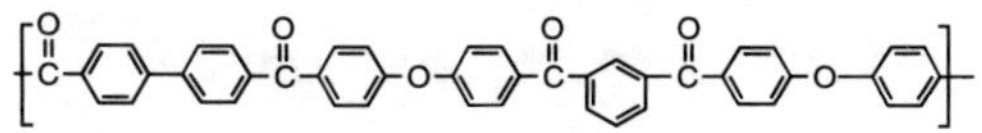	163	387	E-3; $AlCl_3$, $(CH_3)_2SO_2$, CH_2Cl_2	US Pat. 4826947	1989	Raychem
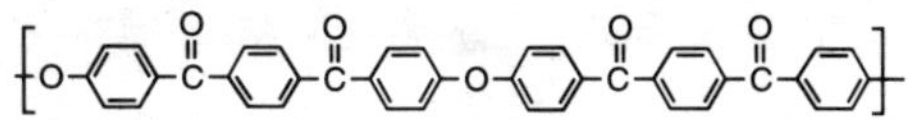	180	389	E-3; $AlCl_3$, $(CH_3)_2SO_2$, CH_2Cl_2	US Pat. 4826947	1989	Raychem
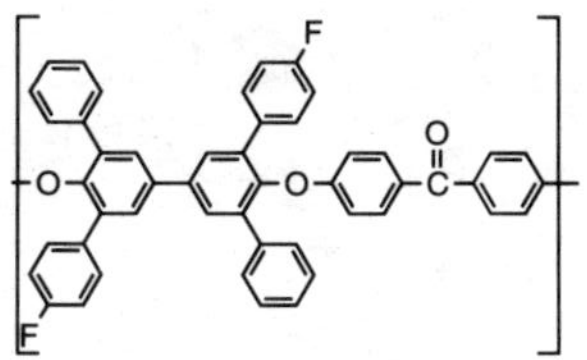		392	N-9; K_2CO_3, NMP; F	*J. Polym. Sci., Polym. Chem.*, **31**, 1907	1993	McGill
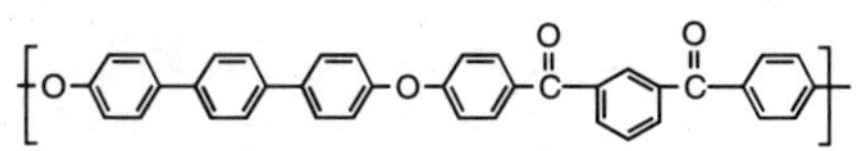	174	394	N-9; K_2CO_3, Diphenyl sulfone; F	*Bull. Soc. Chim. Belg.*, **98**, 667	1989	ICI
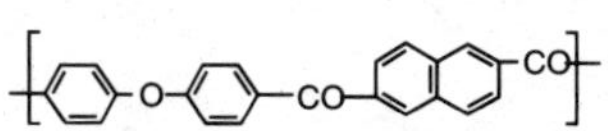	210	395	E-5; TFMSA, P_2O_5	US Pat. 4820794	1989	Eastman Kodak
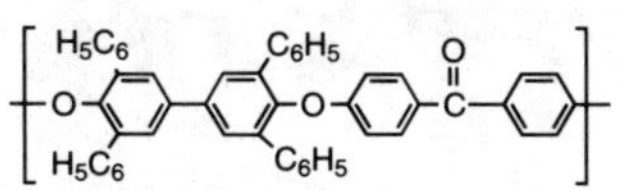	220	400 (>)	N-9; K_2CO_3, NMP; F	*J. Polym Sci., Polym. Chem.*, **31**, 1907	1993	McGill U.
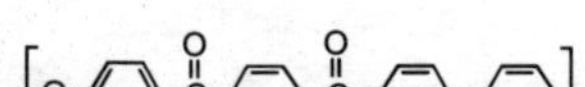		402	E-4; HF, BF_3	Rec. Adv. Mech. & Syn.	1987	ICI

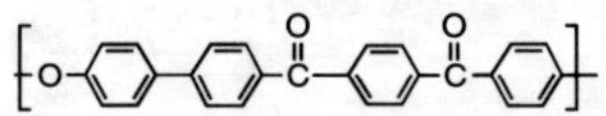		402	E-4; HF, BF_3	US Pat. 3956240	1976	Raychem
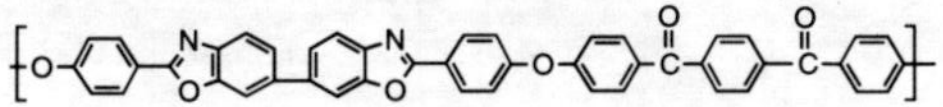	237	405, 446	N-9; K_2CO_3, DMAC, Diphenyl sulfone; F	*Polym. Prep.*, **32**(1), 646	1991	NASA/ Langley
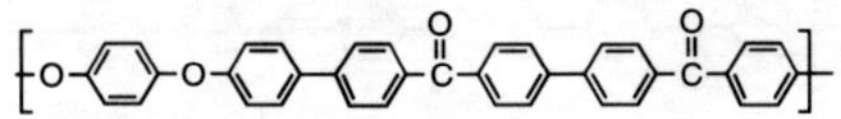	200	407	E-5; TFMSA	GB Pat. 2116990	1983	ICI
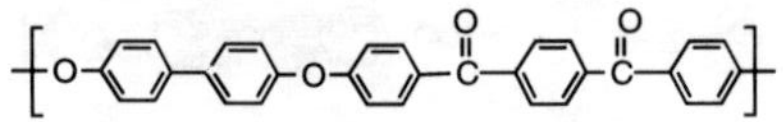	168	409	N -9	*SAMPE*, **32**, 382	1987	ICI
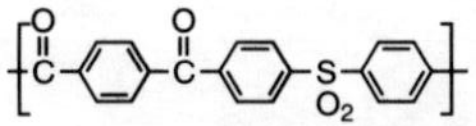	228	414	N-13; NaH, DMF; F	*Polym. Prep.*, **34**(1), 570	1993	Va. P. I. & State U.
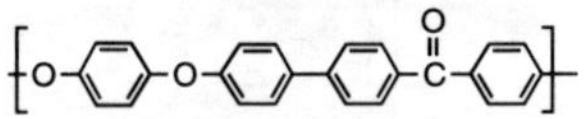	167	416	N-9; K_2CO_3, Diphenyl sulfone; F	*Polym.*, **22**, 1096	1981	ICI
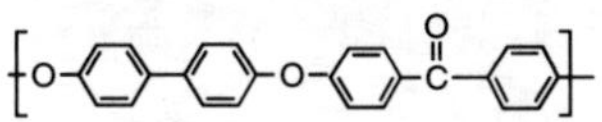		421 422	N-11; Bulk; F	US Pat. 4331798; *Polym.*, **25**(8), 1151	1982 1984	ICI ICI
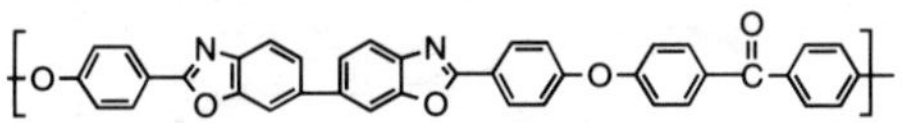	249	425	N-9; K_2CO_3, DMAC, Diphenyl sulfone; F	*Polym. Prep.*, **32**(1), 646	1991	NASA/ Langley
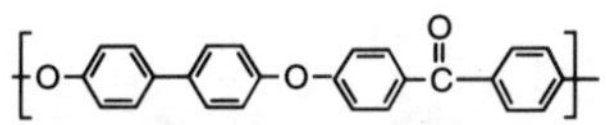	170	428	N-9; K_2CO_3, Diphenyl sulfone; F	*Polym. Prep.*, **32**(2), 166;	1991	Amoco
	167	416		*Polym.*, **18**, 365	1977	ICI
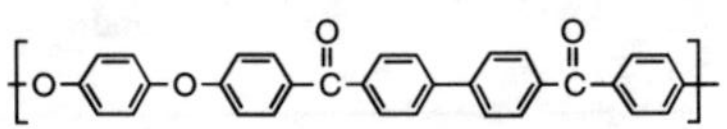	192	428	E-5 ; TFMSA	*Polym. Prep.*, **25**(2),17; *Polym.*, **29**, 1902; US Pat. 4820794	1984 1988	ICI

TABLE 21—*continued*

Polymer structure	Tg((°C)	Tm (°C)	Synthesis	Reference	Year	Organization
	180	411	E-5; TFMSA, P_2O_5		1989	Eastman Kodak
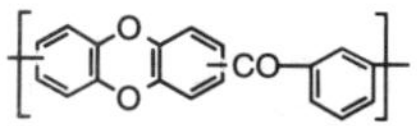	200	430	E-4; BF_3, HF	GB Pat. 1250251	1971	DuPont
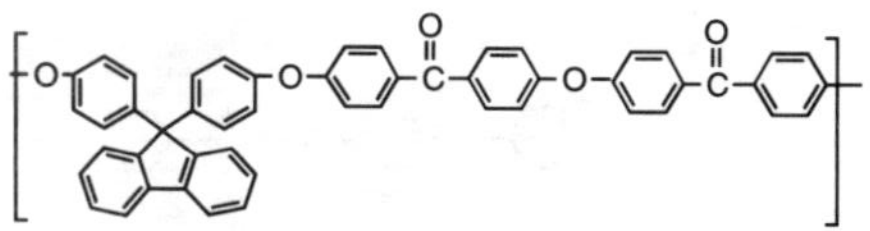	231	430	N-9; K_2CO_3, DMAC; F	*Polym.*,**29**, 358	1988	NASA/ Langley
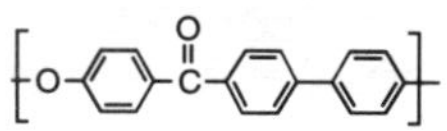		435	E-4; HF, BF_3	US Pats. 3914298; 3965146; 4111908; 4229564	1975 1976 1978 1980	Raychem
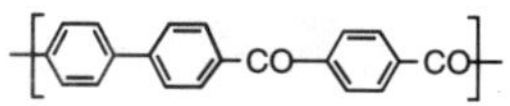		450 (>)	E-5; TFMSA, P_2O_5	US Pat. 4820794	1989	Eastman Kodak
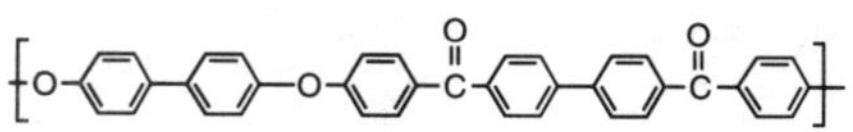	209	469	E-5; TFMSA	*Polym. Prep.*, **25**(2),17; *Polym.*, **29**,1902	1984 1988	ICI
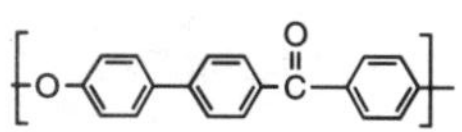	216	478	E-5 ; TFMSA	*Polym. Prep.*, **25**(2),17; *Polym.*, **29**, 1902	1984 1988	ICI
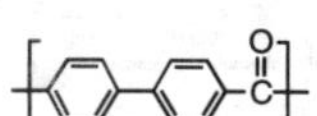		500	E-4; HF, BF_3	US Pat. 3441538	1969	DuPont
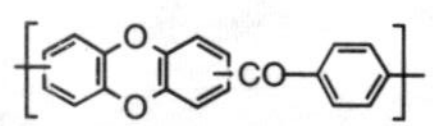	260	525	E-4; BF_3, HF	GB Pat. 1250251	1971	DuPont
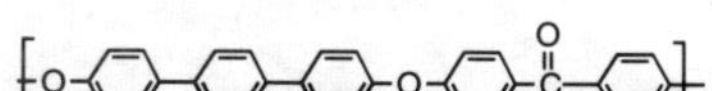		Highly crystalline	N-9; K_2CO_3, Diphenyl sulfone; F	*Bull. Soc. Chim. Belg.*, **98**, 667	1989	ICI

CHAPTER 7

Polyarylethers Containing Heterocyclic Structures (Non-Imide)

One of the most recent thrusts in polyarylether technology is the synthesis of polymers that contain heterocyclic groups in the polymer repeat unit. Table 22 lists those polymers that have been uncovered for inclusion in this book. Polymers that contain the cyclic imide moiety are not included in this table or chapter since they are reported separately in Tables 11–13. Actually, one could say that the original heterocyclic ring-containing polyarylether was prepared from pyromellitic dianhydride and 4,4′-oxydianiline and commercialized around 1960 by the DuPont Co. It is shown in formula VII-1 with its various tradenames. Since this polymer was not made by polyetherification chemistry, there is a tendency to overlook it as a member of the polyarylether family.

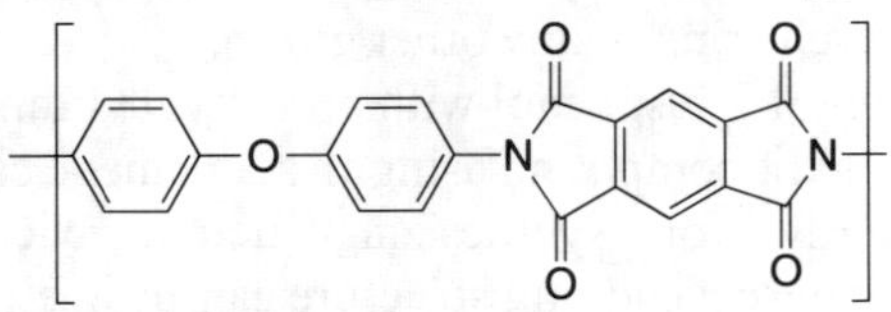

(VII-1)

Kapton™, Pyre M L™, Vespel™, Pyralin™

The majority of the polymers in Table 22 were made by nucleophilic methods, with the carbonate process being used most frequently. A significant number were also made by conducting a (post) reaction on a preformed polyarylether. The precursor polymers most often used by this latter method were polyaryletherketones. A variety of reasons have been put forth for incorporating heterocyclic structures into polyarylethers. Improved thermal stability and solubility properties are mentioned most frequently. Thermally stable heterocyclic ring containing polymers have been known for many years (*1*), but many suffered from intractability. The aryl ether linkage increases the solubility of polymers containing this type of ring. Since polyarylethers are so thermally stable to begin with, incorporation of a heterocyclic structure does little to improve their thermal stability. Of course, increased polymer solubility does allow one to use solution methods to fabricate fibers and films of these polymers. Two major application areas that have been of sustained interest are microelectronic materials and the development of matrix resins for use in thermoplastic composites. The use of heterocyclic rings to incorporate fluorine-containing structures into a polyarylether has been reported (*2*, *3*). The property improvements that were being sought were imparted by the fluorine groups not the heterocyclic rings. These included decreased water absorption, melt viscosity, discoloration tendency, electrical conductivity and increases in Tg and solubility. Thus, heterocyclic structures were being used simply because they provided the easiest way to incorporate the desired fluoro groups.

This strategy can be generalized. Heterocyclic ring-containing polyarylethers that also contain phosphorous groups have been reported (*4*). The property improvements that are obtained include improved resistance to oxygen plasma and atomic oxygen, flame resistance, adhesion to metals, metal ion binding characteristics and thermal stability. Again in this case, these improvements stem from the presence of the phosphine oxide groups, not the heterocyclic ring structures. Certain heterocyclic rings can indirectly lead to improved thermal stability by opening on thermolysis to provide crosslinking sites. This effect will lead to a higher insolubles or char content, and thus, a lower weight loss in a TGA type of thermal stability test. However, polyarylethers are of commercial interest primarily for their thermoplastic, melt processable nature. This property must not be compromised by the incorporation of a heterocyclic ring or any other group.

The polymerization of a bisphenol with an activated dihalo (or other displaceable group) aromatic compound using an alkali metal carbonate has been used most frequently for synthesizing heterocyclic ring-containing polyarylethers. The heterocyclic ring structure can be in either the bisphenol reactant or

Hetero Rings in Bisphenol Reactant	*Hetero Rings in Activated Disubstituted Aromatic Reactant*
Benzimidazole	Benzoxazole
Benzoxazole	Benzthiazole
Cyclic Ketal	1,2,4-Oxadiazole
Lactam	1,2,4-Triazole
Lactone	1,2,5-Oxadiazole (and N-Oxide)
Imidazole	Pyridazine
Oxazole	Pyridine
Piperazine	Quinoxaline
Pyrazole	s-Triazine
Thiazole	Thiophene

the activated dihalo aromatic reactant (*8*). The accompanying table illustrates the types of structures that have been incorporated into these two reactants in a polymerization using the carbonate process. Poly(arylether-pyrazoles) have been prepared from a bisphenol containing the pyrazole ring as shown in the following equation (*6*).

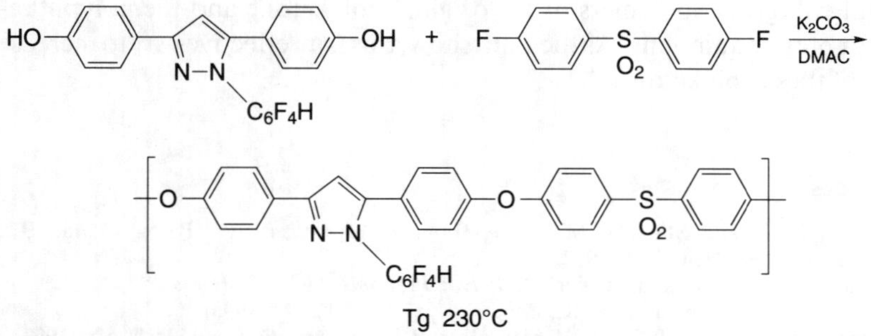

(VII-2)

An example of the use of triazole groups to activate aromatic halogen towards polymerization is shown in equation VII-3 (*7*).

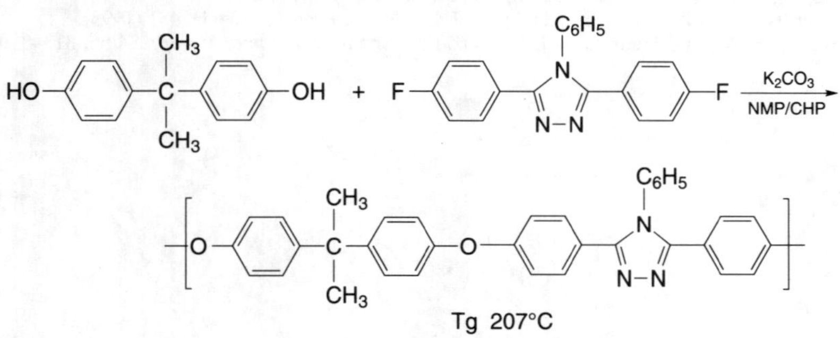

(VII-3)

Polyisoquinolines and polyphthalazines have been synthesized by the post reaction of amines with polyarylethers containing the 1,2-dibenzoylbenzene moiety (*5*). As illustrated in the accompanying equation, reaction of the polyketone with hydrazine monohydrate and acetic acid in refluxing chlorobenzene gave the polyphthalazines as intensely yellow colored polymers. Reaction with benzylamine in the presence of 1,8-diazabicyclo[5.4.0]undecene in refluxing chlorobenzene gave polyisoquinolines as pale yellow polymers. Use of less than stoichiometric amounts of the amine

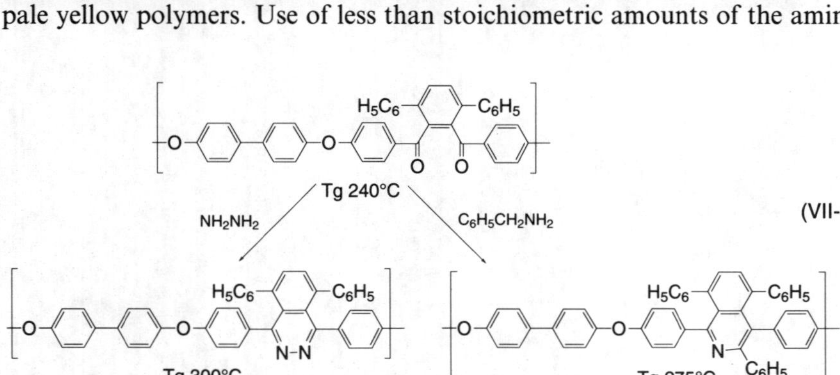

(VII-4)

reactant will give copolymers of the original polyketone and the polyphthalazine or polyisoquinoline. As the data show, this is an effective way to increase the Tg of these polyketones.

References

1. Cotter, R. J., and Matzner, M., "*Ring-Forming Polymerizations*," Parts B,1 and B,2. Academic Press, New York, 1972.
2. Maier, G., Hecht, R., and Nuyken, O., *Polym. Preprints*, **34**(1), 429 (1993).
3. Bass, R. G., and Srinivasan, K. R., *Polym. Preprints*, **34**(1), 441 (1993).
4. Connell, J. W., Smith, J. G., Jr., Hergenrother, P. M., *Polym., Preprints*, **34**(1), 525 (1993).
5. Singh, R., and Hay, A. S., *Polym. Preprints*, **32**(1), 393 (1991); *Macromolecules*, **24**, 2640, 2643 (1991); *Macromolecules*, **25**, 1025, 1033 (1992).
6. Bass, R., and Srinivasan, K. R., *Polym. Preprints*, **34**(1), 441 (1993).
7. Carter, K. R., Miller, R.D., and Hedrick, J. L., *Polym. Preprints*, **34**(1), 481 (1993).
8. Labadie, J. W., and Hedrick, J. L., Makromol. Chem., *Macromol. Symp.*, **54/55**, 313–330 (1992).

TABLE 22
Polyarylethers containing heterocyclic rings (non-imide) in the repeat unit.

Polymer structure	Tg (°C)	Tm (°C)	Synthesis	Reference	Year	Organization
One Oxygen Heteroatom						
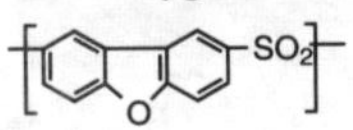	360		E-1; Dimethyl sulfone, Nitrobenzene, $FeCl_3$	*Polym.*, **15**, 456; *J. Polym. Sci.*, A-1, **8**, 2035	1974 1970	ICI 3M
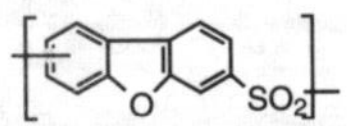		410–440	E-1	*Adv. Macromol. Chem.*, **1**, 335	1968	3M
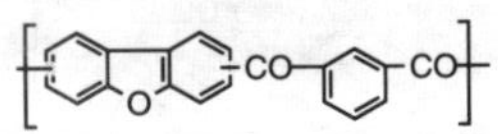			E-4; BF_3, HF	US Pat. 3441538	1969	DuPont
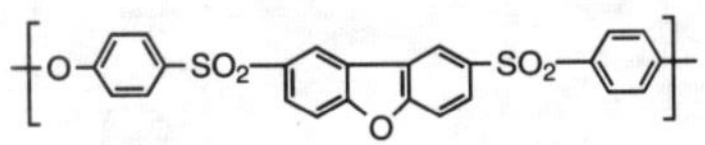	290		E-1; $FeCl_3$, Nitrobenzene, Dimethyl sulfone	*J.Polym. Sci.*, A-1, **8**, 2035	1970	3M
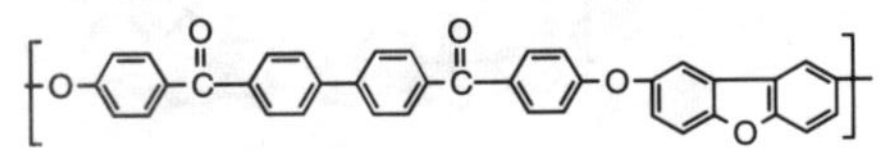	225		A-12; Bulk, CsF	*J. Chem. Soc., Chem. Commun.*, 336	1990	ICI
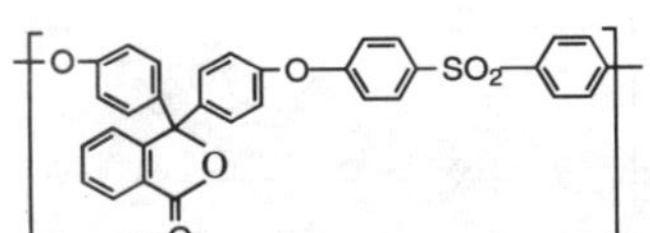			N-6; K_2CO_3, NMP; Cl	*Makromol. Chem.*, **191**, 2027	1990	Hamburg U
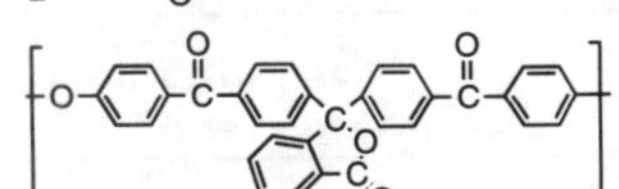			E-3; $AlCl_3$, Dichlorethane	*Acta Polymerica*, **39**, 452	1988	F. Schiller Univ., Jena

TABLE 22—*continued*

Polymer structure	Tg (°C)	Tm (°C)	Synthesis	Reference	Year	Organization
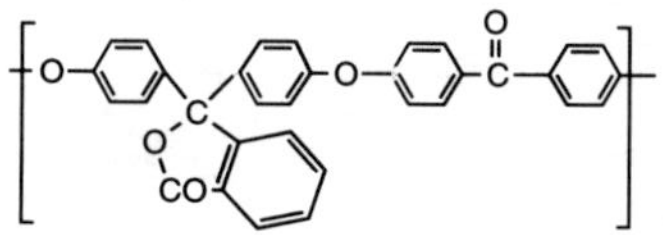	214		N-11; CsF, Bulk; F	*New Polym. Mater.*, **1**, 127	1988	Hamburg U.
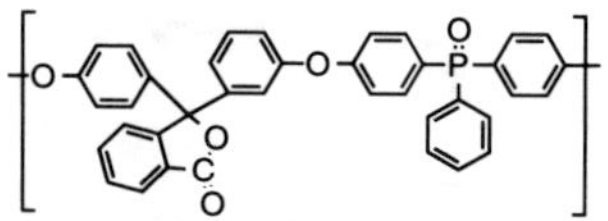	279		N-17; K_2CO_3, NMP, DMAC; F	*Polym. Prep.*, **33**(2), 231; **34**(1), 310	1992 1993	Va. P. I & State U.
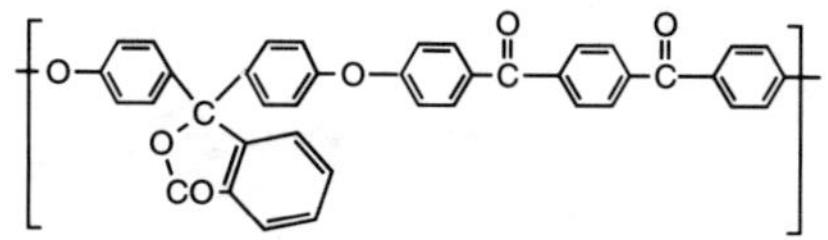	210		N-11; CsF, Bulk; F	*New Polym. Mater.*, **1**, 127	1988	Hamburg U.
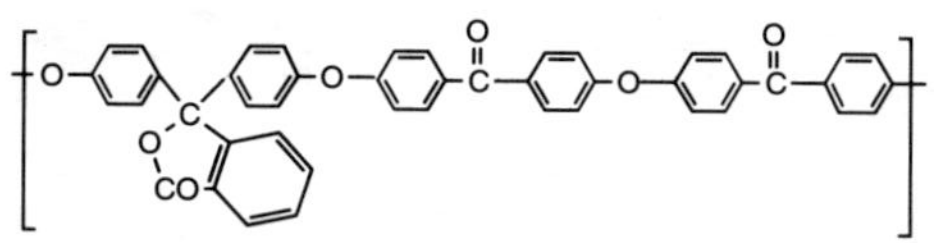	191		N-11; CsF, Bulk; F	*New Polym. Mater.*, **1**, 127	1988	Hamburg U.
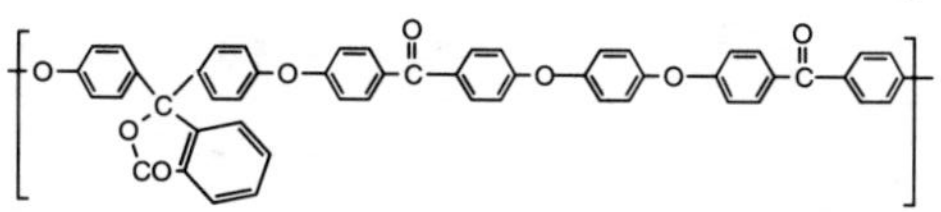	184		N-11; CsF, Bulk; F	*New Polym. Mater.*, **1**, 127	1988	Hamburg U.
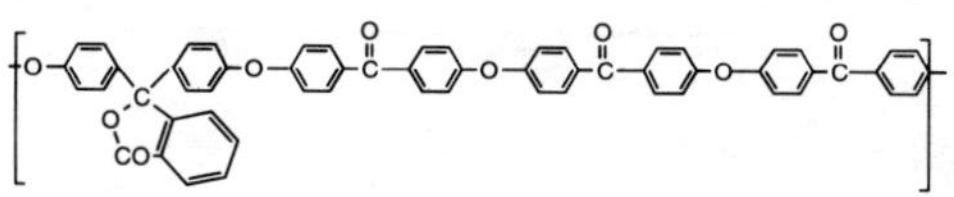	192		N-11; CsF, Bulk; F	*New Polym. Mater.*, **1**, 127	1988	Hamburg U.
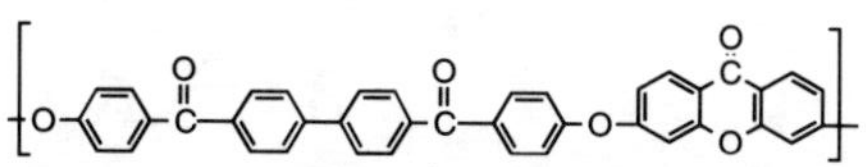			A-12; Bulk, CsF	*J. Chem. Soc., Chem. Commun.*, 336	1990	ICI

One Nitrogen Heteroatom

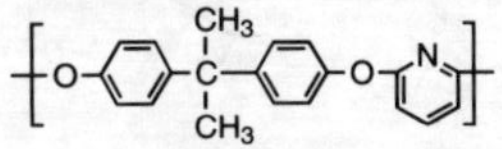		N-19; K_2CO_3, NMP; Cl	*Makromol. Chem.* **191**, 2027	1990	Hamburg U.
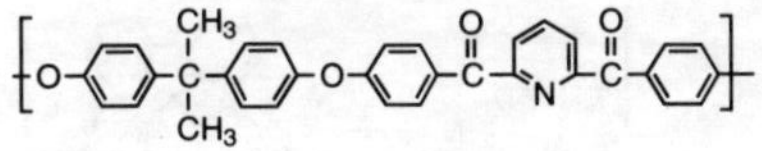	132	N-9; K_2CO_3, DMAC; F	*Polym.*, **29**, 358	1988	NASA/ Langley
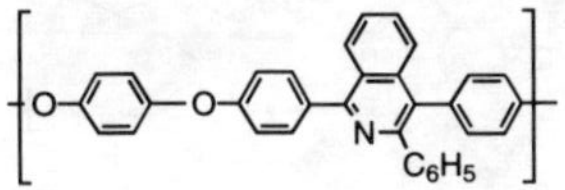	220	Post Reaction on PAEK	*Macromol.*, **25**, 1033	1992	McGill U.
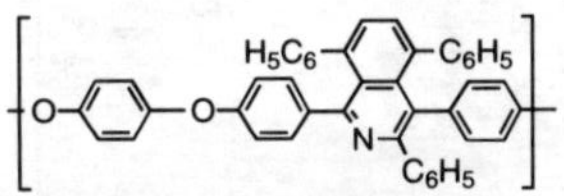	260	Post Reaction on PAEK	*Macromol.*, **25**, 1033	1992	McGill U.
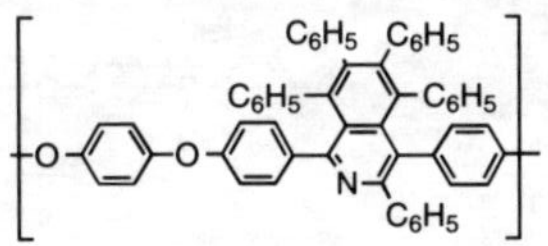	297	Post reaction on PAEK	*Macromol.*, **24**(9), 2643	1991	McGill U.
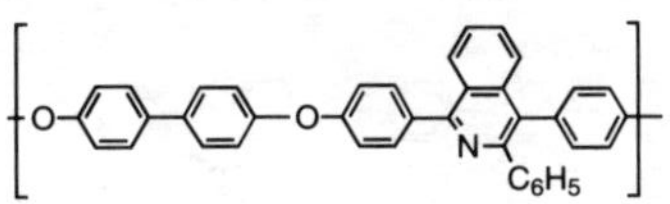	230	Post Reaction on PAEK	*Macromol.*, **25**, 1033	1992	McGill U.
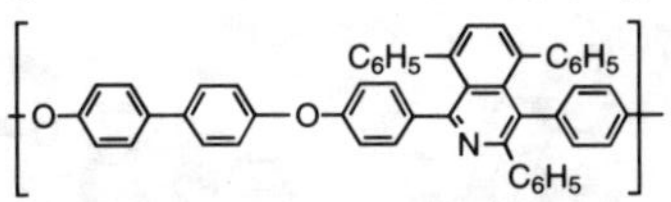	275	Post Reaction on PAEK	*Macromol.*, **25**, 1033	1992	McGill U.
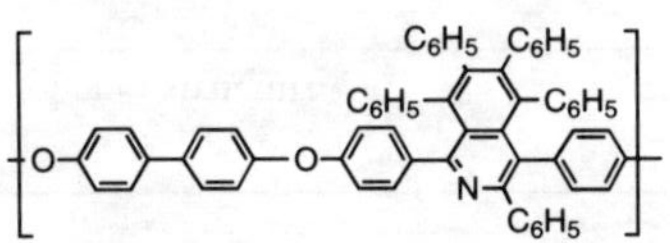	320	Post reaction on PAEK	*Macromol.*, **24**(9), 2643	1991	McGill U.

TABLE 22—*continued*

Polymer structure	Tg (°C)	Tm (°C)	Synthesis	Reference	Year	Organization
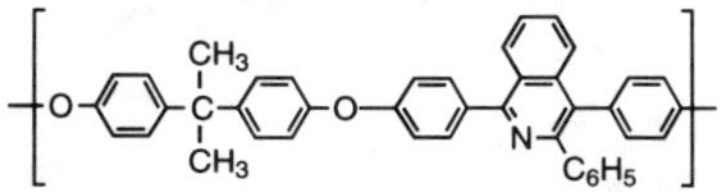	226		Post reaction on PAEK	*Macromol.*, **24**(9), 2643	1991	McGill U.
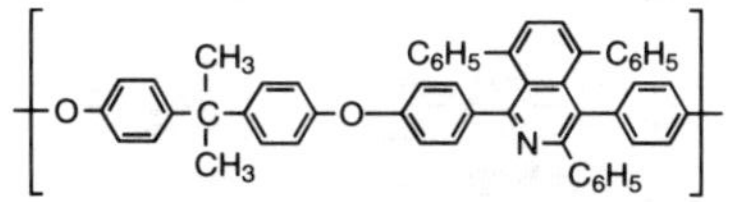	234		Post reaction on PAEK	*Macromol.*, **24**(9), 2643	1991	McGill U.
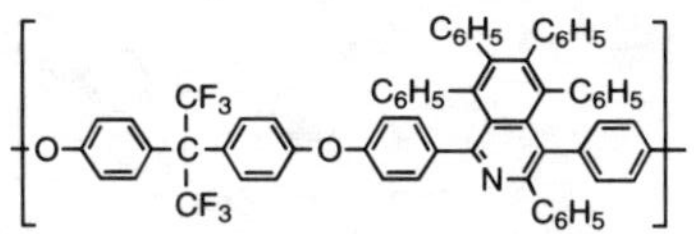	294		Post reaction on PAEK	*Macromol.*, **24**(9), 2643	1991	McGill U.
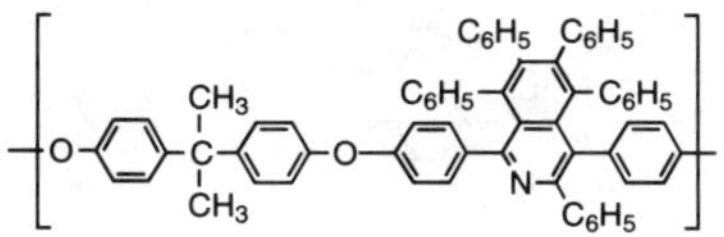	280		Post reaction on PAEK	*Macromol.*, **24**(9), 2643	1991	McGill U.
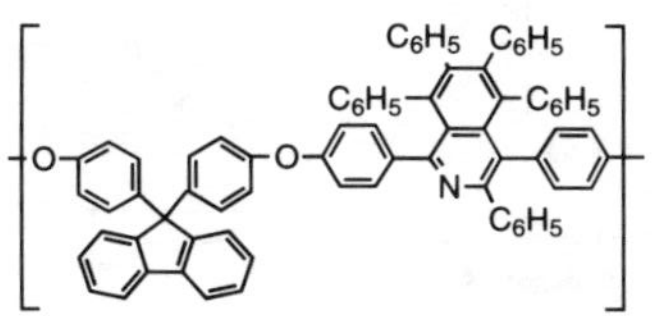	323		Post reaction on PAEK	*Macromol.*, **24**(9), 2643	1991	McGill U.
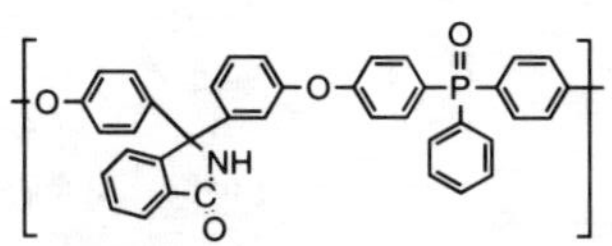	294		N-17; K_2CO_3, NMP, DMAC; F	*Polym. Prep.*, **33**(2), 231; **34**(1), 310	1992 1993	Va. P. I & State U.

Structure						
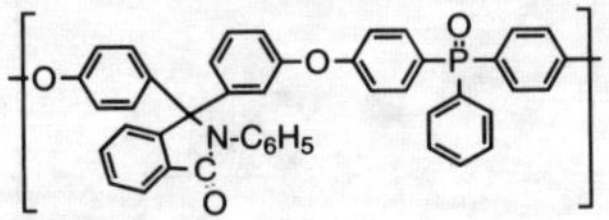	273		N-17; K_2CO_3, NMP, DMAC; F	*Polym. Prep.*, **33**(2), 231; **34**(1), 310	1992 1993	Va. P. I & State U.
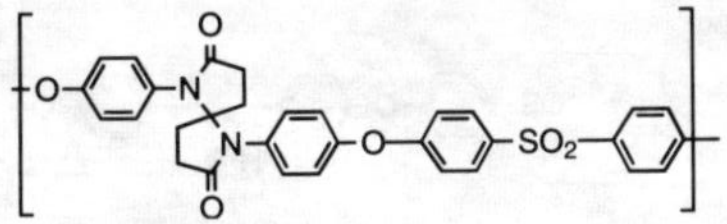	272		N-2; Cl	*Polym. Prep.*, **32**(2), 180	1991	Shell
One Sulfur Heteroatom						
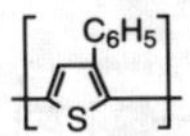	177		A-19; DMAC; Cl	*Macromol.*, **24**, 2694	1991	Yamagata U.
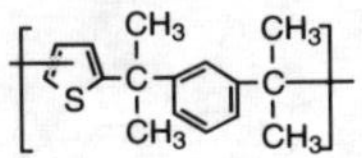	120		E-9; Trityl Cl, $AlCl_3$, Nitrobenzene, CH_2Cl_2	*J. Polym. Sci.*, A-1, **10**, 2365	1972	DuPont
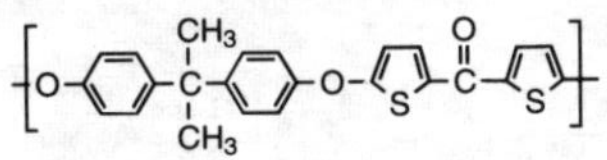	124		N-9; K_2CO_3, NMP; Cl	*Polym. Prep.*, 33(1), 418; **33**(2), 170	1992	U. No. Carolina
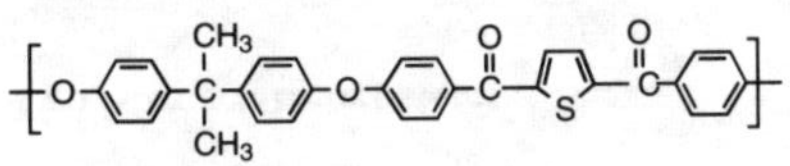	158		N-9; K_2CO_3, NMP; F	*Polym. Prep.*, **32**(2), 172; **33**(1), 485	1991 1992	U. No. Carolina, U. Florida
Two Oxygen Heteroatoms						
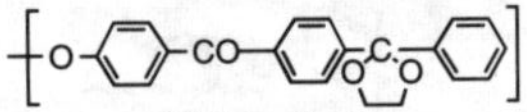 20% Ketalization		315	Post Ketalization on PEKK	US Pat. 3734888	1973	DuPont
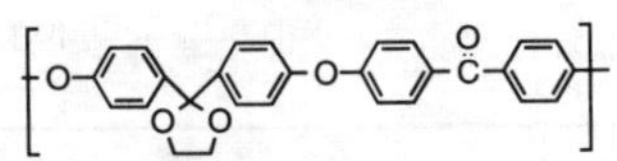	144		N-9; K_2CO_3, DMAC, Sulfolane; F	*Macromol.*, **20**, 1204; US Pat. 4882397	1987 1989	Amoco

TABLE 22—*continued*

Polymer structure	Tg (°C)	Tm (°C)	Synthesis	Reference	Year	Organization
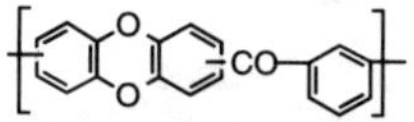	200	430	E-4; BF_3, HF	GB Pat. 1250251	1971	DuPont
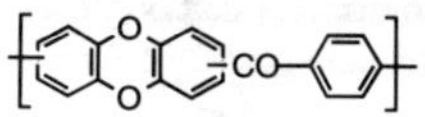	260	525	E-4; BF_3, HF	GB Pat. 1250251	1971	DuPont
Two Nitrogen Heteroatoms						
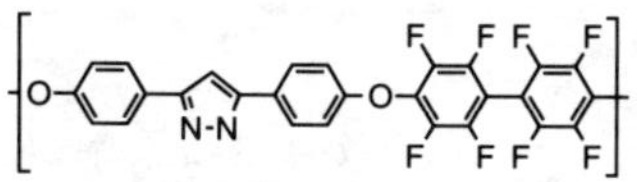	196		N-17; K_2CO_3, DMAC; F	*Polym. Prep.*, **34**(1), 441	1993	Va. Comm. U.
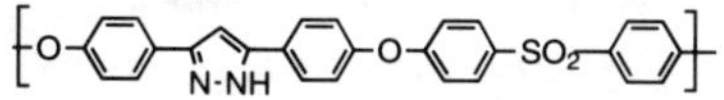	260		N-2; K_2CO_3, DMAC, NMP; Cl	*Polym. Prep.*, **32**(1), 619	1991	Va. Comm. U.
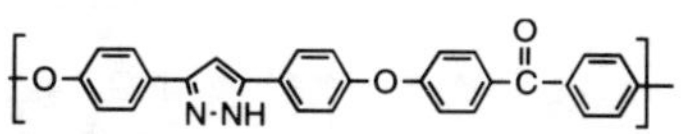	229		N-9; K_2CO_3, DMAC, NMP; Cl	*Polym. Prep.*, **32**(1), 619	1991	Va. Comm. U.
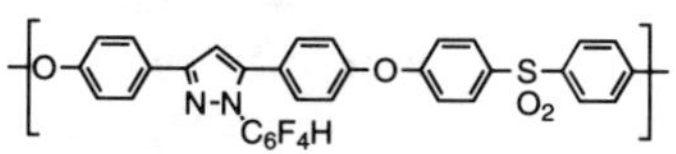	230		N-9; K_2CO_3, DMAC; F	*Polym. Prep.*, **34**(1), 441	1993	Va. Comm. U.
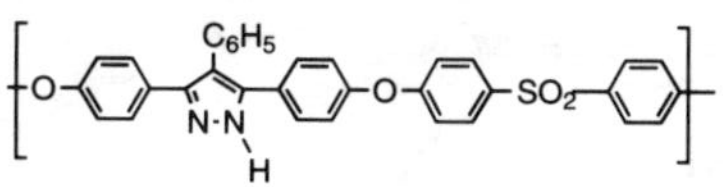	237		N-2; K_2CO_3, DMAC; Cl	*Polym. Prep.*, **32**(2), 160	1991	Va. Comm. U.; NASA/ Langley
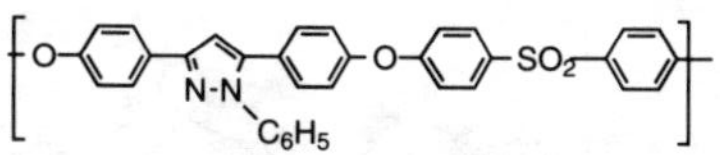	235		N-2; K_2CO_3, DMAC, NMP; F	*Polym. Prep.*, **32**(1), 619	1991	Va. Comm. U.
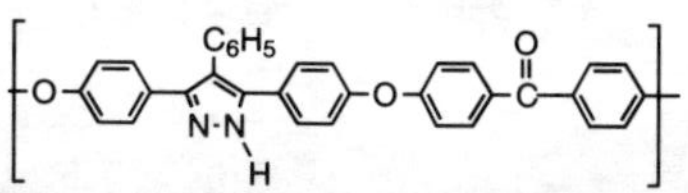	208		N-9; K_2CO_3, DMAC; Cl	*Polym. Prep.*, **32**(2), 160	1991	Va. Comm. U.; NASA/ Langley

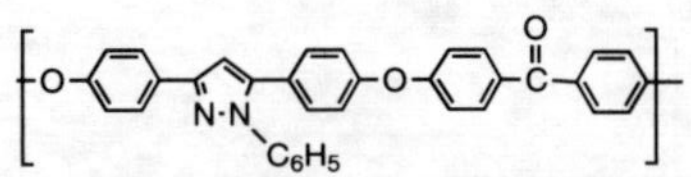	205	N-9; K_2CO_3, DMAC, NMP; F	*Polym. Prep.*, 32(1), 619	1991	Va. Comm. U.
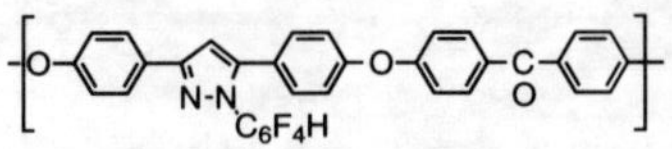	204	N-9; K_2CO_3, DMAC; F	*Polym. Prep.*, **34**(1), 441	1993	Va. Comm. U.
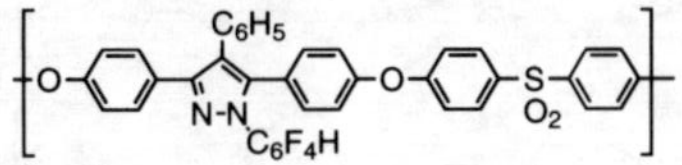	221	N-9; K_2CO_3, DMAC; F	*Polym. Prep.*, **34**(1), 441	1993	Va. Comm. U.
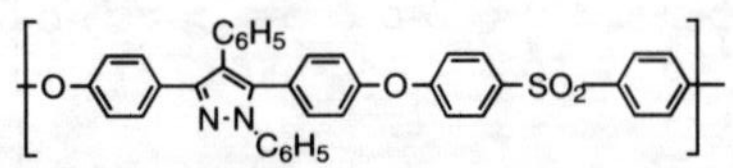	243	N-2; K_2CO_3, DMAC; F	*Polym. Prep.*, **32**(2), 160	1991	Va. Comm. U.; NASA/ Langley
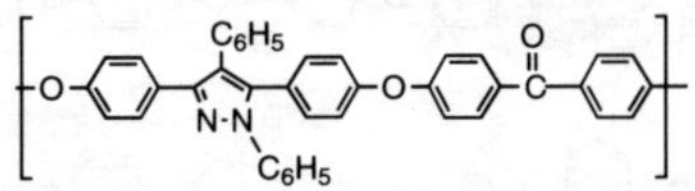	205	N-9; K_2CO_3, DMAC; F	*Polym. Prep.*, **32**(2), 160	1991	Va. Comm. U.; NASA/ Langley
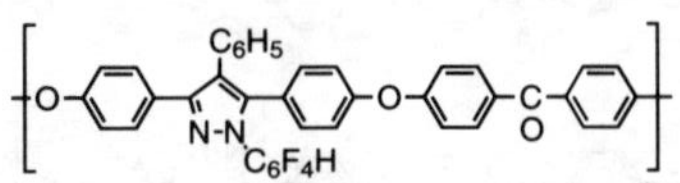	213	N-9; K_2CO_3, DMAC; F	*Polym. Prep.*, 34(1), 441	1993	Va. Comm. U.
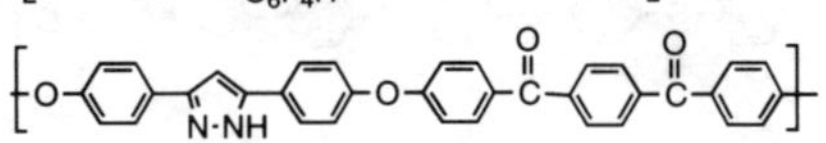	235	N-9; K_2CO_3, DMAC, NMP; F	*Polym. Prep.*, **32**(1), 619	1991	Va. Comm. U.
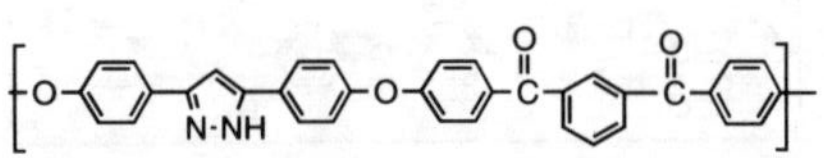	205	N-9; K_2CO_3, DMAC, NMP; Cl	*Polym. Prep.*, **32**(1), 619	1991	Va. Comm. U.
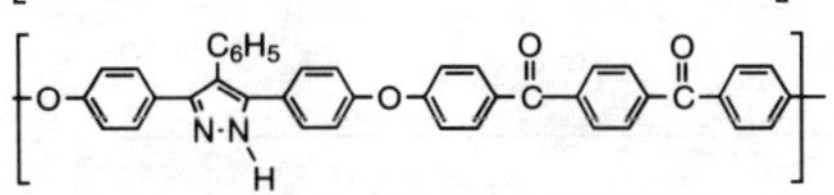	200	N-9; K_2CO_3, DMAC; Cl	*Polym. Prep.*, **32**(2), 160	1991	Va. Comm. U.; NASA/ Langley

TABLE 22—*continued*

Polymer structure	Tg (°C)	Tm (°C)	Synthesis	Reference	Year	Organization
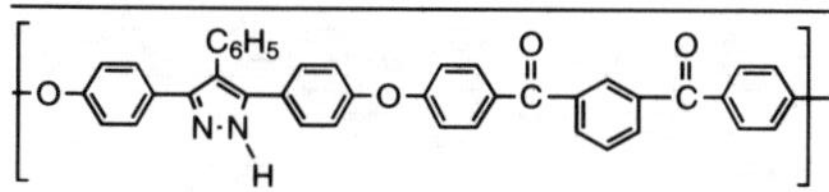	188		N-9; K_2CO_3, DMAC; Cl	*Polym. Prep.*, **32**(2), 160	1991	Va. Comm. U.; NASA/ Langley
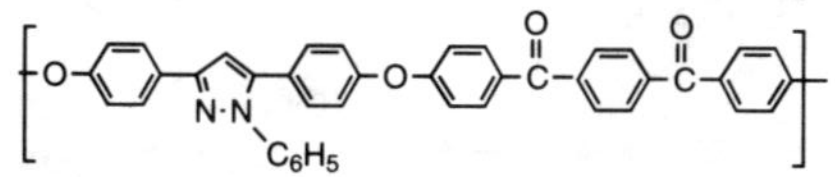	205		N-9; K_2CO_3, DMAC, NMP; F	*Polym. Prep.*, **32**(1), 619	1991	Va. Comm. U.
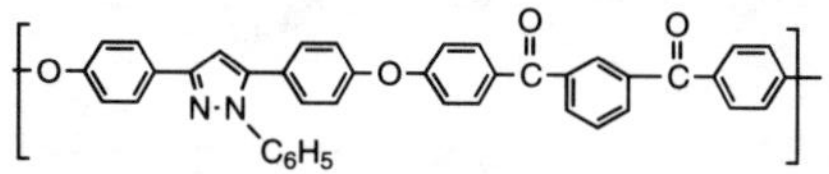	190		N-9; K_2CO_3, DMAC, NMP; F	*Polym. Prep.*, **32**(1), 619	1991	Va. Comm. U.
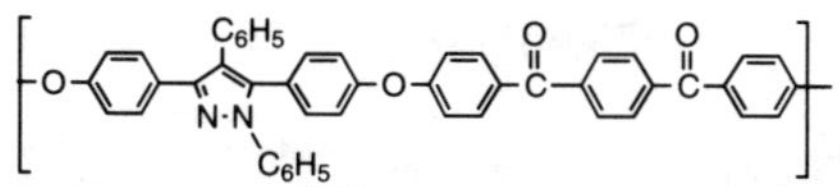	210		N-9; K_2CO_3, DMAC; F	*Polym. Prep.*, **32**(2), 160	1991	Va. Comm. U.; NASA/ Langley
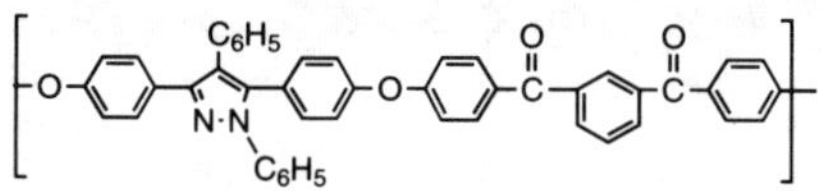	196		N-9; K_2CO_3, DMAC; F	*Polym. Prep.*, **32**(2), 160	1991	Va. Comm. U.; NASA/ Langley
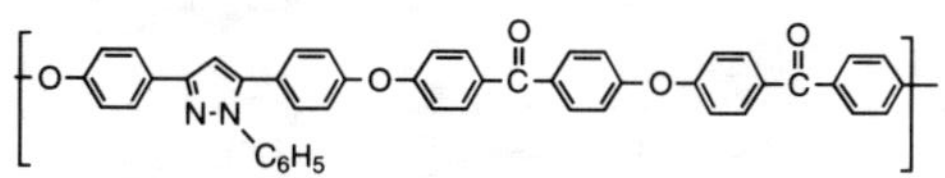	190		N-9; K_2CO_3, DMAC, NMP; F	*Polym. Prep.*, **32**(1), 619	1991	Va. Comm. U.
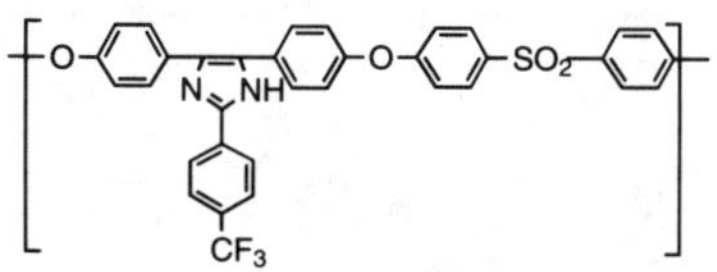	284		N-2; K_2CO_3, DMAC; F	*Polym. Prep.*, **32**(2), 162	1991	NASA/ Langley

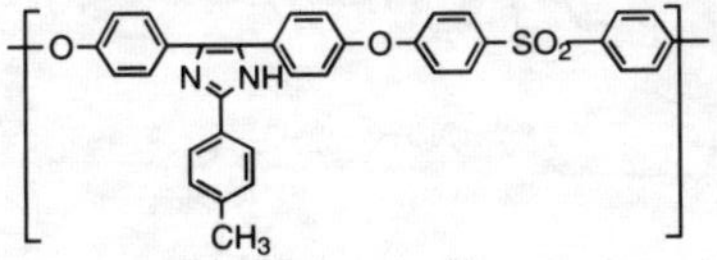	275	N-2; K_2CO_3, DMAC; F	*Polym. Prep.*, **32**(2), 162	1991	NASA/ Langley
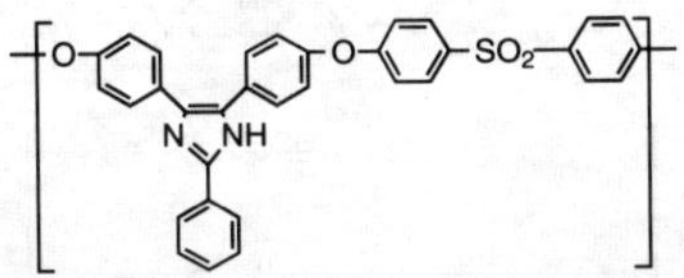	277	N-2; K_2CO_3, DMAC; F	NASA Tech. Briefs, Aug., p. 48	1990	NASA/ Langley
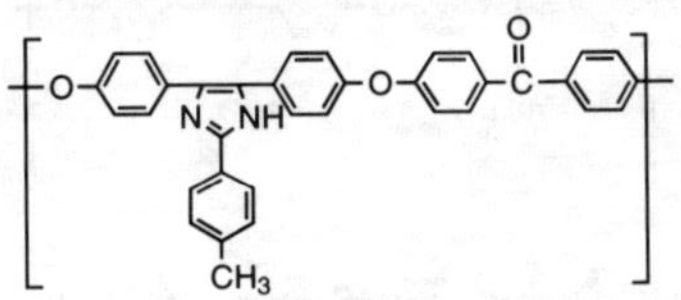	254	N-9; K_2CO_3, DMAC; F	*Polym. Prep.*, **32**(2), 162	1991	NASA/ Langley
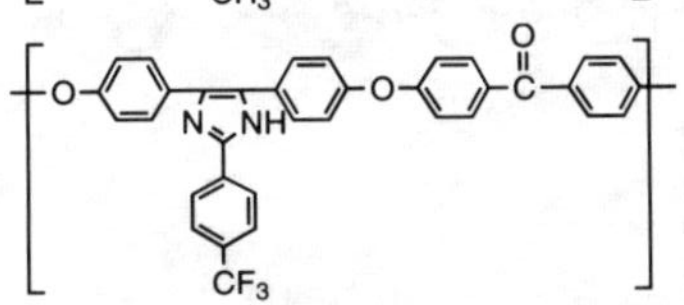	263	N-9; K_2CO_3, DMAC; F	*Polym. Prep.*, **32**(2), 162	1991	NASA/ Langley
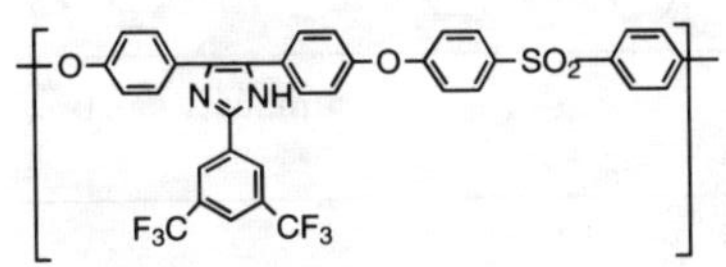	295	N-2; K_2CO_3, DMAC; F	*Polym. Prep.*, **32**(2), 162	1991	NASA/ Langley

TABLE 22—*continued*

Polymer structure	Tg (°C)	Tm (°C)	Synthesis	Reference	Year	Organization
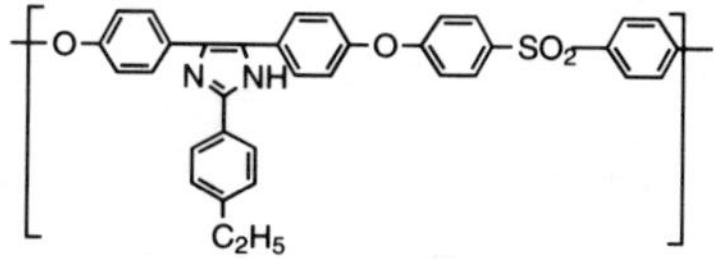	271		N-2; K_2CO_3, DMAC; F	*Polym. Prep.*, **32**(2), 162	1991	NASA/ Langley
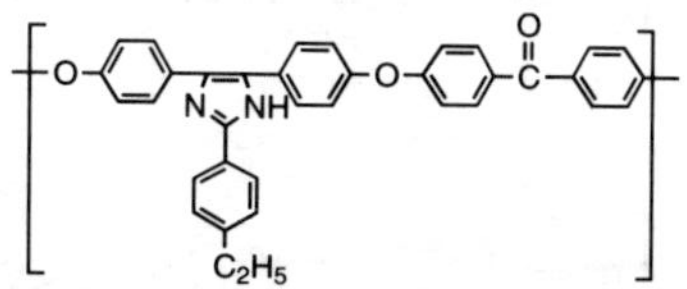	243		N-9; K_2CO_3, DMAC; F	*Polym. Prep.*, **32**(2), 162	1991	NASA/ Langley
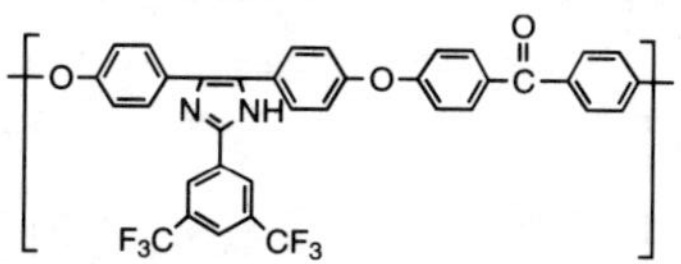	270		N-9; K_2CO_3, DMAC; F	*Polym. Prep.*, **32**(2), 162	1991	NASA/ Langley
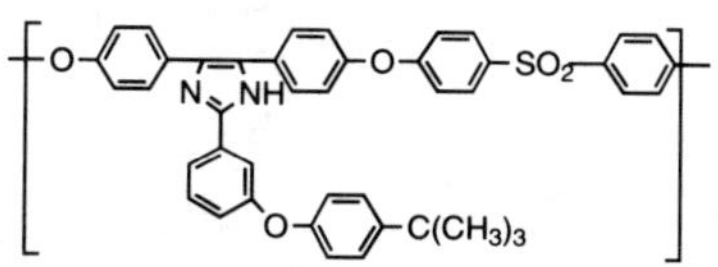	235		N-2; K_2CO_3, DMAC; F	*Polym. Prep.*, **32**(2), 162	1991	NASA/ Langley
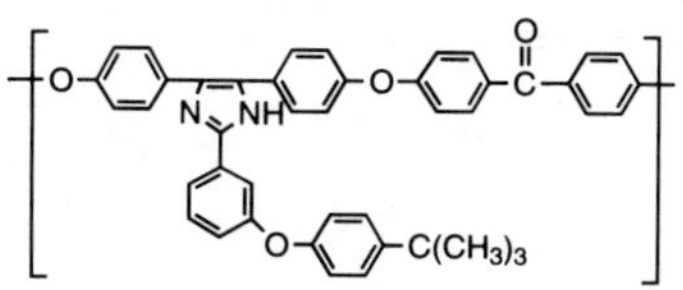	217		N-9; K_2CO_3, DMAC; F	*Polym. Prep.*, **32**(2), 162	1991	NASA/ Langley

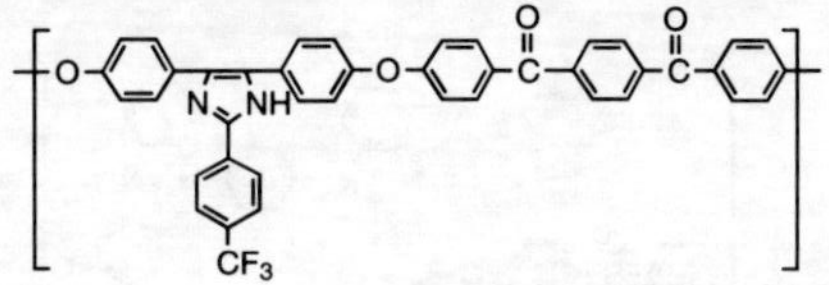	249	N-9; K_2CO_3, DMAC; F	*Polym. Prep.*, **32**(2), 162	1991	NASA/ Langley
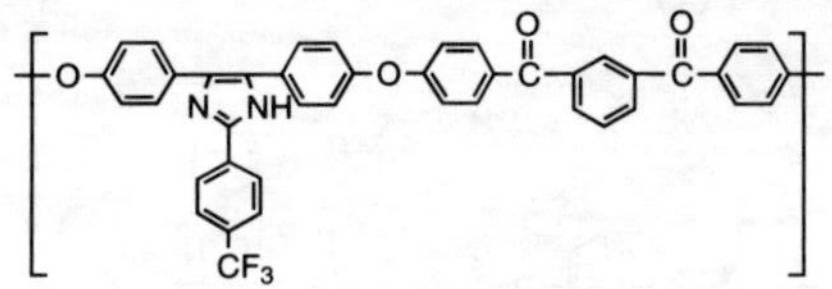	233	N-9; K_2CO_3, DMAC; F	*Polym. Prep.*, **32**(2), 162	1991	NASA/ Langley
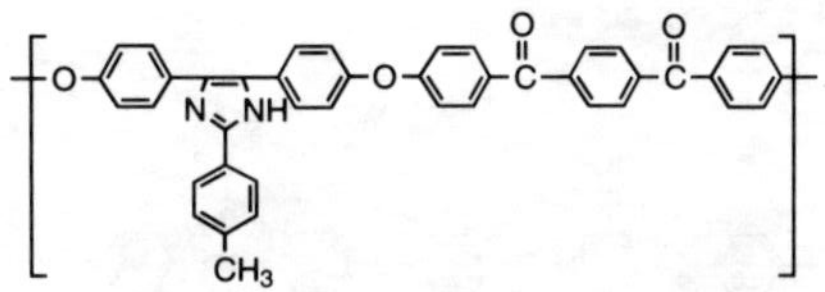	245	N-9; K_2CO_3, DMAC; F	*Polym. Prep.*, **32**(2), 162	1991	NASA/ Langley
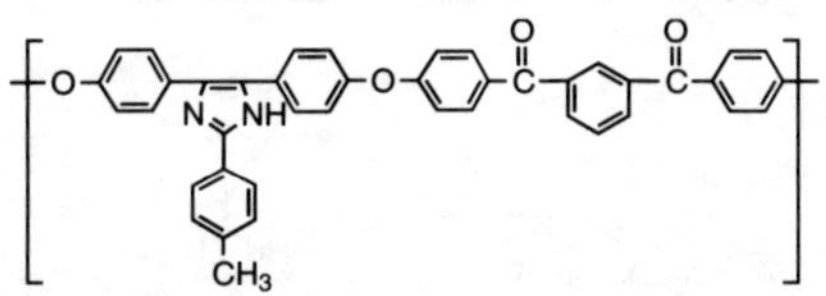	227	N-9; K_2CO_3, DMAC; F	*Polym. Prep.*, **32**(2), 162	1991	NASA/ Langley
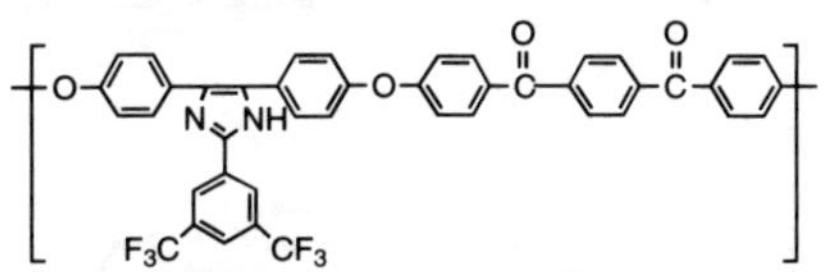	259	N-9; K_2CO_3, DMAC; F	*Polym. Prep.*, **32**(2), 162	1991	NASA/ Langley
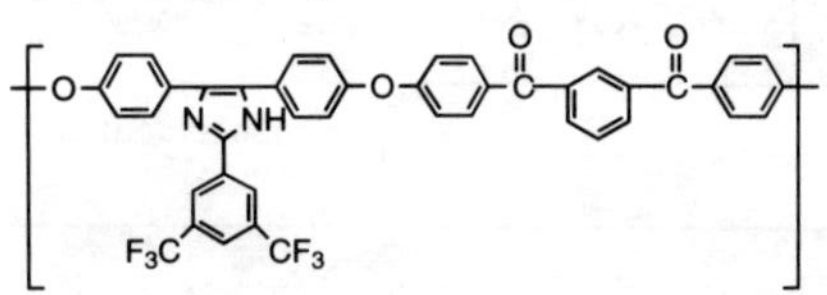	234	N-9; K_2CO_3, DMAC; F	*Polym. Prep.*, **32**(2), 162	1991	NASA/ Langley

TABLE 22—*continued*

Polymer structure	Tg (°C)	Tm (°C)	Synthesis	Reference	Year	Organization
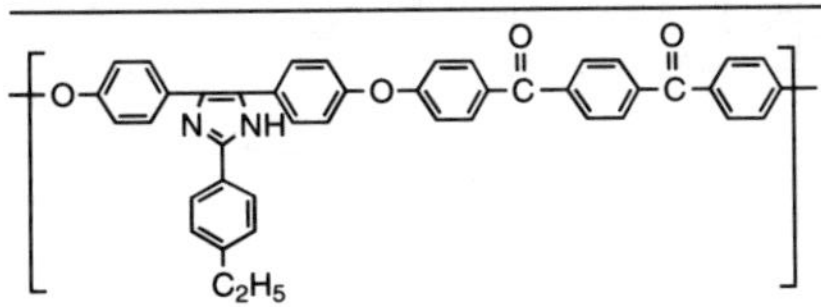	236		N-9; K_2CO_3, DMAC; F	*Polym. Prep.*, **32**(2), 162	1991	NASA/ Langley
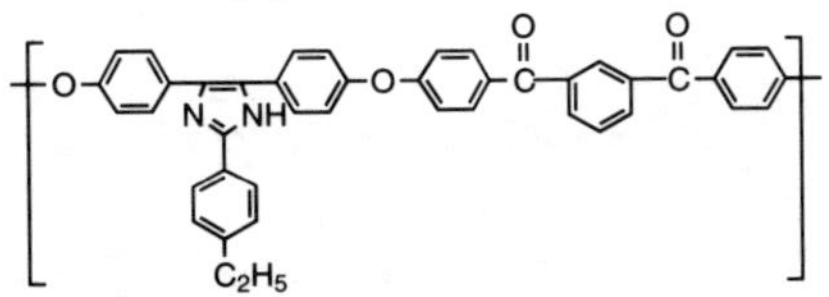	218		N-9; K_2CO_3, DMAC; F	*Polym. Prep.*, **32**(2), 162	1991	NASA/ Langley
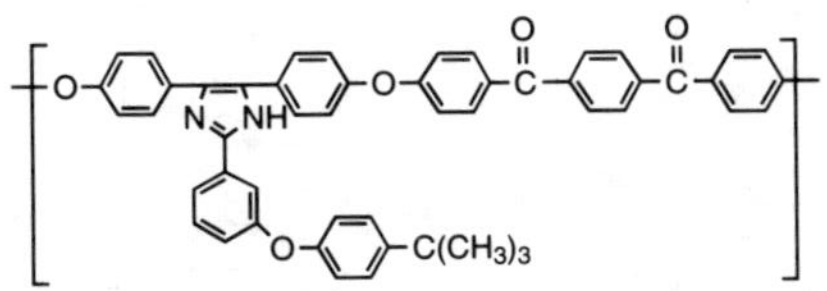	213		N-9; K_2CO_3, DMAC; F	*Polym. Prep.*, **32**(2), 162	1991	NASA/ Langley
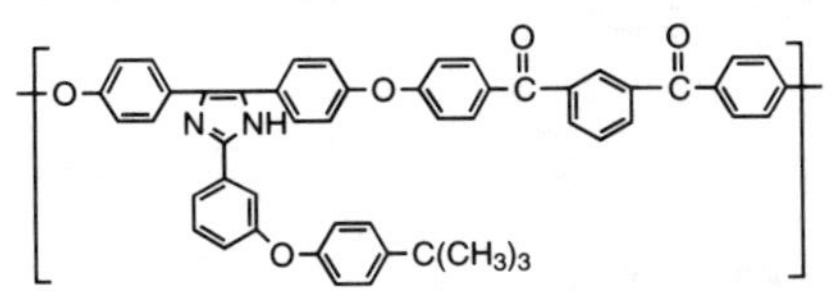	200		N-9; K_2CO_3, DMAC; F	*Polym. Prep.*, **32**(2), 162	1991	NASA/ Langley
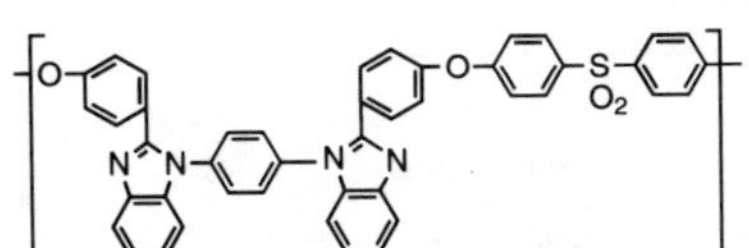	270		N-2; K_2CO_3, Sulfolane; F	*Polym. Prep.*, **33**(1), 1098	1992	NASA/ Langley

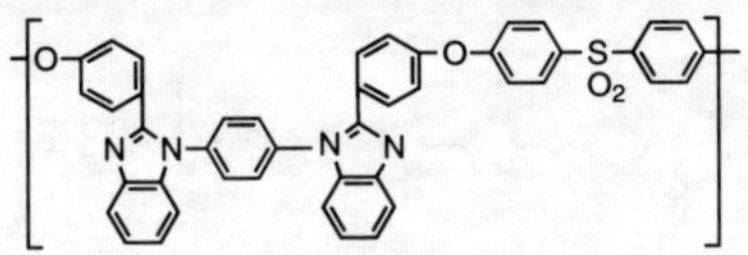	242	N-9; K_2CO_3, Sulfolane; F	*Polym. Prep.*, **33**(1), 1098	1992	NASA/ Langley
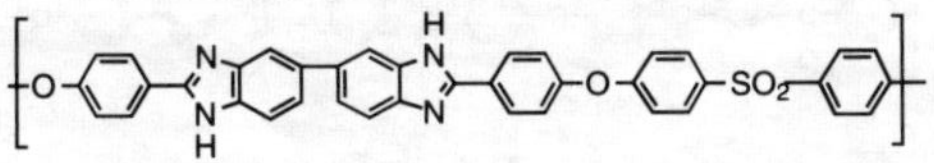	352	N-2; K_2CO_3, DMAC; F	*Polym. Prep.*, **32**(3), 195	1991	NASA/ Langley
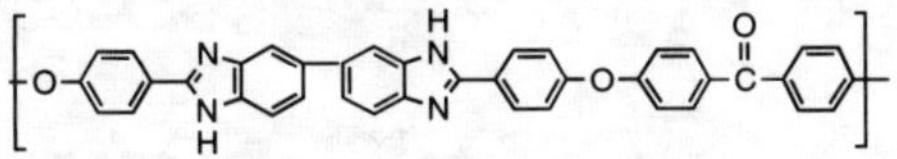	307	N-9; K_2CO_3, DMAC; F	*Polym. Prep.*, **32**(3), 195	1991	NASA/ Langley
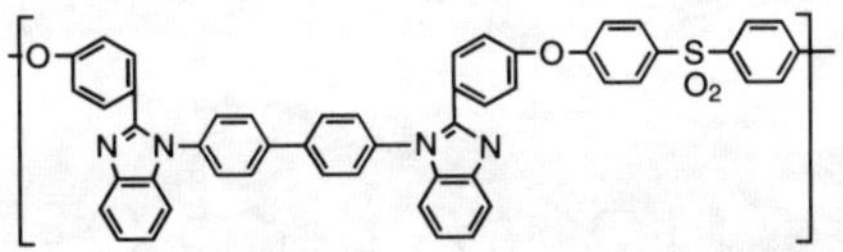	289	N-2; K_2CO_3, Sulfolane; F	*Polym. Prep.*, **33**(1), 1098	1992	NASA/ Langley
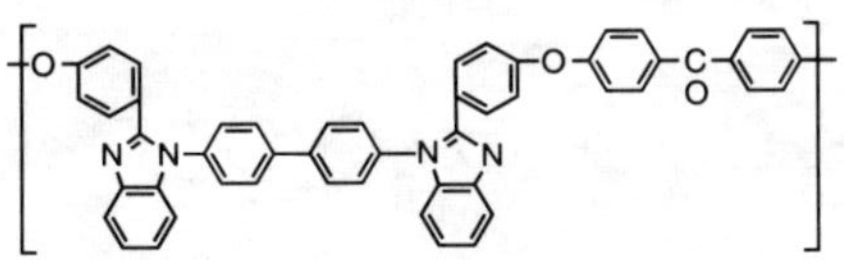	268	N-9; K_2CO_3, Sulfolane; F	*Polym. Prep.*, **33**(1), 1098	1992	NASA/ Langley
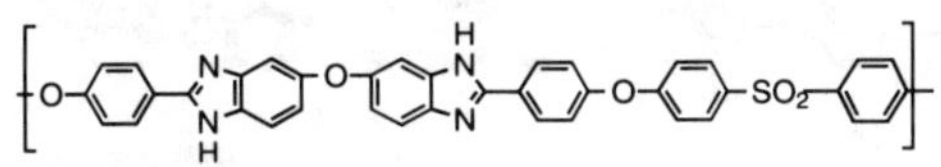	322	N-2; K_2CO_3, DMAC; F	*Polym. Prep.*, **32**(3), 195	1991	NASA/ Langley
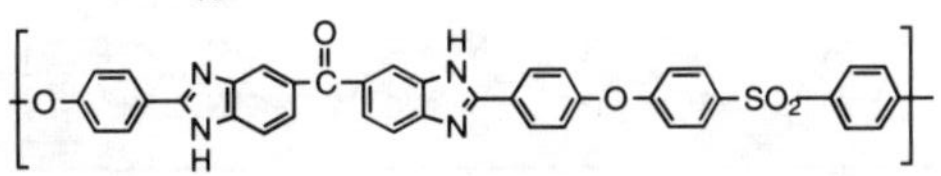		N-2; K_2CO_3, DMAC; F	*Polym. Prep.*, **32**(3), 195	1991	NASA/ Langley
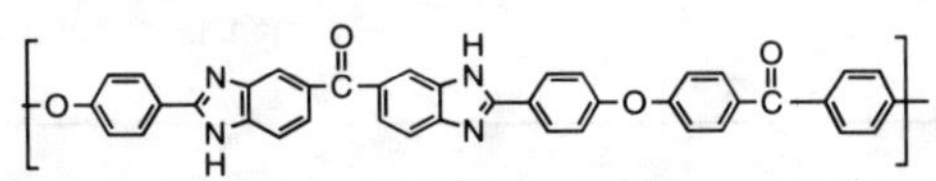		N-9; K_2CO_3, DMAC; F	*Polym. Prep.*, **32**(3), 195	1991	NASA/ Langley

TABLE 22—*continued*

Polymer structure	Tg (°C)	Tm (°C)	Synthesis	Reference	Year	Organization
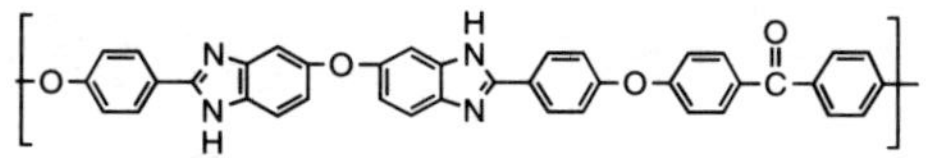	294		N-9; K_2CO_3, DMAC; F	*Polym. Prep.*, **32**(3), 195	1991	NASA/ Langley
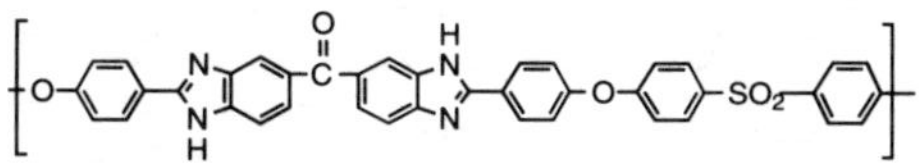			N-9; K_2CO_3, DMAC; F	*Polym. Prep.*, **32**(3), 195	1991	NASA/ Langley
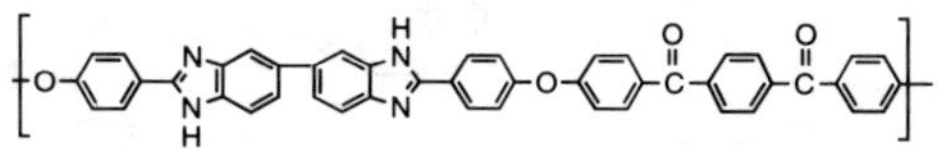	295		N-9; K_2CO_3, DMAC; F	*Polym. Prep.*, **32**(3), 195	1991	NASA/ Langley
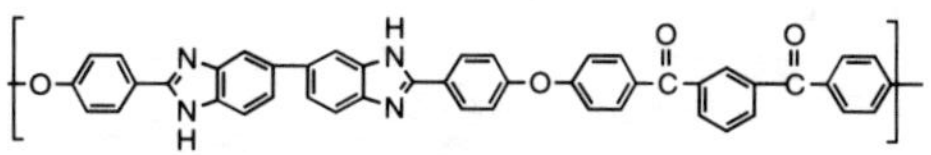	276		N-9; K_2CO_3, DMAC; F	*Polym. Prep.*, **32**(3),195; Polym. *Prep.*, **33**(1), 412; *Polym. Prep.*, **34**(1), 525	1991 1992 1993	NASA/ Langley
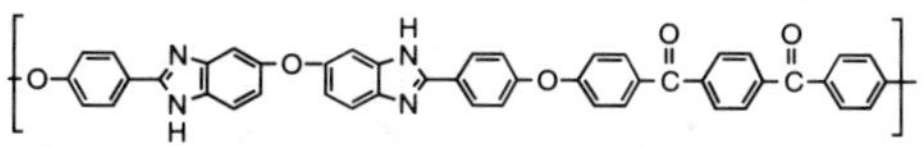	282		N-9; K_2CO_3, DMAC; F	*Polym. Prep.*, **32**(3), 195	1991	NASA/ Langley
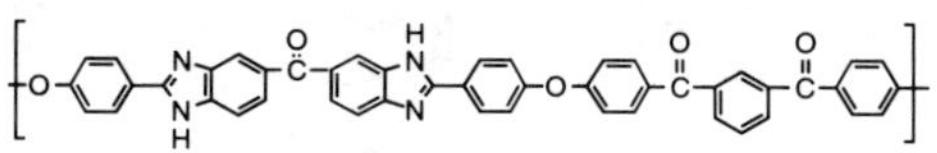	276		N-9; K_2CO_3, DMAC; F	*Polym. Prep.*, **32**(3), 195	1991	NASA/ Langley
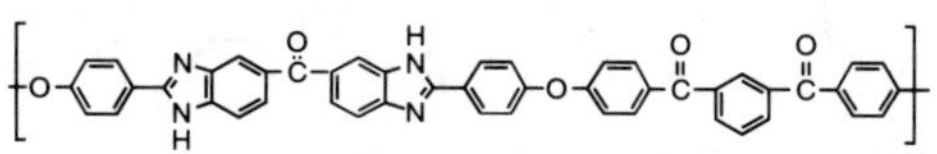	264		N-9; K_2CO_3, DMAC; F	*Polym. Prep.*, **32**(3), 195	1991	NASA/ Langley
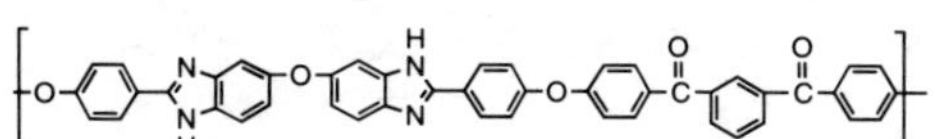	269		N-9; K_2CO_3, DMAC; F	*Polym. Prep.*, **32**(3), 195	1991	NASA/ Langley

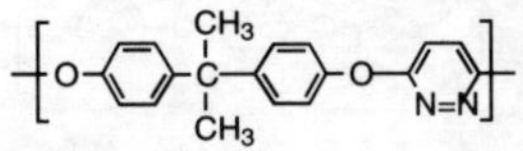	180	250	N-1; NaOH, DMSO, Sulfolane; Cl	*J.Polym.Sci.*, A-1, **5**, 2375	1967	Union Carbide
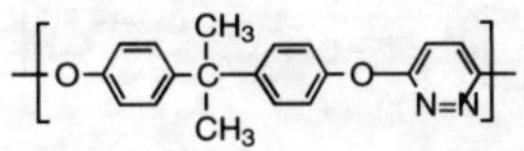			N-19; K_2CO_3, NMP; Cl	*Makromol. Chem.*, **191**, 2027	1990	Hamburg U.
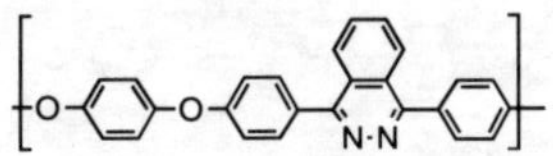	260		Post Reaction on PAEK	*Macromol.*, **25**, 1025	1992	McGill U.
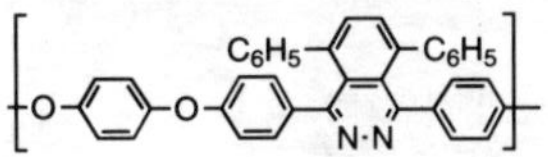	270		Post Reaction on PAEK	*Macromol.*, **25**, 1025	1992	McGill U.
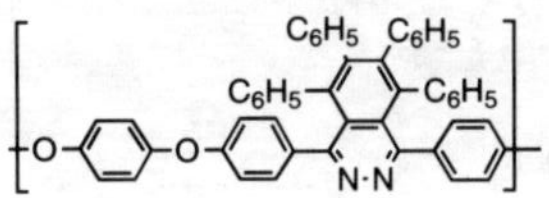	296		Post Reacton on PAEK	*Macromol.*, **24**(9), 2640	1991	McGill U.
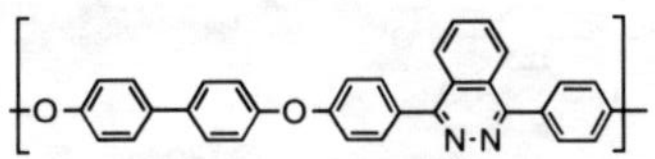	255		Post Reaction on PAEK	*Macromol.*, **25**, 1025	1992	McGill U.
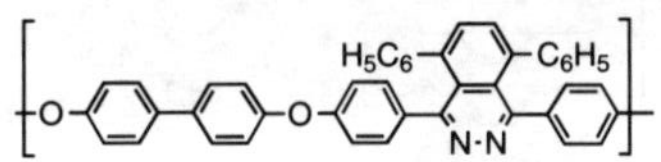	300		Post Reaction on PAEK	*Macromol.*, **25**, 1025	1992	McGill U.
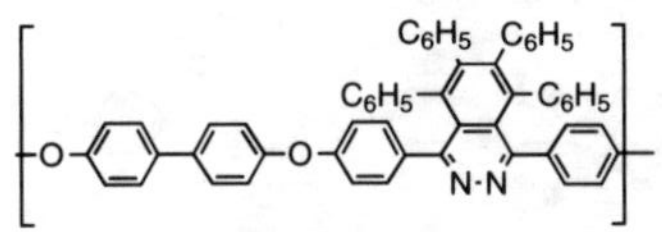	340		Post Reacton on PAEK	*Macromol.*, **24**(9), 2640	1991	McGill U.
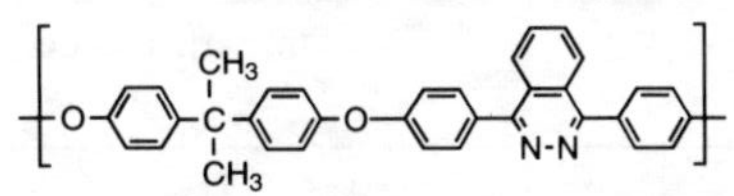	236		Post Reaction on PAEK	*Macromol.*, **24**(9), 2640	1991	McGill U.

TABLE 22—*continued*

Polymer structure	Tg (°C)	Tm (°C)	Synthesis	Reference	Year	Organization
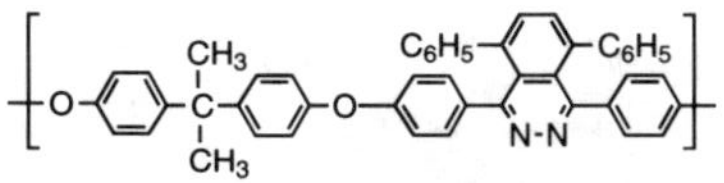	250		Post Reaction on PAEK	*Macromol.*, **24**(9), 2640	1991	McGill U.
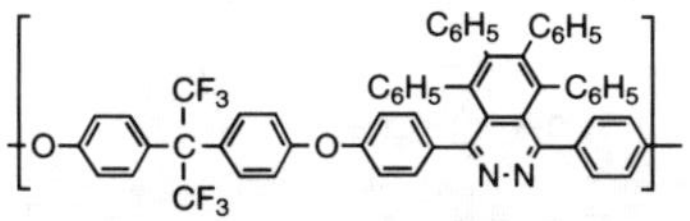	293		Post Reacton on PAEK	*Macromol.*, **24**(9),2640	1991	McGill U.
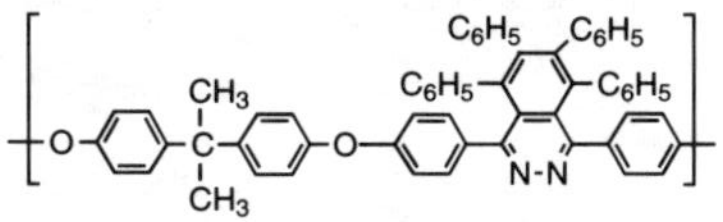	283		Post Reaction on PAEK	*Macromol.*, **24**(9), 2640	1991	McGill U.
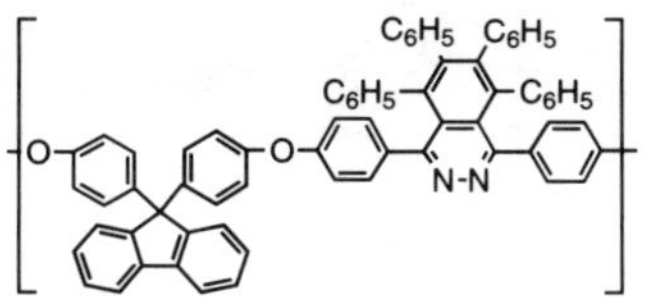	340		Post Reacton on PAEK	*Macromol.*, **24**(9), 2640	1991	McGill U.
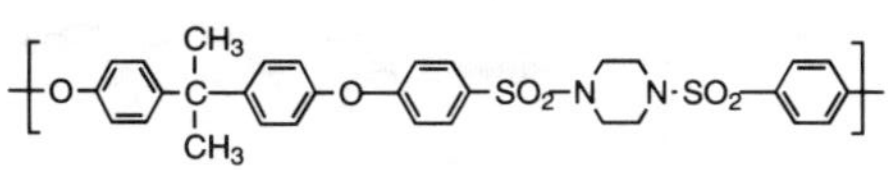		255–260	N-1; NaOH, DMSO; Cl	US Pat. 4108837	1978	Union Carbide
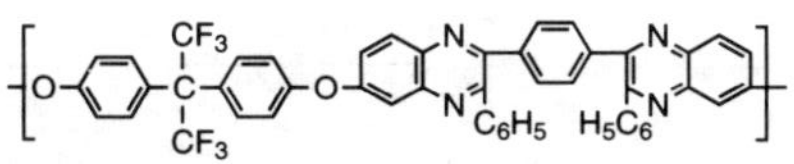	255		N-17; K_2CO_3, NMP; F	*Makromol. Chem., Macromol. Symp.*, **54/55**, 313	1992	IBM
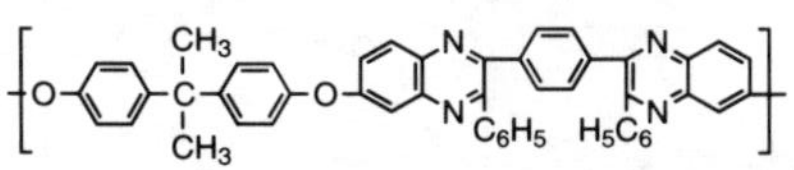	255		N-17; K_2CO_3, NMP; F	*Makromol. Chem., Macromol. Symp.*, **54/55**, 313	1992	IBM

Structure						
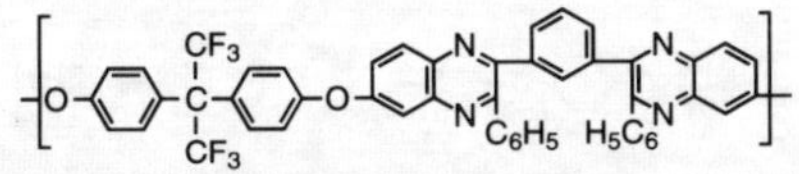	240		N-17; K_2CO_3, NMP; F	*Makromol. Chem., Macromol. Symp.*, **54/55**, 313	1992	IBM
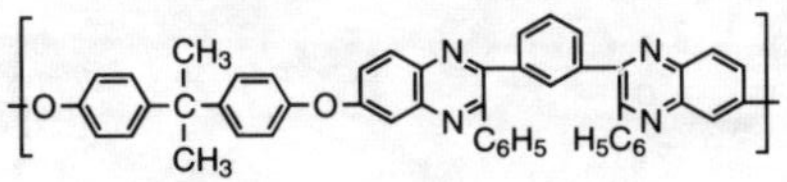	230		N-17; K_2CO_3, NMP; F	*Makromol. Chem., Macromol. Symp.*, **54/55**, 313	1992	IBM
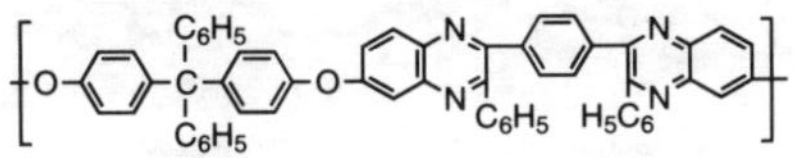	270		N-17; K_2CO_3, NMP; F	*Makromol. Chem., Macromol. Symp.*, **54/55**, 313	1992	IBM
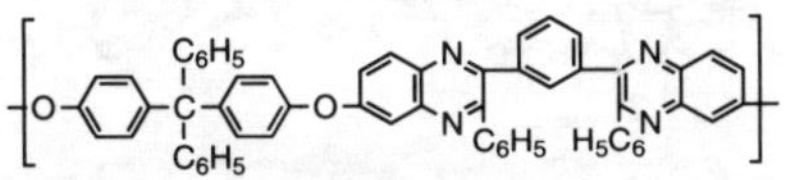	250		N-17; K_2CO_3, NMP; F	*Makromol. Chem., Macromol. Symp.*, **54/55**, 313	1992	IBM
One Oxygen, One Nitrogen Heteroatoms						
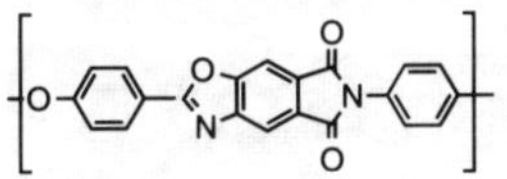		470	N-14; NaH, NMP, CHP; F	*Polym. Prep.*, **33**(2), 142	1992	U. So. Miss.
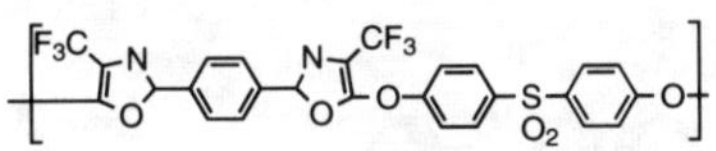	191		N-2; $(Et)_3N$, NMP; F	*Polym. Prep.*, **34**(1), 429	1993	U. Bayreuth
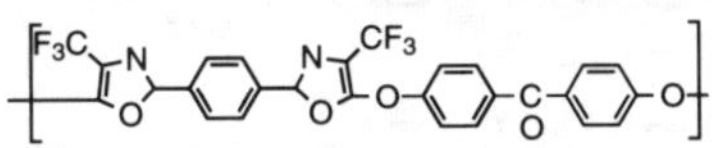	167	278	N-9; $(Et)_3N$, NMP; F	*Polym. Prep.*, **34**(1), 429	1993	U. Bayreuth
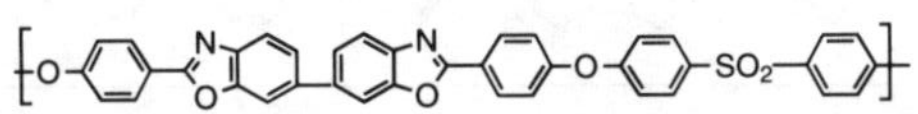	247		N-2; K_2CO_3, DMAC, Diphenyl sulfone; F	*Polym. Prep.*, **32**(1), 646	1991	NASA/ Langley
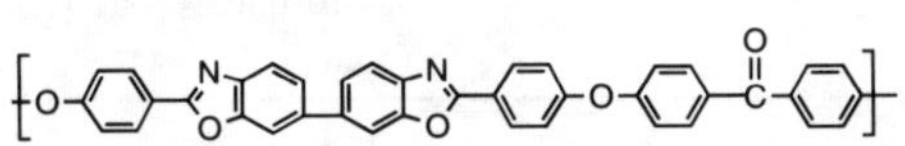	249	425	N-9; K_2CO_3, DMAC, Diphenyl sulfone; F	*Polym. Prep.*, **32**(1), 646	1991	NASA/ Langley

TABLE 22—*continued*

Polymer structure	Tg (°C)	Tm (°C)	Synthesis	Reference	Year	Organization
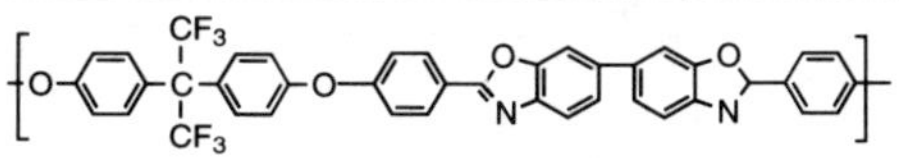	230		N-17; K_2CO_3, NMP; F	*Makromol. Chem., Macromol. Symp.*, **54/55**, 313	1992	IBM
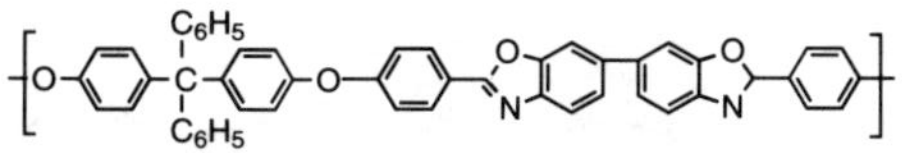	245		N-17; K_2CO_3, NMP; F	*Makromol. Chem., Macromol. Symp.*, **54/55**, 313	1992	IBM
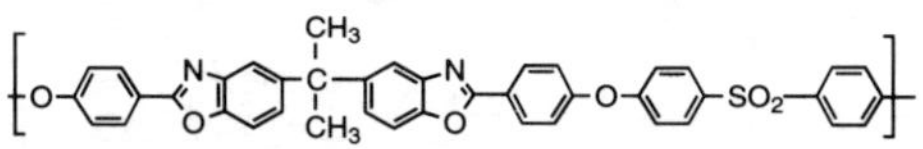	285		N-2; K_2CO_3, DMAC, Diphenyl Sulfone; F	*Polym. Prep.*, **32**(1), 646	1991	NASA/ Langley
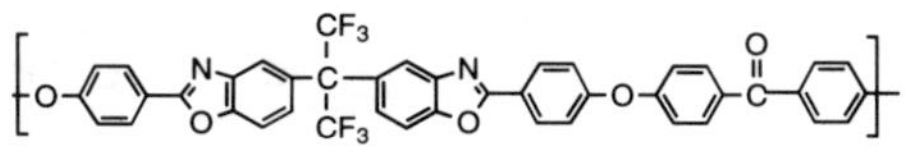	254		N-9; K_2CO_3, DMAC, Diphenyl sulfone; F	*Polym. Prep.*, **32**(1), 646	1991	NASA/ Langley
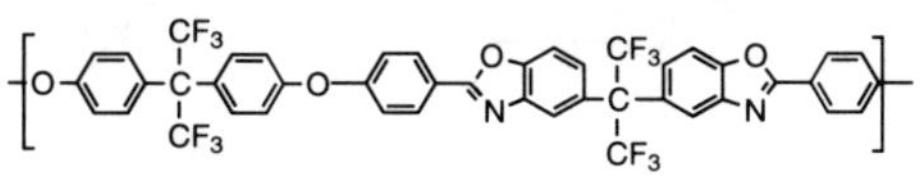	241		N-17; K_2CO_3, NMP; F	*Makromol. Chem., Macromol. Symp.*, **54/55**, 313	1992	IBM
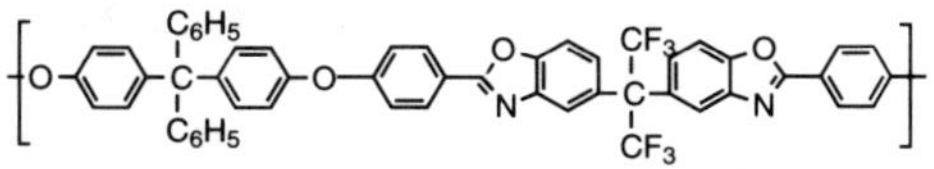	259		N-17; K_2CO_3, NMP; F	*Makromol. Chem., Macromol. Symp.*, **54/55**, 313	1992	IBM
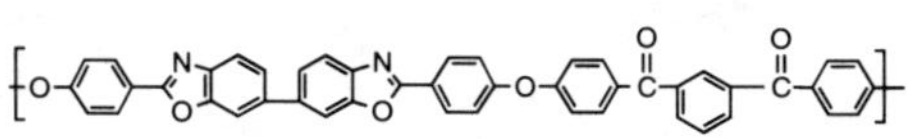	215	373	N-9; K_2CO_3, DMAC, Diphenyl sulfone; F	*Polym. Prep.*, **32**(1), 646	1991	NASA/ Langley
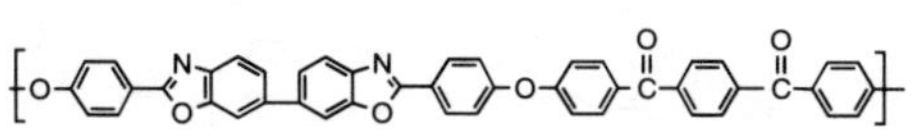	237	405, 446	N-9; K_2CO_3, DMAC, Diphenyl sulfone; F	*Polym. Prep.*, **32**(1), 646	1991	NASA/ Langley
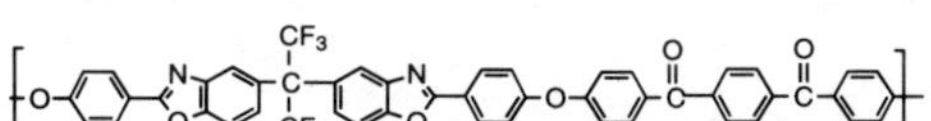	249		N-9; K_2CO_3, DMAC, Diphenyl sulfone; F	*Polym. Prep.*, **32**(1), 646	1991	NASA/ Langley

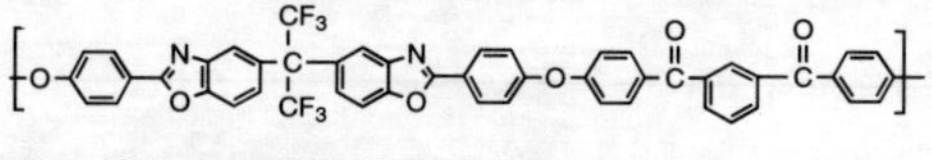	227	N-9; K_2CO_3, DMAC, Diphenyl sulfone; F	*Polym. Prep.*, **32**(1), 646	1991	NASA/ Langley
One Nitrogen, One Sulfur Heteroatoms					
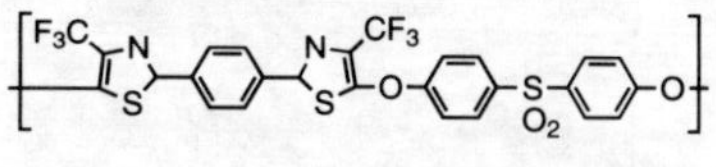	203	N-2; $(Et)_3N$, NMP; F	*Polym. Prep.*, **34**(1), 429	1993	U. Bayreuth
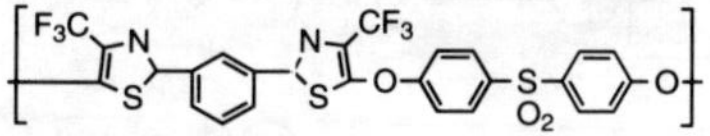		N-2; $(Et)_3N$, NMP; F	*Polym. Prep.*, **34**(1), 429	1993	U. Bayreuth
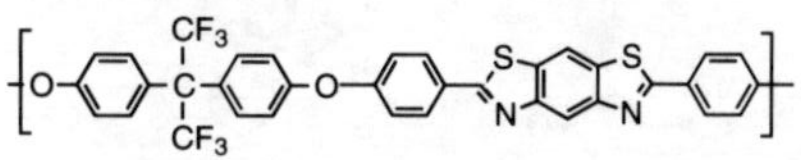		N-17; K_2CO_3, NMP, CHP; F	*Makromol. Chem., Macromol. Symp.*, **54/55**, 313	1992	IBM
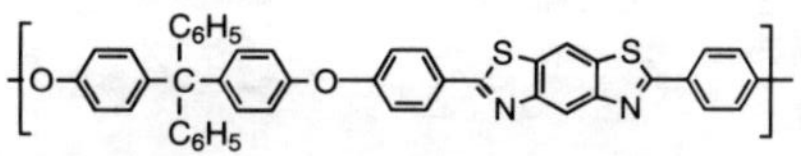		N-17; K_2CO_3, NMP, CHP; F	*Makromol. Chem., Macromol. Symp.*, **54/55**, 313	1992	IBM
Three Nitrogen Heteroatoms					
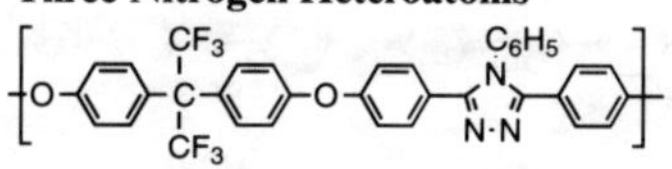	203	N-17; K_2CO_3, DMPU; F	*Polym. Prep.*, **34**(1), 481	1993	IBM
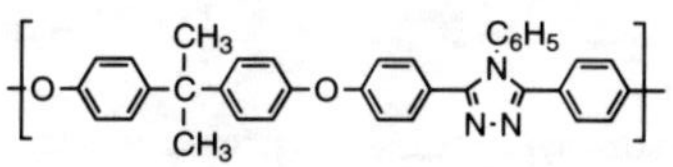	207	N-17; K_2CO_3, DMPU; F	*Polym. Prep.*, **34**(1), 481	1993	IBM
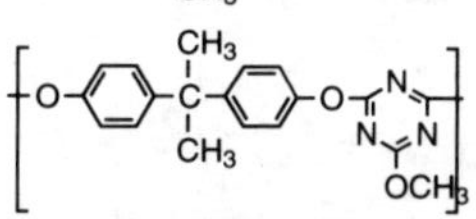		N-1; NaOH, $CHCl_3$; Cl	*Eur. Polym. J.*, **20**, 519	1984	
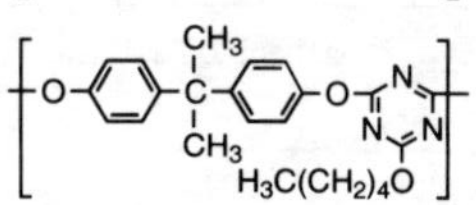		N-1; NaOH, $CHCl_3$; Cl	*J. Polym. Sci.*, A-1, **7**, 2749	1969	

TABLE 22—*continued*

Polymer structure	Tg (°C)	Tm (°C)	Synthesis	Reference	Year	Organization
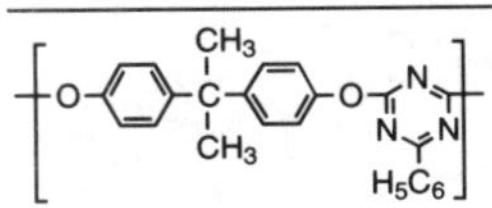			N-1; NaOH, Nitrobenzene; Cl	*J. Polym. Sci.*, A-1, **7**, 3089	1969	
One Oxygen, Two Nitrogen Heteroatoms						
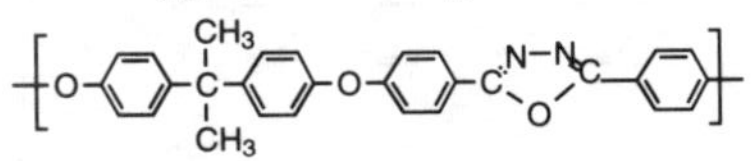	180		N-1; NaOH, DMSO, Sulfolane; Cl,F	*J.Polym.Sci.*, A-1, **5**, 2375	1967	Union Carbide
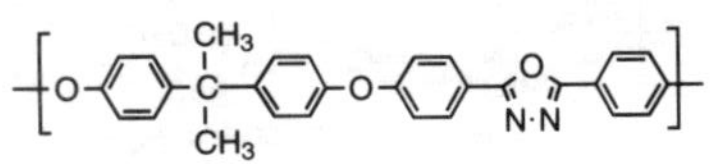	210		N-17; K_2CO_3, NMP, CHP; F	*Polym. Bull.*, **25**, 543	1991	IBM, San Jose
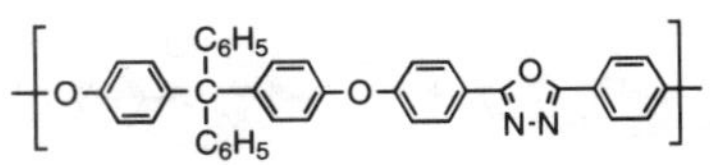	190		N-17; K_2CO_3, NMP, CHP; F	*Polym. Bull.*, **25**, 543	1991	IBM, San Jose
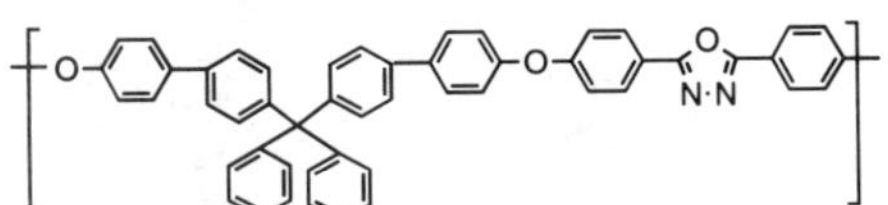	220		N-17; K_2CO_3, NMP, CHP; F	*Polym. Bull.*, **25**, 543	1991	IBM, San Jose
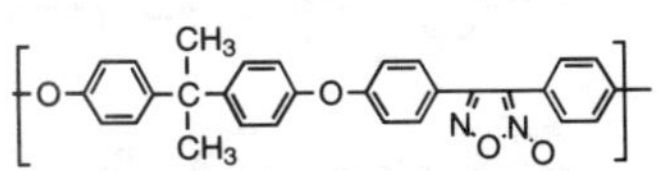			N-1; NaOH, DMSO; Cl	*Eur. Polym. J.*, **22**, 447	1986	
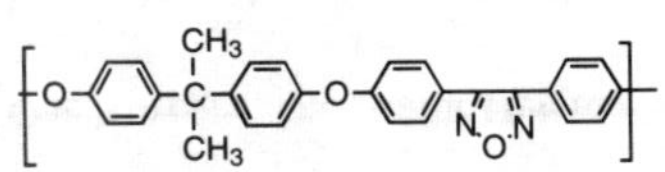			N-1; NaOH, DMSO; Cl	*Eur. Polym. J.*, **22**, 447	1986	

CHAPTER 8

Commercialized Polyarylethers and their Engineering Properties

Introduction

Many individual members of the polyarylether family have been commercialized over the last thirty years. In fact, it is probable that more specific structural entities have been commercialized from this family than from any other polymer class. Some materials that were commercialized early are no longer available, or at least not from the original developers. However, this chapter presents property data for all of the commercialized polymers, irrespective of their current status.

The data that are presented were obtained from manufacturers' brochures and other references. The thermoplastic nature of polyarylethers is one of the main reasons for their success in meeting needs in many application areas. The data in this chapter have been obtained for the most part on injection molded parts by standardized test methods. Thus, these data are more meaningful than those obtained on small laboratory film or compression molded samples. This is the basic reason that no attempt was made to tabulate the scattered property data that exist for many of the non-commercialized polymers that are in the Tables (1–22) in the first seven chapters of this book. The existence of data on injection molded specimens also signify that the polymer has been, or is more than a "laboratory curiosity." It has been scaled up and developed to the point where it possesses enough melt stability to withstand melt processing. The high temperatures required to process the polyarylethers mandates the need to rigorously manage the end group composition and purity of these polymers. Incidentally, even though polyarylethers with Tg's much greater than 275°C can be prepared, their potential is intrinsically limited because of an inability to be melt fabricated. Alternate solution based fabrication methods could be used for these polymers in coatings, fiber and film applications.

In the tables that follow, property data for the neat polymer, a glass fiber reinforced product, and a mineral filled product are presented, when available. For many of these polymers, a much wider variety of products has been formulated and are available for specific end use applications. Of course, many colors are also available for these products. The manufacturers should be contacted for their recommendations and expertise to select a product for an application. The property data in this Chapter serve as an introduction to the range of properties that are available. Additionally, scientists will find the data useful in their ongoing structure vs. property studies.

Properties for the products listed in the following tabulation are presented in this chapter.

POLYARYLETHERSULFONES

Polyethersulfone

VICTREX™ PES—ICI
ULTRASON E™—BASF

Bisphenol A polysulfone

UDEL™—AMOCO
ULTRASON S™—BASF
MINDEL™—AMOCO

Biphenol polysulfone

RADEL R™—AMOCO

Bisphenol S/ Hydroquinone polysulfone

RADEL A™—AMOCO

Biphenylene/Diphenyl Oxide Sulfone Copolymer

ASTREL 360™—3M
PES 720P™—ICI

Biphenyl-based Polyethersulfone

VICTREX™ HTA—ICI

POLYARYLETHERKETONES

Poly(etherketone)-PEK

STILAN™—RAYCHEM
VICTREX™ PEK—ICI

Poly(etheretherketone)-PEEK

VICTREX™ PEEK—ICI
KADEL™—AMOCO

Poly(etheretherketoneketone)—PEEKK

HOSTATEC™—HOECHST

Poly(etherketone)(Etherketoneketone)-PEKEKK

ULTRAPEK™—BASF

Poly(etherketoneketone)-PEKK

DECLAR™—DUPONT

POLYARYLETHERIMIDES

Polyetherimide

ULTEM™ 1000—GE
ULTEM™ 6000—GE

OTHER THERMOPLASTIC POLYIMIDES

Polymide 2080—DOW
XPI™ 182—AMERICAN CYANAMID
XU 218—CIBA-GEIGY
LARC-TPI—MITSUI TOATSU
PISO—NASA/LANGLEY

POLYARYLETHERS

Poly(2,6-Dimethylphenylene oxide)-PPO
PPO™—GE
NORYL™—GE

Poly(Arylene Ether Nitrile)-PEN
PEN™—IDEMITSU

POLYETHERSULFONE

Structure

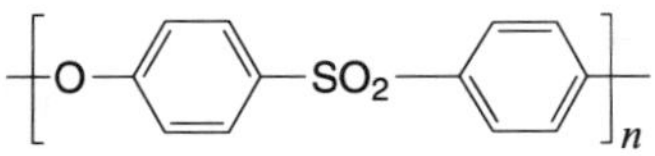

Suppliers/Selected Products

	Neat Products	Modified Products
BASF Corporation Plastics Materials 100 Cherry Hill Road Parsippany, NJ 07054, USA	Ultrason™ E3010	-Ultrason™ E3010G6-30% Glass Fiber. -Ultrason KR4101-30% Mineral Filled
ICI Advanced Materials Business (1) Group Wilmington, DE 19897, USA	Victrex™ PES 4800G	-Victrex™ PES 4101GL30-30% Glass Fiber.

(1) ICI discontinued its PES business in 1991 and may license the technology to Sumitomo Chemical and Mitsui Toatsu. Chemicalweek, July 24, 1991 and December 18/25, 1991.

Key Properties

Excellent Stability to Heat Distortion and Aging—Good Electrical Properties—Transparent—Low Flammability

Typical Applications

Transportation—Medical—Appliances—Mechanical Parts—Electrical/Electronics—Automotive—Hot Fluids—Lighting—Membranes—Coatings

POLYETHERSULFONE

Typical Physical Properties	ASTM Test Method (Other)	English Units (Standard Internat.)	Ultrason™ E3010 (Neat)	Ultrason™ E1010G6 (30% Glass Fiber)	Ultrason™ KR4101 (30% Mineral)	Victrex™ PES 4800G (Neat)	Victrex™ PES-4101GL30 (30% Glass Fiber)
General							
Specific Gravity	D 792	g/cm^3	1.37	1.6	1.62	1.37	
Mold Shrinkage	D 955	in/in (mm/mm)				0.006	
Water Absorption (Immersion at 73°F):	D 570						
24 Hours		%				0.80	
Equilibrium		%	2.1	1.5	1.5	2.3	
FDA						Yes	
Melt Flow Rate:	D 1238 (ISO 1133)						
300°C		g/10 min (cc/10min)	(3)	(6)	(6)		
Optical							
Transmittance	D 1003	%	70–85				
Refractive Index	D 542		1.65	Opaque		1.65	
Mechanical							
Tensile Stress at Yield	D 638	psi (MPa)	13000	22000		12000(83)	
Tensile Stress at Break	D 638	psi (MPa)			13000		25200(140)
Tensile Elongation at Yield	D 638	%	6.7				
Tensile Elongation at Break	D 638	%	15–40	2.1	4.1	40–80	
Tensile Modulus	D 638	psi (MPa)	410000	1550000	680000	345000	
Flexural Strength	D 790	psi (MPa)	18500	28500	21000	18650(129)	
Flexural Modulus (73°F)	D 790	psi (MPa)	370000	1300000	700000	380000 (2600)	1116700 (7700)
Compressive Strength	D 695	psi (MPa)					
Compressive Modulus	D 695	psi (MPa)					
Impact Strength:							

Typical Physical Properties	ASTM Test Method (Other)	English Units (Standard Internat.)	Ultrason™ E3010 (Neat)	Ultrason™ E1010G6 (30% Glass Fiber)	Ultrason™ KR4101 (30% Mineral)	Victrex™ PES 4800G (Neat)	Victrex™ PES-4101GL30 (30% Glass Fiber)
Notched Izod 0.125 in. (3.2mm) Thickness, 73°F	D 256	ft-lb/in (J/m)	1.5	1.7	0.4	1.6(83)	1.7(90)
Unnotched, 0.125 in. Thickness, 73°F	D 4812	ft-lb/in (J/m)	No break	8.3	7.9	No Break	
Tensile Impact Strength	D 1822	ft-lb/in^2 (KJ/m^2)				(337)	
Hardness, Rockwell Shore/Barcol	D 785 D2240/ D2583	M Scale	85	97	84	88	
Thermal							
Melting Temperature							
T_m (crystalline)							
T_g (amorphous)		°F (°C)	428(220)			433(223)	
UL Temp. Rating		°F (°C)				356(180)	
Heat Deflection Temp.	D 648						
264 psi Load(1.84 MPa)		°F (°C)	383(195)	419(215)	403(206)	397(203)	
66 psi Load		°F (°C)	410(210)	430(221)	424(218)		
Coefficient of Linear Thermal Expansion	D 696	in/in-°F (mm/mm-°C)	$3.1x10^{-5}$	1.2×10^{-5}	1.7×10^{-5}	5.5×10^{-5}	
Rate A		°F (°C)					
Rate B		°F (°C)					
Flammability							
Oxygen Index	D 2863	%	38	46	44	38	
Flammability	D-635					Non-burning	
UL94 Flame Class: 0.062 in. (1.57 mm)	(UL 94)	Rating				V-0 (0.02in)	

Thickness						
0.125 in. (3.18 mm)		Rating	V-0	V-0	V-0	V-0
Thickness						
Electrical						
Volume Resistivity (Tinfoil Electrodes)	D 257	(ohm-cm)	$>10^{16}$	$>10^{16}$	$>10^{16}$	2×10^{16}
Surface Resistivity	D 257	ohm	$>10^{14}$	$>10^{14}$	$>10^{14}$	
Dielectric Strength (Short Time)	D 149	V/mil (Kv/mm)				(15.2)
Dielectric Constant	D 150					
60 Hz						3.71
10^3 Hz						3.69
1 MHz			3.5	4.1	4.0	3.71
Dissipation Factor (Tinfoil Electrodes):		D 150				
60 Hz						0.0007
10^3 Hz						0.0022
1 Mhz			0.011	0.01	0.01	0.0109
Arc Resistance (Tungsten)	D 495					

BISPHENOL A POLYSULFONE

Structure

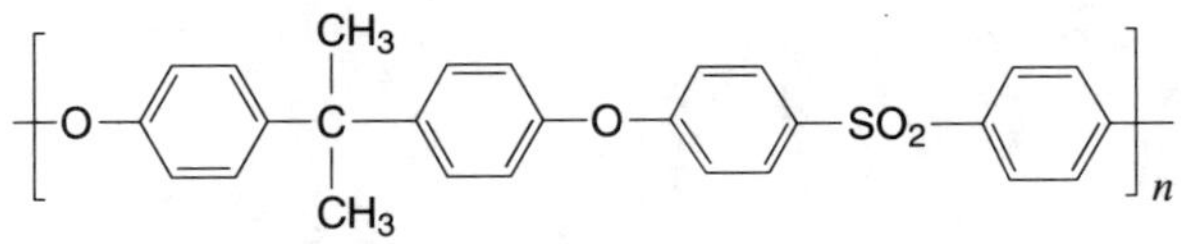

Suppliers/Selected Products

	Neat Products	Modified Products
Amoco Performance Products, Inc. 4500 McGinnis Ferry Road Alpharetta, GA 30202-3944	Udel™ P-1700	—Udel™ GF-130-30% Glass Fiber. —Mindel A-670-ABS modified PSF. —Mindel B-322-PSF alloy with crystalline polymers and glass fibers —Mindel M-800-Mineral filled PSF. —Mindel S-1000-Alloy of PSF(40) and Polycarbonate(60)(1).
BASF Corporation Plastics Materials 100 Cherry Hill Road Parsippany, NJ 07054, USA	Ultrason™ S2010	—Ultrason S2010G6-30% Glass Fiber

(1) Sanchez, P., Remiro, P. M., and Nazabal, *J., Polym. Eng. Sci.*, **32**, 861 (1992).

Key Properties of Udel™ and Ultrason™ S

High Tensile Strength—High Modulus—Low Creep—Excellent Electricals —Good Chemical Resistance—Hydrolytic stability—FDA, USP, 3-A and NSF—UL94 V-0—Transparent

Typical Applications of Udel™ and Ultrason™ S

Medical Devices—Food Contact Uses—Microwave Cookware—Chemical Process Equipment—Electronic Devices—Automotive/Electrical

BISPHENOL A POLYSULFONE

Typical Physical Properties	ASTM Test Method (Other)	English Units (Standard Internat.)	UDEL™ P-1700 (Neat)	UDEL™ GF-130 (30% Glass Fiber)	Ultrason S2010 (Neat)	Ultrason S2010G6 (30% Glass Fiber)
General						
Specific Gravity	D 792 (ISO1183)	g/cm^3	1.24	1.49	1.24	1.49
Density	D 792	lb/in^3	0.0447			
Mold Shrinkage	D 955	in/in (mm/mm)	0.007	0.002		
Water Absorption (Immersion at 73°F):	D 570					
24 Hours		%	0.3	0.49		
Equilibrium		%	0.62		0.8	0.5
FDA			Yes	—		
Melt Flow	D 1238 (ISO 1133)	dgm/min44 psi (cc/min)	6.5(650°F) (343.3°C)	7–8(650°F) (343.3°C)	(50) (300°C)	(20) (300°C)
Optical						
Transmittance	D 1003	%			70–85	
Haze	D 1003	%	5	Opaque		
Refractive Index	D 542				1.63	Opaque
Mechanical						
Tensile Stress at Yield	D 638	psi (MPa)	10200(70)	15600(108)	11500	18000
Tensile Stress at Break	D 638	psi (MPa)				
Tensile Elongation at Yield	D 638	%	5-6		5.7	
Tensile Elongation at Break	D 638	%	50-100		40-70	1.8
Tensile Modulus	D 638	psi (MPa)	360000 (2500)	1070000 (7380)	390000	1450000
Flexural Strength	D 790	psi (MPa)	15400 (106)	22400(155)	17500	23500
Flexural Modulus						
73°F	D 790	psi (MPa)	390000 (2700)	1100000 (7590)	370000	1250000
300°F	D 790	psi (MPa)	300000			

BISPHENOL A POLYSULFONE

Typical Physical Properties	ASTM Test Method (Other)	English Units (Standard Internat.)	UDEL™ P-1700 (Neat)	UDEL™ GF-130 (30% Glass Fiber)	Ultrason S2010 (Neat)	Ultrason S2010G6 (30% Glass Fiber)
Compressive Strength	D 695	psi (MPa)	13900 (96)			
Compressive Modulus	D 695	psi (MPa)	374000 (2579)			
Shear Strength, Ult.	D732	psi (MPa)	6000 (41)			
Impact Strength:						
Notched Izod 0.125 in. (3.2mm) Thickness, 73°F	D 256	ft-lb/in (J/m)	1.3(69)	1.4(75)	1.1	1.5
Unnotched, 0.125 in. Thickness, 73°F	D 4812	ft-lb/in (J/m)	No Break		No Break	6.0
Tensile Impact Strength	D 1822	ft-lb/in^2 (KJ/m^2)	200(421)	52(109)		
Hardness, Rockwell	D 785	M Scale R Scale	M69 R120		69	87
Thermal						
Melting Temperature, °C						
T_m (crystalline)						
T_g (amorphous)			374 (190)			
Heat Deflection Temp.	D 648					
264 psi Load(1.84 MPa)		°F (°C)	345 (174)	358 (181)	336(169)	365(185)
66 psi Load		°F (°C)	358 (181)		354(179)	372(189)
Coefficient of Linear Thermal Expansion	D 696	in/in-°F (mm/mm-°C)	3.1×10^{-5} (5.6×10^{-5})		3.1×10^{-5}	1.1×10^{-5}
Thermal Conductivity	C 177	Btu-in/ft^2-hr-°F (W/m-°C)	1.8 (0.26)			
Vicat Softening Temperature	D 1525	°F (°C)	371 (188)			
Flammability						
Oxygen Index	D 2863	%	30		30	40

UL94 Flame Class:	(UL 94)					
0.062 in. (1.57 mm) Thickness		Rating	HB[60 mil] (1.47mm)	V1[62 mil] (1.57mm)		
0.125 in. (3.18 mm) Thickness		Rating	V-2		V-2	V-0
0.250 in. (6.35 mm) Thickness		Rating	V-0			
Electrical						
Volume Resistivity (Tinfoil Electrodes)	D 257	(ohm-cm)	5×10^{16}		$>10^{16}$	$>10^{16}$
Surface Resistivity	D 257	ohm	3×10^{16}		$>10^{14}$	$>10^{14}$
Dielectric Strength (Short Time under Oil at 73°F)	D 149	V/mil (Kv/mm)	425(17)	475(19)		
Dielectric Constant (Tinfoil Electrodes):	D 150					
60 Hz			3.15			
10^3 Hz			3.14	3.7		
1 MHz			3.10		3.2	3.7
Dissipation Factor (Tinfoil Electrodes):	D 150					
60 Hz			0.0011			
10^3 Hz			0.0013	0.004		
1 Mhz			0.0050		0.0055	0.006
Arc Resistance, Tungsten	D 495	sec	60			

Key Properties of Mindel™ A Modified Polysulfone

Hot Water Resistance—Dimensional Stability—FDA and NSF Recognition —High Heat Deflection—Platable on Conventional ABS Lines

Typical Applications of Mindel™ A Modified Polysulfone

Food Contact Applications—Automotive Parts(painted and plated)—Plumbing Fixtures

Key Properties of Mindel™ B Modified Polysulfone

High Heat Distortion Temperature—Toughness and Rigidity—Chemical Resistance—Warp Resistance

Typical Applications of Mindel™ B Modified Polysulfones

Food Contact —Electrical/Electronic—Connectors—Snap-fit Assemblies

Key Properties of Mindel™ M Modified Polysulfone

High Heat Distortion Temperature—Hydrolytic Stability—Qualified for FDA-Approved Devices—Toughness and Processability—Stress Crack Resistance

Typical Applications of Mindel™ M Modified Polysulfones

Food Handling Equipment—Electrical Devices—Medical Equipment—Miscellaneous Devices

Key Properties of Mindel™ S Polysulfone Alloys

High Heat Distortion Temperature—Hydrolytic Stability—Qualified for FDA-Approved Devices—Toughness and Processability—Stress Crack Resistance

Typical Applications of Mindel™ S Polysulfone Alloys

Food Processing—Medical Devices—Biological Contact— Institutional Feeding Systems

MODIFED BISPHENOL A POLYSULFONE

Typical Physical Properties	ASTM Test Method (Other)	English Units (Standard Internat.)	MINDEL™ A-670	MINDEL™ B-322	MINDEL™ M-800	MINDEL™ S-1000
General						
Specific Gravity	D 792	g/cm^3	1.13	1.47	1.61	1.23
Mold Shrinkage	D 955	in/in (cm/cm)	0.0066	0.003	0.005	0.0066
Water Absorption (Immersion at 73°F):	D 570					
24 Hours		%	0.25	0.14	0.46	0.20
Equilibrium		%			0.7	0.51
FDA			Yes (to 212°F)	–	–	Yes
Melt Flow	D 1238	dgm/min44 psi (0.3MPa)	12 (550°F) (288°C)	4–9 (527°F) (275°C)	8.5 (650°F) (343.3°C)	13 (608°F) (320°C)
Optical						
Haze	D 1003	%	Opaque	Opaque	Opaque	Opaque
Mechanical						
Tensile Stress at Yield	D 638	psi (MPa)	7200(49.6)	15000(103)	9500(65.5)	9500(65.5)
Tensile Stress at Break	D 638	psi (MPa)		2.5		8300 (57)
Tensile Elongation at Yield	D 638	%			5	5–6
Tensile Elongation at Break	D 638	%				100
Tensile Modulus	D 638	psi (MPa)	308000 (2123)	–	650000 (4480)	350000 (2412)
Flexural Strength	D 790	psi (MPa)	12000(83)	23000(159)	14300(98.6)	14100(97.8)
Flexural Modulus (73°F)			316000 (2180)	1000000 (6200)	750000 (5172)	381000 (2620)
Impact Strength:						
Notched Izod 0.125 in. (3.2mm) Thickness, 73°F	D 256	ft-lb/in (J/m)	7.0 (374)	1.0 (53)	.65 (32)	1.6 (85)
Tensile Impact Strength	D 1822	ft-lb/in^2 (KJ/m^2)	90 (189)	40 (84)	–	130 (273)

MODIFED BISPHENOL A POLYSULFONE

Typical Physical Properties	ASTM Test Method (Other)	English Units (Standard Internat.)	MINDEL™ A-670	MINDEL™ B-322	MINDEL™ M-800	MINDEL™ S-1000
Hardness						
Rockwell	D 785	M Scale		M80	M70	
Thermal						
Melting Temperature, ° C						
T_m (crystalline)						
T_g (amorphous)						
Heat Deflection Temp.	D 648					
264 psi Load(1.84 MPa)		°F (°C)	300 (149)	320 (160)	354 (179)	300 (149)
Coefficient of Linear Thermal Expansion	D 696	in/in-°F (mm/mm-°C)				$5.7x10^{-5}$ ($10.3x10^{-5}$ cm/cm-°C)
Flammability						
Oxygen Index	D 2863	%		35		
UL94 Flame Class:	(UL 94)	Rating	–	V-0 (31 mil) (0.79 mm)	V-0 (32 mil) (0.81 mm)	V-0 (133 mil)(3.38 mm)
Electrical						
Volume Resistivity (Tinfoil Electrodes)	D 257	(ohm-cm)		1.0×10^{16}	1.52×10^{16}	
Dielectric Strength (Short Time under Oil at 73°F)	D 149	V/mil (Kv/mm)	427 (17)	500 (20)	450 (17.7)	500 (20)
Dielectric Constant (Tinfoil Electrodes):	D 150					
60 Hz				3.7	3.7	3.12
10^3 Hz			3.28	3.7	3.8	3.1
1 MHz				3.7	3.7	3.5
Dissipation Factor (Tinfoil Electrodes):	D 150					
60 Hz				0.002	0.003	0.0027
10^3 Hz			0.0055	0.003	0.003	0.001
1 Mhz				0.009	0.006	0.007
Arc Resistance:	D 495					
Tungsten Electrodes		sec		125	125	130

BIPHENOL POLYSULFONE

Structure

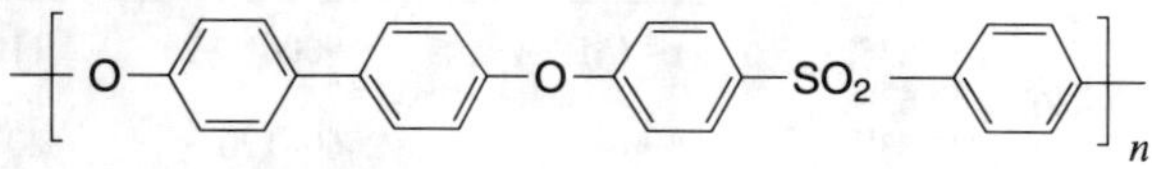

Suppliers/Selected Products

	Neat Products	Modified Products
Amoco Performance Products, Inc. 4500 McGinnis Ferry Road Alpharetta, GA 30202-3944	Radel™ R-5000 Polyphenylsulfone	Radel™ R-7000A

Key Properties

High Impact Resistance—Meets OSU Test Criteria—Excellent Chemical Resistance

Typical Applications

Aircraft Cabin Interior Components—Sterilizable Medical Devices—Electrical/Electronic—Chemical Process Equipment—Automotive

BIPHENOL POLYSULFONE

Typical Physical Properties	ASTM Test Method (Other)	English Units (Standard Internat.)	RADEL™ R-5000	RADEL™ R 7000A
General				
Specific Gravity	D 792	g/cm^3	1.29	1.37
Mold Shrinkage	D 955	in/in (mm/mm)	0.006	0.006
Water Absorption (Immersion at 73°F):	D 570			
24 Hours				0.22
Equilibrium		%	1.1	
FDA				
Melt Flow	D 1238	dgm/min44 psi (0.3MPa)	5–9 (400°C)	
Melt Flow Rate,380°C/44psi	D 1238	g/10 min		10
Optical				
Haze	D 1003	%	Transparent	
Refractive Index			1.6724	
Mechanical				
Tensile Stress at Yield	D 638	psi (MPa)	10100 (70)	

BIPHENOL POLYSULFONE

Typical Physical Properties	ASTM Test Method (Other)	English Units (Standard Internat.)	RADEL™ R-5000	RADEL™ R 7000A
Tensile Stress at Break	D 638	psi(MPa)	9000	11000(76)
Tensile Elongation at Yield	D 638	%	7.2	
Tensile Elongation at Break	D 638	%	60–120	20
Tensile Modulus	D 638	psi (MPa)	340000 (2300)	405000 (2790)
Flexural Strength	D 790	psi (MPa)	13200 (91)	16000 (110)
Flexural Modulus, 73°F	D 790	psi (MPa)	350000 (2400)	400000 (2760)
Compressive Strength	D 695	psi (MPa)	14350(99)	
Compressive Modulus	D 695	psi (MPa)	251000 (1729)	
Shear Strength	D 732	psi (MPa)	9100(62)	
Impact Strength:				
Notched Izod 0.125 in. (3.2mm) Thickness, 73°F	D 256	ft-lb/in (J/m)	13(694)	1.9(101)
Unnotched, 0.125 in. Thickness, 73°F	D 4812	ft-lb/in (J/m)	No Break	No Break
Tensile Impact Strength	D 1822	ft-lb/in^2 (KJ/m^2)	190(400)	
Hardness				
Rockwell	D 785	M Scale R Scale		
Falling Dart Impact (10 lb. Wt., 1 in. Dia. Dart)	(Miles)	ft-lb (J)	(>135) (1/8"specimen)	
Thermal				
Melting Temperature, °C				
T_m (crystalline)				
T_g (amorphous)			220	
Heat Deflection Temp.	D 648			
264 psi Load(1.84 MPa)		°F (°C)	405 (207)	360 (182)
66 psi Load		°F (°C)	417 (214)	380 (193)
Coefficient of Linear Thermal Expansion	D 696	in/in-°F (mm/mm-°C)	3.1×10^{-5} $(5.6 \times 10^{-5}$	
Thermal Conductivity	C 177	Btu-in/ft^2-hr°F (W/m-°C)	2.42(0.35)	
Specific Heat	D 2766	Btu/ lb-°F (J/g°C)	0.28 (1.17)	
Flammability				
Oxygen Index	D 2863	%	38	
UL94 Flame Class:	(UL 94)			
0.062 in. (1.57 mm) Thickness		Rating	V-0 (0.030 in)	
Electrical				
Volume Resistivity (Tinfoil Electrodes)	D 257	(ohm-cm)	$>10^{15}$	

BIPHENOL POLYSULFONE

Typical Physical Properties	ASTM Test Method (Other)	English Units (Standard Internat.)	RADEL™ R-5000	RADEL™ R 7000A
Surface Resistivity	D 257	ohm		
Dielectric Strength (Short Time under Oil at 73°F)	D 149	V/mil (Kv/mm)	360(15)	
Dielectric Constant (Tinfoil Electrodes):	D 150			
60 Hz			3.44	
10^3 Hz			3.45	
1 MHz				
Dissipation Factor (Tinfoil Electrodes):	D 150			
60 Hz			0.00058	
10^3 Hz			-	
1 Mhz			0.0076	
Arc Resistance:	D 495			
Tungsten Electrodes		sec	41	

BISPHENOL S/HYDROQUINONE POLYSULFONE

Structure

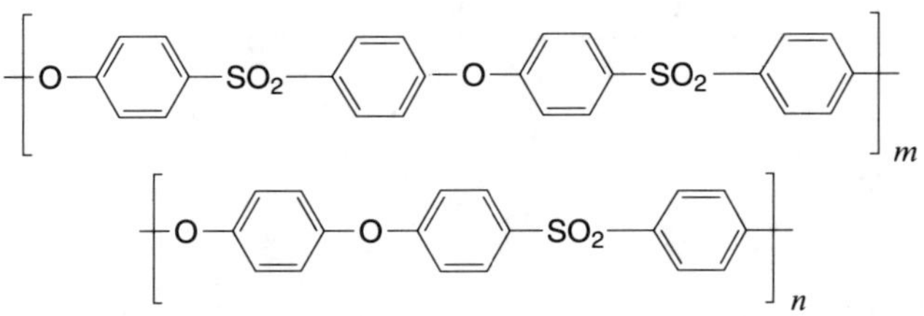

Mole Ratio m/n=4.7(1)

Suppliers/Selected Products

	Neat Products	Modified Products
Amoco Performance Products, Inc. 4500 McGinnis Ferry Road Alpharetta, GA 30202-3944	Radel™ A-200 Polyethersulfone	Radel™ A-230-30% Glass Fiber.

(1) Ghosal, K., Chern, R. T., and Freeman, B. D., J. Polym. Sci.: B: Polym. Physics, 31, 891 (1993).

Key Properties

Toughness at High Temperatures—Hydrolytic Stability—Good Chemical Resistance

Typical Applications

Lighting Fixtures—Electrical/Electronic—Medical Devices—Chemical Process Equipment

BISPHENOL S/HYDROQUINONE POLYSULFONE

Typical Physical Properties	ASTM Test Method (Other)	English Units (Standard Internat.)	RADEL™ A-200	RADEL™ AG-230
General				
Specific Gravity	D 792	g/cm^3	1.37	1.58
Mold Shrinkage	D 955	in/in (mm/mm)	0.006	0.003
Water Absorption (Immersion at 73°F):	D 570			
24 Hours		%	0.4	
Equilibrium		%	1.85	
FDA			Yes	No
Melt Flow	D 1238	dgm/min44 psi (0.3MPa)	20(380°C)	10
Optical				
Haze	D 1003	%		Opaque
Refractive Index	D 542		1.6510	
Mechanical				
Tensile Stress at Yield	D 638	psi(MPa)	12000 (83)	18300 (126)
Tensile Stress at Break	D 638	psi(MPa)		
Tensile Elongation at Yield	D 638	%	6.5	1.9
Tensile Elongation at Break	D 638	%	50–100	1.9
Tensile Modulus	D 638	psi(MPa)	385000 (2600)	1250000 (8600)
Flexural Strength	D 790	psi(MPa)	16100 (111)	26000 (179)
Flexural Modulus- 73°F	D 790	psi(MPa)	420000 (2900)	1170000 (8100)
Compressive Strength	D 695	psi(MPa)	14500(100)	25600(177)
Compressive Modulus	D 695	psi(MPa)	388000 (2680)	1119000 (7720)
Shear Strength	D732	psi(MPa)	7270 (50)	9490(65)
Taber Abrasion, 1kg.	D1044	mg/1000 cy	19	21
Impact Strength:				
Notched Izod 0.125 in. (3.2mm) Thickness, 73°F	D 256	ft-lb/in (J/m)	1.6 (85)	1.4 (75)
Unnotched, 0.125 in. Thickness, 73°F	D 4812	ft-lb/in (J/m)	No Break	12
Tensile Impact Strength	D 1822	ft-lb/in^2 (KJ/m^2)	160(336)	34 (71)
Hardness				
Rockwell	D 785	M Scale	85	
		R Scale	127	121
Thermal				
T_g (amorphous)			428(220)	
Heat Deflection Temp.	D 648			
264 psi Load(1.84 MPa)		°F (°C)	399(204)	420(216)
66 psi Load		°F (°C)	417(214)	428(220)
Coefficient of Linear Thermal Expansion	D 696	in/in-°F (mm/mm-°C)	2.7×10^{-5} (4.9×10^{-5})	1.7×10^{-5} (3.1×10^{-5})

BISPHENOL S/HYDROQUINONE POLYSULFONE

Typical Physical Properties	ASTM Test Method (Other)	English Units (Standard Internat.)	RADEL™ A-200	RADEL™ AG-230
Thermal Conductivity	C 177	Btu-in/ft^2-hr °F (W/m-°C)	1.13(0.16)	1.35(0.20)
Relative Temperature Index: Index:	(UL746B)			
Electrical		°C	180	
Mechanical with Impact		°C	180	
Mechanical without Impact		°C	180	
Vicat Softening Temperature: Rate B	D 1525	°F (°C)	418(215)	424(218)
Flammability				
Oxygen Index	D 2863	%	39.3	40.3
Self Ignition Temperature	D1929	°F (°C)	936(502)	
UL94 Flame Class:	(UL 94)			
0.062 in. (1.57 mm) Thickness		Rating	V-0 @.031" (0.8 mm)	V-0 @.031" (0.8 mm)
Electrical				
Volume Resistivity (Tinfoil Electrodes)	D 257	(ohm-cm)	1.7×10^{15}	$>10^{16}$
Dielectric Strength (Short Time under Oil at 73°F)	D 149	V/mil (Kv/mm)	380(15)	440(17)
Dielectric Constant (Tinfoil Electrodes):	D 150			
60 Hz			3.51	4.11
10^3 Hz			3.50	4.13
1 MHz			3.54	4.17
Dissipation Factor (Tinfoil Electrodes):	D 150			
60 Hz			0.0017	0.0019
10^3 Hz			0.0022	0.0018
1 Mhz			0.0056	0.0094
Arc Resistance:	D 495			
Tungsten Electrodes		sec	81	

BIPHENYLENE/DIPHENYL OXIDE SULFONE COPOLYMER

Structure

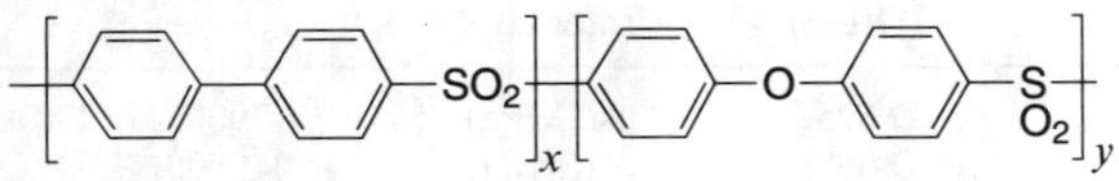

Astrel™ 360- x:y = 60-70: 30-40

Victrex 720P-x«y

Suppliers/Selected Products (1)

	Neat Products
3M Company Chemical Division 3M Center St. Paul, Minnesota 55101, USA	Astrel™ 360
ICI Advanced Materials Business Group Wilmington, DE 19897, USA	Victrex™ 720P

(1) These products are no longer commercially available.

Key Properties

Long Term Service at High Temperatures—Excellent Dielectric Properties—Good Chemical Resistance

Suggested Applications

Lighting Fixtures—Electrical/Electronic—Chemical Process Equipment

BIPHENYLENE/DIPHENYL OXIDE SULFONE COPOLYMER

Typical Physical Properties	ASTM Test Method (Other)	English Units (Standard Internat.)	Astrel™ 360	Victrex™ PES 720P
General				
Specific Gravity	D 792	g/cm^3	1.36	
Mold Shrinkage	D 955	in/in (mm/mm)	0.008	
Water Absorption (Immersion, 24 Hrs.,73°F)	D 570	%	1.8	
Mechanical				
Tensile Stress at Yield	D 638	psi (MPa)		
Tensile Stress at Break	D 638	psi (MPa)	13000	
Tensile Elongation at Yield	D 638	%	13	
Tensile Modulus	D 638	psi (MPa)	370000	
Flexural Strength	D 790	psi (MPa)	17200	
Flexural Modulus, 73°F	D 790	psi (MPa)	395000	

BIPHENYLENE/DIPHENYL OXIDE SULFONE COPOLYMER

Typical Physical Properties	ASTM Test Method (Other)	English Units (Standard Internat.)	Astrel™ 360	Victrex™ PES 720P
Compressive Strength	D 695	psi (MPa)	17900	
Compressive Modulus	D 695	psi (MPa)	340000	
Impact Strength:				
Notched Izod 0.125 in. (3.2mm) Thickness, 73°F	D 256	ft-lb/in (J/m)	5	
Unnotched, 0.125 in. Thickness, 73°F	D 4812	ft-lb/in (J/m)	40	
Hardness: Rockwell	D 785	M Scale	M110	
Thermal				
T_g (amorphous)		°F (°C)	550 (288)	483(250)
Heat Deflection Temp.	D 648			
264 psi Load(1.84 MPa)		°F (°C)	525	
Coefficient of Linear Thermal Expansion	D 696	in/in-°F (mm/mm-°C)	2.6×10^{-5}	
Thermal Conductivity	C 177	Btu-in/ft^2-hr-°F (W/m-°C)	1.32	
Electrical				
Volume Resistivity (Tinfoil Electrodes)	D 257	(ohm-cm)	3.2×10^{16}	
Surface Resistivity	D 257	ohm	6.2×10^{15}	
Dielectric Strength (Short Time under Oil at 73°F)	D 149	V/mil (Kv/mm)	350	
Dielectric Constant (Tinfoil Electrodes): 60Hz	D 150		3.94	
Dissipation Factor(Tin)60hz	D 150		0.003	
Arc Resistance: Tungsten	D 495	sec	67	

BIPHENYL-BASED POLYETHERSULFONE

Structure (1)

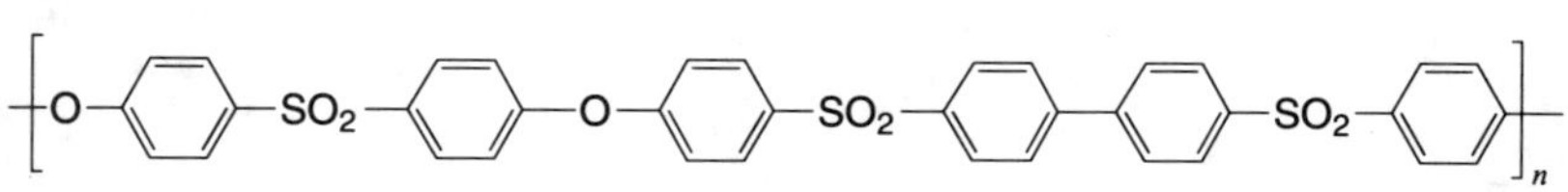

Supplier (2)

	Neat Products	Modified Products
ICI Advanced Materials Business Group Wilmington, DE 19897, USA	Victrex™ HTA	Victrex™ HTA-30% Glass Fiber

(1) Parker, D., et. al., *Ullmann's Encyclopedia of Industrial Chemistry*, A21, **449** (1992). Smith, C. P. (ICI), *Preprints of Ryder Conference on Performance Plastics*, **2/87**, Atlanta, GA.

(2) This product is no longer commercially available.

BIPHENYL-BASED POLYETHERSULFONE

Typical Physical Properties	ASTM Test Method (Other)	English Units (Standard Internat.)	Victrex HTA	Victrex HTA 30% Glass Fiber
General				
Specific Gravity	D 792 (ISO1183)	g/cm^3	1.36	1.58
Water Absorption (Immersion at 73°F):	D 570(ISO 62)			
Equilibrium		%	2.75	
Mechanical				
Tensile Stress at Yield	D 638 (ISO527)	psi (MPa)		
Tensile Stress at Break	D 638 (ISO527)	psi (MPa)	9900(68)	19400(134)
Tensile Elongation at Yield	D 638 (ISO527)	%	8	
Tensile Elongation at Break	D 638 (ISO527)	%	19	4
Tensile Modulus	D 638 (ISO527)	psi (MPa)	333500 (2300)	
Flexural Modulus, 73°F	D 790 (ISO178)	psi (MPa)	360000 (2500)	1233000 (8500)
Impact Strength:				
Notched Izod 0.125 in. (3.2mm) Thickness, 73°F	D 256	ft-lb/in (J/m)	2.2(12.3 KJ/mm^2)	(5.8 KJ/mm^2)
Unnotched, 0.125 in. Thickness, 73°F	D 4812	ft-lb/in (J/m)	60	
Hardness				
Rockwell	D 785	M Scale	M101	
Thermal				
T_g (amorphous)		°F (°C)	500(260)	500(260)
Max. Cont. Service Temp.		°F (°C)	392(200)	392(200)
Heat Deflection Temp.	D 648 (ISO75)			
264 psi Load(1.84 MPa)		°F (°C)	454(234)	486(252)
66 psi Load		°F (°C)		
Coefficient of Linear Thermal Expansion	D 696	in/in-°F (mm/mm-°C)	(5.4×10^{-5})	(2.3×10^{-5})
Flammability				
Oxygen Index	D 2863	%	38	
UL94 Flame Class:	(UL 94)			
0.062 in. (1.57 mm) Thickness		Rating		V-0
0.125 in. (3.18 mm) Thickness		Rating	V-0	
Electrical				
Volume Resistivity (Tinfoil Electrodes)	D 257 (ISO93)	ohm-cm	10^{17}	
Dielectric Constant (Tinfoil Electrodes):	D 150 (ISO250)			
10^3 Hz			3.8	
Dissipation Factor (Tinfoil Electrodes):	D 150 (ISO250)			
10^3 Hz			0.004	

POLY(ETHERKETONE)-PEK

Structure

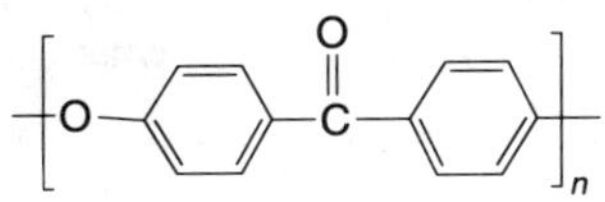

Suppliers/Selected Products (1)

	Neat Products	Modified Products
ICI Advanced Materials Business Group Wilmington, DE 19897, USA	Victrex™ PEK	Victrex™ PEK- 30% Glass Fiber.
Raychem Corporation Materials Division 300 Constitution drive Menlo Park, CA 94025, USA	Stilan™ 1000	

(1) These products are no longer available from these suppliers.

Key Properties of Poly(etherketone)

Superior Temperature Performance—Exceptional Chemical Resistance—Low Toxic Gas and Smoke Emission—Oustanding Radiation Resistance—Low Flammability—Good Electrical Properties—Toughness, Strength—Low Moisture Absorption—Excellent Hydrolysis Resistance.

Suggested Applications of Poly(etherketone)

Round Wire and Extruded Flat Cable Insulation—Bearings—Connectors—Parts for Service in Corrosive Environments.

Typical Physical Properties	ASTM Test Method (Other)	English Units (Standard Internat.)	VICTREX PEK	VICTREX PEK 30% Glass Fiber	Stilan 1000
General					
Specific Gravity (ISO1183)	D 792 g/cm^3		1.3	1.53	1.30
Water Absorption (Immersion at 73°F):	D 570(ISO 62)				
24 Hours		%			0.3
Equilibrium		%			0.7
Mechanical					
Tensile Stress at Yield	D 638 (ISO527)	psi (MPa)			13–17000
Tensile Stress at Break	D 638 (ISO527)	psi (MPa)	16000(110)	24650(170)	11–15000
Tensile Elongation at Yield	D 638 (ISO527)	%			10–11
Tensile Elongation at Break	D 638 (ISO527)	%	5	4	80–120
Tensile Modulus	D 638 (ISO527)	psi (MPa)	580000 (4000)	1522000 (10500)	
Flexural Strength	D 790 (ISO178)	psi (MPa)			11–16000
Flexural Modulus 73°F	D 790 (ISO178)	psi (MPa)	536500 (3700)	1305000 (9000)	240000
Compressive Strength	D 695	psi (MPa)			11–16000
Impact Strength:					
Notched Izod 0.125 in. (3.2mm) Thickness, 73°F	D 256	ft-lb/in (J/m)	1.52 (7 KJ/m^2)	9 (KJ/m^2)	0.9–1.4
Unnotched, 0.125 in. Thickness, 73°F	D 4812	ft-lb/in (J/m)			No break
Hardness					
Rockwell	D 785	M Scale	105	109	M83
		R Scale	126	126	R104
Shore/Barcol	D2240/ D2583				D80-90
Thermal					
Melting Temperature					
T_m (crystalline)		°F (°C)	704(373)	704(373)	692(366)
T_g (amorphous)		°F (°C)	330(162)	330(162)	

POLY(ETHERKETONE)

Typical Physical Properties	ASTM Test Method (Other)	English Units (Standard Internat.)	VICTREX PEK	VICTREX PEK 30% Glass Fiber	Stilan 1000
Max. Cont. Service Temp.		°F (°C)	500(260)	500(260)	
Heat Deflection Temp.	D 648 (ISO75)				
264 psi Load(1.84 MPa)		°F (°C)	367(186)	676(358)	333(167)
66 psi Load		°F (°C)			394(201)
Coefficient of Linear Thermal Expansion	D 696	in/in-°F (mm/mm-°C)	(5.7×10^{-5})	(1.7×10^{-5})	2.3×10^{-5}
Thermal Conductivity	C 177	Btu-in/ft^2-hr-°F (W/m-°C)			1.90
Specific Heat (J/g°C)	D 2766	Btu/lb-°F			0.32
Flammability					
Oxygen Index	D 2863	%	40	46	36
UL94 Flame Class:	(UL 94)				
0.062 in. (1.57 mm) Thickness		Rating	V-0	V-0	
0.125 in. (3.18 mm) Thickness		Rating	V-0	V-0	V-0
Electrical					
Volume Resistivity (Tinfoil Electrodes)	D 257 (ISO93)	ohm-cm	10^{17}	10^{17}	1×10^{16}
Dielectric Strength (Short Time under Oil at 73°F)	D 149 (ISO243)	V/mil (Kv/mm)			660
Dielectric Constant (Tinfoil Electrodes):	D 150 (ISO250)				
60 Hz					3.4
10^3 Hz			3.5	3.9	3.4
1 MHz			3.4	3.9	3.3
Dissipation Factor (Tinfoil Electrodes):	D 150 (ISO250)				
60 Hz					0.002
10^3 Hz			0.002	0.002	0.003
1 Mhz			0.005	0.005	0.005

POLY(ETHERETHERKETONE)-PEEK

Structure

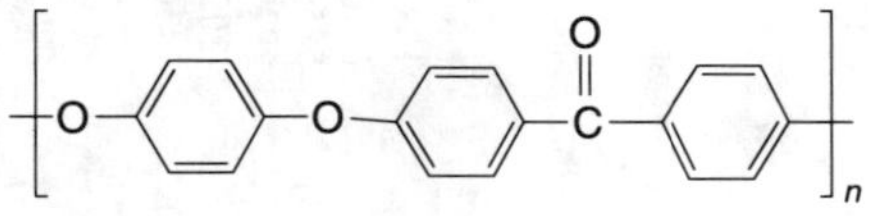

Suppliers/Selected Products

	Neat Products	Modified Products
ICI Advanced Materials Business Group Wilmington, DE 19897, USA	Victrex™ PEEK 450G	Victrex™ PEEK 450GL30-30% Glass Fiber. Victrex™ PEEK 450CA30-30% Carbon Fiber.
Amoco Performance Products, Inc. 4500 McGinnis Ferry Road Alpharetta, GA 30202-3944	Kadel™ E-1000 (1)	Kadel™ E-1130-30% Glass Fiber.

(1) May be a copolymer of the type described in Harris, J. E., et. al., *Macromolecules*, **26**, 2366 (1993).

Key Properties

Excellent Thermal and Oxidative Stability—Superior Chemical Resistance—Strength and Stiffness—Low Smoke and Toxicity—ExcellentWear and Abrasion Resistance—Radiation Resistance

Typical Applications

Chemical Processing—Aviation. Aerospace Composites—Compressor Parts—Bearings and Seals—Fibers and Film—Wire and Cable Coatings—Wear Rings

POLY(ETHERETHERKETONE)

Typical Physical Properties	ASTM Test Method (Other)	English Units (Standard Internat.)	VICTREX™ PEEK 450G	VICTREX™ PEEK 450GL30	VICTREX™ PEEK 450CA30
General					
Specific Gravity	D 792 (ISO1183)	g/cm^3	1.32 (35% Cryst.)	1.49	1.44
Mold Shrinkage	D 955	in/in (mm/mm)	1.1 %	0.5%	0.1–1.4%
Water Absorption (Immersion at 73°F):	D 570(ISO 62)				
24 Hours		%	0.5	0.11	0.06
Equilibrium		%	0.5		
Mechanical					
Tensile Stress at Yield	D 638 (ISO527)	psi (MPa)	13300 (92)		
Tensile Stress at Break	D 638 (ISO527)	psi (MPa)		22800(157)	33800(233)
Tensile Elongation at Yield	D 638 (ISO527)	%	4.9		
Tensile Elongation at Break	D 638 (ISO527)	%	>50	2.0	1.5
Tensile Modulus	D 638 (ISO527)	psi (MPa)	522100 (3600)	1406800 (9700)	1885400 (13000)
Flexural Strength	D 790 (ISO178)	psi (MPa)	24700(170)	33800(233)	51500(355)
Flexural Modulus, 73°F	D 790 (ISO178)	psi (MPa)	594500 (4100)	1495200 (10310)	2929000 (20200)
Shear Strength	D 3846	psi (MPa)	7700(53)	14100(97)	14100(97)
Shear Modulus	-	psi (MPa)	188500 (1300)	348000 (2400)	
Compressive Strength	D 695	psi (MPa)	17100(118)	31200(215)	34800(240)
Impact Strength:					
Notched Izod 0.125 in. (3.2mm) Thickness, 73°F	D 256	ft-lb/in (J/m)	1.55(83)	1.8(96)	1.6(85)
Unnotched, 0.125 in. Thickness, 73°F	D 4812	ft-lb/in (J/m)	No Break	7.7(409)	7.9(421)
Hardness					
Rockwell	D 785	M Scale	99	103	107
		R Scale	126	124	124
Shore/Barcol	D2240/ D2583				

Thermal					
Melting Temperature					
T_m (crystalline)		°F (°C)	633(334)	633(334)	633(334)
T_g (amorphous)		°F (°C)	289(143)	289(143)	289(143)
UL Temp. Rating (estimated)		°F (°C)	480(249)	480(249)	
Heat Deflection Temp.	D 648 (ISO75)				
264 psi Load(1.84 MPa)		°F (°C)	320(160)	600(315)	600(315)
UL Continuous Use Temp.		°F (°C)	482(250)	482(250)	482(250)
Coefficient of Linear Thermal Expansion	D 696	in/in-°F (mm/mm-°C)	(47×10^{-6})	(22×10^{-6})	15×10^{-6}
Thermal Conductivity	C 177	Btu-in/ft^2-hr-°F (W/m-°C)	(1.25)	(0.43)	(0.92)
Specific Heat	D 2766	Btu/ lb-°F (J/g°C)	(0.32 Cal/gm°C)		
Flammability					
Oxygen Index	D 2863	%	35		
UL94 Flame Class:	(UL 94)				
0.062 in. (1.57 mm) Thickness		Rating	V-0	V-0	V-0
0.125 in. (3.18 mm) Thickness		Rating	V-0		V-0
Electrical					
Volume Resistivity (Tinfoil Electrodes)	D 257 (ISO93)	ohm-cm	4.9×10^{16}	10^{16}	1.4×10^{5}
Surface Resistivity	D 257 (ISO93)	ohm	10^{15}	10^{15}	
Dielectric Strength (Short Time under Oil at 73°F)	D 149 (ISO243)	V/mil (Kv/mm)	480	500	
Dielectric Constant (Tinfoil Electrodes):	D 150 (ISO250)				
60 Hz			3.2		
1 MHz				3.61	
Dissipation Factor (Tinfoil Electrodes):	D 150 (ISO250)				
60 Hz			0.001		
1 Mhz				0.0043	
Arc Resistance:	D 495				
Tungsten Electrodes		sec		124	

KADEL™ POLYKETONE

Typical Physical Properties	ASTM Test Method (Other)	English Units (Standard Internat.)	KADEL™ E-1000	KADEL™ E-1130 (30% Glass Filled)
General				
Specific Gravity	D 792 (ISO1183)	g/cm^3	1.30	1.47
Mold Shrinkage	D 955	in/in (mm/mm)	0.008–0.012	0.001–0.009
Water Absorption (Immersion at 73°F):	D 570(ISO 62)			
24 Hours		%	0.10	0.07
Mechanical				
Tensile Stress at Break	D 638 (ISO527)	psi (MPa)	13500	23700
Tensile Elongation at Break	D 638 (ISO527)	%	50	2.2
Tensile Modulus	D 638 (ISO527)	psi (MPa)	520000	1410000
Flexural Strength	D 790 (ISO178)	psi (MPa)	24500	34100
Flexural Modulus	D 790 (ISO178)			
73°F		psi (MPa)	530000	1450000
250°F		psi (MPa)	46000	322400
Compressive Strength	D 695	psi (MPa)	17200	30000
Impact Strength:				
Notched Izod 0.125 in. (3.2mm) Thickness, 73°F	D 256	ft-lb/in (J/m)	1.6	1.8
Unnotched, 0.125 in. Thickness, 73°F	D 256	ft-lb/in (J/m)	No Break	13.5
Tensile Impact Strength	D 1822	ft-lb/in^2 (KJ/m^2)	55	65
Shear Strength	D732	psi (MPa)	15400	13600
Taber Abrasion, 1kg	D1044	mg/1000 cy	8	
Thermal				
Melting Temperature				
T_m (crystalline)		°F (°C)		
T_g (amorphous)		°F (°C)		
Heat Deflection Temp.	D 648 (ISO75)			
264 psi Load(1.84 MPa)		°F (°C)	323 (162)	619(326)
66 psi Load		°F (°C)	582 (305)	644(340)
Coefficient of Linear Thermal Expansion	D 696 E831	in/in-°F (m/m-°C)	24.7×10^{-6}	10.3×10^{-6}
Flammability				
Oxygen Index (LOI)	D 2863	%	34.2	40.5
UL94 Flame Class:	(UL 94)			
0.062 in. (1.57 mm) Thickness		Rating	V-0 @31 mils	V-0 @31 mils
Electrical				
Volume Resistivity (Tinfoil Electrodes)	D 257 (ISO93)	ohm-cm (ohm-m)	1.66×10^{18} ohm-in	$1.64 x 10^{18}$ ohm-in
Dielectric Strength (Short Time under Oil at 73°F)	D 149 (ISO243)	V/mil (kV/mm)	355	369
Dielectric Constant (Tinfoil Electrodes):	D 150 (ISO250)			
60 Hz			3.25	3.72

KADEL™ POLYKETONE

Typical Physical Properties	ASTM Test Method (Other)	English Units (Standard Internat.)	KADEL™ E-1000	KADEL™ E-1130 (30% Glass Filled)
1 MHz			3.28	3.74
Dissipation Factor (Tinfoil Electrodes):		D 150 (ISO250)		
60 Hz			0.0078	0.0068
1 Mhz			0.0033	0.0038

POLY(ETHERETHERKETONEKETONE)-PEEKK

Structure

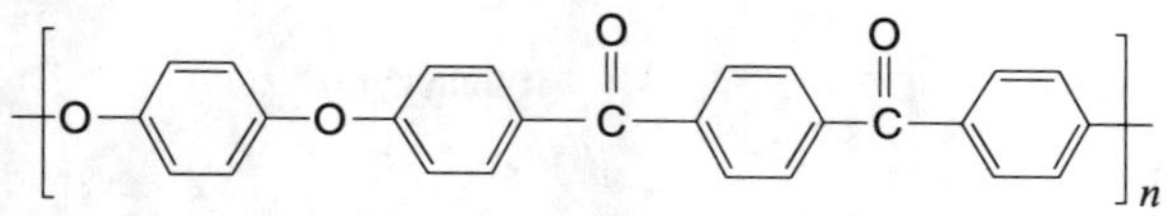

Suppliers/Selected Products

	Neat Products	Modified Products
Hoechst AG 65926 Frankfurt am Main Germany	Hostatec	Hostatec-30% Glass Fiber

(1) Hoechst withdrew from the polyetherketone business in 1992. See also Chemicalweek 11/21/90, p. 10, for polymer structure.

Key Properties

Very High Use Temperature—Good Electrical and Mechanical Properties at Elevated Temperatures—Inherent Flame Resistance—Extraordinary Chemical Resistance.

Typical Applications

Chemical Processing—Aviation—Aerospace Composites—Compressor Parts—Bearings and Seals—Fibers and Film—Wire and Cable Coatings—Wear Rings

POLY(ETHERETHERKETONEKETONE)

Typical Physical Properties	ASTM Test Method (Other)	English Units (Standard Internat.)	HOSTATEC PEK	HOSTATEC PEK (30% Glass Filled)
General				
Specific Gravity	D 792 (ISO1183)	g/cm^3	1.3	1.5
Mold Shrinkage	D 955	in/in (mm/mm)	1.1%	0.5%
Mechanical				
Tensile Stress at Yield	D 638 (ISO527)	psi (MPa)	(100)	
Tensile Stress at Break	D 638 (ISO527)	psi (MPa)	(90)	(155)
Tensile Elongation at Yield	D 638 (ISO527)	%	5.5	
Tensile Elongation at Break	D 638 (ISO527)	%	28	2.3
Tensile Modulus	D 638 (ISO527)	psi (MPa)	(4000)	(11000)
Flexural Strength	D 790 (ISO178)	psi (MPa)		
Flexural Modulus	D 790 (ISO178)			
73°F		psi (MPa)	(3700)	(9500)
Impact Strength:				
Notched Izod 0.125 in. (3.2mm) Thickness, 73°F	D 256	ft-lb/in (J/m)	(80)	(90)
Thermal				
Melting Temperature				
T_m (crystalline)		°F (°C)	680(360)	680(360)
T_g (amorphous)		°F (°C)	333(167)	333(167)
Flammability				
UL94 Flame Class:	(UL 94)			
0.125 in. (3.18 mm) Thickness		Rating	V-0	V-0

POLY(ETHERKETONEETHERKETONEKETONE)-PEKEKK

Structure

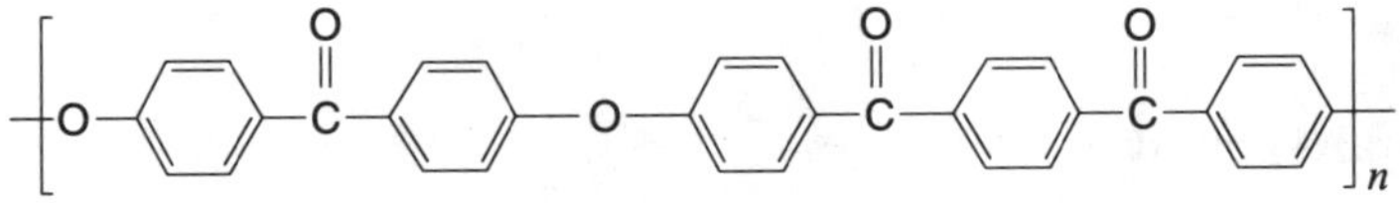

Suppliers/Selected Products

	Neat Products	Modified Products
BASF AG D-6700 Ludwigshafen Germany	Ultrapek™ A-2000	Ultrapek™ A-2000 G6-30% Glass Fiber.

Key Properties

Great Rigidity and Strength—High Level of Mechanical Properties over a Wide Range of Temperatures—Good Tribological Properties—Good Dielectric properties—Excellent Chemical Resistance—Radiation Resistance —Good Stress Cracking Resistance.

Typical Applications

Cable Connectors—Distillation Column Packing—Wire Coatings—Lighting Components—Seals—Medical Protheses.

POLY(ETHERKETONEETHERKETONEKETONE)

Typical Physical Properties	ASTM Test Method (Other)	English Units (Standard Internat.)	Ultrapek A 2000	Ultrapek A 2000 G6 30% GF
General				
Specific Gravity	D 792 (ISO1183)	g/cm^3	1.3	1.53
Water Absorption (Immersion at 73°F):	D 570(ISO 62)			
Equilibrium		%	0.8	0.5
Mechanical				
Tensile Stress at Yield	D 638 (ISO527)	psi (MPa)	(118)	
Tensile Stress at Break	D 638 (ISO527)	psi (MPa)		(185)
Tensile Elongation at Yield	D 638 (ISO527)	%	5.2	
Tensile Elongation at Break	D 638 (ISO527)	%		>2.5
Tensile Modulus	D 638 (ISO527)	psi (MPa)	(4000)	(12000)
Flexural Strength	D 790 (ISO178)	psi (MPa)	(130)	(250)
Impact Strength:				
Notched Izod 0.125 in. (3.2mm) Thickness, 73°F	D 256	ft-lb/in (J/m)	5.6 KJ/m^2	11 KJ/m^2
Shore/Barcol	D2240/ D2583		D86	D90
Thermal				
Melting Temperature				
T_m (crystalline)		°F (°C)	718(381)	718(381)
T_g (amorphous)		°F (°C)	338(170)	338(170)
Heat Deflection Temp.	D 648 (ISO75)			
264 psi Load(1.84 MPa)		°F (°C)	338(170)	>482(>250)
66 psi Load		°F (°C)	>482(>250)	662(350)
Coefficient of Linear Thermal Expansion	D 696	in/in-°F (mm/mm-°C)	(4.1×10^{-5})	
Thermal Conductivity	C 177	Btu-in/ft^2-hr-°F (W/m-°C)	(0.22)	(0.42)
Specific Heat	D 2766	Btu/ lb-°F (J/g°C)	(1.4)	(1.1)
Vicat Softening Temperature: Rate B	D 1525 (ISO306)	°F (°C)	464(>240)	464(>240)

POLY(ETHERKETONEETHERKETONEKETONE)

Typical Physical Properties	ASTM Test Method (Other)	English Units (Standard Internat.)	Ultrapek A 2000	Ultrapek A 2000 G6 30% GF
Flammability				
UL94 Flame Class:	(UL 94)			
0.062 in. (1.57 mm) Thickness		Rating	V-0	V-0
Electrical				
Volume Resistivity (Tinfoil Electrodes)	D 257 (ISO93)	ohm-cm	$>10^{16}$	$>10^{16}$
Surface Resistivity	D 257 (ISO93)	ohm	$>10^{13}$	$>10^{14}$
Dielectric Strength (Short Time under Oil at 73°F)	D 149 (ISO243)	V/mil (Kv/mm)	(83)	(88)
Dielectric Constant (Tinfoil Electrodes):	D 150 (ISO250)			
1 MHz			3.3	3.8
Dissipation Factor (Tinfoil Electrodes):	D 150 (ISO250)			
1 Mhz			0.0053	0.005

POLY(ETHERKETONEKETONE)-PEKK

Structure

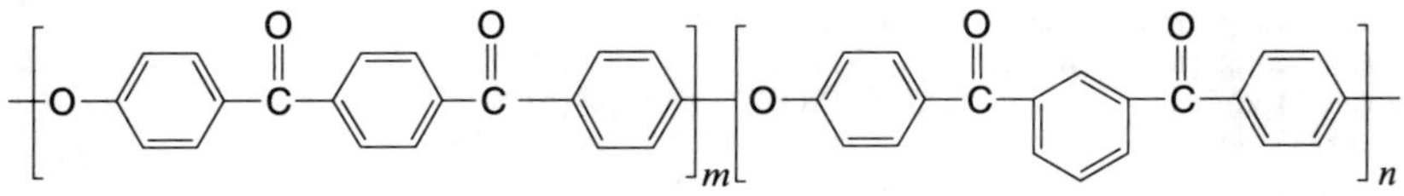

Supplier/Selected Products (1)

	Neat Products	Modified Products
DuPont Polymers Specialty Polymers Wilmington, DE 19898	PEKK	Declar™ Sheet-12.5% TiO_2

(1) The commercial product may be based on 60/40 tere/iso (m/n) copolymer. See Sauer, B. B., and Hsiao, B. S., *Polym.*, **34**, 3315 (1993) and Gardner, K. H., Hsiao, B. S., Matheson, R. R., Jr., and Wood, B. A., *Polym.*, **33**, 2483 (1992).

Key Properties

Heat resistance—Minimal Smoke and Gas Generation—Good Formability—Chemical Inertness—Resistance to Fatigue.

Typical Applications

Aircraft Cabin Interiors—Thermoplastic Matrix for Composites.

POLY(ETHERKETONEKETONE)

Typical Physical Properties	ASTM Test Method (Other)	English Units (Standard Internat.)	PEKK (26% cryst.)	DECLAR™ (12.5% TiO_2)
General				
Specific Gravity	D 792 (ISO1183)	g/cm^3	1.3	1.3–1.49
Mold Shrinkage	D 955	in/in (mm/mm)		0.004
Water Absorption (Immersion at 73°F):	D 570(ISO 62)			
Equilibrium		%	0.3	< 0.2
Mechanical				
Tensile Stress at Yield	D 638 (ISO527)	psi (MPa)		
Tensile Stress at Break	D 638 (ISO527)	psi (MPa)	14800(102)	10–14500 (69–100)
Tensile Elongation at Yield	D 638 (ISO527)	%		5
Tensile Elongation at Break	D 638 (ISO527)	%	4	45–130
Tensile Modulus	D 638 (ISO527)	psi (MPa)	650000 (4500)	400–610000 (2760–4210)
Flexural Strength	D 790 (ISO178)	psi (MPa)		16–22000 (110–152)
Flexural Modulus, 73°F	D 790 (ISO178)			400–610000 (2760–4210)
Compressive Strength	D 695	psi (MPa)		16100 (111)
Gardner Impact	D3029	ft-lbs (J)		(>320) (text. side up)
Thermal				
Melting Temperature				
T_m (crystalline)		°F (°C)	641(338)	581(305)
T_g (amorphous)		°F (°C)	313(156)	310(155)
Heat Deflection Temp.	D 648 (ISO75)			
264 psi Load(1.84 MPa)		°F (°C)		299(148)
Flammability				
Oxygen Index (LOI)	D 2863	%	40	
NBS Smoke Density	E662-79	Ds at 4 min		10
UL94 Flame Class:	(UL 94)			
0.125 in. (3.18 mm) Thickness		Rating	V-0	

ULTEM™ POLYETHERIMIDES

Structure

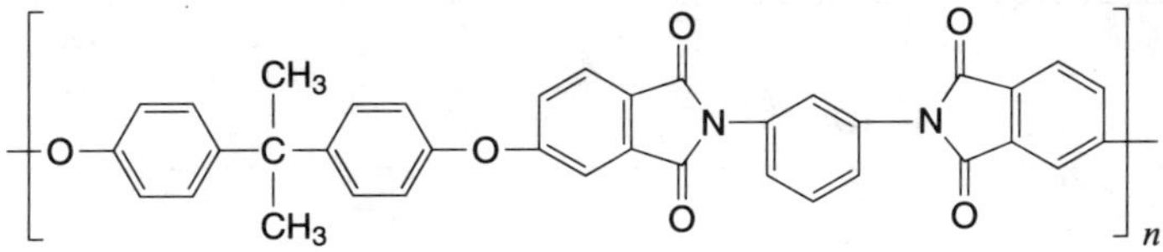

Ultem™ 1000

Supplier/Selected Products

	Neat Products	Modified Products
GE Plastics One Plastics Avenue Pittsfield, MA 01201, USA	ULTEM™ 1000 ULTEM™ 6000 (1)	ULTEM™ 2300–30% Glass Fiber. ULTEM™ 3254–20% Mineral Filler

(1) Described as being based on a "polyetherimide copolymer."

Key Properties

High Heat Resistance—Exceptional Strength and Modulus—Inherent Flame Resistance with Low Smoke Emission—High Dielectric Strength—Stable Dielectric Constant and Dissipation Factor over a Wide Range of Temperatures and Frequencies—Broad Chemical Resistance—Outstanding Processability on Conventional Molding Equipment—Excellent Machinability and Finishing Characteristics.

Typical Applications

Aircraft Cabin Interiors—Circuit Boards—Connectors—Bearings—Small Appliances—Automotive Under Hood Applications—Medical Instruments and Devices—Computer Disks and Components—Microwave Oven Hardware—Food Packaging Containers.

ULTEM POLYETHERIMIDES

Typical Physical Properties	ASTM Test Method (Other)	English Units (Standard Internat.)	ULTEM™ 1000	ULTEM™ 2300 (30% Glass Fiber)	ULTEM™ 3254 (20% Mineral Filled)	ULTEM™ 6000
General						
Specific Gravity	D 792 (ISO1183)	g/cm^3	1.27	1.51	1.42	1.29
Mold Shrinkage	D 955	in/in (mm/mm)	0.007	0.003		0.006
Water Absorption (Immersion at 73°F):	D 570(ISO 62)					
24 Hours		%	0.25	0.16		0.28
Equilibrium		%	1.25	0.9		
FDA						
Melt Flow Rate: 300°C/1.2 kg Load	D 1238	g/10 min			11.9 (337°C)	
Mechanical						
Tensile Stress at Yield	D 638 (ISO527)	psi (MPa)	15200(105)	24500 (170)		15000(105)
Tensile Stress at Break	D 638 (ISO527)	psi (MPa)	(90)	(160)	14600 (100)	
Tensile Elongation at Yield	D 638 (ISO527)	%	7			7-8
Tensile Elongation at Break	D 638 (ISO527)	%	60	3	3	30
Tensile Modulus	D 638 (ISO527)	psi (MPa)	430000 (3000)	1300000 (9000)		
Flexural Strength	D 790 (ISO178)	psi (MPa)	22000(150)	33000 (230)	21000 (140)	21000(145)
Flexural Modulus	D 790 (ISO178)				895000 (6200)	440000 (3000)
73°F		psi (MPa)	480000 (3300)	1300000 (9000)		
Compressive Strength	D 695	psi (MPa)	21900(150)	30700 (210)	18700 (130)	20500(140)
Compressive Modulus	D 695	psi (MPa)	480000 (3300)	938000 (6500)	500000 (3400)	
Impact Strength: Notched Izod 0.125 in. (3.2mm)	D 256	ft-lb/in (J/m)	1 (50)	2 (110)	0.6 (30)	0.8(40)

ULTEM POLYETHERIMIDES

Typical Physical Properties	ASTM Test Method (Other)	English Units (Standard Internat.)	ULTEM™ 1000	ULTEM™ 2300 (30% Glass Fiber)	ULTEM™ 3254 (20% Mineral Filled)	ULTEM™ 6000
Thickness, 73°F						
Unnotched, 0.125 in.	D 256	ft-lb/in (J/m)	25(1300)	8 (430)		25(1300)
Thickness, 73°F						
Gardner Impact	D3029	ft-lbs (J)	27(37)			
Shear Strength	D732	psi (MPa)	15000(100)	14000 (100)		15000(105)
Taber Abrasion, 1kg	D1044	mg/1000 cy	10			8
Hardness						
Rockwell	D 785	M Scale	109	114		110
		R Scale				
Thermal						
Melting Temperature						
T_g (amorphous)		°F (°C)				
UL Temp. Rating (estimated)		°F (°C)	338 (170)	356 (180)		
Heat Deflection Temp.	D 648 (ISO75)					
264 psi Load(1.84 MPa)		°F (°C)	392 (200)	410 (210)	401 (205)	420(216)
66 psi Load		°F (°C)	410 (210)	414 (212)		430(221)
Coefficient of Linear	D 696	in/in-°F	3.1×10^{-5}	1.1×10^{-5}	1.7×10^{-5}	2.9×10^{-5}
Thermal Expansion	E 831	(m/m-°C)	$(5.6x10^{-5})$	(2×10^{-5})	(3.1×10^{-5})	(5.2×10^{-5})
Relative Temperature Index:	(UL746B)					
Index						
Electrical		°C	170	180		
Mechanical with Impact		°C	170	170		
Mechanical without Impact		°C	170	180		
Rate B		°F (°C)	426 (219)	442 (228)		453(234)
Flammability						
Oxygen Index (LOI)	D 2863	%	47	50		44
UL94 Flame Class:	(UL 94)					

0.062 in. (1.57 mm) Thickness		Rating	V-0@0.016" 5V@0.075"	V-0@0.10"	V-0	V-0
Electrical						
Volume Resistivity (Tinfoil Electrodes)	D 257 (ISO93)	ohm-cm (ohm-m)	6.7×10^{17} (6.7×10^{15})	3×10^{16} (3×10^{14})	9.4×10^{15}	1×10^{17} (1×10^{15})
Surface Resistivity	D 257 (ISO93)	ohm				
Dielectric Strength (Short Time under Oil at 73°F)	D 149 (ISO243)	V/mil (kV/mm)	710 (28)	770 (30)	475	750(30)
Dielectric Constant (Tinfoil Electrodes):	D 150 (ISO250)					
10^3 Hz			3.15	3.7		3.0
Dissipation Factor (Tinfoil Electrodes):		D 150 (ISO250)				
10^3 Hz			0.0013	0.0015		0.0010
Arc Resistance:	D 495					
Tungsten Electrodes		sec	128	85		

OTHER THERMOPLASTIC POLYIMIDES

Polyimide 2080

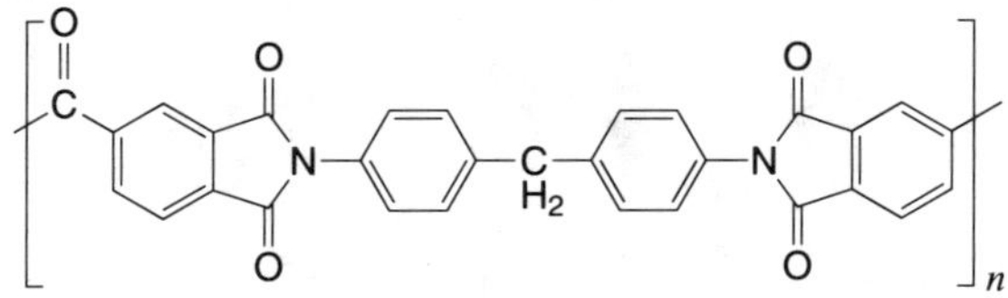

Supplier—Dow U. S. A., 2040 Willard H. Dow Center, Midland MI 48674. This product is no longer commercially available.

XPI™ 182

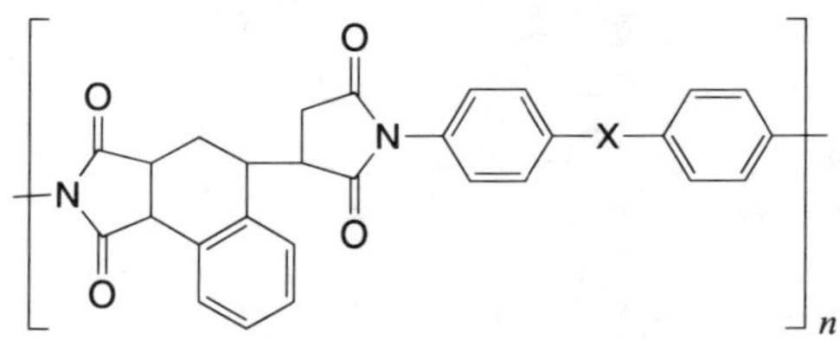

Supplier—American Cyanamid Co., One Cyanamid Plaza, Wayne, NJ 07470, USA. This product is no longer commercially available.

XU 218

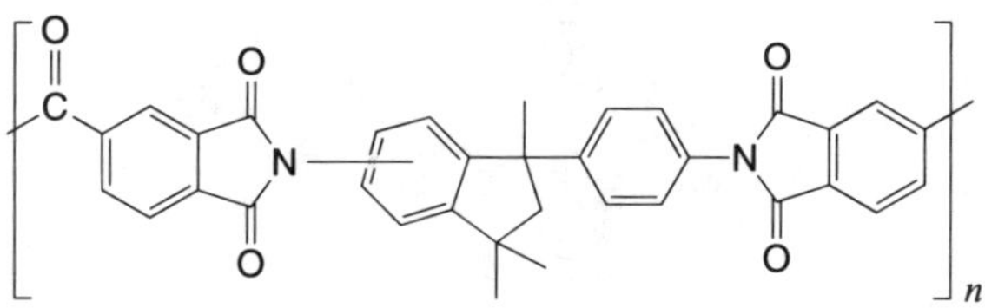

Supplier—Ciba-Geigy Corp., Polymers Division, Seven Skyline Drive, Hawthorne, NY, 10532.

LARC-TPI

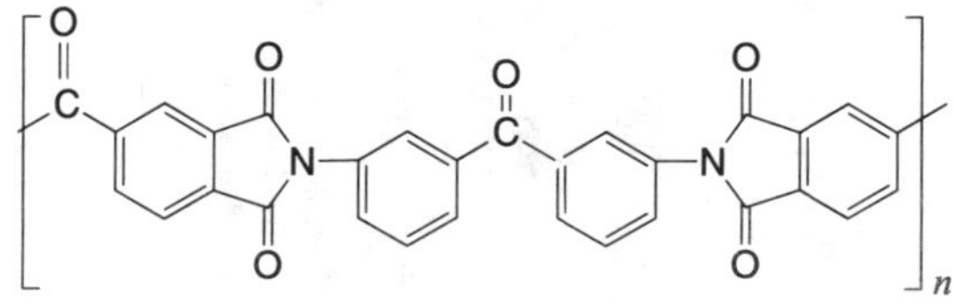

Supplier—Mitsui Toatsu Chemicals, Inc. New York, NY.

PISO

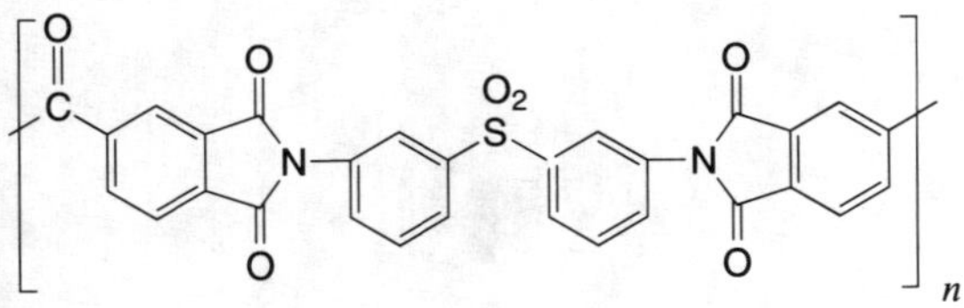

Supplier— NASA Langley Research Center, USA. This product is not commercially available.

OTHER THERMOPLASTIC POLYIMIDES

Typical Physical Properties	ASTM Test Method (Other)	English Units (Standard Internat.)	Polyimide 2080 (2)	XPI™ 182 (2)	XU 218 (1,2)	LARC-TPI (1,3)	PISO (1,2)
General							
Specific Gravity	D 792	g/cm^3	1.40	1.28	1.2	1.38	
Mold Shrinkage	D 955	in/in (cm/cm)		0.006			
Water Absorption (Immersion at 73°F):	D 570						
24 Hours		%		0.75			
Melt Flow	D 1238	dgm/min44 psi (0.3MPa)					
Mechanical							
Tensile Stress at Yield	D 638	psi (MPa)				17300	
Tensile Stress at Break	D 638	psi (MPa)	17100	11000		19700	9200
Tensile Elongation at Break	D 638	%	10	7		4.8	
Tensile Modulus	D 638	psi (MPa)	300000			540000	660000
Flexural Strength	D 790	psi (MPa)	28000	18000		23000	
Flexural Modulus73°F	D 790	psi (MPa)	500000	480000			700000
Compressive Strength	D 695	psi (MPa)	29900	22300			
Compressive Modulus	D 695	psi (MPa)	295000				
Impact Strength:							
Notched Izod 0.125 in. (3.2mm) Thickness, 73°F	D 256	ft-lb/in (J/m)	1.45	0.6–0.8			
Unnotched, 0.125 in. Thickness, 73°F	D 4812	ft-lb/in (J/m)	6.0	3.0–5.0			
Hardness, Rockwell	D 785	M Scale	E99	119			
Thermal							
T_g (amorphous)			590(310)		617(325)	500(260)	524(273)
Heat Deflection Temp. 264 psi Load(1.84 MPa)	D 648	°F (°C)	527(275)	>500			
Coefficient of Linear	D 696	in/in-°F	28×10^{-6}	2.85×10^{-5}			

Thermal Expansion		(mm/mm-°C)		
Thermal Conductivity	C 177	Btu-in/ft^2-hr-°F (W/m-°C)		0.087
Flammability				
Oxygen Index	D 2863	%	44	
Electrical				
Dielectric Constant (Tinfoil Electrodes):	D 150			
60 Hz			3.43	3.45
10^3 Hz			3.42	3.40
1 MHz				3.21
Dissipation Factor (Tinfoil Electrodes):		D 150		
60 Hz			0.0055	0.0014
10^3 Hz			0.0018	0.0028
1 Mhz				0.0102

(1) Data may not have been obtained by the method indicated in Column 2.
(2) Product is not currently available.
(3) Product may be available from Mitsui Toatsu Chemicals, NY, NY.

POLY(2,6-DIMETHYLPHENYLENE OXIDE)-PPO

Structure

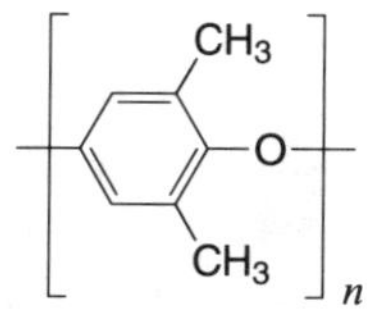

Supplier/Selected Products

	Neat Products (1)	Modified Products (2)
GE Plastics One Plastics Avenue Pittsfield, MA 01201, USA	PPO™ C-1001	NORYL™ Grades PC180X, N225, N300, N190X, 731, SE100, SE1.

(1) This product is not currently commercially available neat. Small amounts are available from lab supply houses.
(2) Noryl™ products are blends of PPO with polystyrene, including impact modified polystyrenes. Since these two polymers are thermodynamically compatible, the heat distortion temperatures (and glass transition temperatures) of the mixtures are averages of the two components. Prevex™ resins are based on similar technologies. Noryl GTX™ resins are nylon blends with PPO™.

Key Properties of PPO™

Useful Temperature Range of More than 600°F — Outstanding Resistance to Aqueous Chemicals and Steam — Excellent Mechanical Properties — Excellent Dielectric Characteristics Over a Wide Temperature Range.

Key Properties of Noryl™ Products

Good Processability—Impact Resistance —Heat Resistance—Flame Retardance—Colorability and Aesthetic Appearance — Excellent Electrical Properties—Low Water Absorption—Glass Reinforced Composites.

Typical Applications of Noryls™

Large and Small Home Appliances—Automotive Under Hood Applications—Medical Instruments and Devices—Food Contact Applications—Computer Components—Hospital and Office Furniture Interior and Exterior Automotive Applications—Fluid Handling Applications.

Typical Physical Properties (Other)	ASTM Test Method Internat.)	English Units (Standard)	PPO™ C-1001	NORYL™ Resin (Unfilled grades)
General				
Specific Gravity	D 792 (ISO1183)	g/cm^3	1.06	1.06–1.12
Mold Shrinkage	D 955	in/in (mm/mm)	0.007	0.006
Water Absorption	D 570(ISO 62)			
(Immersion at 73°F):				
24 Hours		%	0.06	0.06–0.08
Equilibrium		%	0.1	
Mechanical				
Tensile Stress at Yield	D 638 (ISO527)	psi (MPa)	10500	6900(48)–10500(72)
Tensile Stress at Break	D 638 (ISO527)	psi (MPa)	9500	6000(41)–900(48)
Tensile Elongation at Yield	D 638 (ISO527)	%	6–7	3.6–7
Tensile Elongation at Break	D 638 (ISO527)	%	50-80	18–45
Tensile Modulus	D 638 (ISO527)	psi (MPa)	370000	374500(2582)
Flexural Strength	D 790 (ISO178)	psi (MPa)	15000	11000(76)–16500(114)
Flexural Modulus 73°F	D 790 (ISO178)	psi (MPa)	370000	345000(2379)-
380000(2620)				
Compressive Strength	D 695	psi (MPa)		11700(81)
Impact Strength:				
Notched Izod 0.125 in. (3.2mm) Thickness, 73°F	D 256	ft-lb/in (J/m)	1.5–1.9	3.5(187)–5.8(309)
Unnotched, 0.125 in. Thickness, 73°F	D 256	ft-lb/in (J/m)		13.5(720)–22(1173)
Tensile Impact Strength	D 1822	ft-lb/in^2 (KJ/m^2)	350–450	
Gardner Impact	D3029	ft-lbs (J)		11(15)–19(26)
Shear Strength	D732	psi (MPa)		7700 (53)
Taber Abrasion, 1kg	D1044	mg/1000 cy	17	64(H-18)–76(CS-17)
Hardness				
Rockwell	D 785	R Scale	120	115-120
Thermal				
T_g (amorphous)		°F (°C)	428(220)	212–428(100–220)

Typical Physical Properties (Other)	ASTM Test Method Internat.)	English Units (Standard	PPO™ C-1001	NORYL™ Resin (Unfilled grades)
UL Temp. Rating (estimated)		°F (°C)		
Heat Deflection Temp.	D 648 (ISO75)			
264 psi Load(1.84 MPa)		°F (°C)	375(190)	175(79)–295(146)
66 psi Load		°F (°C)		195(203)–315(157)
Coefficient of Linear Thermal Expansion	D 696 E831	in/in-°F (m/m-°C)	2.9×10^{-5}	$3x10^{-5}(5.4 \times 10^{-5})$- $4.8 \times 10^{-5}(8.6 \times 10^{-5})$
Thermal Conductivity	C 177	Btu-in/ft^2-hr-°F (W/m-°C)	1.33	(0.24)
Relative Temperature Index: Index	(UL746B)			
Electrical		°C		50–110
Mechanical with Impact		°C		50–105
Mechanical without Impact		°C		50–110
Vicat Softening Temperature: Rate B	D 1525 (ISO306)	°F (°C)		235(113)
Flammability				
Oxygen Index (LOI)	D 2863	%		22–39
UL94 Flame Class: 0.062 in. (1.57 mm) Thickness	(UL 94)	Rating		HB, V-0, V-1 (depends on prod. composition)
Electrical				
Volume Resistivity (Tinfoil Electrodes)	D 257 (ISO93)	ohm-cm (ohm-m)	10^{17}	$1.8 \times 10^{16}(1.8 \times 10^{14})$
Dielectric Strength (Short Time under Oil at 73°F)	D 149 (ISO243)	V/mil (kV/mm)	500	400(16)–630(25)
Dielectric Constant (Tinfoil Electrodes):	D 150 (ISO250)			
60 Hz			2.58	2.65–2.79
1 MHz			2.58	
Dissipation Factor (Tinfoil Electrodes):	D 150 (ISO250)			
60 Hz			0.00035	0.0007–0.0031
1 Mhz			0.0009	
Arc Resistance:	D 495			
Tungsten Electrodes		sec	75	

*** Noryl™ data is the range for unfilled grades PC180X, N225, N300, N190X, 731, SE100, SE1

POLY(ARYLENE ETHER NITRILE)-PEN

Structure

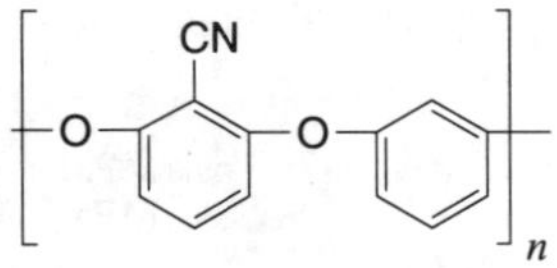

Supplier/Selected Products

	Neat Products	Modified Products
Idemitsu Material Corp. 3-1-1 Marunouchi, Chiyoda-ku 100 Tokyo, Japan	PEN™	PEN™- 305 Glass Filled

Key Properties of PEN™

Excellent Thermal and Mechanical Stability Over a Wide Temperature Range—Excellent Adhesive Properties—Wear Resistance—Chemical Resistance—Creep Resistance.

Suggested Applications

Glass and Carbon Fiber Reinforced Composites—Molded Parts for Engineering Applications.

POLY(ARYLENE ETHER NITRILE)-PEN™

Typical Physical Properties	ASTM Test Method (Other)	English Units (Standard Internat.)	PEN™	PEN™ (30% Glass Filled)
Mechanical				
Tensile Stress at Yield	D 638	psi (MPa)		
Tensile Stress at Break	D 638	psi (MPa)	(137)	(204)
Tensile Elongation at Yield	D 638	%		
Tensile Elongation at Break	D 638	%	10	3
Tensile Modulus	D 638	psi (MPa)	(3600)	(8900)
Flexural Strength	D 790	psi (MPa)	(194)	(265)
Flexural Modulus				
73°F	D 790	psi (MPa)	(4000)	(11200)
Tensile Impact Strength	D 1822	ft-lb/in^2 (KJ/m^2)		
Hardness				
Rockwell	D 785	M Scale	114	
Thermal				
Melting Temperature, ° C				
T_m (crystalline)			340	
T_g (amorphous)			148	
Heat Deflection Temp.	D 648			
264 psi Load(1.84 MPa)		°F (°C)	165	
Flammability				
Oxygen Index	D 2863	%	42	

REFERENCES FOR TABLES 1–17

This is a list of the complete references that are cited in Tables 1–17, including authors. It is arranged alphabetically by reference for ready access from the tables. Page citations for references in Tables 18–22 (all of which were taken from tables 1–17) are not included in this list.

REFERENCE	YEAR	AUTHORS	PAGE
ACS Symp.Ser., **4**, 63;	1972	Rose, J. B.	106
Acta Polymerica, **39**, 452	1988	Gileva, N. G., Solotuchin, M. G., Salaskin, S. N., Sultanova, V. S., Horhold, H. H., Raabe, D.	143
Adv. Chem. Ser., **176**, 559	1979	Olabisi, O., Farnham, A. G.	118
Adv. Macromol. Chem., **1**, 335	1968	Ivin, K. J., Rose, J. B.	108
Angew. Makromol. Chemie, **188**, 143, 155	1991	Podkoscielny, W., Dethloff, J., Dethloff, M., Brunn, J.	122,123
Appl. Polym. Symp., No. **26**, 373	1975	Robeson, L. M., Farnham, A. G., McGrath, J. E.	113,116, 118,119, 121,194, 198
Brit. Polym. J., **4**, 391	1972	Attwood, T. E., Newton, A. B., Rose, J. B.	106,107, 133
Brit. Polym. J., **17**, 4	1985	Jin, X. J., Bishop, M. T., Ellis, T. S., Karasz, F. E.	137
Bull. Soc. Chim. Belg., **98**, 667	1989	Staniland, P. A.	124,127, 128,130, 136,141, 145,147, 150,154
Can. Pat. 946091	1974	O'Shea, F. X., Cornell, R. J.	109
Chem. Ind. (15), 461	1968	Rose, J. B.	106,107
Chimia, **28**(9), 561	1974	Rose, J., B.	106,107
Comp. Polym. Sci. (Allen), **5**, 483	1989	Staniland, P. A.	141
Ency. Chem. Tech., **18**, 594	1982	White, D. M., Cooper, G. D.	183
Ency. Polym. Sci. & Eng., **13**,196	1988	Harris, J. E., Johnson, R. N.	108,111, 114–117, 119,121, 141,142, 189,195
Ency. Polym. Sci. & Eng., **12**, 364	1988	Verbicky, J. W., Jr.	167,173–175
Eur. Pat. 0161576	1986	Sumitomo Chemical Co.	109
Eur. Pat. 192260	1986	Gay, F. P., Brunette, C. M.	159
Eur. Pat. 218064	1987	Mayske, P. J., Tresper, E.	106
Eur. Pat. 254431	1988	Sweeney, W.	133
Eur. Pat. 282096	1988	Percec, V., Nicholas, P. P.	113,139
Eur. Pat. 327501	1989	Ciba Geigy AG	123
Eur. Pat. 331305	1989	ICI	128
Eur. Polym. J., **4**, 275	1968	van Dort, H. M., Hoefs, C. A. M., Magre, E. P., Schopf, A. J., Yntema, K.	183
Eur. Polym. J., **20**, 519	1984	Shah, P. P.	191
Eur. Polym. J., **22**, 447	1986	Mandal, B. K., Maiti, S.	198
Gaofenzi Tongxun, **6**, 442	1984	Qu, X., Zhang, S., Liu, L., Bin, Z.	136

REFERENCE	YEAR	AUTHORS	PAGE
GB Pat. 971227	1964	Goodman, I., McIntyre, J. E., Russell, W.	141
GB Pat. 1078234	1967	Johnson, R. N., Farnham, A. G.	109,124, 190
GB Pat. 1250251	1971	DuPont Co.	138
GB Pat. 2116990	1983	Colquhoun, H. M., Lewis, D. F.	150,154, 158,162
GB Pat. 2549529; *CA*, **87**, 136615	1977	Blinne, G., Cordes, C.	129
GB Pat. 2557652; *CA*, **87**, 102832	1977	Blinne, G., Cordes, C.	129
GB Pat.. 2940190; *CA*, **95**, 8035	1981	Blinne, G., Wurmb, R., Schmidt, F.	127
GB Pat.. 3014230; *CA*, **96**, 7266	1982	Blinne, G., Wurmb, R., Schmidt, F.	127,150
GB Pat.. 3829520	1990	BASF-AG	149
High Perf. Polym.: Their Origin & Devel., Elsevier, p. 171	1986	Rose, J. B.	107
High Perf. Polym.: Their Origin & Devel., Elsevier, p.157;	1986	Clendinning, R. A., Farnham, A. G., Johnson, R. N.	107,111
High Perf. Polym.: Their Origin & Devel., Elsevier, p. 317	1986	King, J. J., Lee, B. M.	174
J. Appl. Polym. Sci., **16**, 461	1972	Shultz, A. L., Gendron, B. M.	184
J. Appl. Polym. Sci., **24**,1145	1979	Port, A. B., Still, R. H.	183,184
J. Appl. Polym. Sci., **28**, 2925	1983	Robeson, L. M., Crisafulli, S. T.	116
J. Appl. Polym. Sci., **47**, 753	1993	Li, S., Karasz, F. E.	171
J. Appl. Polym. Sci., **47**, 1961	1993	Matsumoto, K., Xu, P.	172
J. Chem. Soc., Chem. Commun., (19), 1338	1985	Segal, J. A.	135
J. Chem. Soc., Chem. Commun., 336	1990	Colquhoun, H. M., Dudman, C. C., Thomas, M., O'Mahoney, C. A., Williams, D. J.	128,157, 162
J. M. S.-Pure Appl. Chem., **A29**(12), 1141	1992	Kim, W., Hay, A. S.	110,137
J. Memb. Sci., **72**, 91	1992	Ghosal, K., Chern, R. T.	116,184
J. Memb. Sci., **74**, 193	1992	Kamps, K. M. P., Teunis, H. A., Wessling, M., Smolders, C. A.	107,116
J. Memb. Sci., **78**, 123	1993	Tam, C. M., Dal-Cin, M., Guiver, M. D.	107
J. Org. Chem., **29**, 207	1964	Gipstein, E., Wellisch, E., Sweeting, O. J.	108
J. Polym. Mater., **2**, 115	1985	Mandal, B. K., Maiti, S.	167
J. Polym. Sci., **58**, 469	1972	Blanchard, H. S., Finkbeiner, H. L., Russell, G. A.	183
J. Polym. Sci., A, **58**, 581	1962	Hay, A. S.	183,184
J. Polym. Sci., A-1, **4**, 722	1966	Cohen, S. M., Young, R. H.	106
J. Polym. Sci., A-1, **5**, 2375	1967	Johnson, R. N., Farnham, A. G., Clendinning, R. A., Hale, W. F., Merriam, C. N.	108,114–121,123, 127,129, 135,141, 142,148, 151,189, 190,195
J. Polym. Sci., A-1, **5**, 2399	1967	Hale, W. F., Farnham, A. G., Johnson, R. N., Clendinning, R. A.,	116

REFERENCE	YEAR	AUTHORS	PAGE
J. Polym. Sci., A-1, **6**, 3345	1968	Iwakura, Y., Uno, K., Takiguchi, T.	132,133
J. Polym. Sci., A-1, **7**, 2749	1969	Breed, L. W., Elliott, R.L.	191
J. Polym. Sci., A-1, **7**, 3089	1969	Nakamura, Y., Mori, K., Tamura, K., Saito, Y.	192
J. Polym. Sci., A-1, **8**, 2035	1970	Vogel, H. A.	111,112, 115,124, 128
J. Polym. Sci., A-1, **10**, 2365	1972	Fritz, A., Rees, R. W.	187–189, 191
J. Polym. Sci., B, Polym. Phys., **25**, 311	1987	Harris, J. E., Robeson, L. M.	135,141, 149,150
J. Polym. Sci., B, Polym. Phys., **31**, 891	1993	Ghosal, K., Chern, R. T., Freeman, B. D.	107
J. Polym. Sci., B, Polym. Phys., **31**, 983	1993	Aitken, C. L., Paul, D. R.,	128,129, 157,158
J. Polym. Sci., B, Polym.Phys., **31**, 1061	1993	Aitken, C. L., Paul, D. R., Mohanty, D. K.	112, 113
J. Polym. Sci., C, No. **16**, 715	1967	Jennings, B. E., Jones, M. E. B., Rose, J. B.	108,109, 112,115, 127,129
J. Polym. Sci., Polym. Chem., **11**, 561	1973	Relles, H. M., Schluenz, R. W.	167,168, 171
J. Polym. Sci., Polym. Chem., **14**, 2275	1976	Bell, V. L., Stump, B. L., Gager, H.	170
J. Polym. Sci., Polym. Chem., **18**, 3069	1980	Takeoshi, T., Wirth, J. G., Heath, D. R., Kochanowski, J. E., Manello, J. S., Webber, M. J.	173, 175–180, 189 190
J. Polym. Sci., Polym. Chem., **19**, 1635	1981	White, D. M., Takeoshi, T., Williams, F. J., Relles, H. M., Donahue, P. E., Klopfer, H. J., Loucks, G. R., Manello, J. S., Matthews, R. O., Schluenz, R. W.	173,176–180
J. Polym. Sci., Polym. Chem., **20**, 609	1982	Sutter, A., Schmutz, P., Marvel, C. S.	143
J. Polym. Sci., Polym. Chem., **23**, 287	1986	Hedrick, J. L., Mohanty, D. K., Johnson, B. C., Viswanathan, R., Hinkley, J. A., McGrath, J. E.	108,111
J. Polym. Sci., Polym. Chem., **24**, 2447	1986	Mandal, B. K., Maiti, S.	187,195, 196
J. Polym. Sci., Polym. Chem., **25**, 1093	1987	Hergenrother, P. M., Wakelyn, N. T., Havens, S. J.	178,180, 182
J. Polym. Sci., Polym. Chem., **27**, 661	1989	Durvasula, V. R., Stuber, F. A., Bhattacharjee, D.	133,150
J. Polym. Sci., Polym. Chem., **28**, 367	1990	Colon, I., Kwiatkowski, G. T.	111
J. Polym. Sci., Polym. Chem., **29**, 965	1991	Percec, V., Wang, J. H., Oishi, Y., Feiring, A. E., Nicholas, P. P.	113,123
J. Polym. Sci., Polym. Chem., **29**, 1045	1991	Wang, Z. Y., Hay, A. S.	122,143, 144,149, 153
J. Polym. Sci., Polym. Chem., **30**, 221	1992	Andrews, S. M.	112

REFERENCE	YEAR	AUTHORS	PAGE
J. Polym. Sci., Polym. Chem., **30**, 2511	1992	Bochmann, M., Kelly, K., Lu, J.	111,138, 139
J. Polym. Sci., Polym. Chem., **31**, 141	1993	Giesa, R., Keller, U., Eiselt, P., Schmidt, H.	170
J. Polym. Sci., Polym. Chem., **31**, 853	1993	Ueda, M., Toyota, H., Ouchi, T., Sugiyama, J., Yonetake, K., Masuko, T., Teramoto, T.	117,121
J. Polym. Sci., Polym. Chem., **31**, 923	1993	Percec, V., Okita, S.	194
J. Polym. Sci., Polym. Chem., **31**, 1261	1993	Yang, H., Hay, A. S.	185–187
J. Polym. Sci., Polym. Chem., **31**, 1907	1993	Yang, H., Hay, A. S.	139,140, 184–186
J. Polym. Sci., Polym. Chem., **31**, 2015	1993	Yang, H., Hay, A. S.	184–186
J. Polym. Sci., Polym. Lett. Ed., **17**, 85	1979	Imai, Y., Ueda, M., Ii, M.	117
J. Polym. Sci., Polym. Symp., **74**, 93	1986	Takeoshi, T., Kochanowski, J. E., Manello, J. S., Webber, M. J.	171,174–182
Jap. Pat. 60-14739	1985		188
Jap. Pat. 60-228541	1985	Ueda, M.	106
Jap. Pat. 61-275326	1986	S. Matsuo, R. Takasawa	112
Jap. Pat. 1065130	1989	Asahi Chemical Ind KK	163
Jap. Pat. 1087628	1989	Toyo Soda Mfg. KK	116
Jap. Pat. 1098629	1989	Idemitsu Kosan KK	122
Jap. Pat. 02011634	1990	New Japan Chemical Co.	106
Kunst. Ger. Plastics, **79**, 75	1989	Munstedt, H., Zeiner, H.	150
Macromol., **20**, 2675	1987	Ueda, M., Sato, M.	132,133, 150,159, 160
Macromol., **20**, 1204	1987	Kelsey, D. R., Robeson, L. M., Clendinning, R. A.,	132,142
Macromol., **22**, 517	1989	Kricheldorf, H. R., Delius, U.	144,161, 163–166
Macromol., **23**, 926	1990	Ueda, M., Ichikawa, F.	139,147, 148
Macromol., **23**, 5371	1990	Labadie, J. W., Hedrick, J. L.	192,196
Macromol., **24**, 2637	1991	Singh, R., Hay, A. S.	151,153
Macromol., **24**, 2640	1991	Singh, R., Hay, A. S.	194,197, 198,200
Macromol., **24**, 2643	1991	Singh, R., Hay, A. S.	194,198–201
Macromol., **24**, 2694	1991	Ueda, M., Miyaji, Y., Ito, T.	183
Macromol., **25**, 64	1992	Percec, V., Okita, S., Wang, J. H.	189,193
Macromol., **25**, 1017	1992	Singh, R., Hay, A. S.	141,143, 147,149
Macromol., **25**, 1025	1992	Singh, R., Hay, A. S.	193,194, 197
Macromol., **25**, 1033	1992	Singh, R., Hay, A. S.	193,194, 197
Macromol., **26**, 1777	1993	Strukelj, M., Paventi, M., Hay, A. S.	196
Makromol. Chem. Rapid Comm., **13**, 91	1992	Moore, J. S.	133

REFERENCE	YEAR	AUTHORS	PAGE
Makromol. Chem., **130**, 45	1969	Radlmann, E., Schmidt, W., Nischkl, G. E.	114, 138 142
Makromol. Chem., **140**, 83	1970	Baron, A. L., Blank, D. R.	111, 114
Makromol. Chem., **179**, 2989	1978	Imai, Y., Ueda, M., II, M.	117
Makromol. Chem., **191**, 2027	1990	Kricheldorf, H. R., Jahnke, P.,	109, 116, 120, 190
Makromol. Chem., Macromol. Symp., **54/55**, 313	1992	Labadie, J. W., Hedrick, J. L.	201–203
Makromol. Chem., Rapid Comm., **3**, 821	1982	Sato, M., Kondo, H., Yokoyama, M.	116
Makromol. Chem., Rapid Comm., **11**, 535	1990	Rehahn, M., Schluter, A., Wegner, G.	133, 134 188
Mod. Plast., p. 86, 11/81	1981	Stening, T. C., Smith, C. P., Kimber, P. J.	135
NASA Tech. Briefs, Aug., p. 48	1990	Connell, J. W., Hergenrother, P. M.	125
New Polym. Mater., **1**, 127	1988	Kricheldorf, H. R., Delius, U., Tonnes, K. U.	137, 139, 142, 143, 148, 151, 152, 158, 160, 161, 163, 165
Polym. Bull., **25**, 543	1991	Hedrick, J. L.	198, 199, 202
Polym. Bull., **29**, 477	1992	Pak, S. J., Lyle, G. D., Mercier, R., McGrath, J. E.	121
Polym. Bull., **29**, 521	1992	Goh, S. H., Lau, W. W. Y.,Lee, C. S.	118
Polym. J., **23**, 297	1991	Ueda, M., Ito, T.	111, 112
Polym. News **11,** 359	1986	Robeson, L. M., Dickinson, B. L., Crisafulli, S. T.	106, 116
Polym. Prep., **16**(1), 476	1975	Robeson, L. M., Farnham, A. G., McGrath, J. E.	118
Polym. Prep., **20**(1), 211	1979	Reinhardt, B. A., Arnold, F. E.	123
Polym. Prep., **20**(1), 191	1979	Attwood, T. E., Dawson, P. C., Freeman, J. L., Hoy, L. R. J., Rose, J. B., Staniland, P. A.	135, 136
Polym. Prep., **21**(2), 164	1980	Kellman, R., Gerbi, D. J., Williams, R. F., Morgan, J. L.	108, 189, 190
Polym. Prep., **22**(2), 383	1981	Kellman, R., McPheeters, J. C., Gerbi, D. J., Williams, R. F.	188, 189
Polym. Prep., **22**(2), 385	1980	Kellman, R., Gerbi, D. J., Williams, R. F., Morgan, J. L.	190
Polym. Prep., **23**(1), 284	1982	Mohanty, D. K., Hedrick, J. L., Gobetz, K., Johnson, B. C., Yilgor, I., Yilgor, E., Yang, R., McGrath, J. E.	109,114–116, 189, 190
Polym. Prep., **24**(2), 312	1983	Takeoshi, T., Kochanowski, J. E., Manello, J. S., Webber, M. J.	178–182
Polym. Prep., **25**(1), 160	1984	Harris, F. W., Beltz, M. W., Das, S., Gupta, R. K.	172, 173
Polym. Prep., **25**(2), 19	1984	Mohanty, D. K., Sachdeva, Y., Hedrick, J. L., Wolfe, J. F., McGrath, J. E.	120, 143, 191

REFERENCE	YEAR	AUTHORS	PAGE
Polym. Prep., **25**(2),17	1984	Colquhoun, H. M.	134, 135, 147, 154
Polym. Prep., **26**(2), 287	1985	Harris, F. W., Karnavas, A. J., Cucuras, C. N., Das, S.	172, 173
Polym. Prep., **31**(1), 342	1990	Harris, F. W., Hsu, S., Tso, C. C.	170
Polym. Prep., **32**(1), 93	1991	Smith, C. D., Gungor, A., Keister, K. M., Marand, H. A., McGrath, J. E.	191
Polym. Prep., **32**(1), 389	1991	Kim, W. G., Hay, A. S.	113, 139, 140
Polym. Prep., **32**(1), 391	1991	Percec, V., Wang, J. H.	148, 149, 187
Polym. Prep., **32**(1), 392	1991	Percec, V., Wang, J. H.	139
Polym. Prep., **32**(1), 619	1991	Bass, R. G., Srinivasan, K. R.	124, 125, 145, 154, 155, 162
Polym. Prep., **32**(1), 646	1991	Smith, J. G., Jr., Connell, J., W., Hergenrother, P. M.	130, 131, 164–166
Polym. Prep., **32**(2), 160	1991	Bass, R. G., Srinivasan, K. R.	125, 126, 145, 146, 154, 156, 157
Polym. Prep., **32**(2), 162	1991	Connell, J. W., Croall, C. I.	125, 126, 145–147, 155–157
Polym. Prep., **32**(2), 164	1991	Strukelj, M., Hay, A. S.	125–127, 146, 147, 156, 157, 168, 169, 172
Polym. Prep., **32**(2), 166	1991	Clendinning, R. A., El-Hibri, M. J., Matzner, M., Kwiatkowski, G. T.	111,127–129, 138
Polym. Prep., **32**(2), 172	1991	DeSimone, J. M., Stompel, S., Samulski, E. T.	151
Polym. Prep., **32**(2), 180	1991	Wang, P. C., Asbell, W. J., Chuah, H. H.	128
Polym. Prep., **32**(2), 258	1991	Yokelson, H. B., Morland, R. B., Palka, A. M., Hunt, C. L.	119, 159
Polym. Prep., **32**(2), 262	1991	Rogers, J. H., Jr.	135
Polym. Prep., **32**(2), 264	1991	Hsiao, B. S., Chang, I. Y.	135
Polym. Prep., **32**(3), 195	1991	McKean, D. R., Labadie, J. W.	130, 131 165, 166
Polym. Prep., **32**(3), 255	1991	Gardner, K. H., Hsiao, B. S., Matheson, R. R., Jr.	159
Polym. Prep., **32**(3), 376	1991	Chern, R. T., Ghosal, K., Semar, S.	183
Polym. Prep., **32**(3),195	1991	McKean, D. R., Labadie, J. W.	164
Polym. Prep., **33**(1) 409	1992	Muellerleile, J. T., Risch, B. G., Brandom, D. K., Wilkes, G. L.	180
Polym. Prep., **33**(1), 411	1992	Hergenrother, P. M., Smith, J. G., Jr., Connell, J. W.	165
Polym. Prep., **33**(1), 418	1992	DeSimone, J. M., Sheares, V. V., Samulski, E. T.	142

REFERENCE	YEAR	AUTHORS	PAGE
Polym. Prep., **33**(1), 485	1992	Brennan, A. B., Wang, Y. Q.	152
Polym. Prep., **33**(1), 1014	1992	Lucas, M., Hedrick, J. L.	169
Polym. Prep., **33**(1), 1098	1992	Smith, J. G., Jr., Connell, J. W., Hergenrother, P. M.	130, 162, 164
Polym. Prep., **33**(2), 142	1992	Sundar, R. A., Mathias, L. J.	167
Polym. Prep., **33**(2), 170	1992	DeSimone, J. M., Sheares, V. V., Berndt, S. C.	142
Polym. Prep., **33**(2), 231	1992	Priddy, D. B., McGrath, J. E.	195
Polym. Prep., **33**(2), 241	1992	Smith, J. G., Jr., Connell, J. W., Hergenrother, P. M.	129, 161, 163, 202
Polym. Prep., **33**(2), 351	1992	Wang, Z. Y., Zhang, C.	126, 129, 148, 154, 158, 159, 193, 197, 199
Polym. Prep., **34**(1), 58	1993	Miller, T. M., Neenan, T. X., Kwock, E. W., Stein, S. M.	134, 187
Polym. Prep., **34**(1), 310	1993	Priddy, D. B., Franks, M., Konas, M., Vrana, M. A., Yoon, T. H., McGrath, J. E.	195
Polym. Prep., **34**(1), 413	1993	Babb, D. A., Ezzell, B. R., Clement, K. S., Richey, W. F., Kennedy, A. P., Frye, D. R.	188
Polym. Prep., **34**(1), 415	1993	Carter, K. R., Kim, S. Y., Labadie, J. W.	193–195
Polym. Prep., **34**(1), 417	1993	Nakamura, S., Suzuki, Y., Tamura, E., Kuriki, M., Saegusa, Y.	187
Polym. Prep., **34**(1), 429	1993	Maier, G., Hecht, R., Nuyken, O.	124, 145
Polym. Prep., **34**(1), 441	1993	Bass, R. G., Srinivasan, K. R.	125, 126, 145, 147, 197
Polym. Prep., **34**(1), 461	1993	Jensen, B. J. Hergenrother, P. M., Nwokogu, G.	143
Polym. Prep., **34**(1), 481	1993	Carter, K. R., Miller, R. D., Hedrick, J. L.	198, 199
Polym. Prep., **34**(1), 521	1993	Carter, K. R., Hedrick, J. L.	123, 144, 196
Polym. Prep., **34**(1), 525	1993	Connell, J. W., Smith, J. G., Jr., Hergenrother, P. M.	165, 167
Polym. Prep., **34**(1), 570	1993	Gibson, H. W., Wang, J., Pandya, A.	109, 136
Polym. Rxns. & New Polymers, ACS No.129, 131	1973	Cornell, R., J.	127
Polym., **6**, 589	1965	Cudby, M. E., Feasey, R. G., Jennings, B. E., Jones, M. E. B., Rose, J. B.	106, 107
Polym., **15**, 456	1974	Rose, J. B.	107, 108, 130
Polym., **18**, 359	1977	Atwood, T. E., Barr, D. A., King, T., Newton, A. B., Rose, J. B.	106, 107, 112, 128, 131
Polym., **18**, 365	1977	Atwood, T. E., King, T., McKenzie, I. D., Rose, J. B.	132, 138

REFERENCE	YEAR	AUTHORS	PAGE
Polym., **18**, 370	1977	Atwood, T. E., King, T., Leslie, V. J., Rose, J. B.	106, 107, 112,127–131
Polym., **21**, 577	1980	Dawson, P. C., Blundell, D. J.	132, 135, 183
Polym., **22**, 1096	1981	Atwood, T. E., Dawson, P. C., Freeman, J. L., Hoy, L. R. J., Rose, J. B., Staniland, P. A.	108, 132, 135, 136, 138, 141, 149
Polym., **24**, 953	1983	Blundell, D. J., Osborn, B. N.	135
Polym., **25**, 1827	1984	Viswanathan, R., Johnson, B. C., McGrath, J. E.	108, 114
Polym., **25**, 1151	1984	Kricheldorf, H. R., Bier, G.	114, 135, 138,141–143
Polym., **26**, 174	1985	Hergenrother, P. M., Jensen, B. J., Havens, S. J.	142, 151
Polym., **29**, 358	1988	Hergenrother, P. M., Jensen, B. J., Havens, S. J.	121, 122, 142,150–153,158–163
Polym., **29**, 1902	1988	Colquhoun, H. M., Lewis, D. F.	134, 135, 148, 154
Polym., **30**, 1137	1989	Guiver, M. D., Kutowy, O., ApSimon, J. W.	117
Polym., **32**, 840	1991	McHattie, J. S., Koros, W. J., Paul, D. R.	116, 118, 120
Polym., **34**, 3315	1993	Sauer, B. B., Hsiao, B. S.	174
Polym., **34**, 1512	1993	Sakaguchi, Y., Tokai, M., Kato, Y.	
Polym.,**13**, 465	1974	Newton, A.B., Rose, J. B.	106
Preprints, Japan-US Pol. Symp.	1985	Sroog, C. E.	135
Proc. Int. Membrane Conf. p. 187	1986	Guiver, M. D.	121
Prog.Polym. Sci., **12**,111	1986	Maiti, S., Mandal, B. K.	122, 183
Prog. Polym. Sci., **14**, 173	1989	Imai, Y., Oishi, Y.	109, 111, 115
Rec. Adv. Mech. & Syn. Asp. of Polym., p. 207. Reidel.	1987	Rose, J. B.	139
SAMPE, **32**, 382	1987	Cogswell, F. N., Leach, D. C., McGrail, P. T., Colquhoun, H. M., MacKenzie, P., Turner, R. M.	131, 147, 165
US Pat. 3332909	1967	Farnham, A. G., Johnson, R. N.	190, 194, 195
US Pat. 3441538	1969	Marks, B. M.	106, 132, 134, 136, 138, 139
US Pat. 3734888	1973	Darms, R.	137
US Pat. 3914298	1975	Dahl, K. J.	134
US Pat. 3953400	1976	Dahl, K. J.	132
US Pat. 3956240	1976	Dahl, K. J., Jansons, V.	134, 136, 139, 141, 148–150
US Pat. 3965146	1976	Dahl, K. J.	134
US Pat. 4108837	1978	Johnson, R. N., Farnham, A. G.	127

REFERENCE	YEAR	AUTHORS	PAGE
US Pat. 4111908	1978	Dahl, K. J.	134
US Pat. 4156068	1979	Hartmann, L. A.	118
US Pat. 4229564	1980	Dahl, K. J.	132, 134
US Pat. 4247682	1981	Dahl, K. J.	132
US Pat. 4331798	1982	Staniland, P. A.	135, 138, 142, 143
US Pat. 4361693	1982	Jansons, V.	132, 135, 136
US Pat. 4398021	1983	St. Clair, T. L., Yamaki, T.	170
US Pat. 4473684	1984	Maresca, L. M., Chao, H. S.	118
US Pat. 4489027	1984	St. Clair, T. L., Yamaki, D. A.	170
US Pat. 4645819	1987	Sterzel, H.	135
US Pat. 4794155	1988	Woo, E. R., Tuinstra, H. E., Mullins, M. J.	111, 132, 147
US Pat. 4797457	1989	Guiver, M. D., ApSimon, J. W., Kutowy, O.	113,117–122
US Pat. 4806618	1989	Imai, Y., Kakimoto, M., Maruyama, Y., Koishi, T.	121, 149
US Pat. 4816556	1989	Gay, F. P., Brunette, C. M.	159
US Pat. 4820794	1989	Darnell, W. R., Jackson, W. J., Jr.	132, 134, 136–141, 147, 148
US Pat. 4826947	1989	Jansons, V., Mazzanti, J.B., Moore, S.	135, 136, 159, 160, 162
US Pat. 4833219	1989	Guiver, M. D., ApSimon, J. W., Kutowy, O.	113,117–121
US Pat. 4839459	1989	Darnell, W. R., Jackson, W. J., Jr.	160
US Pat. 4843131	1989	Becker, P., Edwards, L. M., Horner, P. J., Macknick, A. B., Moore, S., Mosso, R. J.	149
US Pat. 4849503	1989	Cotter, R. J., Rimsa, S. B., Barclay, R. B., Jr.	109
US Pat. 4861856	1989	Darnell, W. R., Jackson, W. J., Jr.	124, 127, 145, 149, 150
US Pat. 4882397	1989	Kelsey, D. R.	132, 142
US Pat. 4912181	1990	Becker, P., Edwards, L. M., Horner, P. J., Macknick, A. B., Moore, S., Mosso, R. J., White, L. J.	149

Author Index

This index lists the authors whose work is cited in the body of the chapters.

F

G

H

I

J

K

R

S

T

U

V

W

X

Y

Z

Subject Index

D

E

F

R